Inclusion Compounds
Volume 3

Inclusion Compounds

Volume 3
Physical Properties and Applications

Edited by

J. L. Atwood
University of Alabama, USA

J. E. D. Davies
University of Lancaster, UK

D. D. MacNicol
University of Glasgow, UK

1984

ACADEMIC PRESS

(*Harcourt Brace Jovanovich, Publishers*)

London   Orlando   San Diego   San Francisco   New York
Toronto   Montreal   Sydney   Tokyo   São Paulo

ACADEMIC PRESS INC. (LONDON) LTD.
24–28 Oval Road,
London NW1 7DX

United States Edition Published by
ACADEMIC PRESS INC.
(Harcourt Brace Jovanovich, Inc.)
Orlando, Florida 32887

British Library Cataloguing in Publication Data

Inclusion compounds.
  Vol. 3
  1. Molecular structure—Congresses
  2. Chemistry, Physical organic—Congresses
  I. Atwood, J. L.   II. Davies, J. E. D.
  III. MacNicol, D. D.
  547,1'22   QD461
  ISBN 0-12-067103-4

Printed in Great Britain by J. W. Arrowsmith Ltd.
Bristol BS3 2NT

Contributors to Volume 3

ANDREETTI, G. D., *Istituto di Strutturistica Chimica, Università di Parma, Via M. D'Azeglio 85, 43100 Parma, Italy.*

ARAD-YELLIN, R., *Department of Structural Chemistry, The Weizmann Institute of Science, 76100 Rehovot, Israel*

BERGERON, R. J., *J. Hillis Miller Health Center, College of Pharmacy, University of Florida, Gainesville, FL 32610, USA*

BRESLOW, R., *Department of Chemistry, Columbia University, New York, NY 10027, USA*

CHRISTENSEN, J. J., *Department of Chemistry, Brigham Young University, Provo, UT 84602, USA*

DAVIDSON, D. W., *Division of Chemistry, National Research Council, Ottawa, Canada, K1A 0R9*

DAVIES, J. E. D., *Department of Chemistry, University of Lancaster, Lancaster LA1 4YA, UK*

FARINA, M., *Istituto di Chimica Industriale, Università di Milano, Via Venezian 21, 20133 Milan, Italy*

GREEN, B. S., *Department of Structural Chemistry, The Weizmann Institute of Science, 76100 Rehovot, Israel*

GUARINO, A., *Istituto di Chimica Nucleare, C.N.R., Area della Ricerca di Roma, 00016 Monterotondo Stazione, C.P.10, Rome, Italy*

IZATT, R. M., *Department of Chemistry, Brigham Young University, Provo, UT 84602, USA*

JOHNSON, L. N., *Department of Zoology, Laboratory of Molecular Biophysics, University of Oxford, South Parks Road, Oxford OX1 3PS, UK*

KNOSSOW, M., *Laboratoire de Physique, Centre Pharmaceutique, 92290 Chatenay-Malabry, France*

LAMB, J. D., *Department of Chemistry, Brigham Young University, Provo, UT 84602, USA*

McBRIDE, D. W. Jr., *Department of Chemistry, Brigham Young University, Provo, UT 84602, USA*

PARSONAGE, N. G., *Department of Chemistry, Imperial College of Science and Technology, South Kensington, London SW7 2AY, UK*

RIPMEESTER, J. A., *Division of Chemistry, National Research Council, Ottawa, Canada K1A 0R9*

ŠINGLIAR, M., *Research Institute for Petrochemistry, 972 71 Nováky, Czechoslovakia*

SMITH, N. O., *Department of Chemistry, Fordham University, Bronx, NY 10485, USA*

SMOLKOVÁ-KEULEMANSOVÁ, E., *Department of Analytical Chemistry, Charles University, Albertov 2030, Prague 2, Czechoslovakia*

SOPKOVÁ, A., *Department of Analytical Chemistry, P. J. Šafarik's University, Moyzesova 11, 04167 Košice, Czechoslovakia*

STAVELEY, L. A. K., *Department of Inorganic Chemistry, University of Oxford, South Parks Road, Oxford OX1 3QR, UK*

SYBILSKA, D., *Institute of Physical Chemistry, Polish Academy of Sciences, ul. Kasprzaka 44/52, 01-224 Warsaw, Poland*

SZEJTLI, J., *Chinoin Pharmaceutical and Chemical Works Ltd, Biochemical Research Laboratory, Újpest 1, Budapest 1325, Hungary*

TABUSHI, I., *Department of Synthetic Chemistry, Faculty of Engineering, Kyoto University, Yoshida, Kyoto, Japan*

TSOUCARIS, G., *Laboratoire de Physique, Centre Pharmaceutique, 92290 Chatenay-Malabry, France*

PREFACE

In September 1980 the Institute of Physical Chemistry of the Polish Academy of Sciences hosted the First International Symposium on 'Clathrate Compounds and Molecular Inclusion Phenomena' at Jachranka, near Warsaw. At this timely meeting, the first devoted entirely to all types of inclusion behaviour, the unanimous opinion of the participants was that every effort should be made to draw together in print the various threads from which the rich tapestry of Inclusion Chemistry is currently being woven.

As a first step in this direction, the proceedings of the conference were published in special issues of the *Journal of Molecular Structure* (Volume 75, Number 1, 1981) and the *Polish Journal of Chemistry* (Volume 56, Number 2, 1982). However, to obtain a more global modern picture of Inclusion Chemistry it was apparent that an up-to-date Comprehensive Treatise would be necessary. In view of the rapid advances being made at present, it was clear that such a work could only be produced on an acceptable timescale, and with a sufficient depth of treatment of recent work, by inviting recognized international authorities to write on their own particular fields of interest. Accordingly, this was the plan chosen for the present work.

Earlier useful books, in English, have appeared on inclusion compounds over the years, each reflecting the state of knowledge at the time of publication, three being *Clathrate Inclusion Compounds*, Reinhold, 1962, by M. Hagen; *Non-Stoichiometric Compounds*, Academic Press, 1964, edited by L. Mandelcorn; and *Clathrate Compounds*, Chemical Publishing Company, 1970 by V. M. Bhatnagar. The most comprehensive of these is undoubtedly the book edited by L. Mandelcorn (1964) and in some ways the present treatise may be regarded as complementary to that work.

The editors note, with pleasure, the greatly increasing interest in inclusion phenomena, as evidenced by recent relevant publications on *specific* aspects of Inclusion Chemistry: *Cyclodextrin Chemistry*, by M. L. Bender and M. Komiyama, Springer-Verlag, 1977; *Host–Guest Complex Chemistry I and II*, edited by F. Vögtle, Springer-Verlag, 1981; *Ionophores and their Structures*, by M. Dobler, Wiley, 1981; *Cyclodextrins and their Inclusion Complexes*, by J. Szejtli, Akadémiai Kiadó, Budapest, 1982; and *Intercalation Chemistry*, edited by M. S. Whittingham and R. J. Jacobson, Academic Press, 1982. Also a new journal devoted to inclusion compounds *The Journal of Inclusion Phenomena* has been launched by the Reidel Publishing Company.

We have great pleasure in dedicating these three volumes to Professor H. M. Powell, FRS, whose pioneering crystallographic work laid firm foundations for subsequent work in Inclusion Chemistry.

We wish to thank Professor Powell for kindly agreeing to write the important introductory chapter; and we are indebted also to all our other contributors for their help and participation in writing this book. We must also thank the staff of Academic Press for the efficient way in which the book has been produced.

The present volume is the third of a three volume series designed to provide comprehensive coverage of all aspects of inclusion compounds. Volumes 1 and 2 reviewed structural and design aspects of inclusion compounds whilst the present volume is mainly concerned with their physical properties, applications, and biological significance.

March 1984

J. L. Atwood
J. E. D. Davies
D. D. MacNicol

Contents

Chapter 3. NMR, NQR and dielectric properties of clathrates
    D. W. Davidson and J. A. Ripmeester

Chapter 4. Crystallographic studies of inclusion compounds
    G. D. Andreeti

Chapter 5. Host lattice—guest molecule energy transfer processes
    A. Guarino

Chapter 12. Cycloamylose-substrate binding *R. J. Bergeron*

Contents of Volume 1

Structural aspects of inclusion compounds formed by inorganic and organometallic hosts

Contents of Volume 2

Structural aspects of inclusion compounds formed by organic host lattices

Dedicated to
H. M. Powell, FRS
who laid the firm foundation on
which this book is based

1 · THERMODYNAMIC STUDIES OF CLATHRATES AND INCLUSION COMPOUNDS

N. G. PARSONAGE
Imperial College of Science and Technology, London, UK

L. A. K. STAVELEY
University of Oxford, Oxford, UK

1. General introduction

In embarking on the study of the thermodynamic properties of inclusion compounds the initial assumption which is made is that the host-lattice behaves as if it were inert. This leads to the conclusion that each thermo-dynamic property may be split into contributions from the host and guest substances. This assumption, which is almost universally made, has been questioned by Hazony and Ruby,[1] who have pointed out that for a β-quinol clathrate with all of the cages occupied by krypton atoms the mass of guest substance is far from negligible compared with that of the quinol associated with it, so that coupling of the motions of the components would be expected. However, a stronger case for the importance of host–guest coupling would arise if the frequency of the rattling mode of the guest molecules was close to that of important modes of the host lattice, and this generally does not

INCLUSION COMPOUNDS III
ISBN 0-12-067103-4

seem to be the case. Moreover, the rattling modes are of such low wavenumber value ($\sim 35\ cm^{-1}$ for Kr) that lattice modes of that value would be behaving almost classically above ~ 120 K in any case. Thus for measurements above that temperature the effect of such coupling would be expected to be unimportant. The best support for the above assumption is, however, the remarkable success which has been achieved by theories in which independence of the contributions is invoked.

Another assumption, which also is fundamental to the simplicity of the picture generally adopted, is that there are no correlations between the behaviour of guest molecules in the same sample, even when they are in adjacent guest sites. This is a good approximation for most purposes, and lies at the root of some of the studies of inclusion compounds in which they have been employed as a type of organized matrix isolation system. Consideration of the inter-guest potential and the distances involved shows that there are several systems in which this assumption should fail. This can lead to the occurrence of transitions, which are discussed in the final section of this chapter.

The last assumption which is commonly made in simple theories of these systems is that there is orientational disorder of the guest molecules. This is probably the worst of the three assumptions, although it can, nevertheless, provide a useful starting-point for the understanding of the properties of these compounds. In many real systems, then, we shall expect to find hindered, rather than free, rotation of the guest molecules. A considerable amount of work has been devoted to determining the number of such preferred orientations and the barriers between them. These considerations will feature prominently in the penultimate section of this chapter.

2. Decomposition pressures

2.1. Experimental methods and results

A general difficulty in making these measurements on the heterogeneous equilibria is the slowness with which equilibrium is reached. Indeed, for the clathrates of β-quinol such a measurement can be essentially ruled out as being unfeasible, at least at ordinary temperatures. For the clathrate hydrates the situation is not so unfavourable, but considerable care (and much waiting) is necessary for valid results to be obtained. For the inclusion compounds of urea and thiourea the corresponding measuring variables are usually the concentrations (or activities) of the guest and host substances in solutions in equilibrium with the inclusion compound; less often the decomposition pressure can be measured directly.

2.1.1. β-quinol clathrates

The empty β-quinol host structure, which is unstable and slowly reverts to α-quinol, can be stabilized by the predominently attractive interaction with guest molecules provided that a sufficiently large fraction of the cages is filled. It follows that, at a given temperature, it is not possible to have a clathrate with less than this fraction of cages occupied as the stable phase. A schematic phase diagram for a gas-clathrate system at a particular temperature is shown in Fig. 1.

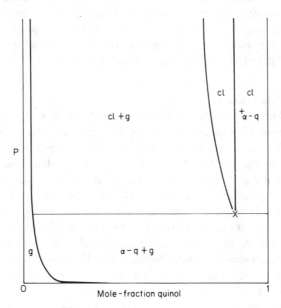

Fig. 1. Schematic P–x diagram for quinol $\rightarrow$ gas system at a fixed temperature. cl = clathrate; α-q = α-quinol; g = gas.

Careful measurements of the univariant triple-point ($\times$) (clathrate + α-quinol + gas) were first made by van der Waals and Platteeuw[2] using n-propanol as an auxiliary solvent. The latter has molecules which are too large to be encaged, but it is a good solvent for quinol and greatly facilitates the attainment of equilibrium. In this method the composition of the clathrate is calculated from the amount of gas used in converting a known amount of α-quinol to the clathrate. Measurements were made for argon

Table 1. Equilibrium pressures and compositions for the β-quinol + clathrate + gas equilibrium at 298.15 K

	Ar[3]	Kr[3]	Xe[4]	N$_2^3$	HCl[3]
P obs (atm)	3.4	0.4	0.058	5.8	0.01
P calc (atm)	3.4	0.4	0.06	5.2	0.02
y^* (obs)	0.34	0.34	0.34	—	—

* Fraction of cavities occupied.

(at 25, 60 and 120° C), krypton (at 25° C) and xenon (at 25° C).[3,4] Table 1 shows some of the results obtained for the equilibrium pressure.

For guest substances which are liquid under normal conditions observation of the pressure cannot be used to indicate the formation of a new phase, and so some other observable quantity must be employed. Van der Waals and Platteeuw have used the refractive index of the supernatant liquor to indicate the entry to a new phase region.[3]

2.1.2. Clathrate hydrates

As discussed earlier (Chapter 5, Volume 1) the hydrates exist in two common structural forms (I and II), as well as at least one other rare form (the tetragonal form adopted by Br$_2$ and (CH$_2$)$_3$O). None of these host lattice structures is stable in the absence of guest molecules. The situation is, therefore, similar in this respect to that of β-quinol discussed above. An extra complexity of all these hydrate structures is that they each contain two kinds of cavity which can be used for clathration. This raises the possibility of simultaneously including two guest substances, one in each type of cavity, although most fundamental studies have been made with a single guest substance. Clathrate hydrates can be formed with liquid as well as gaseous guest substances, and these may or may not be water-soluble. Initially, we shall discuss the water-insoluble gaseous guest substances.

The problem of maintaining equilibrium between gas and hydrate has been overcome by the technique originated by Barrer and Edge[5] of grinding the crystals by rolling steel ball-bearings over them in the presence of the gas. This is thought to work by breaking up layers of reacted material covering the surface of the solid.

Where the guest molecules are clearly too large to occupy the smaller cavities in a structure, analysis of the results can proceed as for the β-quinol clathrates. If, however, both kinds of cavity are in use a problem arises which has not been satisfactorily solved: there is no general way to decide on the distribution of the guest between the two kinds of cavity. An exception is the Structure I hydrate of xenon. Ripmeester and Davidson[6] have noted that in the ^{129}Xe resonance spectrum two lines are observed, which are assigned to xenon atoms in the two types of cage. From the relative intensities of these lines they have found that $y_1/y_2 = 0.77 \pm 0.02$, where y_1 and y_2 are the fractional occupancies of the small and large cages, respectively. This value is to be compared with those yielded by the usual theories, which are much closer to unity. Barrer and Edge[5] in their study of Structure II hydrates adopted a method of simplification: they effectively filled the large cavities with chloroform and then studied the sorption of simple gases (argon, krypton) in the smaller cages.

A p–T diagram for a clathrate hydrate system is shown in Fig. 2.[7a] Although two quadruple points are shown, not all systems display the upper

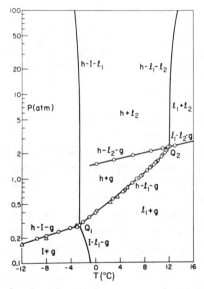

Fig. 2. Phase diagram for the $SO_2 + H_2O$ system. I = ice Ih; h = clathrate hydrate; l_1 and l_2 are water- and guest-rich liquids, respectively. (Adapted from *Water, a comprehensive treatise*, (ed. F. Franks) Plenum Press, New York and London, 1973, Vol. 2, with permission).

one. In particular, highly volatile or gaseous guest substances or those which are miscible with water do not have the upper quadruple point, since an aqueous and a guest-rich liquid phase would need to be present at a temperature which is too high for the existence of ice. Platteeuw and van der Waals[8] have made measurements of the univariant $h + I + g$ equilibrium for the Structure I hydrates of nine simple gases. They obtained values for the pressure, but not for the occupancy, at 273 K (Table 2).

Table 2. *Decomposition pressures for hydrate + ice Ih + gas at 273.15 K*

	P (atm)		
	Expt.[8,31,14]	Cell model with Lennard-Jones potential	Cell model with Kihara[31] potential
Ar	95.5	95.5	a
Kr	14.5	15.4	a
Xe	1.15	1.0	a
CH_4	26	19.0	13.0, 19.0[b]
N_2	140	90.0	115, 115[b]
O_2	100	63.0	120
CO_2	12.47	0.71, 1.70[b]	9.0
N_2O	10.0	0.6, 1.52[b]	8.2
C_2H_6	5.20	1.1	8.4
C_2H_4	5.44	0.50	1.3, 0.82[b]
CF_4	41.5	1.6	0.6

[a] Parameters chosen to give correct pressure for these substances.
[b] Choice of force or core parameters.

Dimethyl ether, which forms a tetragonal clathrate similar to that of bromine, and at higher temperatures a Structure II hydrate, has a complicated phase diagram containing no fewer than four stable quadruple points, as well as two other quadruple points which involve metastable phases.[9] Cyclopropane can also form two hydrates according to the conditions of temperature and pressure,[10,11] indicating that it is at the borderline with respect to size between those substances forming Structure I hydrates and those preferring Structure II. By extending the measurements on the tetrahydrofuran hydrate (Structure II) to high pressures (~3.6 kbar) Gough and Davidson[12] were able to observe transitions to the ice forms II and III, as well as to I. In contrast with the variety of ices which can occur at elevated pressures, no evidence was found for any "new" clathrate form under these conditions.

Largely because of the sluggishness with which equilibrium is approached, especially when no liquid phase is present, some very inaccurate data have been reported. Thus, for the three-phase $h + I + g$ equilibrium for methane and ethane Falabella and Vanpee[13] found the temperature at which $p = 1$ atm to be 193.2 K and 240.8 K, respectively, as compared with the previously accepted values of 244 K and 257 K. Again, for the corresponding argon system an early value for the temperature at which $p = 1$ atm of 230 K has been corrected to 149.6 K.[5] Again, the widely quoted value for the decomposition pressure of the CF_4 hydrate of ~1 atm at 273.2 K is far too low: a subsequent study yielded the value 41.5 atm[14] and this has been confirmed by Garg *et al.*[15]

Difficulty in ensuring attainment of equilibrium was probably the cause also of some incorrect suggestions that in many Structure II hydrates only one-half of the cages were used. This led to an idealized formula of $X.34H_2O$, instead of $X.17H_2O$ in the normal Structure II hydrates.[16-19] Phase diagram studies on one of the systems concerned, 1,3 dioxane + water, by Morcom and Smith[20] conflicted with these results and conclusions, only the normal Structure II hydrate being found. Davidson *et al.*[21] have subsequently shown beyond doubt by X-ray, dielectric and NMR means that for 1,3 dioxane only the normal hydrate is obtained. (See Chapter 3, Volume 3).

A quite different type of error with respect to the phase diagrams of hydrates has arisen from a proposal to detect the formation of a hydrate by studying the complete phase diagram for hydrates of binary mixtures with a known hydrate-former.[22] As a result of a false thermodynamic argument it was concluded that methanol could form both Structure I and II hydrates, a result which would have important consequences in the oil industry because of the widespread use of methanol to suppress hydrate formation. The false argument has been exposed, and it has also been demonstrated that all the dielectric and NMR features found in the double hydrates investigated in the previous work could be explained without any need to postulate clathration of the methanol.[23]

2.1.3. Channel compounds of urea and thiourea

Once again, the host structure is unstable when empty, or even when only partly filled. As the guest is held in open channels, rather than in cages, it might be thought that measurements of decomposition pressure would be more straightforward, because it would be easier for equilibrium to be reached with the vapour phase. In fact, reliable measurements could only be made for inclusion compounds of the more volatile guest substances, e.g. thiourea + cyclopentane, which has been studied by a dew-point method. For many inclusion compounds the pressures would be very small indeed and so the method would be unsuitable. Estimates of the free energy of formation have come instead from measurements on the solution equilibrium:

$$\text{urea (solid or solution)} + \text{guest (solution)} \rightleftharpoons \text{inclusion compound} \quad (1)$$

From such studies Redlich and co-workers[24] arrived at the following formulae for three homologous series of urea inclusion compounds:

$$\Delta G_f^\ominus \text{ (kJ (mol. guest)}^{-1}) = -9.11 + 1.59x \qquad \textit{n-alkanes} \qquad (2a)$$

$$= -8.39 + 1.62x \qquad \textit{n-acids} \qquad (2b)$$

$$= -10.47 + 1.62x \qquad \textit{n-alcohols} \qquad (2c)$$

where x is the number of carbon atoms in each guest molecule. Corresponding measurements for a number of thiourea inclusion compounds have also been made, some by the dew-point method (Fig. 3)[25]

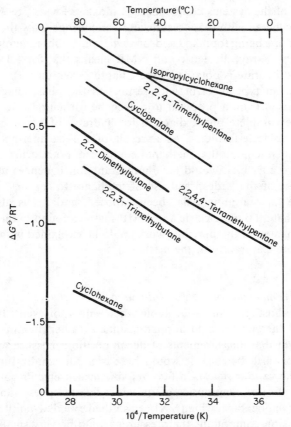

Fig. 3. Standard free energy of formation against temperature for several thiourea inclusion compounds[25]. (Adapted from *Non-stoichiometric Compounds*, (ed. L. Mandelcorn), Academic Press, New York and London, Chapter 8, with permission).

2.2. Theoretical treatments

Successful theories for both the β-quinol clathrates and the clathrate hydrates have been developed from a theory of J. H. van der Waals.[26] That for the β-quinol clathrates is the easier because there is only one type of cavity to be considered, as against two for either of the two common forms of clathrate hydrate. We shall, therefore, start by considering the β-quinol compounds. For the two-phase $h + g$ equilibrium the statistical mechanics proceeds exactly as for the Langmuir adsorption. By equating the chemical potential of the guest in the two phases there is obtained the equation:

$$y = Cp/(1 + Cp) \text{ or } p = C^{-1}y/(1 - y) \tag{3}$$

where y is the fractional occupancy of the cavities, p is the pressure of the gas (which is taken to be perfect), $C = Z_c(T)/kTZ_g(T)$ and Z_c and Z_g are the configuration integrals per guest molecule when enclathrated and when in the gas phase (with the volume factor removed), respectively. When the three-phase equilibrium

$$\alpha\text{-quinol} + \text{guest} \rightleftharpoons \text{clathrate} \tag{4}$$

is considered there is an additional equation, which arises from equating the chemical potential of quinol in the two solid phases

$$\mu_Q^\alpha = \mu_Q^c = \mu_Q^\beta + kT\nu \ln(1 - y) \tag{5}$$

where μ_Q^α, μ_Q^c and μ_Q^β are the chemical potentials of quinol in the pure α-form, the clathrate, and the pure (empty) β-form, respectively, and ν is the number of cavities per host molecule ($= \frac{1}{3}$).

A conclusion which follows immediately from Equation 5 is that, since μ_Q^α and μ_Q^β are independent of the guest substance, y should likewise be independent of the guest, but should depend upon the temperature. This has been confirmed experimentally for the rare gas clathrates (argon, krypton, xenon), for all of which y was found to be 0.34, within experimental error.[2,4]

To obtain information on the p–y equilibrium it is necessary to evaluate C. For this, van der Waals employed the cell model, which had first been proposed by Lennard-Jones and Devonshire for the representation of dense fluids.[27] By using this model van der Waals assumed that for the purpose of calculating the potential energy of interaction between each encaged molecule and the wall of its cavity it was adequate to replace the discrete centres of force in the wall by a spherically smeared out model. In view of the large number of atoms comprising the wall of each cavity, and its near-spherical shape, this seemed to be a reasonable simplification. At the time at which the theory was put forward such a simplification was, indeed, necessary for further progress to be made. Van der Waals also followed Lennard-Jones and Devonshire in assuming that the guest molecules interacted with each element of the wall according to the Lennard-Jones 12–6 potential

$$\varepsilon = 4\varepsilon^*((\sigma/r)^{12} - (\sigma/r)^6) \tag{6}$$

where ε is the potential energy for a distance apart r, ε^* is the potential energy minimum and σ is the distance at which the potential energy is zero. It remained to choose the two interaction parameters (ε^*, σ) of the potential. The molecule-wall parameters were first related to those for a pair of molecules (mm) and a pair of wall elements (ww) by means of the Lorentz–Berthelot combining rules:

$$\varepsilon_{mw}^* = (\varepsilon_{mm}^* \varepsilon_{ww}^*)^{1/2} \tag{7a}$$

$$\sigma_{mw} = \tfrac{1}{2}(\sigma_{mm} + \sigma_{ww}) \tag{7b}$$

where ε_{mw}^*, ε_{mm}^*, ε_{ww}^* are the energy parameters for a molecule–wall, molecule–molecule and wall–wall interaction, respectively, and σ_{mw}, σ_{mm}, σ_{ww} are the corresponding distance parameters. The first of these relationships is now known to give a value which is rather too large for the simpler case of a gaseous mixture; the second equation seems to be very nearly correct. ε_{mm}^* and σ_{mm} are known fairly well for most of the simple guest substances, but there is no direct way of determining ε_{ww}^* and σ_{ww}. σ_{ww} was chosen to be 0.330 nm from a knowledge of the distance of closest approach of aromatic molecules in crystals, whilst $z\varepsilon_{ww}^{*\,1/2}$, the quantity actually appearing in the equations (z being the number of discrete force centres in the wall) was chosen so as to fit the results for the enthalpy of formation found by Evans and Richards[28] (Section 3.2). The predictions for the pressure of the α-quinol + clathrate + gas equilibrium at 25° C were in very good agreement with experiment for the rare gas clathrates, but were less so for the compounds of diatomic guests. Subsequently, van der Waals has proposed that the totality of data could be better represented if σ_{ww} was reduced to 0.300 nm (Section 3.2 and Fig. 4).[3] This alteration does not greatly affect the agreement for the above pressure values.

Following this very considerable success, an extension of the theory to deal with the clathrate hydrates was natural. Barrer and Stuart[29] and Platteeuw and van der Waals[8] have each formulated such theories. In place of Equation 5 we now have

$$\nu_1 \ln(1 - y_1) + \nu_2 \ln(1 - y_2) = -(\mu_{H_2O}^{0(c)} - \mu_{H_2O}^{ice})/kT \tag{8}$$

where ν_1 and ν_2 are stoichiometric coefficients for cells of types 1 and 2, y_1 and y_2 are the fractional occupancies of type 1 and 2 cells, and $\mu_{H_2O}^{0(c)}$ and $\mu_{H_2O}^{ice}$ are the chemical potentials of water in the empty clathrate and the ice phases, respectively. It will be noticed that the triple-point coverages (y_1, y_2) are not fixed when the temperature is fixed. Thus, on changing the guest substance, y_1 would change, there being a corresponding change in y_2 so as to obey Equation 8. In fact, there are several difficulties in testing Equation 8 experimentally, because of the problem of determining y_1 and y_2, rather than just the overall fractional occupancy ($=(\nu_1 y_1 + \nu_2 y_2)/$

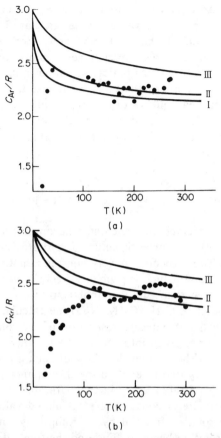

Fig. 4. Comparison of experimental values and values calculated from classical theory for the heat capacity contributions of (a) argon and (b) krypton in their quinol clathrates. ●: experimental values. Calculated values: I for $\sigma_{ww} = 0.29$ nm; II for $\sigma_{ww} = 0.30$ nm; III for $\sigma_{ww} = 0.33$ nm.[45]. (Reproduced with permission, *Mol. Phys.*, 1961, **4**, 153.)

$(\nu_1 + \nu_2))$. In the case of the Structure II hydrates it is usually true that only the large cages are occupied, since if the guest molecules could be accommodated in the smaller cages of Structure II then a type I hydrate (in which the small cages are very similar to the small cages of Structure II, but the large cages are smaller than those of Structure II) would probably form. If, indeed, only the larger cages of a Structure II hydrate are used then the problem reduces to that of the β-quinol clathrates and it loses much of its interest. The value of $\Delta\mu = \mu_{H_2O}^{0(c)} - \mu_{H_2O}^{ice}$ for the Structure I and II hydrates were chosen by van der Waals and Plattteeuw[3] as 700 and 820 J mol^{-1}, respectively. More recently the value for the Structure I hydrates has been estimated by Davidson[7b] to be 1110 ± 60 J mol^{-1} from a consideration of the large number of hydrates which have a composition $X.nH_2O$ with $n \approx 6.1$. For the Structure II host lattice, which has angles which are closer to those of ice Ih than are those of Structure I, he suggested $\Delta\mu <$ 1110 J mol^{-1}. From the frequency of occurrence of nearly full large cavities he suggested $\Delta\mu > 520$ J mol^{-1}, thereby bracketing $\Delta\mu$.

The equilibrium pressure for either two- or three-phase equilibrium is given by the equation

$$p = kT\frac{Z_g(T)}{Z_{c,1}(T)} \cdot \frac{y_1}{1 - y_1 - y_2} = kT\frac{Z_g(T)}{Z_{c,2}(T)} \cdot \frac{y_2}{1 - y_1 - y_2} \qquad (9)$$

Table 2 shows the observed and calculated decomposition pressures (from Equations 8 and 9) for some Structure I hydrates at 273 K based on $\Delta\mu = 700$ J mol^{-1}. The agreement is very good for the three rare gases and quite good for methane, which is probably freely rotating (see Section 4.4 for the rotation of methane in β-quinol clathrates). For the elongated molecules (N_2, O_2, CO_2, C_2H_4, C_2H_6, N_2O) the calculated pressure is too low. At first this was attributed to restriction of the rotation of the guest molecules, which is not taken into account by the theory. However, it has been pointed out by Mazo[30] that the effect of such a restriction would be the reverse of that required to explain the discrepancies. The argument is that if the molecules rotated freely in the clathrate the observed pressure would be (approximately) equal to the calculated value. The molecules would only cease rotating freely if by so doing they could lower the free energy. Hence if rotation becomes hindered the free energy must be lower than for free rotation. Equating a lower free energy with a lowered pressure leads to Mazo's conclusion. The reason for the discrepancy between the calculated and observed pressures must therefore be sought elsewhere. This was found by McKoy and Sinanoglu[31] to be the use of a single point centre of force for the guest molecule. They showed that account could be taken satisfactorily of the finite size of the molecule by using a Kihara core model. In this treatment the distance r in the Lennard-Jones potential is replaced by the shortest distance between the core of the guest molecule and that of the element of wall. For tetrahedral molecules the core was taken to be spherical, so that no difficulty arose in obtaining the shortest distance of separation of the cores. This was not the case for the more interesting elongated molecules, for which a linear core was used, since the appropriate distance is a fairly complicated function of the orientation of the molecule as well as of the distance of its centre from the wall. McKoy and Sinanoglu overcame this problem by a preliminary averaging over molecular orientations. This is, of course, an approximation, but results show a considerable improvement for the calculated pressures so obtained (Table 2). The most serious error remaining is in the prediction for the CF_4 hydrate. Indeed, the agreement was good when the old (and incorrect) experimental value of ~1 atm was accepted! The very large discrepancy observed now is matched by that for the CCl_4 hydrate, for which the predicted decomposition pressure at $0°C$ is only ~0.04 times the vapour-pressure of liquid CCl_4 although it is known that the hydrate only forms in the presence of a *hilfgas*.[7c]

This model, using the Kihara core potential, has been applied by Parrish and Prausnitz[32] and Holder et al.[33] to hydrates of mixed gases. In each case only spherical cores were considered. In these later treatments account is taken of the effect of change of pressure on $\Delta\mu$. (It may be noted that since the pressure at the α-quinol + clathrate + gas or ice + clathrate + gas triple points depends upon the gas concerned, $\Delta\mu$ should also show a similar dependence. In turn, this would slightly modify the observation that, if only one type of cavity were in use, then y at a given temperature would not vary with change of the enclathrated gas.) For comparison with their theoretical predictions Holder et al. made measurements of the $p-T$ curve for the clathrate + liquid + gas equilibrium for the binary systems involving argon, krypton and methane. There was very good agreement in the pressure values for the pure hydrates, but the discrepancies for the mixed clathrates were quite large (up to ~25%) with the calculated values being too low for argon + methane and too high for krypton + methane. One source of error is the theoretical treatment of the chemical potential of the liquid phase, but it is difficult to see why this should be more troublesome for a mixed, rather than a single, clathrate. With regard to the experiments, the attainment of equilibrium within the clathrate for a mixture of guest substances may well be slower than for a pure guest substance, because of the need to allow time for redistribution of the substances to take place. However, if this were the only source of error it would be expected that the discrepancies for the argon + methane and krypton + methane systems would be in the same sense. The work of Holder et al.[33] led to results for the occupancy of the smaller cages in the pure methane clathrate which appeared to be far too low ($y_1 \approx 0.05$, whereas the composition of the hydrate indicates that $y_1 \approx 1$). This was attributed to uncertainty in the value to be chosen for the σ parameter for water, to which the results for y_1 are very sensitive. A similar sensitivity to the choice of σ for the wall molecules, for some properties at least, has been found for the β-quinol clathrates (Section 4). The possibility that the predictions of Holder et al. may not in fact be as bad as suggested here arises as a result of Ripmeester and Davidson's determination of y_1/y_2 for the xenon hydrate[6] (Section 2.1.2).

The next stage in the improvement of theories of clathrate hydrates and quinol clathrates would appear to be the use of numerical methods so as to avoid some of the simplifications which are made in the methods described above. An attempt to do this for a clathrate hydrate, using the Kihara core potential between atoms, has been made by Tester *et al.*[34] However, faced with the problem of evaluating the configuration integral (Z_c), they made the approximation of replacing it by

$$Z_c \approx \exp[-\langle U \rangle / kT] \cdot V_{cell} \tag{10}$$

where $\langle U \rangle$ was obtained by the usual Metropolis Monte Carlo procedure, and so is Boltzmann-averaged, and V_{cell} is the volume of a cavity (an ill-defined quantity). This approximation does not appear to be "safe", and it is therefore not surprising that, agreement for the argon hydrate having been "forced" by the choice of parameter, the predicted pressure for the xenon clathrate was in error by as much as 43%, with the krypton clathrate showing better accord, with only 7% error. The computational problem here is very similar to that encountered with molecules sorbed in molecular sieves, for which the configuration integral has been obtained by a coupling parameter integration.[35,36] The latter work was concerned with systems in which up to eight molecules could be present in the same cavity, and furthermore it was not assumed that there would be the same number of particles in each cage. It is clear, therefore, that the rather easier problem presented by the clathrates should be susceptible to attack by this method.

A by-product of the development of these theories is an understanding of the stabilization of a clathrate hydrate by means of a *hilfgas.* It is clear from the theories that filling of the cavities in the host structure stabilizes it with respect to ice. Since many would-be hydrate-formers could, on account of their size, only use the larger cavities the smaller are in a sense being "wasted". Introduction of a second component with smaller molecules suitable for the smaller cavities would clearly stabilize the complete structure. Many examples are known of the use of molecules such as methane and hydrogen sulphide for this purpose. An important example is the formation of Structure II hydrates by *n*-butane in the presence of methane, which occupies the smaller cages. Without the methane as *hilfgas n*-butane does not form a hydrate.[37]

The cell theories of the clathrates involve a number of simplifications, and we now examine some of these. The models assume that the only interactions are of the Lennard-Jones 12-6 type. In view of the polarity of the host material (water and quinol) it may have been expected that there would be important interactions arising from the electric fields and, possibly, the electric field gradients inside the cages. Davidson[38] has calculated the electric fields arising from the dipole and quadrupole moments of the water molecules in cages similar to those which occur in the hydrates. He found that if, of the four water molecules hydrogen-bonded in a regular tetrahedral manner to each molecule of the cage wall, three (tangentially bonded) had their centres on the surface of the spherical cavity and the other hydrogen-bond was directed radially outwards (radially bonded), then the electric field at the centre of the cage arising from the water *dipoles* would be identically zero provided that the Bernal–Fowler rules for the placing of the protons were obeyed. In fact, the O–O–O angles are known to be not quite tetrahedral, and this causes there to be a small residual field. For a regular dodecahedron, for which the angles between the radial and the three tangential bonds would be $110° \, 44.3'$ the electric field would be given by

$$\mathbf{E} = -0.0705 \, \mathbf{P} r^{-3} = 0.268 \, m_{\mathrm{OH}} r^{-3} \sum_{k=1}^{10} \mathbf{r}_k, \tag{11}$$

where $\mathbf{P}$ is the vector sum of the dipole moments of the water molecules of the wall, m_{OH} is the dipole moment associated with a single OH bond, r is the distance from the cell centre to any of the water molecules comprising the wall and $\mathbf{r}_k$ is the unit vector in the direction of the k-th radial bond. This still represents almost complete cancellation of the fields. The pentagonal dodecahedra of the hydrates are nearly, but not quite, regular dodecahedra. The 14- and 16-hedra, which correspond to the larger cells of Structures I and II, respectively, also have fairly small residual fields. When the fields arising from the quadrupoles of the water molecules were calculated the amount of cancellation was found to be less. Nevertheless, the electric fields inside the cavities would be expected to be small, with the main contribution coming from the water quadrupoles. As the distribution of protons is believed to be random (or nearly so), subject to Bernal–Fowler constraints,[39] these fields would vary from cage to cage, leading to a broadening of the energy spectrum. Hazony and Herber[40] and Holk[41] have also questioned the nature of the potential inside the β-quinol cage. They suggest that in the krypton clathrates there may also be an appreciable interaction with the oxygen atoms of the wall by a charge-transfer process, similar to that which leads to the formation of KrF_2 and XeF_4. Some support for this contention has come from the observation that the shielding of the nuclei of xenon atoms is sufficiently altered by clathration that two NMR lines are found for the Structure I hydrate of ^{129}Xe.[6] The application of this has been discussed earlier (Section 2.1.2).

Free rotation of the guest molecules is normally assumed. The extent to which this assumption is correct for several β-quinol clathrates will be examined in Section 4. Rapid isotropic reorientation (on the NMR time-scale) of the guest molecules in the hydrates of CH_4 and CF_4 (Structure I) and SF_6 (Structure II) even at temperatures as low as 4 K, 22 K and 13 K, respectively, are indicated by the NMR second moments of those hydrates.[15] Although for rotation to be effective thermodynamically the reorientation must be much more frequent than is nececessary for narrowing of the NMR lines, nevertheless it appears to be quite safe to assume that all of these molecules would be reorientating almost freely at temperatures near 283 K, at which the pressure measurements were made. The possibility of coupling between the rotational motion of the guest molecules and the rattling motion has been considered by Gough *et al.*[42] For the somewhat unusual tetragonal hydrate of dimethylether they concluded that rotation was only possible if accompanied by a suitable translational motion. As well as the evidence for restricted rotation in some β-quinol clathrates (Section 4), other evidence indicates the existence of preferred orientations in the clathrates.

3. Enthalpies of clathration

3.1. Introduction

From the thermodynamic and statistical mechanical point of view there are close similarities between inclusion and monolayer adsorption. Thus, when a guest molecule leaves the gaseous phase to become trapped in a void in the host lattice, there is inevitably an entropy decrease, just as there is when it becomes adsorbed on a solid surface. Spontaneous inclusion and adsorption, therefore, both require that the associated enthalpy changes should be negative. There are two methods by which enthalpies of clathration can be determined, namely either using direct calorimetry, or indirectly from a set of isotherms at different temperatures. For clathrates in which the pure guest is a gas, these isotherms link the equilibrium pressure p with y, the fraction of the cavities in the host lattice which are occupied by guest molecules. This relation between p and y has exactly the same form as the Langmuir isotherm for monolayer adsorption (Equation 3). For less volatile guests, such as those involved in the urea inclusion compounds, the experimental study becomes a matter of measuring the equilibrium constant for the formation of the inclusion compound.

Just as conformity of a Langmuir isotherm for adsorption requires that the enthalpy of adsorption must be independent of the fraction of the surface covered, so also the same kind of relation between p and y means that ΔH_c, the enthalpy of clathration per mole of guest, should be independent of y. This is of course consistent with the commonly made assumption mentioned in Section I that guest–guest interaction is negligible. Accordingly, ΔH_c is essentially determined by guest–host interaction, and its chief interest is that it is a source of information about this interaction. Experimental values of ΔH_c have also been used to evaluate interaction parameters involved in the statistical mechanical theories of clathrates. Values of ΔH_c for other clathrates of the same kind can then be calculated and compared with the experimental estimates.

At present, ΔH_c values are available for the following types of system: (1) the clathrates of β-quinol and β-phenol; (2) ice hydrates; (3) the inclusion compounds of urea with straight-chain aliphatic compounds as the guests; (4) inclusion compounds in which thiourea acts as the host. We will briefly consider each of these in turn.

3.2. Quinol and phenol clathrates

Quinol is an unusual host in that the unstable β-form which gives the clathrates can in fact be prepared. Using a differential technique involving twin calorimeters to find the difference in the enthalpies of solution of the β-form and the stable α-form, Evans and Richards[43] found $\Delta H = 0.55 \text{ kJ mol}^{-1}$ for the process α-quinol $\rightarrow \beta$-quinol. With the same technique, they then determined ΔH_c for the β-quinol clathrates of Ar, N_2, O_2, HCl, HBr, CH_3OH, and HCOOH (Evans and Richards[28]). In one calorimeter, a known mass of α-quinol was dissolved (usually in water or a very dilute acid solution), and in the other, using the same solvent, a known mass of a clathrate of known composition and containing about the same mass of quinol. Experiments with the argon clathrate were made with y varying from 0.14 to 0.74, and with the oxygen clathrate with y between 0.12 and 0.60. The β-form only achieves thermodynamic stability when y exceeds ~ 0.34. For both clathrates, the difference in the enthalpies of solution was a linear function of y, as it should be if ΔH_c is independent of y. Extrapolation to $y = 0$ gave values of 0.70 kJ mol^{-1} (from the argon system) and 0.75 kJ mol^{-1} (from the oxygen system) for the enthalpy difference between β-quinol and α-quinol, in reasonable agreement with the directly determined figure.

Direct measurements of ΔH_c for the methane clathrate were made in the same way by Parsonage and Staveley,[44] and for the krypton clathrate by Grey *et al.*[45] Indirect estimates of ΔH_c for krypton and xenon clathrates were made by Allison and Barrer[46] from p, y isotherms.

ΔH_c results for β-quinol clathrates are given in Table 3. ΔH_c is the enthalpy gain when one mole of gaseous guest is enclathrated in 3 moles of β-quinol. Evans and Richards noted that ΔH_c is approximately proportional to the polarizability of the guest molecules. In view of the connection between polarizability and dispersion forces, this observation about ΔH_c is consistent with the conclusion reached by J. H. van der Waals[26] that the energy of the interaction between host and guest in these clathrates derives virtually completely from dispersion forces. More recently, however, Hirokawa[47] has investigated the anisotropic potential for nitrogen and oxygen molecules trapped in the β-quinol lattice by combining an atom–atom Lennard-Jones potential with electrostatic multipole–multipole (EMM) interaction. This study led to the result that for nitrogen, the dominant contribution to the overall potential comes from the EMM interaction, while for oxygen this and the Lennard-Jones potential contribute on about equal terms.

Table 3. Values of ΔH_c, the enthalpy gain for the process $3Q + M(g) \rightarrow (Q)_3.M$, where $Q = quinol$ and $M = guest$. The calculated values are those given by van der Waals and Platteeuw[3]

Enthalpy gain $\Delta H_c(\text{kJ mol}^{-1})$		
M	Expt.	Calc.
Ar	25.1[28]	21.3
Kr	26.4[45]	28.0
	25.5[46]	
Xe	40.6	36.0
CH_4	30.2[44]	27.2
	25.5[46]	
N_2	24.3[28]	21.3
O_2	23.1[28]	22.6
HCl	38.5[28]	34.3
HBr	42.7[28]	
CH_3OH	46[28]	
HCOOH	51[28]	

Van der Waals[26] calculated the energy of clathration for several quinol clathrates, using a value for the energy parameter ε^*_{ww} which gave the best agreement with the experimental values for the clathration energies. Later, van der Waals and Platteeuw[3] presented slightly different values, having chosen the energy parameter to give the best overall fit with "the aggregate of experimental data at present available". Their results, expressed as ΔH_c values, are compared with experiment in Table 3. The agreement is very fair.

Small molecules can also form clathrates with phenol in its unstable β-form, which offers the guest molecule two kinds of cavity. For the systems with Kr, Xe, CH_4, C_2H_4, C_2H_6, CO_2 and SO_2 as the guest, Allison and Barrer[46] measured dissociation pressures over a range of temperature, and so derived ΔH_c values. These varied from $22.4\,kJ\,mol^{-1}$ for the krypton clathrate to $44.8\,kJ\,mol^{-1}$ for the sulphur dioxide clathrate. These values refer to the formation of an amount of the β-phenol clathrate containing one mole of guest from the gaseous guest and α-phenol, the normal form of the host. They also obtained isotherms for the first six of the above gases in β-phenol, and hence derived ΔH_c values for the process

$$M(g) + (n/y)\beta\text{-phenol} \rightarrow M.(n/y)\beta\text{-phenol}.$$

By combining these latter ΔH_c values with those resulting from the dissociation pressure studies, they obtained ΔH for the process α-phenol $\rightarrow$ β-phenol. This quantity proved to be dependent on the guest molecule, increasing from $0.25\,kJ\,mol^{-1}$ for methane to $13\,kJ\,mol^{-1}$ for ethane. In other words, the guest can cause the phenol lattice to be perturbed, with a consequential energy change which may not be negligible in comparison with the host–guest interaction energy. The values of ΔH_c, the enthalpy increase on forming the clathrate from the β-form and the gaseous guest, are as follows:

guest	Kr	Xe	CH_4	C_2H_6	C_2H_4	CO_2
$-\Delta H_c(kJ\,mol^{-1})$	24.5	44.5	27	47.5	60.5	65

3.3. Clathrate hydrates

Both type I and type II hydrates have cavities of two sizes, the difference in size being rather larger for type II than for type I. Chloroform forms a type II hydrate in which the $CHCl_3$ molecules are necessarily restricted to the larger holes, leaving the smaller cavities empty. As already mentioned (Section 2) Barrer and Edge[5] used chloroform hydrate to study the clathration of rare gas atoms in the smaller holes. For ΔH_c for the process

$$CHCl_3\text{-hydrate} + \text{one mole rare gas} \rightarrow CHCl_3\text{-rare gas-hydrate}$$

they obtained the following values, in $kJ\,mol^{-1}$: -26 (Ar); -28 (Kr); -31 (Xe). Using this source of information about the interaction of a rare gas atom with the water molecules lining the cavity, they were able to analyse their results for the enthalpy of formation of the rare gas type I hydrates, that is the enthalpy gain for the process

$$\text{normal ice} + \text{one mole rare gas} \rightarrow \text{type I hydrate}$$

They thus arrived at ΔH_c values for the clathration of guest molecules in each type of cavity. The following ΔH_c values refer to the process

empty hydrate structure + one mole rare gas → hydrate with water molecules in one kind of hole only.

	Ar	Kr	Xe
$-\Delta H_c$(kJ mol^{-1}) for (smaller) dodecahedral holes	24.6	27.2	34.0
$-\Delta H_c$(kJ mol^{-1}) for (larger) tetrakai decahedral holes	22.7	24.5	29.8

It will be noted that the energy liberated on clathration is somewhat larger when the rare gas atom enters the smaller cavity.

In carrying out this analysis it was necessary to have a value for the enthalpy change for the process

normal ice → empty type I hydrate structure

The estimate made by Barrer and Edge of this quantity was ~1.5 kJ mol^{-1}. Platteeuw and J. H. van der Waals[8] had previously concluded that this enthalpy change is small, since, on the assumption that it is zero, they obtained good agreement between the experimental values and calculated values of ΔH_c for some type I hydrates, as the following figures show:

	Kr	Xe	CH$_4$
$-\Delta H_c$(calc.)(kJ mol^{-1})	60.2	68.1	60.2
$-\Delta H_c$(expt.)(kJ mol^{-1})	50	70	61

Here, ΔH_c refers to the process

$$M(g) + nH_2O(l) \rightarrow M.nH_2O(hydrate)$$

The calculated values were obtained from the theory of Platteeuw and van der Waals.

Thus, for quinol, phenol and ice clathrates, the enthalpy difference between the stable solid form of the host and the clathrate form with the cavities empty, but undistorted, is quite small in all three cases. But of course, were this otherwise, the clathrates could only be stabilized by much stronger host-guest interaction.

3.4. Channel compounds of urea and thiourea

Redlich et al.[24] studied a considerable number of channel compounds of urea, the guests being several n-alkanes with between 7 and 16 carbon atoms, n-alkenes, halides, alcohols, ketones, acids, methyl esters, ether and dioxane. Measurements of the equilibrium constant for the formation of an inclusion compound were carried out over a temperature range (usually of ~25–40 K), and yielded ΔH_c values which, for members of a homologous series as the guests, proved to be an approximately linear function of the number of carbon atoms in the chain. As is to be expected, the enthalpies of clathration are numerically greater than in the systems so far considered. Thus, ΔH_c for the formation of the inclusion compound from the gaseous hydrocarbon is $-67\ \text{kJ mol}^{-1}$ for $n\text{-}C_7H_{16}$, and $-169\ \text{kJ mol}^{-1}$ for $n\text{-}C_{16}H_{32}$, the average increment in ΔH_c per CH_2 group being $11.3\ \text{kJ mol}^{-1}$. In their theoretical treatment of the inclusion compounds of urea with n-alkanes, Parsonage and Pemberton[48] obtained estimates of 10.9 and $14.2\ \text{kJ mol}^{-1}$ for this increment, thus confirming their assumption that the energy of interaction of host and guest atoms can be adequately represented by a Lennard-Jones 6–12 potential.

A similar but less comprehensive experimental study of the inclusion compounds of thiourea was carried out by Redlich and his co-workers.[49] Here the channels are wider, and the guest molecules generally broader. The guest molecules used by Redlich *et al.* included branched-chain alkanes and cycloparaffins. The clathration enthalpies are again considerable, ΔH_c for the formation of the inclusion compound from a mole of gaseous cyclohexane being $-48.1\ \text{kJ mol}^{-1}$, for example.

4. Heat capacity measurements

4.1. Introduction

The determination of the heat capacity of clathrates over an extended range of temperature serves two main purposes. One is to establish whether or not the clathrate has transitions, and if it has, to investigate their thermodynamic characteristics. This aspect is discussed and illustrated in Section 5. The other objective is to obtain information which can lead to some insight into the motion of the guest molecules in the cavities, and it is this which is discussed in this Section.

The quantity which has to be determined is the contribution C_M which the guest molecules M make to the measured overall heat capacity C_p of the clathrate. This contribution will in general be only a relatively small proportion of C_p. Thus, for the β-quinol clathrate with 78% of the cavities filled with argon atoms, the latter contribute ~4 per cent to C_p at 240 K. This therefore limits the precision with which C_M can be assessed. With clathrates of other kinds, with a less favourable guest/host ratio, C_M could be an even smaller fraction of C_p, to such an extent that the approach we are about to discuss would scarcely be feasible or worthwhile.

4.2. Monatomic guest species

All of the work to be reviewed in this Section involves clathrates of β-quinol. The simplest are those in which the guest is a rare gas. The first experiments were carried out by Parsonage and Staveley[50] on the argon clathrate, and by Grey *et al.*[45] on the krypton clathrate. For both systems, C_p at any one temperature proved to be a linear function of y. (The ranges of y were 0.20 to 0.78 for the argon clathrate, and 0.24 to 0.80 for the krypton system). This is reassuring evidence that guest–guest interaction plays a negligible role. Accordingly, the contribution to C_p made by a mole of guest (C_{Ar} or C_{Kr}) was readily obtained.

For monatomic guest species, the only kind of movement which has to be considered is the 'rattling' of the atom in the cavity. It has already been

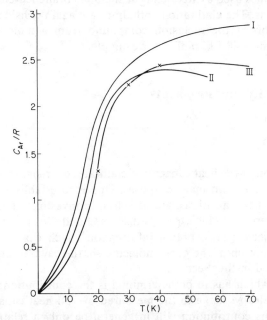

Fig. 5. Comparison of experimental values and values calculated from quantum mechanical theories for the heat capacity contribution (C_{Ar}) of argon in the quinol clathrate. I; harmonic oscillator approximation: II; anharmonic oscillator approximation: III; experimental values.[45] (Reproduced with permission, Mol. Phys., 1961, **4**, 153.)

pointed out in the introductory Section that the rattling frequency is low (e.g. 35 cm^{-1} for krypton), and so from, say, 120 K upwards it should be entirely reasonable to compare the experimental values of C_{Ar} and C_{Kr} with those calculated from a classically-based theory such as that of J. H. van der Waals. Such comparisons are shown graphically in Fig. 4 for the argon and krypton clathrates. Agreement between experiment and theory from about 100 K upwards is good, especially if the parameter σ_{ww} is reduced from the value of 0.33 nm used by van der Waals to 0.29 nm.

For calculations of the heat capacity contribution of the guest at lower temperatures, the simplest model one can adopt is, of course, the harmonic oscillator. In the circumstances, this might be expected to be a rather poor approximation, and to take account of anharmonicity, Parsonage[51] expanded the Lennard-Jones and Devonshire potential in powers of r^2 as far as r^4. The resulting values of C_{Ar} calculated with this improvement are compared with experiment in Fig. 5. The agreement is reasonably good for argon, but less so for krypton (Fig. 6). Fig. 6 also embodies the results of calculations by Grey and Staveley,[52] who constructed a partition function from the first 200 energy levels for the krypton atoms obtained by solving

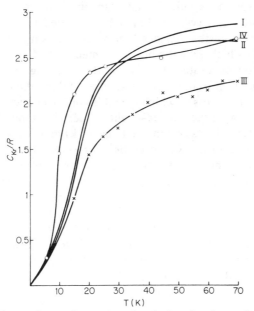

Fig. 6. Comparison of experimental and calculated values of the heat capacity contribution (C_{Kr}) of krypton in the quinol clathrate. I, II and III have the same significance as in Fig. 5. IV is for the quantum-mechanical cell model.[45] (Reproduced with permission, *Mol. Phys.*, 1963, **7**, 83.)

the Schrödinger equation using the Lennard-Jones and Devonshire potential. The C_{Kr} values so calculated, however, do not show improved agreement with experiment. Hazony and Ruby[1] suggested that the discrepancies between calculation and experiment for the krypton clathrate might be attributed to changes in the quinol phonon spectrum due to host–guest coupling, (see Parsonage and Staveley[53a]).

4.3. Diatomic guest molecules

The three most thoroughly studied β-quinol clathrates in this category are those with carbon monoxide, nitrogen and oxygen as the guests. For all three systems C_p was found to have a linear dependence on y. None of these three molecules could be expected to rotate freely, in the literal sense, in its cavity, and it was clearly of interest to attempt to estimate V_0, the height of the potential barrier inhibiting free rotation. The contribution C_m which each mole of guest makes to C_p may now be considered to have two components, namely $C(vib)$ (that due to the rattling), and $C(rot)$ (that due to the hindered rotation). The contribution from the intramolecular vibration of the diatomic species is quite insignificant up to room temperature. To obtain $C(rot)$ from C_M, $C(vib)$ has first to be estimated. Grey and Staveley[52] calculated $C(vib)$ from the van der Waals theory, using values for the σ and ε parameters which gave the best agreement with C_{Ar} and C_{Kr} for the argon and krypton clathrates. To analyse the resulting values of $C(rot)$, each diatomic guest molecule was regarded as making the contribution of two one-dimensional rotors to which the tables of Pitzer and Gwinn[54] could be applied. In this way, the barrier height for carbon monoxide molecules in the β-quinol clathrate was estimated to be 4.6 kJ mol^{-1} (Fig. 7). Stepakoff and Coulter[55,56] used their own experimental results for C_p for the carbon monoxide clathrate (which gave somewhat higher results for C_{CO} than those of Grey and Staveley), and treated them in two ways, according to the temperature. At high temperatures they used the classical cell model approach, but adjusted the parameters ε and σ and also V_0 to give the best fit with C_{CO}. This gave the considerably higher figure of 12.0 kJ mol^{-1} for V_0. They interpreted the C_{CO} values from 15 K to 80 K, however, on the basis that the carbon monoxide molecules undergo three-dimensional harmonic oscillation with a frequency of 97.3 cm^{-1}, and also two-dimensional hindered rotation with a frequency $\nu(lib)$ of 43.1 cm^{-1}. Using the relation

$$\nu(lib) = (V_0/2\pi^2 I)^{1/2}$$

this leads to a value of 3.1 kJ mol^{-1} for V_0, which is much nearer the estimate of Grey and Staveley. A relatively low barrier of 2.9 kJ mol^{-1} was also

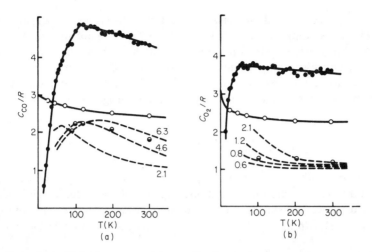

Fig. 7. Analysis of the heat capacity contribution versus temperature curves for (a) carbon monoxide and (b) oxygen in their quinol clathrates. ●, experimental values; ○, calculated values of C_{vib}/R; ◐; derived values of $C_{rot}/R = (C_{CO} - C_{vib,CO})/R$ ˙ or $(C_{O_2} - C_{vib,O_2})/R$. Broken curves, calculated values of C_{rot}/R for hindering barriers with values (in kJ mol^{-1}) shown.[52] (Reproduced with permission, *Mol. Phys.*, 1963, 7, 83.)

suggested by Ball and McKean[57] from a study of the infrared spectrum of the clathrate, but there may be some uncertainty in the derivation of this value of V_0 (see Parsonage and Staveley[53b]).

Grey and Staveley[52] also studied the nitrogen clathrate. Application of the same approach as that just outlined for the carbon monoxide clathrate gave $V_0 = 4.6$ kJ mol^{-1}. This is in reasonable agreement with the value of 3.93 kJ mol^{-1} derived by Meyer and Scott[58] from a nuclear quadrupole resonance investigation. Coulter *et al.*[59] fitted their own results for C_{N_2} with a rattling frequency of 83.4 cm^{-1} and a librational frequency ν(lib) of 37.9 cm^{-1}, whence $V_0 = 2.1$ kJ mol^{-1}.

For the oxygen clathrate, Grey and Staveley[52] obtained a lower barrier height of 0.84 kJ mol^{-1}, (Fig. 7). There are two other independent estimates of this quantity, one of 0.54 kJ mol^{-1} from the temperature dependence of the magnetic susceptibility (Cooke *et al.*;[60] Meyer *et al.*[61]) and the other of 1.05 kJ mol^{-1} from an electron spin resonance study by Foner *et al.*[62] There seems little doubt, therefore, that the librational motion of the oxygen molecule in the cavities in β-quinol is much less restricted than that of the nitrogen or carbon monoxide molecules. Although the bond in the oxygen molecule is longer than those in the nitrogen and carbon monoxide molecules, the oxygen molecule appears to be effectively the smallest of the three. Its collision diameter as evaluated from viscosity measurement is 0.343 nm, as compared with 0.359 nm for CO and 0.368 nm for N_2.

Roper[63] measured C_p for one sample of the hydrogen chloride β-quinol clathrate, and by finding values of ν(vib) and ν(lib) which fitted the C_p results up to 100 K, estimated that V_0 was 1.3 kJ mol^{-1}. Stepakoff[55] made similar measurements on one specimen of the hydrogen bromide clathrate.

4.4. Polyatomic guest molecules

The methane clathrate was studied by Parsonage and Staveley.[44] Once again, C_p was a linear function of y, and once again J. H. van der Waals' theory was used to evaluate the rattling contribution. The results of the analysis of C_{CH_4} are shown in Fig. 8. It will be seen that above about 150 K, C(rot)

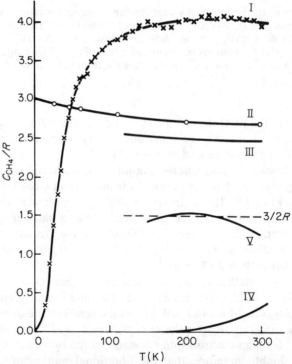

Fig. 8. Analysis of the contribution of methane to the heat capacity of the quinol clathrate (C_{CH_4}).I; C_{CH_4} (expt.): II; C_{vib} from the classical theory of van der Waals: III; corrected values of C_{vib}: IV; the contribution of the internal vibrational modes of methane (C_{int}): V; value deduced for the rotational contribution[44] (C_{rot}). (Reproduced with permission, *Mol. Phys.*, 1960, **3**, 59.)

is $3R/2$ within experimental error, the classical value for free three-dimensional rotation. It may be noted that $C(\text{rot})$ is only ~2.8% of C_p at 300 K. The experimental results up to 50 K were used by Coulter *et al.*,[59] who obtained a good fit over this range with $\nu(\text{vib}) = 160 \text{ cm}^{-1}$ and $\nu(\text{lib}) = 56.5 \text{ cm}^{-1}$. Use of the relation between V_0 and $\nu(\text{lib})$ quoted above gave $V_0 = 0.8 \text{ kJ mol}^{-1}$.

From this discussion of the clathrates having methane or a diatomic molecule as the guest, it should be evident that there are rather severe limitations on the precision with which these relatively low barrier heights can be estimated from heat capacity data. This is partly because the relevant quantity C_M is only a few per cent of the measured C_p, and partly because of the assumptions or approximations which have to be made in analysing C_M. Perhaps one of the most important of these is that the contributions from the rattling and the librational movement are regarded as being quite separate.

The transitions in the β-quinol clathrates of hydrogen cyanide and methanol, which have been studied by heat capacity measurements, are considered in Section 5. With larger molecules such as these as guests, it is doubtful whether an analysis of the heat capacity of the kind applied to the clathrates with smaller guest molecules would be meaningful.

5. Thermal anomalies

5.1. Introduction

So far it has been assumed that there is no coupling between the motions of guest molecules in different cages. By "no coupling" is meant that the coupling must be small compared with kT. It follows that it is at low temperatures that any evidence of such coupling is most likely to be seen.

An early observation of this kind was that of Meyer and Scott on the ^{14}N nuclear quadrupole resonance spectra of nitrogen β-quinol clathrates.[58] They found that the spectrum showed splitting into seven lines and that the relative intensities of these were dependent on the fraction of cavities occupied. Their conclusion was that the splitting was related to the occupation of adjacent cages, the coupling being transmitted by a slight distortion of the walls of the cages. No such distortion was, however, observable by X-ray diffraction. Very similar effects were found in the low-temperature (1.5–4.2 K) ESR spectra of O_2–N_2 mixed clathrates.[62] The splitting observed was, in this case, attributed to the effect on the resonating O_2 spin of N_2 molecules in neighbouring cavities.

Much larger effects, in fact large thermal anomalies, have subsequently been found in the inclusion compounds of urea, thiourea and quinol, and these are discussed below.

5.2. Urea and thiourea inclusion compounds

The guest aliphatic molecules lie along the channels in the host lattice and are almost completely in the all-*trans* configuration. At first sight it would appear that the only interaction between guest molecules would be that between the end-groups of the adjacent included molecules in the same channel (longitudinal interactions) (Fig. 9). This interaction would tend to lead to orientation of the planes of the hydrocarbon skeletons. However,

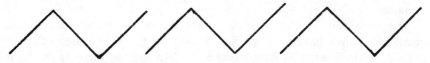

Fig. 9. End-interactions of alkane molecules in urea channel compounds.

such a system, if there were no interactions between molecules in different channels, would just correspond to a set of one-dimensional systems. Now it is well-known that a one-dimensional system cannot possess long-range order at non-zero temperatures unless the interactions are of infinite range. However, it is found experimentally that the inclusion compounds of the simple n-alkanes do show gradual transitions, with most of the heat absorption occurring over ~ 20 K (Fig. 10).[64] Since the transition is gradual (with a rounded maximum) and does not involve non-analyticity of the free energy it is not strictly forbidden by the above rules. Furthermore, the longitudinal interaction is also long-range, although very weakly so. Nevertheless, bearing in mind the "spirit" of the rule, the experimental observation is very surprising indeed and suggests that the assumption of no interaction between guest molecules in different channels should be re-examined. The transverse interaction might be thought to be far too weak to have any noticeable effect at temperatures as high as those at which the anomalies are observed (100–180 K). However, it was found that if the system was represented by an Ising model with strong interactions in one direction (along the channel) and much weaker interactions along the other two, and values for the Ising parameters were chosen on the basis of the intermolecular potentials, then disordering transitions were found at temperatures similar to those at which the experimental effects were observed (Fig. 11).[48] This treatment involved the application of a mean field approximation in the transverse directions,

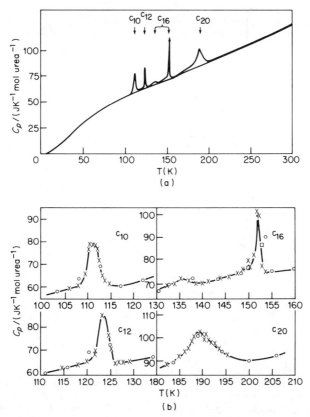

Fig. 10. Heat capacity versus temperature for four urea *n*-alkane inclusion compounds. The quantity of inclusion compound referred to is that which contains 1 mole of urea. (a) Overall plot for the four inclusion compounds; (b) detailed plots in the transition region for each inclusion compound.[64] (Reproduced with permission, *Trans. Faraday Soc.*, 1965, **61**, 2112.)

and this probably accounts for the shape predicted for the heat capacity anomaly, which was of the classical Ehrenfest second-order type. Had an exact treatment of the Ising model been possible the shape would probably have been similar to the second-order transitions discussed by Pippard,[65] with an infinite divergence. The rounding of the transition could be attributed to the guest-host interaction which acts in a manner analogous to a field of variable direction in the magnetic case. As a result of the insight provided by this theory two main reasons can be found why the transitions occur at such high temperatures. Firstly, the lateral interactions are not as weak as might have been thought because they arise from the summing of

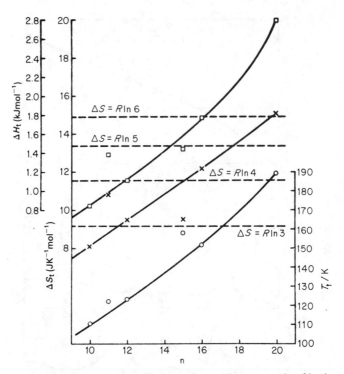

Fig. 11. The enthalpies (□) and entropies (×) of transition per mole of hydrocarbon and the transition temperature (○) of the urea n-alkane inclusion compounds; n = the number of carbon atoms in the n-alkane molecule. Also shown are the ordinates corresponding to $\Delta S = R \ln 3$, $R \ln 4$, $R \ln 5$, and $R \ln 6$.[64] (Reproduced with permission, *Trans. Faraday Soc.*, 1965, **61**, 2112.)

a large number of atom–atom potentials which act in concert. Secondly, the strong, longitudinal interactions tend to cause the molecules within a channel to be strongly correlated, which further increases the effectiveness of the lateral interactions. In view of the reorientational nature of the transition it is not surprising that when a branched-chain alkane, such as 2-methylpentadecane, is used as the guest the anomaly disappears entirely.[66] It was also noted that the behaviour of inclusion compounds of "even" and "odd" n-alkanes showed the alternating effect which is so common amongst solids composed of long-chain molecules (Fig. 11).[48] A by-product of this theory was the realization that the end-groups of the guest molecules should approach more closely (0.374 nm against ~0.41 nm) than they would in the absence of the host lattice. The closing-up of the chains of guest molecules is a consequence of the attempt to avoid "wasting" the attractive guest-host interactions by having too large a gap between successive guest molecules. Observation of the closing-up had been made just previously by X-ray diffraction.[67] A consequence of the closing-up is a marked increase in the strength of the orientational interaction between end-groups, which in turn raises the predicted transition temperature. These transitions have subsequently been studied by X-ray diffraction[68] and this has confirmed the picture presented above.

On the basis of the above model similar transitions would be expected for many other inclusion compounds, and indeed for a series of six even alk-1-enes (C_{10}–C_{20}) only the lowest did not show a thermal anomaly.[69,70] (It is a general observation for both alkane and alkene inclusion compounds that the entropy of the transition decreases as the chain is shortened.) The alkene inclusion compound behaved in a generally similar manner to the alkane compounds, although the transition temperature of the $C_{14}H_{28}$ adduct was approximately twice that which would be expected by interpolation between the data for its neighbours. As the two ends of the alkene molecules are different (ane, ene) an extra type of disorder (head-tail) is possible in these compounds. Since the interaction between end-groups would be dependent upon whether the pair concerned were ane-ane, ene-ene or ane-ene, the transition temperature would be expected to be sensitive to the nature of the end contacts. The arrangement of end contacts would of course be frozen-in at the time of crystallization, and no subsequent change in the head-tail ordering could occur. Monte Carlo simulations of the Ising models corresponding to the various possible types of head-tail ordering (-head head-tail tail-head head-, head-tail head-tail head-, and random) have been carried out.[71] The Monte Carlo method suffers from very slow convergence near to transitions, but the existence of a transition and its approximate location (to ±20 K) could be easily determined. The problem of the unusually high transition temperature of the $C_{14}H_{28}$ inclusion compound, however, remains to be solved.

The urea inclusion compound of trioxane

$$\begin{array}{c} CH_2-O \\ \diagup \qquad \diagdown \\ O \qquad\qquad CH_2 \\ \diagdown \qquad \diagup \\ CH_2-O \end{array}$$

has been shown by DTA to have three thermal anomalies between 113 K and room temperature, the entropies of the transitions being 7.65, 2.25 and 3.11 J K^{-1} mol. guest^{-1} for the transitions at 189.9 K, 200.8 K and 242.5 K, respectively.[72] X-ray diffraction patterns indicated that, whereas at room temperature the periodicity of the guest lattice in the c-direction was twice that of the host lattice, on cooling through the transition at 240 K this changed to a periodicity equal to that of the host lattice; no further change in this periodicity occurs on cooling through the lower transitions, at which some distortion of the host lattice takes place. The existence of two kinds of resonance line in the ^{14}N NQR spectrum for each of the phases suggested that there were two kinds of domain in the crystal, with the trioxane molecules occupying only one of the two kinds of site in each domain. There is a great similarity between this inclusion compound and the cyclohexane inclusion compound of thiourea, which has the same room temperature structure and shows at least two transitions on cooling, their entropies being similar to those of the urea trioxane inclusion compound. Corresponding to the thermal anomalies, wide-line NMR on the urea trioxane inclusion compound showed narrowing to occur in three steps (at approximately 128 K, 198 K and 243 K).[73]

Following the observations of transitions in the urea inclusion compounds, a natural next step was to investigate the corresponding thiourea inclusion compounds. These have yielded a rich sequence of anomalies, but probably because of the generally greater complexity of the guest substances involved it has not so far been possible to provide a theoretical treatment for any of these effects. The first compounds to be examined were those of the cycloparaffins (C_6–C_8) and tetrahydropyran.[71,74–76] Transitions were found in all these compounds (Table 4; Fig. 12). DTA studies[76] failed to disclose peaks in either the C_7 or C_8 inclusion compounds, and located

Table 4. *Transition temperatures and enthalpies and entropies of transition of thiourea inclusion compounds. (The quantities referred to are those which contain one mole of hydrocarbon.)*

Guest	T_t (K)	ΔH_t (kJ mol^{-1})	ΔS_t (J K^{-1} mol^{-1})
c-$C_6H_{12}^{75}$	128.8	1.142	8.87
	130–150	0.331	2.34
	153–161	0.036	0.25
	170.8	0.142	0.84
	210–240	0.084	0.38
c-$C_7H_{14}^{75}$	162.4	0.123	0.75
	241	0.190	0.79
	262	0.031	0.12
	Shallow hump	1.606	11.54
c-$C_8H_{16}^{75}$	187.2	0.708	3.78
	240	7.186	31.68
	265	0.236	0.89
$(CH_3)_3C.C_2H_5^{75}$	69.9	0.310	4.35
	89.5	1.160	12.93
	169.6	0.240	1.42
$(CH_2)_5O^{16}$	170	5.9	34
	184	0.63	3.4

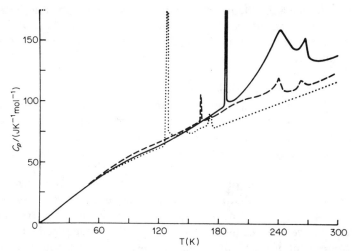

Fig. 12. Heat capacity versus temperature for thiourea inclusion compounds. Dotted line; c-C_6H_{12}: dashed line; c-C_7H_{14}: full-line; c-C_8H_{16}.[75] (Reproduced with permission, *J. Chem. Thermodyn.*, 1972, **4**, 829.)

only one for C_6 (at $\sim$129 K). An X-ray study by the same workers, however, located two transitions for the latter compound ($\sim$148 K and $\sim$126 K). The large hump found for the C_8 inclusion compound has been attributed on the basis of its size ($\Delta S_t = R \ln 45.1$), together with a consideration of geometric possibilities, to conformational changes of the guest molecules.[75] Some evidence for the effect of inclusion on the relative stability of conformational isomers in a related compound has been reported by Fukushima.[77] He found that in the thiourea c-$C_6H_{11}Cl$ inclusion compound the substituent was predominantly *axial*, whereas in the pure guest solid it is entirely *equatorial* and in the pure liquid it is largely so.

The sandwich compound ferrocene($Fe(C_5H_5)_2$) can also form an inclusion compound with thiourea and this has been found by DTA to have a thermal anomaly at $\sim$162 K.[78] Although it has been studied by X-ray diffraction the nature of the process involved is still unclear.[79]

5.3. β-quinol clathrates

Transitions have also been found in the β-quinol clathrates of some guests having large dipole moments. Since these compounds can be prepared with a wide range of filling of the cavities it would be expected that any cooperative transition would be very dependent upon this degree of occupancy. Indeed, the theory of such processes would appear to be a form of the site Percolation Problem of statistical mechanics[80] in which the probable extent of clusters of adjacent occupied sites is investigated. It is therefore unfortunate that in some of the work on this topic the occupancy has not been reported.

The first report of a transition of this kind was for a methanol clathrate with $y = 0.974$.[81] A DTA run showed a transition at 66 K, and a sharp change in the dielectric loss was found at the same temperature. A low-temperature calorimetric study was then made by Belliveau:[82] a sample with $y = 0.552$ showed no anomaly, but others, with $y = 0.644$, 0.750 and 0.989, displayed transitions for which the transition temperature varied linearly with y from 35 K for $y = 0.644$ to 71 K for $y = 0.989$. More recent work[83] on the methanol clathrate has confirmed that $\mu = 1.3$ D, the value used by Matsuo et al.[81] to explain their results, is also appropriate to explain the dielectric data. A favoured arrangement for the methanol molecules is suggested which leads to the component of the dipole moment along the c-axis being 1.28 D. This model is consistent with the observed small activation barrier against rotational jumps $(2.59 \pm 1.3 \text{ kJ mol}^{-1})$.

The β-quinol HCN clathrate with $y \approx 1$ was also examined calorimetrically, and C_p was found to have a sharp peak at 178.1 K with $\Delta S_t = (0.687\text{--}0.730)R$, within experimental error equal to $R \ln 2 (= 0.693\,R)$, suggesting that the process is one of simple dipole ordering; a change in the behaviour of the permittivity was found at the same temperature.[84] Davies and Williams,[85] using a sample with $y = 0.8$, found a break in the permittivity at ~ 225 K, but nothing else down to their lowest temperature (113 K). It could well be that the transition equivalent to that found by Matsuo et al.[84] would occur at <113 K for a crystal with $y = 0.8$.

References

1. Y. Hazony and S. L. Ruby, *J. Chem. Phys.*, 1968, **49**, 1478.
2. J. H. van der Waals and J. C. Platteeuw, *Recl. Trav. Chim. Pays-Bas*, 1956, **75**, 912.
3. J. H. van der Waals and J. C. Platteeuw, *Adv. Chem. Phys.*, 1959, **2**, 1.
4. J. N. Helle, D. Kok, J. C. Platteeuw and J. H. van der Waals, *Recl. Trav. Chim. Pays-Bas*, 1962, **81**, 1068.
5. R. M. Barrer and A. V. J. Edge, *Proc. R. Soc. London, Ser. A*, 1967, **300**, 1.
6. J. A. Ripmeester and D. W. Davidson, *J. Mol. Struct.*, 1981, **75**, 67.
7. D. W. Davidson in *Water, a comprehensive treatise*, (ed. F. Franks), Plenum Press, New York and London, 1973. Vol. 2, Ch. 3. (a) p. 116; (b) p. 154; (c) p. 164.
8. J. C. Platteeuw and J. H. van der Waals, *Mol. Phys.*, 1958, **1**, 91.
9. S. L. Miller, S. R. Gough and D. W. Davidson, *J. Phys. Chem.*, 1977, **81**, 2154.

10. D. R. Hafemann and S. L. Miller, *J. Phys. Chem.*, 1969, **73**, 1392, 1398.
11. Y. A. Majid, S. K. Garg and D. W. Davidson, *Can. J. Chem.*, 1969, **47**, 4697.
12. S. R. Gough and D. W. Davidson, *Can. J. Chem.*, 1971, **49**, 2691.
13. B. J. Falabella and M. Vanpee, *Ind. Eng. Chem., Fundam.*, 1974, **13**, 228.
14. S. L. Miller, E. J. Eger and C. Lundgren, *Nature*, 1969, **221**, 468.
15. S. K. Garg, S. R. Gough and D. W. Davidson, *J. Chem. Phys.*, 1975, **63**, 1646.
16. J. C. Rosso and L. Carbonnel, *C.R. Hebd. Séances Acad. Sci.*, 1971, **C272**, 136, 713.
17. J. Kaloustian, J. C. Rosso and L. Carbonnel, *C.R. Hebd. Séances Acad. Sci.*, 1972, **C275**, 249.
18. L. Carbonel, J. C. Rosso and J. Kaloustian, *C.R. Hebd. Séances Acad. Sci.*, 1974, **C279**, 243.
19. J. Kaloustian, J. C. Rosso, C. Caranoni and L. Carbonnel, *Rev. Chim. Mineral.*, 1976, **13**, 334.
20. K. W. Morcom and R. W. Smith, *J. Chem. Thermodyn.*, 1971, **3**, 507.
21. D. W. Davidson, S. R. Gough, F. Lee and J. A. Ripmeester, *Rev. Chim. Mineral.*, 1977, **14**, 447.
22. H. Nakayama and M. Hashimoto, *Bull. Chem. Soc. Jpn.*, 1980, **53**, 2427.
23. D. W. Davidson, S. R. Gough, J. A. Ripmeester and H. Nakayama, *Can. J. Chem.*, 1981, **59**, 2587.
24. O. Redlich, C. M. Gable, A. K. Dunlop and R. W. Millar, *J. Am. Chem. Soc.*, 1950, **72**, 4153.
25. L. C. Fetterly in *Non-stoichiometric Compounds*, (ed. L. Mandelcorn), Academic Press, New York and London, 1964, Ch. 8.
26. J. H. van der Waals, *Trans. Faraday Soc.*, 1956, **52**, 184.
27. J. E. Lennard-Jones and A. F. Devonshire, *Proc. R. Soc. London, Ser. A*, 1937, **163**, 53; 1938, **165**, 1.
28. D. F. Evans and R. E. Richards, *Proc. R. Soc. London, Ser. A*, 1954, **223**, 238.
29. R. M. Barrer and W. I. Stuart, *Proc. R. Soc. London, Ser. A*, 1957, **243**, 172.
30. R. M. Mazo, *Mol. Phys.*, 1964, **8**, 515.
31. V. McKoy and O. Sinanoglu, *J. Chem. Phys.*, 1963, **38**, 2946.
32. W. R. Parrish and J. M. Prausnitz, *Ind. Eng. Chem. Proc. Res. Dev.*, 1972, **11**, 26.
33. G. D. Holder, G. Corbin and K. D. Papadopoulos, *Ind. Eng. Chem. Fundam.*, 1980, **19**, 282.
34. J. W. Tester, R. L. Bivins and C. C. Herrick, *Am. Inst. Chem. Eng. J.*, 1972, **18**, 1220.
35. H. J. F. Stroud and N. G. Parsonage, *Proc. 2nd Int. Conf. Molec. Sieve Zeolites, Worcester, Mass.*, 1971. (Advances in Chemistry Series No. 102) Vol. 2, p. 138.
36. H. J. F. Stroud, E. Richards, P. Limcharoen and N. G. Parsonage, *J. Chem. Soc. Faraday Trans. I*, 1976, **72**, 942.
37. D. W. Davidson and J. A. Ripmeester, *J. Glaciol.*, 1978, **21**, 33.
38. D. W. Davidson, *Can. J. Chem.*, 1971, **49**, 1224.
39. F. Hollander and G. A. Jeffrey, *J. Chem. Phys.*, 1977, **66**, 4699.
40. Y. Hazony and R. H. Herber, *J. Inorg. Nucl. Chem.*, 1971, **33**, 961.
41. B. Holk, *Phys. Rev.*, 1975, **B12**, 1620.
42. S. R. Gough, S. K. Garg, J. A. Ripmeester and D. W. Davidson, *J. Phys. Chem.*, 1977, **81**, 2158.
43. D. F. Evans and R. E. Richards, *J. Chem. Soc.*, 1952, 3932.
44. N. G. Parsonage and L. A. K. Staveley, *Mol. Phys.*, 1960, **3**, 59.
45. N. R. Grey, N. G. Parsonage and L. A. K. Staveley, *Mol. Phys.*, 1961, **4**, 153.

46. S. A. Allison and R. M. Barrer, *Trans. Faraday Soc.*, 1968, **64**, 549.
47. S. Hirokawa, *Mol. Phys.*, 1978, **36**, 29.
48. N. G. Parsonage and R. C. Pemberton, *Trans. Faraday Soc.*, 1967, **63**, 311.
49. O. Redlich, C. M. Gable, L. R. Beason and R. W. Millar, *J. Am. Chem. Soc.*, 1950, **72**, 4161.
50. N. G. Parsonage and L. A. K. Staveley, *Mol. Phys.*, 1959, **2**, 212.
51. N. G. Parsonage, D. Phil. Thesis, Oxford, 1959.
52. N. R. Grey and L. A. K. Staveley, *Mol. Phys.*, 1963, **7**, 83.
53. N. G. Parsonage and L. A. K. Staveley, *Disorder in Crystals*, Clarendon Press, Oxford, 1978. (a) p. 728; (b) p. 731.
54. K. S. Pitzer and W. D. Gwinn, *J. Chem. Phys.*, 1942, **10**, 428.
55. G. L. Stepakoff, *Diss. Abstr.*, 1963, B24, 1424.
56. G. L. Stepakoff and L. V. Coulter, *J. Phys. Chem. Solids*, 1963, **24**, 1435.
57. D. F. Ball and D. C. McKean, *Spectrochim. Acta*, 1962, **18**, 933.
58. H. Meyer and T. A. Scott, *J. Phys. Chem. Solids*, 1959, **11**, 215.
59. L. V. Coulter, G. L. Stepakoff and G. C. Roper, *J. Phys. Chem. Solids*, 1963, **24**, 171.
60. A. H. Cooke, H. Meyer, W. P. Wolf, D. F. Evans and R. E. Richards, *Proc. R. Soc. London, Ser. A*, 1954, **225**, 112.
61. H. Meyer, M. C. M. O'Brien and J. H. van Vleck, *Proc. R. Soc. London, Ser. A*, 1957, **243**, 414.
62. S. Foner, H. Meyer and W. H. Kleiner, *J. Phys. Chem. Solids*, 1961, **18**, 273.
63. G. C. Roper, *Diss. Abstr.*, 1966, B27, 1444.
64. R. C. Pemberton and N. G. Parsonage, *Trans. Faraday Soc.*, 1965, **61**, 2112.
65. A. B. Pippard, *The Elements of Classical Thermodynamics*, Cambridge University Press, 1954, Ch. 9.
66. R. C. Pemberton and N. G. Parsonage, *Trans. Faraday Soc.*, 1966, **62**, 553.
67. F. Laves, N. Nicolaides and K. C. Peng, *Z. Kristallogr.*, 1965, **121**, 258.
68. Y. Chatani, Y. Taki and H. Tadokoro, *Acta Crystallogr.*, 1977, B33, 309.
69. A. F. G. Cope and N. G. Parsonage, *J. Chem. Thermodyn.*, 1969, **1**, 99.
70. D. J. Gannon and N. G. Parsonage, *J. Chem. Thermodyn.*, 1972, **4**, 745.
71. N. G. Parsonage, *Disc. Faraday Soc. No. 48*, 1969, 215.
72. (a) R. Clément, C. Mazieres and L. Guibé, *J. Solid State Chem.*, 1972, **5**, 436; (b) R. Claude, R. Clément and A. Dworkin, *J. Chem. Thermodyn.*, 1977, **9**, 1199.
73. R. Clément, *J. Magn. Reson.*, 1975, **20**, 345.
74. R. Clément, M. Gourdji and L. Guibé, *Mol. Phys.*, 1971, **21**, 247.
75. A. F. G. Cope, D. J. Gannon and N. G. Parsonage, *J. Chem. Thermodyn.*, 1972, **4**, 829.
76. R. Clément, J. Jegoudez and C. Mazieres, *J. Solid State Chem.*, 1974, **10**, 46.
77. K. Fukushima, *J. Mol. Struct.*, 1976, **34**, 67.
78. R. Clément, R. Claude and C. Mazieres, *J. Chem. Soc., Chem. Commun.*, 1974, 654.
79. E. Hough and D. G. Nicholson, *J. Chem. Soc., Dalton Trans.*, 1978, 15.
80. V. K. S. Shante and S. Kirkpatrick, *Adv. Phys.*, 1971, **20**, 325.
81. T. Matsuo, H. Suga and S. Seki, *J. Phys. Soc. Jpn.*, 1967, **22**, 677.
82. J. F. Belliveau, *Diss. Abstr.*, 1970, B31, 2591.
83. M. Massalska-Arodz, *Acta Phys. Polonica*, 1977, **52**, 555.
84. T. Matsuo, H. Suga and S. Seki, *J. Phys. Soc. Jpn.*, 1968, **25**, 641.
85. M. Davies and K. Williams, *Trans. Faraday Soc.*, 1968, **64**, 529.

2 · SPECTROSCOPIC STUDIES OF INCLUSION COMPOUNDS

J. E. D. DAVIES

University of Lancaster, Lancaster, UK

1. Introduction

For the purposes of this chapter the term spectroscopy is taken in its widest possible definition to include the techniques of infrared and Raman spectroscopy, inelastic neutron scattering, electron spin resonance, Mössbauer spectroscopy, photoelectron spectroscopy, and nuclear magnetic resonance spectroscopy, in order to illustrate the wide range of techniques which can be applied to the study of solid state inclusion compounds. Wherever possible the results obtained from the different techniques will be compared.

2. Infrared and Raman spectroscopic studies

The most recent comprehensive review on this topic was published[1] in 1978 and thus in order to minimize overlap with the previous review, this chapter will concentrate mainly on the results published since 1976.

INCLUSION COMPOUNDS III
ISBN 0-12-067103-4

2.1. Comparison of infrared and Raman spectroscopy

This section compares and contrasts the two techniques with particular reference to their use in the study of inclusion compounds.

Infrared spectroscopy is a transmission technique i.e. the infrared radiation is analysed after it has passed through the sample to determine which wavelengths have been absorbed. The fact that it is a transmission technique has one important consequence for the examination of solids, *viz* they have to be examined as mulls or discs in order to reduce the scattering of radiation by the solid particles and thus increase the amount of radiation transmitted through the sample.

The preparation of mulls or discs does require quite drastic treatment of the sample—it has to be ground to a fine powder, and, if a disc is being prepared, it is subsequently compressed with a pressure of about 10–15 tons. This therefore is one disadvantage of infrared spectroscopy in the study of inclusion compounds—they may be partially decomposed during the preparation of the mull or disc, and this effect has been observed in β-quinol clathrates with large guest molecules, and in Hofmann[2] and Werner inclusion compounds.

The experimental arrangement for Raman spectroscopy is very different to that of infrared spectroscopy. Firstly, the sample is irradiated with monochromatic radiation which is usually provided by a laser working in the visible region of the spectrum. Secondly, it is the scattered radiation which is analysed rather than the transmitted radiation as in infrared spectroscopy. This means that there is no need to prepare solid samples in any way. It is possible to examine the *neat solids* and this circumvents the disadvantage of infrared spectroscopy. Another advantage of Raman spectroscopy arises from the fact that visible radiation is used which means that glass sample cells can be used rather than the alkali halide cells which must be used in infrared spectroscopy.

This is particularly useful when studying unstable inclusion compounds. The procedure then is to put some of the inclusion compound in a glass sample tube, add a drop or two of the liquid guest and seal the tube. The inclusion compound is then under an atmosphere of the gaseous guest component which can prevent its decomposition. It is thus possible to record its Raman spectrum, but impossible to record its infrared spectrum.

A disadvantage of having to use alkali halide cells in infrared spectroscopy has recently been reported[115] when examining species containing 18-Crown-6. This particular crown ether is highly selective towards potassium ions (see Chapter 9, Volume 2) and thus potassium bromide cells should not be used when examining species containing 18-Crown-6, since the recorded spectrum is likely to be that of the K.[18-C-6]$^+$ complex!

The final difference between infrared and Raman spectroscopy arises from the different selection rules governing the activity of vibrations in the two techniques. To summarize these rules:

(i) A diatomic molecule must possess a permanent dipole moment for its vibration to be infrared active. Thus whilst it would not be possible to observe the vibration of the N_2 guest molecule in β-quinol using infrared spectroscopy, it can be observed using Raman spectroscopy.[3]

(ii) For polyatomic molecules, vibrations will be infrared active if there is a change in the dipole moment during the vibration; for Raman activity there must be a change of polarizability. For centrosymmetric polyatomic molecules, vibrations which are infrared active are Raman inactive and *vice versa.*

Thus in order to obtain the maximum amount of information about inclusion compounds using vibrational spectroscopy, it is necessary to use both infrared and Raman spectroscopy since the two techniques give complementary rather than identical information.

Even when a vibration is both infrared and Raman active, there can be a big difference in the intensity of the resulting band in the two spectra. A good example of this is the O—H bond, which gives rise to an intense band in the infrared spectrum but a very much weaker band in the Raman spectrum. This difference is illustrated in Fig. 1 which compares the infrared and Raman spectra of the β-quinol/HCl clathrate. The infrared spectrum is dominated by the broad, intense band arising from the hydrogen bonded O–H groups of the host lattice. This band obscures not only the C–H bands of the host lattice but also the band arising from the HCl guest molecule. In the Raman spectrum, on the other hand, the OH band is very much weaker and it is now possible to observe not only the aromatic C–H bands, but also the vibrational band of the HCl guest molecule.

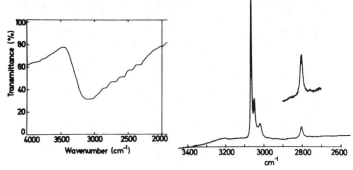

Fig. 1. Infrared (HCBD mull) and Raman (solid) spectra of the β-quinol HCl clathrate. Reproduced with permission from *Molecular Spectroscopy*, Volume 5 page 60.

2.2. Information which is potentially available from the spectra of inclusion compounds

The infrared and Raman spectra of inclusion compounds potentially contain a great deal of information. A brief description is given below under the various headings and more detailed discussion will appear later in the chapter when individual inclusion compounds are discussed.

2.2.1. Confirmation of inclusion compound formation
Confirmation that an inclusion compound has been prepared can be obtained by comparing the spectrum obtained from the product of the attempted preparation with the spectra of the individual components. The presence of bands arising from both the host lattice and the guest species in the spectrum of the product is usually definite proof of the formation of the inclusion compound.

The converse is not necessarily true since the absence of any bands attributable to the guest species need not necessarily mean that an inclusion compound has not been formed. This situation is illustrated in Fig. 2 which

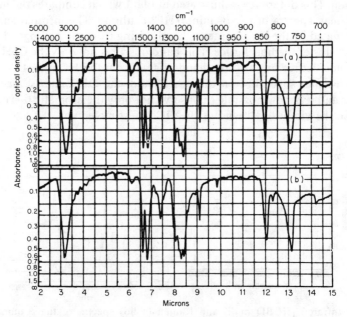

Fig. 2. Infrared spectra of KBr discs of (a) β-quinol HCl clathrate (b) α-quinol. Reproduced with permission from *J. Inorg. Nucl. Chem.*, 1957, **4**, 171.

is taken from the very first publication[4] to report infrared spectra of clathrates in 1957. As mentioned previously the band arising from the HCl guest molecule is obscured by the host lattice band. This overlapping of host lattice and guest molecule bands is one of the biggest disadvantages in the study of the spectra of inclusion compounds—the host lattice does give a vibrational spectrum which will inevitably obscure some of the guest molecule bands thus restricting the amount of information which can be obtained from the spectrum.

The technique of spectral subtraction, which can help to overcome this problem, will be discussed later in Section 2.5.

2.2.2. The state of aggregation and configuration of the guest molecule

Information about the state of aggregation of the guest molecule can be obtained by comparing the band positions in the spectra of the molecule when examined in the vapour, liquid, and solid phases and as the guest molecule, e.g. the position of the ν(O–H) band in the infrared and Raman spectra of CH_3OH and HCO_2H as guest molecules in the β-quinol host lattice is very similar to the vapour phase values.[3] This confirms the existence of a single isolated guest molecule in the almost spherical cavity formed in the β-quinol host lattice. (See Chapter 1, Volume 2.)

If a molecule can exist in different configurations (rotamers, conformers etc.), then by studying its vibrational spectrum in the gaseous, liquid and solid phases it is usually possible to assign a set of spectral bands to the different configurations. Because of the constraints imposed by the host lattice the configuration of the guest molecule may not always be the same as that of the free molecule and information on the configuration can be obtained by comparing the spectrum of the guest molecule with the spectra of the individual configurations. (See the example of the thiourea cyclohexyl chloride inclusion compound described in ref. 1.)

2.2.3. The rotational and translational motions of the guest molecule

In addition to its vibrational degrees of freedom, a guest molecule also possesses rotational and translational degrees of freedom. The rotational motions can be quantized or hindered (librational) depending on the space available for the guest molecule. The translational motions are usually referred to as the "rattling" modes. All of these motions will be low energy motions giving rise to absorption bands in the far-infrared region or the low wavenumber shift Raman region.

Figure 3 illustrates the far-infrared spectra (20–90 cm^{-1}) of α-quinol and of several β-quinol clathrates.[5] Several bands are observed in the spectra

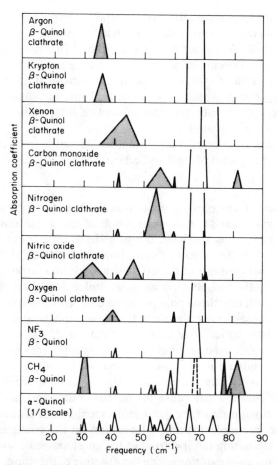

Fig. 3. The far-infrared spectra of various β-quinol clathrates. The shaded bands represent absorptions attributable to the guest molecule, whilst the unshaded bands arise from quinol. Adapted by permission from *J. Chem. Phys.*, 1965, **43**, 4291.

and a detailed discussion of the spectra can be found elsewhere.[1,5] One problem with the far-infrared spectra of β-quinol clathrates is the poor agreement between different studies[1] which may be due to partial decomposition of the samples during sample preparation. Since the far-infrared spectrum of α-quinol is much more complicated than that of the β-polymorph (see Fig. 3), the presence of small amounts of α-quinol would give rise to extra bands in the spectrum.

Although no sample preparation is required in Raman spectroscopy, the low wavenumber shift Raman study suffers from the problem of overlapping host and guest bands since the β-quinol host gives rise to two very intense bands in this region.[6]

Information about the rotational motions of the guest molecule can also be obtained from a study of the temperature dependence of the widths of its vibrational bands. The topic was discussed in detail for β-quinol clathrates in the previous review[1] and the results obtained from the study of zeolite inclusion compounds will be discussed later in the chapter (Section 2.4).

2.2.4. Restricted internal rotations of the guest molecule

The torsional motions of guest molecules give rise to bands appearing in the far-infrared region of the spectrum. It is interesting to compare the torsional frequency and the derived barrier height of the guest molecule with the corresponding values for the free molecule.

One such study has been reported for the benzaldehyde guest molecule, C_6H_5CHO, in thio-Dianin's compound.[7] Corresponding values for the frequency and barrier height of the aldehyde group rotation in benzaldehyde as a gas, liquid and guest molecule are 111 cm^{-1}, 20.5 kJ mol^{-1}; 133 cm^{-1}, 28.0 kJ mol^{-1}; and 126 cm^{-1}, 25.1 kJ mol^{-1} respectively. The guest molecule values are closer to the liquid phase values than to the gaseous phase values indicating a degree of host–guest interaction.

2.2.5. Host lattice–guest molecule interactions

Evidence for host lattice–guest molecule interactions can be obtained by comparing the wavenumber value for a vibration of a molecule in the gaseous phase (ν_{gas}) with the corresponding value when acting as a guest molecule (ν_{guest}). A small value for ($\nu_{gas} - \nu_{guest}$) would indicate very little interaction and such a situation has been reported for the CH_3OH and HCO_2H guest molecules in the β-quinol host lattice.[3]

Marked deviations from the above situations would indicate a degree of interaction between the host lattice and the guest molecule and such behaviour has been observed for the methyl halide guest molecules in β-quinol,[8] and for aromatic guest molecules in Hofmann-type host lattices.[2,9]

A large difference between the gas phase and guest molecule spectrum would indicate a considerable degree of interaction between the host lattice and the guest molecule and in this situation the compound might be better classified as a charge transfer complex rather than as an inclusion compound. Such big differences have been observed in the "inclusion compounds" formed between the xylenes and the [Ni(NCS)$_2$(α-arylalkylamine)$_4$] host lattices.[10] (See Chapter 4, Volume 1.)

2.2.6. Spectra of host lattices and guest molecules in unusual configurations
Inclusion compounds can, on occasion, present us with examples of the host lattice or the guest molecule existing in an unusual configuration whose vibrational spectrum can then be recorded.

The hexagonal inclusion-forming polymorphs of urea and thiourea are stable only in the presence of a guest molecule, and the empty hexagonal polymorphs cannot be obtained. A study of urea and thiourea inclusion compounds is thus the only way in which spectra of the hexagonal polymorphs can be obtained.

The thiourea monohalocyclohexane inclusion compounds are unique in that both the host lattice and the guest molecule exist in unusual configurations. Infrared[11] and Raman[12] vibrational spectroscopic studies have shown that the thiourea host lattice displays a selectivity for the axial conformer of the guest molecule, a unique situation since in the gaseous and liquid phases the equatorial conformer is the more abundant, whilst the solid phase consists of the equatorial conformer alone.

2.3. Intercalates

Intercalates are inclusion compounds where the host lattice has a layered structure and the guest species is accommodated in the interlayer voids. (See Chapter 7, Volume 1.)

2.3.1. Graphite intercalates

Graphite intercalates present a problem for Raman spectroscopy because of their high absorption in the visible region. A back scattering sampling geometry must be used and to prevent guest desorption the measurements are best made at low temperatures and low laser power.[13] Despite this difficulty Raman studies of graphite intercalates are more numerous than infrared studies and all the results here will be derived from Raman studies.

Much attention has been focussed in recent years on the structure of graphite intercalates[14] i.e. whether they are neutral or charged (CM_n; $C^+M_n^-$; $C^-M_n^+$ etc) depending on the nature of the intercalated species. Vibrational spectroscopy should provide such structural information but the majority of the reported Raman studies have concentrated on the lattice modes of the graphite host lattice,[13] and these studies will not be reviewed here. This review will concentrate on the few studies which have reported the spectra of the guest species and the structural information derived therefrom.

The bromine intercalate displays a resonance Raman spectrum of the intercalate when excited with 514.5 nm radiation[15] (Fig. 4), with the band positions being invariant with guest concentration. The stretching mode of the guest gives rise to the fundamental band at 242 cm^{-1} and according to the authors[13] the "correspondence of this value to the molecular stretch mode in solid Br_2 at 300 cm^{-1} and in the free Br_2 molecule at 323 cm^{-1} provides support for the *molecular* identity of the intercalate". The value of 242 cm^{-1} should however be compared with a value of 318 cm^{-1} for

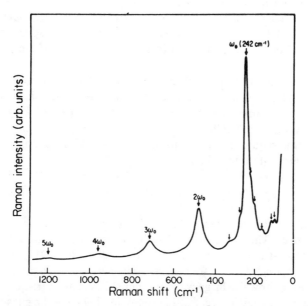

Fig. 4. The resonance Raman spectrum of the bromine intercalate of graphite. Reproduced with permission from *Phys. Rev.*, 1978, **18B**, 7069.

bromine in its clathrate hydrate,[16] and a value of 307 cm^{-1} for bromine[17] in zeolite 5A. Furthermore the Br_2^- species is reported[18] to have a stretching mode value of 160 cm^{-1} in matrix isolated $K^+Br_2^-$ and the value of 242 cm^{-1} observed in the graphite intercalate must surely indicate a large degree of charge transfer character in accordance with the known electron acceptor properties of bromine.

The IBr and ICl intercalates were also studied,[15] with the highest wavenumber band for IBr appearing at 230 cm^{-1} in comparison with a gas phase ν(I–Br) value of 268 cm^{-1}. The behaviour of the ICl intercalate was the complete opposite to that of the Br_2 intercalate with the band position shifting from 98 cm^{-1} to 186 cm^{-1} as the ICl mol.% concentration was increased from 0.92 to 10.7. Even the highest value of 186 cm^{-1} is far removed from the gas phase $\nu(\text{I–Cl})$ value of 384 cm^{-1}, and the interhalogen intercalates are surely worthy of further study.

The Raman spectrum of $FeCl_3$ intercalated in graphite is identical[19] with that of crystalline $FeCl_3$. This suggests that intercalated $FeCl_3$ retains its layer structure, the Raman spectrum providing no evidence for the existence of monomeric $FeCl_3$ or dimeric Fe_2Cl_6. It is also worthwhile referring to the remarkable high resolution electron microscope photographs of the $FeCl_3$ intercalate published by Thomas et al.[20]

The Raman spectrum of the HSO_3F intercalate has been interpreted in terms of a molecular guest species,[21] but the Raman spectrum of the sulphuric acid intercalate has been interpreted in terms of the presence of the $S_2O_7^{2-}$ ion.[22]

The few Raman studies which have reported vibrational values for the guest species have thus provided some structural information. It is however worth pointing out that guest bands have not been observed for guests such as HNO_3, AsF_5, and $SbCl_5$[117] and it has been suggested that resonance enhancement is required in order to observe the Raman bands of the guest species[23] in graphite intercalates.

2.3.2. Other intercalates

The infrared spectra of a range of FeOCl intercalates have been reported by Herber et al. There are marked differences between the intensities of some of the guest bands in $3FeOCl.py$ and $4FeOCl.imid.2H_2O$ when compared with the spectra of the free bases.[24] The bands of intercalated pyridine do not however show the large wavenumber shifts characteristic of coordinated pyridine,[25] suggesting very little interaction between the guest molecule and the iron atom.

The infrared spectra of the guest species in the $6FeOCl.[P(C_2H_5)_3]$ and $6FeOCl.[P(OCH_3)_3]$ intercalates do however show some differences when compared with the spectra of the phosphorus compounds in the vapour phases.[26] In particular, significant shifts are observed in the $\nu(\text{P–O})$ and $\nu(\text{P–C})$ bands suggesting that a direct involvement of the phosphorus lone pairs is involved in the host–guest interaction.

The infrared spectrum[27] of Kryptofix 22 when intercalated into FeOCl shows none of the changes observed when this cyclic polyether is complexed with rare earth ions, where the ion is situated at the centre of the cavity. These spectroscopic data are in agreement with the model deduced from the Mössbauer data which indicate an interaction between the nitrogen lone pairs and the iron atoms.

The infrared and Raman spectra of the intercalates $MPS_3.0.33G[M = Mn,$ Cd; $G = Co(\eta^5\text{-}C_5H_5)_2^+$, $Cr(\eta^6\text{-}C_6H_6)_2^+]$ have been reported by Mathey *et al.*[28] The differences between the spectra of the empty and intercalated host lattices and between the spectra of the guest species and of their iodide salts are minimal, indicating very weak host–guest interactions. Polarized infrared spectra of platelets show that the guest rings are oriented perpendicular to the host lattice layer planes.

The Raman spectra of TaS_2 and $TaSe_2$ intercalated with 1,2-diaminoethane are also superpositions of the spectra of the individual components,[29] again indicative of very weak host–guest interactions.

2.4. Zeolite inclusion compounds

In view of the vast numbers of papers in the literature dealing with infrared and Raman studies of zeolite complexes the reviewer has been forced to limit this section to those papers dealing with the study of the motions of the guest molecule as indicated in Section 2.2.3.

2.4.1. Diatomic guest molecules

The vibrational mode of a gaseous homonuclear diatomic is infrared inactive but Raman active whilst for a gaseous heteronuclear diatomic the mode is both infrared and Raman active. Considerable information about the rotational freedom of a diatomic molecule can be obtained from a study of its vibrational band shape as outlined by Brown and Davies[30] and verified recently by calculations.[31]

A diatomic guest molecule undergoing free rotation would give a P, R envelope with the Q branch being absent in the infrared spectrum, but the Q branch would appear if the rotation is in any way hindered. The Raman spectrum, on the other hand, would show an O, Q, S envelope irrespective of whether the rotation is free or hindered.

Raman spectra have been reported for Br_2 and I_2 adsorbed on different zeolites.[32,33] In view of the known structures of the zeolites (see Chapter 6, Volume 1) it is not surprising that the band positions vary from one zeolite structure to another (e.g. NaA, NaX and NaY) and also as the cation is varied for a given zeolite structure (e.g. LiX, NaX, KX, and CsX).

The Raman spectra of N_2 and O_2 adsorbed on zeolite 4A display[34] bands at $2324\ cm^{-1}$ and $1548\ cm^{-1}$ compared with the gas phase values of 2331 and $1556\ cm^{-1}$ respectively. When examined as guest molecules in β-quinol clathrates[3] the band positions are at the similar values of $2323\ cm^{-1}$ and $1547\ cm^{-1}$ respectively. However, unlike the situation with β-quinol

48   *J. E. D. Davies*

clathrates where the vibrational mode remains infrared inactive, the vibrational mode of homonuclear diatomics such as deuterium,[35–37] nitrogen,[35–39] oxygen[35,36] and hydrogen[36] becomes infrared active when present as guest molecules in zeolites. The appearance of the infrared forbidden band is indicative of an induced dipole moment in the molecule, and the appearance of a prominent Q branch in the observed spectrum indicates hindered rotation of the guest molecule (Fig. 5).

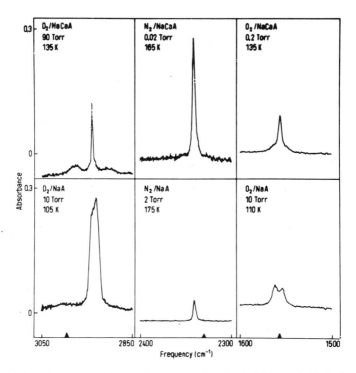

Fig. 5. Infrared spectra of diatomic guest molecules in NaA and NaCaA zeolites. Gas phase fundamental values marked by triangles. Reproduced with permission from *J. Chem. Phys.*, 1977, **66**, 5237.

The main features of the induced infrared spectra are a shift to a lower wavenumber value than the gas phase value for H_2 and D_2, but a shift to a higher value for N_2 and O_2. The magnitudes of the shifts are again dependent on the cations.

The spectrum of guest O_2 is of interest since it consists of a single band in NaCaA but a doublet in NaA (Fig. 5). Proposed explanations for the doublet structure are the presence of dimer species in the cavities or a freely rotating molecule giving the expected P, R structure.

The infrared spectrum[40] of adsorbed CO is also different in NaA and NaCaA, and also very different from the infrared spectrum of CO in β-quinol.[41] In the zeolites the main band position lies above the gas phase value, but at a lower value in β-quinol. The β-quinol and zeolite spectra both contain weaker bands which can be assigned to combination and difference bands of the vibrational mode with translational or librational modes, but at some coverage values the CO/NaCaA spectrum resembles that of O_2/NaA in consisting of a doublet.

Diatomic guest molecules, particularly homonuclear diatomics, can thus provide useful information about the absorption sites and electric fields within zeolites, but further work remains to be done before a complete understanding of the spectra will be obtained.

2.4.2. Triatomic guest molecules

The value of the ν_2 vibration of guest N_2O in NaA lies above[42,43] the gas phase value of 2224 cm^{-1}, once again in complete contrast to the situation with N_2O in β-quinol.[30] The band width decreases with decreasing temperature and the band position moves to higher wavenumber values, as the guest molecule becomes more localized near a cation site. The ν_1 band appears as a doublet due to the two possible orientations of the N_2O molecule.

Similar results are observed[44,45] with CO_2 in NaA, and it seems that the infrared inactive ν_1 mode of this centrosymmetric molecule may become activated in the zeolite.

The ν_3 band of H_2S and D_2S displays a large downward shift from the gas phase value on adsorption[46] in NaY, NaA and NaCaA. The shift and the $\Delta N_{1/2}$ values are greater in NaA than in NaCaA and these differences point to different adsorption sites in the two zeolites.

The infrared[47] and Raman[48] spectra of SO_2 absorbed on a variety of zeolites have been interpreted in terms of two guest species, one being more firmly bound than the other. The Raman study also detected the presence of a variety of decomposition products of SO_2 such as S_3^-, S_2, S_4, S_xO, SO_3^{2-} and SO_4^{2-} generated by surface reactions.

2.4.3. The methane guest molecule

For the gaseous methane molecule, the ν_3 and ν_4 vibrational modes are infrared active whilst the ν_1 and ν_2 modes are inactive. When adsorbed in NaA, the ν_1 mode becomes active[49] (Fig. 6), with a shift of 22 cm^{-1} below

50   J. E. D. Davies

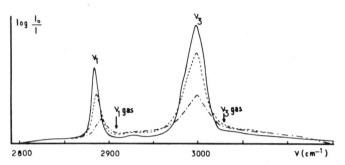

Fig. 6. Infrared spectra of CH_4 adsorbed in NaA zeolite. $-\cdot-\cdot-$ 260 K $----$ 230 K
—— 210 K. Reproduced with permission from *J. Phys.*, 1981, **42**, 1029.

the gas phase value (cf. with a shift of 9 cm^{-1} as a guest in β-quinol[50]). The profile of the ν_3 band shows no variation with temperature indicating that CH_4 retains some rotational freedom within the zeolite cavity. Heat capacity studies of the CH_4/β-quinol clathrate also indicate rotational freedom of the guest molecule. (See Chapter 1, Volume 3.)

2.5. Difference spectra

One of the problems encountered in studying the spectra of inclusion compounds, particularly with organic and organometallic host lattices, is the overlapping of guest molecule bands by the generally much stronger host lattice bands as mentioned in Section 2.2.1. This overlapping of bands reduces the amount of information obtainable from the spectrum and is the reason for including the word "potentially" in the title of Section 2.2.

The availability of any techniques which could remove the host lattice bands from the spectrum would thus increase the amount of information obtainable from the spectrum. The simplest method is to take advantage of the double beam arrangement in an infrared spectrometer by placing a sample of the empty host lattice in the reference beam. Such a method has been used with quinol clathrates with some degree of success.[4,51]

The main disadvantage of this method is the lack of control over the amount of subtraction; this could be achieved by using samples of different concentrations in the reference beam. The advent of computer controlled and computer coupled spectrometers overcomes this problem by providing a variable degree of host band subtraction. The use of this technique in studying some clathrates of quinol and of Dianin's compound is illustrated[52] in Figs. 7–9.

It is quite easy to identify the ν(O–H) band of the guest molecule at 3622 cm^{-1} in the infrared spectrum of the β-quinol/CH$_3$OH clathrate, but more difficult to identify the bands arising from the two guest ν(C–H) modes (Fig. 7b). These two bands can be identified quite readily in the difference spectrum (Fig. 7c). Perfect subtraction of the host lattice spectrum cannot be achieved and it is helpful to compare the difference spectrum with the Raman spectrum of the clathrate (Fig. 7d) to confirm the assignment.

The difference spectrum of the β-quinol/SO$_2$ clathrate (Fig. 8c) highlights the bands at 1337 cm^{-1} arising from the ν_3(b$_2$) mode of the guest. The

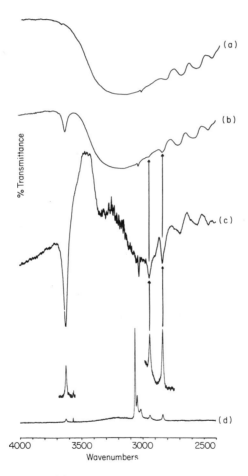

Fig. 7. Infrared spectra of (a) β-quinol/KBr disc (b) β-quinol/CH$_3$OH/HCBD mull/CsI plates (c) Expanded difference spectrum, difference factor = 0.66 (d) Raman spectrum of β-quinol/CH$_3$OH clathrate.

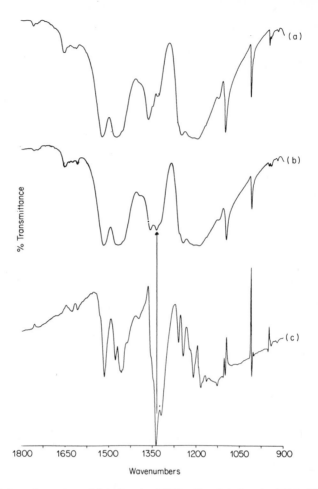

Fig. 8. Infrared spectra of (a) β-quinol/KBr disc (b) β-quinol/SO₂/KBr disc (c) Expanded difference spectrum, difference factor = 0.98.

observation of some derivative type bands in the 900–1100 cm⁻¹ region of the difference spectrum indicates that the band positions of the host lattice do change when a guest molecule is introduced into the host lattice.

The usefulness of the difference technique can be seen from Fig. 9, where the presence of the extra band in the spectrum of the formic acid clathrate of Dianin's compound (Fig. 9b) tends to be masked by the numerous other bands of the host lattice. The difference spectrum (Fig. 9c) picks out the ν(C=O) band of the guest molecule very readily.

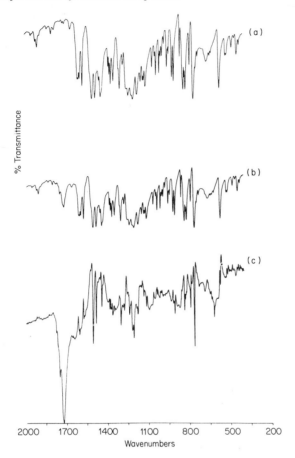

Fig. 9. Infrared spectra of (a) Dianin's compound/KBr disc (b) The formic acid clathrate/KBr disc (c) Expanded difference spectrum, difference factor = 1.00.

3. Inelastic neutron scattering studies

3.1. Introduction

Inelastic neutron scattering (INS)[53,54] can provide information similar to that obtained from infrared and Raman spectroscopy as listed in Section 2.2. The similarity of these two techniques is heightened by the fact that

INS data are quite often presented as plots of intensity vs energy transfer, presented in units of cm^{-1}. The INS data are however complementary, rather than identical, to the spectroscopic data since there are no spectroscopic selection rules in INS and band intensities can differ in the two techniques.

A useful feature of INS is the fact that low frequency, large amplitude vibrations are experimentally the easiest to observe in the neutron spectrum. This is the converse of the case with infrared and Raman spectroscopy, but the resolution obtainable in a neutron spectrum is poorer than that obtainable in a vibrational spectrum.

3.2. Clathrates and intercalates

The far-infrared (30–130 cm^{-1}) and neutron scattering (30–120 cm^{-1}) spectra of the empty β-quinol lattice and of its D_2S clathrate are compared[55] in Fig. 10. Comparison of the β-quinol spectra shows the 63 cm^{-1} peak to be common to both spectra, but there is no coincidence between the higher energy band at 102 cm^{-1} in the neutron spectrum and at 117 cm^{-1} in the far-infrared spectrum. The D_2S guest molecule gives rise to two distinct bands at 57 and 75 cm^{-1} in the far-infrared spectrum but the neutron spectrum consists of a broad ill defined band illustrating the lower resolution obtainable in a neutron spectrum.

The neutron spectra of the β-$C_6D_4(OD)_2$ clathrates containing CH_3OH and CH_3CN have also been reported.[56] The guest molecules themselves give rise to elastic scattering which indicates that their translational motions are restricted. A broad peak at 190 cm^{-1} is assigned to the CH_3OH guest molecule, but a sharp peak at 460 cm^{-1}, due to the $C–C\equiv N$ bending vibration, is observed for the CH_3CN guest molecule. The sharpness of the band indicates that the guest molecule is firmly held in the cage.

The neutron spectrum of the ethylene oxide clathrate deuterate contains bands at 29, 39 and 57 cm^{-1} which are assigned to librational motions of the guest molecule around the z, x, and y axis respectively.[57] This is in good agreement with the far-infrared spectrum.[58]

The neutron spectra of several intercalates of niobium and tantalum disulphide have been reported. The vibrational modes of the guest in $2NbS_2.py$ and $2TaS_2.py$ do not show[59] the large wavenumber shifts characteristic of coordinated pyridine.[25] Since the nitrogen lone pair is *parallel* to the host layers, the dominant interaction seems to be between the lone pair and the sulphur atoms rather than between the lone pair and the transition metal. The neutron spectrum[60] of $TaS_2.NH_3$ has been interpreted in terms of an ionic model $(NH_4^+)_x(NH_3)_{1-x}[TaS_2]^{x-}$ containing small amounts of NH_4^+ cations.

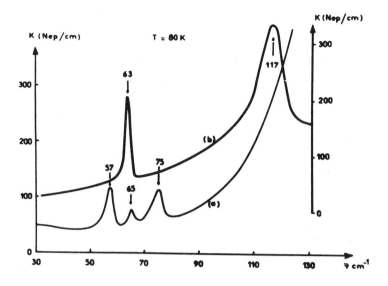

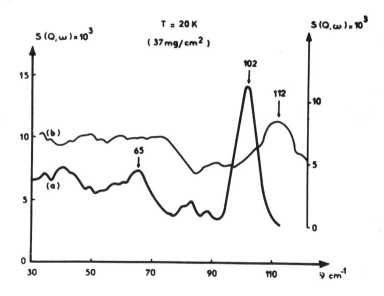

Fig. 10. Top: Far-infrared spectra at 80 K of (a) the β-quinol/D_2S clathrate (b) the empty β-quinol host lattice. Bottom: Neutron scattering spectra at 20 K of (a) the empty β-quinol host lattice (b) the β-quinol/D_2S clathrate. Reproduced with permission from *J. Phys.*, 1976, **37**, 1453.

Figure 11 compares the neutron, Raman and infrared spectra of the 18-Crown-6/KSCN complex.[61] It illustrates quite nicely the complementary nature of the INS data since the neutron spectrum contains bands arising from vibrations which are infrared active only (e.g. the 150 cm^{-1} band), Raman active only (e.g. the 350 cm^{-1} band), both infrared and Raman active (e.g. the 77 cm^{-1} band) as well as a band (300 cm^{-1}) which has no counterpart in the infrared or Raman spectrum.

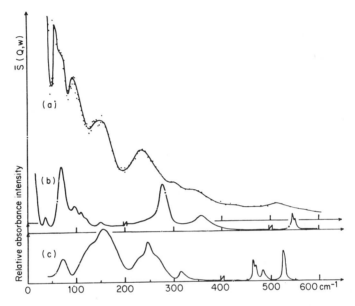

Fig. 11. Spectra of the 18-Crown-6/KSCN complex. (a) the neutron spectrum (b) the Raman spectrum (c) the infrared spectrum. Reproduced with permission from *J. Chim. Phys.*, 1978, **75**, 865.

3.3. Zeolite inclusion compounds

Several studies have been reported of the INS spectra of molecular guests in zeolites concentrating mainly on the low energy motions, the translational and librational modes.

Infrared,[62] Raman,[62] and neutron[63,64] spectra have been reported for C_2H_2 adsorbed in a number of different zeolites, and a comparison of the results is very interesting.

All of the guest molecule fundamentals can be observed for C_2H_2 in NaA and CaA with the $\nu_2\{\nu(C\equiv C)\}$ mode being observed in the infrared spectra.

This mode, which is infrared inactive for the gaseous linear molecule, is also observed in the infrared spectra[62] of C_2H_2 in NaX, KX and CaX. The results were interpreted in terms of a "side-on" interaction between C_2H_2 and the cation, with ethyne retaining its *linear* geometry.

Neutron studies on C_2H_2 and C_2D_2 in AgX[63] and AgA[64] reveal a very different picture, where the ethyne molecule has a *non linear* geometry. The different geometry arises from the fact that silver can back donate the electrons in its filled d orbitals to ethyne, thus giving a stronger interaction than is possible with the alkali and alkaline earth cations.

The adsorption of ethylene on a variety of 13X zeolites has been studied both spectroscopically and calorimetrically.[51] The ν_2 $\{\nu(C{=}C)\}$ and ν_3 $\{\delta CH_2\}$ vibrations, which are both infrared inactive for the gaseous molecule, become activated in the zeolite, and the shifts from the gas phase Raman values are dependent on the nature of the cation. The largest shifts and the largest heats of adsorption are observed with the AgX and CdX zeolites which indicates the presence of a π-bonding interaction between the ethylene and the cation. For the other zeolites (with Li, Na, K, Ba and Ca counter cations) the guest–cation interaction is very much weaker and the guest band widths indicate considerable rotational freedom of the ethylene guest molecule.

Neutron studies of C_2H_4 and C_2D_2 in AgX[65] and AgA[66] also reveal a strong cation–guest interaction leading to the observation of bands which can be assigned to librational and hindered translational modes of the guest relative to the surface.

4. Electron spin resonance studies

Electron spin resonance (ESR) spectroscopy detects the presence of paramagnetic guest species in inclusion compounds. The paramagnetic species can be the host lattice, the parent guest species or the paramagnetic species produced when a diamagnetic guest species is irradiated.

Very few studies seem to have been reported of paramagnetic parent guest species. Di-*t*-butyl nitroxide has been studied as a guest in Dianin's compound[67] and in thiourea.[68] A study of a single crystal of the thiourea inclusion compound at room temperature indicates a rapid, large amplitude motion about the molecular y axis with the guests well oriented along the channel direction. The single crystal spectrum of the guest in Dianin's compound at 293 K indicates the presence of two magnetically different species in equal concentrations. One species rotates rapidly and randomly whilst species 2 rotates about the $\bar{3}$ axis of the crystal.

The biradical (1) also displays a rapid rotational motion although one of the molecular axes is aligned parallel to the channels in the thiourea host lattice.[69]

1

There have been several studies of the radicals produced during the irradiation of a diamagnetic guest species. Inclusion compounds turn out to be attractive systems for studying free radicals since the isolation of a guest molecule from its neighbours can lead to the formation of long lived free radicals.

The majority of the studies have used urea as the host lattice with either X-ray or γ-ray irradiation. Some of the detected radicals are:

(a) $RCH_2^\bullet CHCO_2R'$ produced from long chain alkyl esters.[70]
(b) $RCH_2^\bullet CHCO_2H$ produced from long chain carboxylic acids.[70]
(c) $RCH_2^\bullet CHCOR'$ produced from aliphatic ketones.[71a]
(d) $R\overset{\bullet}{C}HOR'$ produced from aliphatic ethers.[71b]
(e) $R\overset{\bullet}{C}HSR$ produced from $R_2S\{R = \text{ethyl or } n\text{-hexyl}\}$.[72]
(f) $C_2H_5O_2C\overset{\bullet}{C}HCH=CHCO_2C_2H_5$ and $C_2H_5O_2C\overset{\bullet}{C}HCH_2CH_2CO_2C_2H_5$ produced from[73] $C_2H_5O_2CCH_2CH=CHCO_2C_2H_5$.
(g) $CH_3(CH_2)_4\overset{\bullet}{C}HC\equiv CCO_2CH_3$ produced from methyl-2-nonynoate in both urea and perhydrotriphenylene, but with the radical adopting different conformations in the two hosts.[74]
(h) X-irradiation of both 6-undecanol and 1-decanol result in the removal of the hydrogen atom from the carbon atom attached to the hydroxyl group.[75]
(i) $\sim CH_2\overset{\bullet}{C}HCH_2\sim$ and $\sim CH_2\overset{\bullet}{C}(O^-)OH$ from palmitic acid.[76]
(j) $CH_3(CH_2)_5CH=NOH$ produces[77] an iminoxy radical $RCH=NO^\bullet$ and two secondary radicals of the type $R\overset{\bullet}{N}R'$ and $RR'NO^\bullet$.
(k) γ-irradiation of both polyethylene[78] and $n\text{-}C_{24}H_{50}$[79] in the presence of O_2 produces peroxy radicals.

Many olefinic guests in urea undergo polymerization on irradiation. (See Chapter 10, Volume 3.) This polymerization can be prevented by isolating the olefinic guests using an auxiliary guest as a diluent. Such a study[80] on olefinic ester guests $RCH=CR'CO_2R''$ using n-decane as a diluent prevented their polymerization and resulted in the observation of hydrogen addition free radicals of the type $RCH_2\overset{\bullet}{C}R'CO_2R''$.

X-irradiation of β-quinol clathrates of HCl, SO_2, CO_2, HCO_2H, CH_3OH and CH_3CN produced the same ESR spectrum arising from the p-hydroxyphenoxyl radical.[81] The CH_3CN clathrate produced an additional spectrum arising from the CH_2CN radical possessing rapid rotational motion around the radical C–C bond parallel to the crystal c axis even at 77 K.

Irradiation of Y type zeolites produces the H_2O^- radical[82] whilst irradiation of guest CF_2BrCH_2Br in Dianin's compound produces the Br_2^- radical anion.[83]

The Hofmann-type host lattices (see Chapter 2, Volume 1) turn out to be excellent systems for stabilizing free radicals.[84] γ-irradiation of the $Cd(en)M(CN)_4.2C_6H_6$ (M = Cd, Hg) inclusion compounds leads to the formation of the C_6H_7 radical which can survive for a few hours at 423 K. Since irradiation of $Cd(en)Ni(CN)_4.2C_6H_6$ does not lead to the formation of the C_6H_7 radical, it seems that the radical is stabilized in the biprismatically shaped cavity and not in the tetragonally shaped cavity.

The γ-irradiation of guests in cyclodextrins produces hydrogen addition type radicals and hydrogen abstraction type radicals depending on the nature of the guest.[85] The former are produced from benzene (C_6H_7 radical) and acetone (1-hydroxy-1-methylethyl radical), whilst the latter are produced from toluene (benzyl radical), m-fluorotoluene (m-fluorobenzyl radical), cyclohexa-1,4-diene (cyclohexadienyl radical) and cyclohepta-1,3,5-triene (cycloheptatrienyl radical). The radical anions were produced from hexafluorobenzene and cyclo-octa-1,3,5,7-tetraene.

5. Mössbauer and photoelectron spectroscopic studies

5.1. Graphite intercalates

Much interest centres around the exact nature of the guest species in graphite intercalates i.e. whether a neutral guest species retains its neutrality or whether there is a degree of charge transfer between the host lattice and the guest species. In addition to wideline magnetic resonance studies,[86] Mössbauer spectroscopy and photoelectron spectroscopy (PES) have also been used and some of the observed results are discussed here.

The antimony (V) pentahalides have been studied using both Mössbauer spectroscopy and PES and the results obtained from the two techniques can be compared.

An early Mössbauer study[87] on the SbF_5 and $SbCl_5$ intercalates was interpreted as indicating the presence of Sb(V) and Sb(III) species with the formation of a carbon–halogen bond to preserve the $1:5$ Sb: halogen ratio.

A more recent study[88] of the $SbCl_5$ intercalate also reports evidence for the presence of Sb(V) as $SbCl_6^-$ and Sb(III) as $SbCl_3$ together with weak evidence for neutral intercalated $SbCl_5$. It is also suggested that there is a second Sb(III) species present, $SbCl_4^-$, with the $[SbCl_4^-]/[SbCl_3]$ ratio being ~ 0.25. A PES study[89] of the SbF_5 intercalate reports the observation of bands assignable to neutral intercalated SbF_5, SbF_6^- and possibly SbF_3. A PES study of the AsF_5 intercalate[90] indicated that less than 30% of the AsF_5 is converted to AsF_6^- according to the scheme:

$$3AsF_5 + 2e \rightarrow 2AsF_6^- + AsF_3.$$

Mössbauer studies of the $FeCl_2$ and $FeCl_3$ intercalates are in general agreement.[91,92] The $FeCl_3$ intercalate gives a single band slightly shifted from that of anhydrous $FeCl_3$ indicating a partial transfer of π electrons from the graphite to empty d orbitals. For the $FeCl_2$ intercalate there have been reports of both two[92] and three[91] distinct Fe sites with shifts similar to that of anhydrous $FeCl_2$ indicating the absence of any π electron transfer. The mixed $FeCl_3/AlCl_3$ (1 : 1 or less) intercalate gives a Mössbauer band characteristic of Fe(II) arising[93] from the species $C_n^+ Cl^- . FeCl_2.3AlCl_3$.

A PES study[94] of the Stage 1 H_2SO_4 intercalate has been interpreted in terms of the presence of $S_nO_{4n}^{2-}$ ($n \geqslant 1$) ions hydrogen bonded together. This is in complete disagreement with the Raman study[22] (Section 2.3.1) which was interpreted in terms of the presence of $S_2O_7^{2-}$ ions. See also ref. 118.

5.2. Intercalates of FeOCl

Evidence for electron transfer from the guest species to the host lattice has been obtained from Mössbauer studies of intercalates of FeOCl.

Such evidence has been obtained for a series of pyridine intercalates:[95] $4FeOCl.py$, $4FeOCl.2,6\text{-}Me_2py$, $6FeOCl.2,4,6\text{-}Me_3py$ and $4FeOCl.4\text{-}NH_2py$; and for ferrocene,[96,97] substituted ferrocene[98] and cobaltocene[96] intercalates: $Fe(C_5H_5)_2^+ .(6FeOCl)^-$, $Fe(C_5Me_5)_2^+ .(13.4FeOCl)^-$, and $Co(C_5H_5)_2^+ .(6FeOCl)^-$.

5.3. Other inclusion compounds

The PES of the HCl guest molecule in the β-quinol host lattice differs markedly from that of gaseous HCl (Fig. 12), and this is taken to be indicative of a substantial host–guest interaction.[99]

^{119}Sn Mössbauer studies have been reported for a variety of Sn(II) complexes with crown ethers. With 15-Crown-5, both $SnCl_2$ and $Sn(NCS)_2$

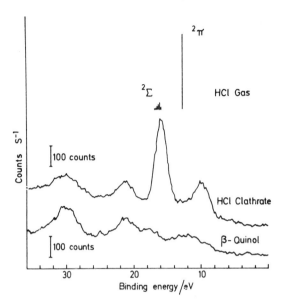

Fig. 12. The photoelectron spectra of β-quinol, (bottom) the HCl clathrate, (middle) and gaseous HCl (top). Reproduced with permission from *J. Chem. Soc., Chem. Commun.*, 1976, 708.

form 2:3 complexes which contain two different kinds of metal atoms in a 2:1 ratio.[100] One of the metal atoms interacts directly with the crown ether whilst the other metal atoms are present as SnX_3^- counterions. With $Sn(ClO_4)_2$, a 2:1 complex is formed with a unique Sn^{2+} ion in the structure, sandwiched between two ring polyether moieties (Fig. 13).

With 18-Crown-6 both $SnCl_2$ and $Sn(NCS)_2$ form 1:2 complexes[101] which contain two distinct tin sites, containing a SnX^+ cation bonded to the crown ether and a SnX_3^- counterion. With $Sn(ClO_4)_2$, a 2:1 complex is again formed containing a unique Sn^{2+} ion.

There seems to be no doubt that Mössbauer and photoelectron spectroscopy will find increasing use in the study of inclusion compounds because of the information about host–guest interactions which they can provide.[116]

6. High resolution solid state NMR studies

High resolution NMR spectra can now be obtained from solid samples using the magic angle spinning technique.[102] The technique turns out to be

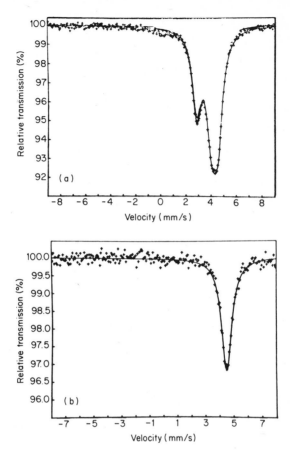

Fig. 13. The ^{119}Sn Mössbauer spectra of (a) (15-Crown-5)$_2$ (SnCl$_2$)$_3$ at 80 K (b) (15-Crown-5)$_2$Sn(ClO$_4$)$_2$ at 78 K. Adapted with permission from *Inorg. Chem.*, 1981, **20**, 3693.

a valuable new structural tool with some advantages over the more conventional X-ray diffraction technique[103] since it can be applied to finely crystalline solids and also to materials composed of atoms of roughly comparable X-ray scattering power.

6.1. Zeolite inclusion compounds

Most of the applications of the technique to the area of inclusion compounds have been to zeolites and their inclusion compounds. A great deal of work, using ^{29}Si and ^{27}Al NMR, has been done on the zeolite hosts yielding new structural information.[103] In this section we will be concerned with studies on inclusion compounds.

A ^{13}C study of CO and CO_2 guest molecules in NaX, NaY and NaA revealed chemical shifts similar to those observed for the gaseous molecules.[104] The CO guest in decationated zeolites showed a large downfield shift indicating an interaction with the extra lattice aluminium ions. The CO_2 guest in decationated zeolites showed no such effect.

The ^{13}C NMR spectrum of the tetrapropylammonium ion in a ZSM-5 type zeolite displays a splitting of the methyl signal which indicates the presence of at least two dissimilar sites within the zeolite.[105] This shows that the NMR spectrum is sensitive to weak interactions between organic molecules and the zeolite framework, again illustrating the fact that useful structural information can be obtained using this technique.

^{13}C NMR spectroscopy has also been used to characterize the carbonaceous residues arising from zeolite catalyzed reactions.[106] An example of an observed spectrum is given in Fig. 14. This is of the residue obtained

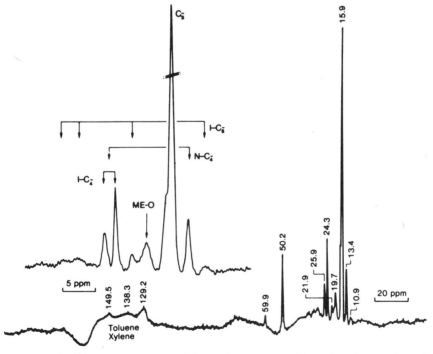

Fig. 14. The ^{13}C NMR spectrum of the carbonaceous residue after the reaction of methanol with the H-ZSM-5 zeolite. Reproduced with permission from *Zeolites*, 1982, **2**, 42.

after the reaction of methanol over H-ZSM-5. The excellent quality of the spectrum allows the identification of alkane products [propane (16.8 ppm), *n*-butane (13.4 and 25.9 ppm), isobutane (24.3 and 25.9 ppm), isopentane (10.9, 21.9, 29.9 and 12.2 ppm)], alkene products [propene (15.9 ppm), linear and branched olefins (~150 ppm)] and aromatic products [benzene, toluene, xylenes (129.2, 138.3 ppm)].

6.2. Quinol clathrates

^{13}C NMR spectroscopy has been found to be a valuable structural technique in its ability to distinguish between the different polymorphs of quinol[107]

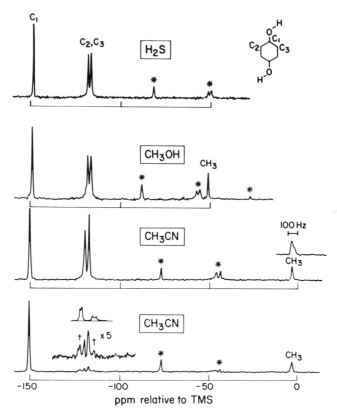

Fig. 15. ^{13}C NMR spectra of β-quinol clathrates. *Spinning side bands. †Features ascribed to [13]CN in the bottom spectrum which was obtained with the C_2 and C_3 lines partly suppressed. Reproduced with permission from *Chem. Phys. Lett.*, 1982, **86**, 428.

(see Chapter 1, Volume 2). Bands arising from the guest molecules can also be observed (Fig. 15), and this study revealed that formamide is an interesting molecule in that it can form either a clathrate or a 2:1 complex with quinol depending on the method of preparation.

^{1}H, ^{2}H and ^{13}C NMR studies of single crystals of the methanol[108,109] clathrate have been reported. The preferred orientation of the guest molecule has the COH plane parallel to the crystal c axis and the angle of 32° between the CO bond and c below 100 K increases to ~40° at 300 K. The ^{13}C spectrum of the acetonitrile clathrate[110] confirms the distorted host lattice structure in this clathrate.

6.3. Other inclusion compounds

The ^{129}Xe chemical shift has been found to be very sensitive to the environment of the atom and it is possible to observe separate signals from the xenon atoms in the small and large cages of the xenon clathrate deuteriohydrate.[111] ^{129}Xe spectra have also been recorded from the guest atom in other host lattices such as zeolites,[112,113] β-quinol,[112] and phenol.[112] There are distinct changes in the ^{13}C chemical shifts of both components, as a result of host guest interaction, when cyclodextrin inclusion compounds are formed.[114,119,120]

There seems to be no doubt that high resolution solid state NMR spectroscopy will be a valuable additional technique in the study of inclusion compounds.

References

1. J. E. D. Davies in *Molecular Spectroscopy* (ed. R. F. Barrow, D. A. Long and J. Sheridan), The Chemical Society, London, 1978, Vol. 5, Ch. 2, p. 60.
2. (a) S. Akyüz, A. B. Dempster, R. L. Morehouse and N. Zengin, *J. Chem. Soc., Chem. Commun.*, 1972, 307; (b) S. Akyüz, A. B. Dempster and R. L. Morehouse, *Spectrochim. Acta*, 1974, **30A**, 1989.
3. J. E. D. Davies, *J. Chem. Soc., Dalton Trans.*, 1972, 1182.
4. R. M. Hexter and T. D. Goldfarb, *J. Inorg. Nucl. Chem.*, 1957, **4**, 171.
5. J. C. Burgiel, H. Meyer and P. L. Richards, *J. Chem. Phys.*, 1965, **43**, 4291.
6. J. W. Anthonsen, *Acta Chem. Scand., Ser. A*, 1975, **29**, 179.
7. D. D. MacNicol, *J. Chem. Soc., Chem. Commun.*, 1973, 621.
8. K. D. Cleaver and J. E. D. Davies, *J. Mol. Struct.*, 1977, **36**, 61.
9. J. E. D. Davies, A. B. Dempster and S. Suzuki, *Spectrochim. Acta*, 1974, **30A**, 1183.
10. R. Leysen and J. van Rysselberge, *Spectrochim. Acta*, 1963, **19**, 237.
11. J. E. Gustaven, P. Klaboe and H. Kvila, *Acta Chem. Scand., Ser. A*, 1978, **32**, 25.
12. A. Allen, V. Fawcett and D. A. Long, *J. Raman Spectrosc.*, 1976, **4**, 285.

13. (a) M. S. Dresselhaus and G. Dresselhaus in *Intercalated Layered Materials* (ed. F. Lévy), D. Reidel Publishing Co., Dordrecht, 1979, Vol. 6, p. 423; (b) M. S. Dresselhaus and G. Dresselhaus, *Adv. Phys.*, 1981, **30**, 139.
14. S. A. Solin, *Adv. Chem. Phys.*, 1982, **49**, 455.
15. J. J. Song, D. D. L. Chung, P. C. Eklund and M. S. Dresselhaus, *Solid State Commun.*, 1976, **20**, 1111.
16. J. W. Anthonsen, *Acta Chem. Scand., Ser. A*, 1975, **29**, 175.
17. J. C. Rubim and O. Sala, *J. Raman Spectrosc.*, 1980, **9**, 155.
18. C. A. Wight, B. S. Ault and L. Andrews, *Inorg. Chem.*, 1976, **15**, 2147.
19. N. Caswell and S. A. Solin, *Solid State Commun.*, 1978, **27**, 961.
20. J. M. Thomas, G. R. Millward, R. Schögl and H. P. Boehm, *Mat. Res. Bull.*, 1980, **15**, 671.
21. B. Iskander, P. Vast, A. Lorriaux-Rubbens, M. L. Dele-Dubois and Ph. Touzain, *Mat. Sci. Eng.*, 1980, **43**, 59.
22. B. Iskander and P. Vast, *J. Raman Spectrosc.*, 1981, **11**, 247.
23. S. A. Solin, *J. Phys. Colloq.*, 1981, **42**, C6-283.
24. R. H. Herber and Y. Maeda, *Inorg. Chem.*, 1981, **20**, 1409.
25. S. Akyüz, A. B. Dempster, J. E. D. Davies and K. T. Holmes, *J. Chem. Soc., Dalton Trans.*, 1976, 1746.
26. R. H. Herber and Y. Maeda, *Physica*, 1981, **105B**, 243.
27. R. H. Herber and R. A. Cassell, *J. Chem. Phys.*, 1981, **75**, 4669.
28. Y. Mathey, R. Clément, C. Sourisseau and G. Lucazeau, *Inorg. Chem.*, 1980, **19**, 2773.
29. J. C. Tsang and M. W. Shafer, *Solid State Commun.*, 1978, **25**, 999.
30. J. D. Brown and J. E. D. Davies, *Spectrochim. Acta*, 1979, **35A**, 73.
31. H. Förster, W. Frede and M. Schuldt, *J. Mol. Struct.*, 1982, **80**, 195.
32. R. P. Cooney and P. Tsai, *J. Raman Spectrosc.*, 1979, **8**, 195.
33. J. C. Rubim and O. Sala, *J. Raman Spectrosc.*, 1980, **9**, 155.
34. D. D. Saperstein and A. J. Rein, *J. Phys. Chem.*, 1977, **81**, 2134.
35. H. Förster and M. Schuldt, *J. Chem. Phys.*, 1977, **66**, 5237.
36. H. Förster and M. Schuldt, *J. Mol. Struct.*, 1978, **47**, 339.
37. H. Förster and M. Schuldt, *J. Mol. Struct.*, 1980, **61**, 361.
38. E. Cohen de Lara and Y. Delaval, *J. Chem. Soc., Faraday Trans. 2*, 1978, **74**, 790.
39. E. Cohen de Lara, *J. Chem. Soc., Faraday Trans. 2*, 1981, **77**, 355.
40. H. Förster, W. Frede and M. Schuldt, *J. Mol. Struct.*, 1980, **61**, 75.
41. D. F. Ball and D. C. McKean, *Spectrochim. Acta*, 1962, **18**, 933.
42. E. Cohen de Lara, *Mol. Phys.*, 1972, **23**, 555.
43. E. Cohen de Lara and J. Vincent-Geisse, *J. Phys. Chem.*, 1976, **80**, 1922.
44. H. Förster, M. Schuldt and R. Seeleman, *Z. Phys. Chem. N.F.*, 1975, **97**, 329.
45. Y. Delaval and E. Cohen de Lara, *J. Chem. Soc., Faraday Trans. 1*, 1981, **77**, 869, 879.
46. H. Förster and M. Schuldt, *J. Colloid Interface Sci.*, 1975, **52**, 380.
47. H. Förster and R. Seelemann, *Ber. Bunsenges. Phys. Chem.*, 1976, **80**, 153.
48. P. Tsai and R. F. Cooney, *J. Raman Spectrosc.*, 1980, **9**, 39.
49. E. Cohen de Lara and R. Kahn, *J. Phys.*, 1981, **42**, 1029.
50. J. E. D. Davies and W. J. Wood, *J. Chem. Soc., Dalton Trans.*, 1975, 674.
51. J. L. Carter, D. J. C. Yates, P. J. Lucchesi, J. J. Elliot and V. Kevorkian, *J. Phys. Chem.*, 1966, **70**, 1126.
52. J. E. D. Davies, unpublished results.
53. R. K. Thomas, in *Molecular Spectroscopy* (ed. R. F. Barrow, D. A. Long, and J. Sheridan), The Chemical Society, London, 1979, Vol. 6, Ch. 6, p. 232.

54. J. Howard and T. C. Waddington, in *Advances in Infrared and Raman Spectroscopy* (ed. R. J. H. Clark and R. E. Hester), Heyden, London, 1980, Vol. 7, Ch. 3, p. 86.
55. J. S. Higgins, X. Gerbaux, C. Barthel and A. Hadni, *J. Phys.*, 1976, **37**, 1453.
56. J. S. Downes, J. W. White, P. A. Egelstaff and V. Rainey, *Phys. Rev. Lett.*, 1966, **17**, 533.
57. W. Wegener, J. Vanderhaeghen, S. Hautecler and L. van Gerven, *Physica* (*B + C*), 1978, **95**, 71.
58. J. E. Bertie and S. M. Jacobs, *Can. J. Chem.*, 1977, **55**, 1777.
59. B. C. Tofield and C. J. Wright, *Solid State Commun.*, 1977, **22**, 715.
60. C. Riekel, R. Schöllhorn and J. Tomkinson, *Z. Naturforsch, Teil A*, 1980, **35**, 590.
61. M. Fouassier and J.-C. Lassegues, *J. Chim. Phys.*, 1978, **75**, 865.
62. N. T. Tam, R. P. Cooney and G. Curthoys, *J. Chem. Soc., Faraday Trans. 1*, 1976, **72**, (a) 2577, (b) 2592.
63. J. Howard and T. C. Waddington, *Surf. Sci.*, 1977, **68**, 86.
64. J. Howard, K. Robson and T. C. Waddington, *Zeolites*, 1981, **1**, 175.
65. J. Howard, T. C. Waddington and C. J. Wright, *J. Chem. Soc., Faraday Trans. 2*, 1977, **73**, 1768.
66. J. Howard, K. Robson, T. C. Waddington and Z. A. Kadir, *Zeolites*, 1982, **2**, 2.
67. A. A. McConnell, D. D. MacNicol and A. L. Porte, *J. Chem. Soc.* (A), 1971, 3516.
68. G. B. Birrell, S. P. Van and O. H. Griffith, *J. Am. Chem. Soc.*, 1973, **95**, 2451.
69. G. R. Luckhurst and M. Setaka, *J. Magn. Reson.*, 1977, **25**, 539.
70. O. H. Griffith, *J. Chem. Phys.*, 1964, **41**, 1093.
71. O. H. Griffith, *J. Chem. Phys.*, 1965, **42**, (a) 2644, (b) 2651.
72. O. H. Griffith and M. M. Mallon, *J. Chem. Phys.*, 1967, **47**, 837.
73. O. H. Griffith and E. E. Wedum, *J. Am. Chem. Soc.*, 1967, **89**, 787.
74. G. B. Birrell, A. A. Lai and O. H. Griffith, *J. Chem. Phys.*, 1971, **54**, 1630.
75. G. B. Birrell and O. H. Griffith, *J. Phys. Chem.*, 1971, **75**, 3489.
76. A. Faucitano, A. Perotti, G. Allara and F. F. Martinotti, *J. Phys. Chem.*, 1972, **76**, 801.
77. G. B. Birrell, Z. Ciecierska-Tworek and O. H. Griffith, *J. Phys. Chem.*, 1972, **76**, 1819.
78. Y. Hori, S. Shimada and H. Kashiwabara, *Polymer*, 1977, **18**, 1143.
79. Y. Hori, S. Aoyama and H. Kashiwabara, *J. Chem. Phys.*, 1981, **75**, 1582.
80. E. E. Wedum and O. H. Griffith, *Trans. Faraday Soc.*, 1967, **63**, 819.
81. H. Ohigashi and Y. Kurita, *J. Magn. Reson.*, 1969, **1**, 464.
82. J. C. Vedrine, *Chem. Phys. Lett.*, 1977, **45**, 117.
83. L. D. Kispert and J. Pearson, *J. Phys. Chem.*, 1972, **76**, 133.
84. T. Iwamoto, M. Kiyoki and N. Matsuura, *Bull. Chem. Soc. Jpn.*, 1978, **51**, 390.
85. P. J. Baugh, J. I. Goodhall and J. Bardsley, *J. Chem. Soc., Perkin Trans. 2*, 1978, 700.
86. (*a*) L. B. Ebert, D. R. Mills, J. C. Scanlon and H. Selig, *Mat. Res. Bull.*, 1981, **16**, 831; (*b*) L. B. Ebert and H. Selig, *Syn. Met.*, 1981, **3**, 53.
87. J. G. Ballard and T. Birchall, *J. Chem. Soc., Dalton Trans.*, 1976, 1859.
88. P. Boolchand, W. J. Bresser, D. McDaniel, K. Sisson, V. Yeh and P. C. Eklund, *Solid State Commun.*, 1981, **40**, 1049.
89. R. W. Joyner and F. L. Vogel, *Syn. Met.*, 1981, **4**, 85.
90. M. J. Moran, J. E. Fisher and W. R. Salaneck, *J. Chem. Phys.*, 1980, **73**, 629.
91. K. Ohhashi, T. Shingo, T. Takada and I. Tsujikawa, *J. Phys. Colloq.*, 1979, **40**, C2–269.
92. R. H. Herber and M. Katada, *J. Inorg. Nucl. Chem.*, 1979, **41**, 1097.

93. T. Tominaga, T. Sakai and T. Kimura, *Bull. Chem. Soc. Jpn.*, 1976, **49**, 2755.
94. W. R. Salaneck, C. F. Brucker, J. E. Fischer and A. Metrot, *Phys. Rev.*, 1981, **24B**, 5037.
95. S. Kikkawa, F. Kanamaru and M. Koizumi, *Physica*, 1981, **105B**, 249.
96. T. R. Halbert, D. C. Johnston, L. E. McCandlish, A. H. Thompson, J. C. Scanlon and J. A. Dumesic, *Physica*, 1980, **99B**, 128.
97. H. Schäfer-Stahl, *Mat. Res. Bull.*, 1980, **15**, 1091.
98. H. Stahl, *Inorg. Nucl. Chem. Lett.*, 1980, **16**, 271.
99. R. G. Copperthwaite, *J. Chem. Soc., Chem. Commun.*, 1976, 707.
100. R. H. Herber and G. Carrasquillo, *Inorg. Chem.*, 1981, **20**, 3693.
101. R. H. Herber and A. E. Smelkinson, *Inorg. Chem.*, 1978, **17**, 1023.
102. E. R. Andrew, *Int. Rev. Phys. Chem.*, 1981, **1**, 195.
103. J. M. Thomas, S. Ramdas, G. R. Millward, J. Klinowski, A. Audier, J. Gonzalez-Calbet and C. A. Fyfe, *J. Solid State Chem.*, 1982, **45**, 368.
104. A. Michael, W. Meiler, D. Michel and B. Pfeifer, *Chem. Phys. Lett.*, 1981, **84**, 30.
105. G. Boxhoorn, R. A. van Santen, W. A. van Erp, G. R. Hays, R. Huis and D. Claque, *J. Chem. Soc., Chem. Commun.*, 1982, 264.
106. E. G. Derouane, J.-P. Gilson and J. B. Nagy, *Zeolites*, 1982, **2**, 42.
107. J. A. Ripmeester, *Chem. Phys. Lett.*, 1980, **74**, 536.
108. T. Terao, S. Matsui and A. Saika, *Chem. Phys. Lett.*, 1979, **64**, 582.
109. J. A. Ripmeester, R. E. Hawkins and D. W. Davidson, *J. Chem. Phys.*, 1979, **71**, 1889.
110. J. A. Ripmeester, J. S. Tse and D. W. Davidson, *Chem. Phys. Lett.*, 1982, **86**, 428.
111. J. A. Ripmeester and D. W. Davidson, *J. Mol. Struct.*, 1981, **75**, 67.
112. J. A. Ripmeester, *J. Am. Chem. Soc.*, 1982, **104**, 289.
113. L.-C. de Menorval, J. P. Fraissard and T. Ito, *J. Chem. Soc. Faraday Trans. 1*, 1982, **78**, 403.
114. H. Ueda and T. Nagai, *Chem. Pharm. Bull.*, 1981, **29**, 2710.
115. H. S. Gold and M. R. Rice, *Talanta*, 1982, **29**, 637.
116. R. H. Herber, *Acc. Chem. Res.*, 1982, **15**, 216.
117. Raman spectra of the guest species in $SbCl_5$ intercalates have recently been reported: (a) W. Jones, P. Korgul, R. Schlögl and J. M. Thomas, *J. Chem. Soc., Chem. Commun.*, 1983, 468; (b) R. Schlögl, W. Jones and J. M. Thomas, *ibid.*, 1983, 1330.
118. (a) L. B. Ebert and E. H. Appelman, *Phys. Rev.*, 1983, **28B**, 1637; (b) W. R. Salaneck, C. F. Brucker, J. E. Fischer and A. Metrot, *Phys. Rev.*, 1983, **28B**, 1639.
119. H. Saito, G. Izumi, T. Mamizuka, S. Suzuki and R. Tabeta, *J. Chem. Soc., Chem. Commun.*, 1982, 1386.
120. Y. Inoue, T. Okuda and R. Chujo, *Carbohydr. Res.*, 1983, **116**, C5.

3 · NMR, NQR AND DIELECTRIC PROPERTIES OF CLATHRATES

D. W. DAVIDSON and J. A. RIPMEESTER

National Research Council, Ottawa, Canada

1. Introduction

As this chapter will illustrate, studies of the NMR and dielectric properties have provided considerable information about large-scale molecular motion, especially reorientation, as well as about certain structural and other parameters of the two most frequently-studied clathrate systems, the clathrate hydrates and the β-quinols. Studies of other inclusion compounds by these techniques are just beginning.

Some aspects of the NMR and dielectric behaviour of clathrate hydrates[1,2,3] and quinol clathrates[3] have been previously reviewed but there is now a clear need for evaluation of the present state of affairs.

This chapter is mainly restricted to illustrations of the most significant results. The reader should consult the references cited for more details of experimental techniques and the relevant NMR and dielectric theory of the solid state.

INCLUSION COMPOUNDS III
ISBN 0-12-067103-4

2. Identification and characterization of clathrates

As clathrates are solids consisting of weakly interacting host and guest molecules, NMR and dielectric methods of analysis, which are sensitive to the considerable orientational mobility of guest molecules, are particularly useful in establishing the presence of clathrates. Moreover, the high resolution possible with modern solid state NMR can provide valuable information about clathrate structures.

2.1. Proton and ^{19}F broadline NMR

Classical broadline NMR techniques[4] have been used extensively to characterize previously unknown clathrate hydrates.[5] In the case of proton or fluorine containing molecules held rigidly in a crystal lattice, NMR linewidths of several gauss to several tens of gauss are determined mainly by nuclear dipole–dipole interactions. Sufficiently rapid reorientational motions reduce linewidths to ~1 Gauss. Liquids give much smaller linewidths, usually determined by instrumental resolution.

Clathrate hydrate samples may contain considerable quantities of unencaged guest material, liquid or solid, if the sample has not been conditioned sufficiently. If heavy water is used to form the hydrate, the NMR signal may consist of contributions from guest molecules in the liquid or solid form as well as from enclathrated species. In the temperature range of ~100 to 270 K, guest molecules in the nearly spherical cages normally reorientate

Table 1. Clathrate hydrates first identified by NMR and dielectric[a] techniques

---

Structure I
    SO_2F_2, ClO_3F, GeH_4
Structure I and Structure II
    $(CH_2)_3O$[a], $(CH_2)_2S$
Structure II
    SeF_6, TeF_6, SO_2ClF, CF_3Cl, CF_3I, CF_3NO, $CHFBr_2$, CH_3CF_3, CF_3CF_3, $CH_2=$
    CF_2, cis-$CHF=CHF$, CF_3CH_2Cl, $(CH_3)_2CHCl$, $(CH_3)_2CF_2$, cyclobutane,
    methylcyclopropane, acetone[a], propylene oxide[a], propylene sulphide,
    CH_3CH_2SH, $(CH_3)_2S$, cyclobutanone[a], 1,3-dioxolane[a], 2,5-dihydrofuran[a],
    2,3-dihydrofuran, isoxazole[a], 1,4-dioxane[a], 1,3-dioxane[a], acetaldehyde[a],
    propionaldehyde[a], isobutyraldehyde[a], $(CH_2)_3S$, $(CH_2)_4S$
Structure II, with H_2S help gas
    n-$CH_3(CH_2)_2CH_3$[a], $(CH_3)_4C$, $(CH_3)_2CCl_2$, γ-butyrolactone
Tetragonal structure
    $(CH_3)_2O$

---

nearly isotropically, giving an NMR signal with a linewidth of ~1 G. Structure II deuteriohydrates show narrower lines than Structure I clathrates*.[5] The quality of the hydrate sample, or for that matter whether hydrate exists, can be assessed. Table 1 lists new clathrate hydrates identified by this method.

In the case of quinol clathrates, the narrow, liquid-like ^1H or ^{19}F lines observed in the NMR spectrum of some samples[6,7] have sometimes been incorrectly assigned to enclathrated molecules. Only diffusion of the centre of mass of the guest molecules can lead to such narrow lines: diffusion is impossible in proper clathrates except during decomposition. Lines much narrower than ~1 G wide must then be assigned to liquid inclusions in the clathrate crystals, as demonstrated explicitly for the narrow line observed for methanol-β-quinol.[8]

2.2. Solid state ^{13}C NMR

Clathrates such as those of quinol, Dianin's compound and the cyclodextrins do not lend themselves well to study by broadline NMR techniques, since fully deuterated host material is not usually easily available. However, with the use of such NMR techniques as cross-polarization[9] and magic-angle spinning[10] it has become feasible to record reasonably well-resolved ^{13}C NMR spectra of solid clathrate samples.

In solids, the nuclear shielding is not only affected by chemical inequivalence, as in solution, but also by crystallographic inequivalence. For instance, quinol in solution gives two lines, one attributed to the hydroxyl carbon atom, and the other to the remaining carbon atoms. About 20 solid quinol samples have been examined by ^{13}C NMR spectroscopy.[11,12] The pattern shown in Fig. 1a is typical of the β-quinol lattice. The inequivalence of the two non-substituted quinol carbon atoms varies from a few tenths ppm for cyclopropane-β-quinol to 2.3 ppm for acetonitrile-β-quinol, and depends both on the nature of the guest and the degree of cage occupancy θ. In Table 2 Δ_1 gives the splitting of these two carbon lines and Δ_2, the mean separation of this doublet from the hydroxyl carbon line. The ^{13}C NMR spectrum of the α-quinol polymorph (Fig. 1b) can be understood in terms of the complex unit cell,[13] which has three crystallographically independent quinol molecules. The line from the hydroxyl substituted carbon atom is split into three components of relative intensity 2:3:1, while the other band shows partially resolved features from 12 inequivalent carbon atoms. Mixed α and β phases are also readily identified (Fig. 1c).

---

* Details of the structural aspects of clathrate hydrates can be found in Chapter 5, Volume 1.

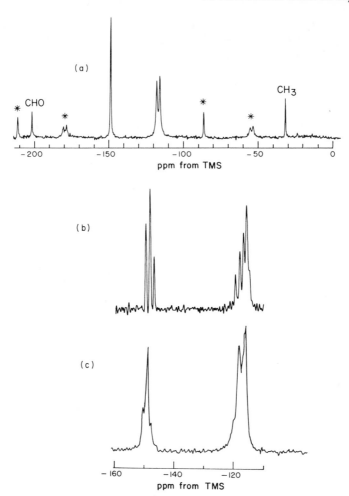

Fig. 1. Proton-enhanced, proton-decoupled (PE) ^{13}C spectra with magic-angle spinning of (a) $CH_3CHO-\beta$-quinol, (b) α-quinol, (c) mixed α- and β-quinol.

Other kinds of clathrate have not been investigated extensively. However, in cases where the clathrate and the normal (non-clathrate) form of the host lattice have different structures, differences similar to those obtained for the quinols can be expected.

In the case of Dianin's compound, these structures are basically the same,[14] but even here some of the spectral lines may be used to indicate the presence or absence of clathrate. Figure 2 shows the ^{13}C NMR spectra of the empty host lattice and of the acetonitrile and octanoic acid clathrates.[137]

Most of the 18 chemically inequivalent lattice carbon atoms are resolved. Although the spectra are similar, some differences are apparent. The largest change occurs in the position of the methyl group resonances at high field. In the empty clathrate, the C_{18} and C_{19} methyl groups, being situated on a flexible ring, can probably occupy at least some of the empty cage spaces. As it has been shown[14] that removal of the C_{18} methyl group markedly affects the van der Waals' surface at the neck of the hourglass shaped cage, (see Chapter 1, Volume 2) we tentatively assign the resonance at highest field to the C_{18} carbon. A useful aid in assigning ^{13}C resonances is the dipolar dephasing technique proposed by Opella et al.[15] By inserting an

Table 2. Guest dependence of ^{13}C NMR spectrum of hydroquinone in β-quinol clathrates

Guest molecule	Δ_1 (ppm)	Guest molecule	Δ_2 (ppm)
CH_2ClF	1.61	Xe	30.95
HCl	1.62	HCl	31.11
H_2S	1.62	$TMO^{(c)}$	31.22
COS	1.62	$TMS^{(c)}$	31.22
Ethylene oxide	1.73	H_2S	31.22
Formaldehyde	1.73	$CH_3SH^{(a)}$	31.22
"Air"	1.83	Formaldehyde	31.27
Trimethylene oxide (TMO)	1.83	"Air"	31.32
Trimethylene sulphide (TMS)	1.83	CH_3F	31.49
		$Formamide^{(c)}$	31.54
Formamide	1.83	$CH_3Cl^{(a)}$	31.49
Methanol	1.83	$COS^{(c)}$	31.55
CH_3F	1.94	Ethylene oxide$^{(c)}$	31.60
$SO_2^{(a)}(\theta = 0.35)$	1.94		
Acetaldehyde	2.05	Methanol	31.65
$SO_2^{(b)}(\theta = 0.93)$	2.20	Formic acid	31.64
		$CH_3Cl^{(b)}$	31.66
Xe	1.29	$CH_3Cl^{(b)}$	31.70
$CH_3Cl^{(a)}$	1.29	$CH_3SH^{(b)}$	31.70
$CH_3Cl^{(b)}$	1.19	CH_3Br	31.75
$CH_3SH^{(a)}$	1.62	CH_3CN	31.76
$CH_3SH^{(b)}$	1.08	$CH_2ClF^{(c)}$	31.86
CH_3Br	0.7	Acetaldehyde	31.92
Ethylene sulphide	0.6	Ethylene sulphide$^{(c)}$	32.11
Cyclopropane	0.5	Cyclopropane$^{(c)}$	32.14

(a) Low occupancy sample.
(b) High occupancy sample.
(c) β-quinol clathrate identified for the first time by ^{13}C NMR.

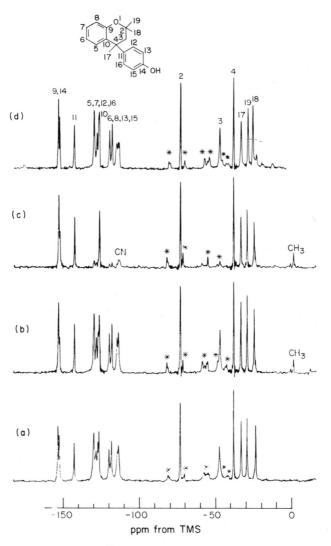

Fig. 2. PE magic-angle spinning ^{13}C spectra of Dianin's compound. (a) empty lattice, (b) and (c) with CH_3CN guest, without and with suppression of signals from C atoms strongly coupled to protons, (d) with octanoic acid guest.

acquisition delay of ~40 μs without proton decoupling, the resonances due to carbons strongly coupled to protons are suppressed (Fig. 2c). In practice, this means that the remaining spectrum contains only lines due to quaternary carbon atoms and carbon atoms such as those in methyl groups and guest species for which dipolar coupling has been reduced by molecular motion.

2.3. ^{129}Xe NMR

The large polarizable electron cloud of xenon leads to a large density-dependent nuclear shielding effect in the ^{129}Xe NMR of atomic xenon.[16,17] The average nuclear shielding and the shielding anisotropy of ^{129}Xe depend on the size and shape of the clathrate cage.[18,19]

Figure 3 shows (a) the proton-decoupled ^{129}Xe NMR spectrum obtained for the structure I xenon hydrate and (b) the spectrum for the corresponding D_2O hydrate. The asymmetric high field line can be assigned to the slightly

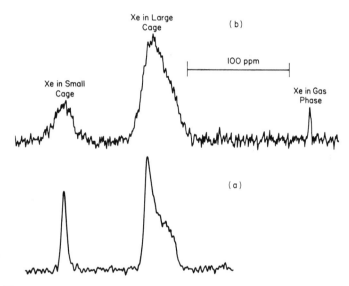

Fig. 3. ^{129}Xe NMR spectra at 275 K of (a) xenon hydrate with proton decoupling, (b) xenon deuteriohydrate.

flattened structure I large cage (symmetry $\bar{4}2m = D_2^d$), the symmetric low field line to xenon in the small, nearly spherical cage (symmetry $m3 = T_h$). With allowance for the presence of three times as many large as small cages in structure I hydrates, the relative intensities of the lines show that the degree of occupancy of the small cage is only 0.74 times that of the large cage, with important implications[20] for the choice of cage-guest interaction potential. No other technique for "seeing" directly the guest molecules in the two kinds of clathrate hydrate cage is presently available.

Cross polarization and magic angle spinning may also be used to good advantage in obtaining ^{129}Xe NMR spectra. Figure 4 shows that the highly anisotropic shielding pattern characteristic of ^{129}Xe trapped in the β-quinol

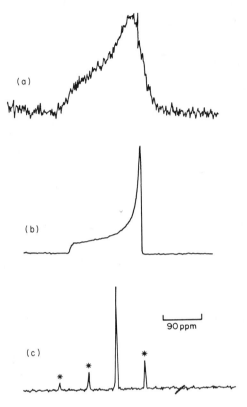

Fig. 4. ^{129}Xe NMR spectra of Xe–β-quinol (a) without proton enhancement, (b) with enhancement, (c) with enhancement and magic-angle spinning.

cage reduces to a single line at the isotropic shielding value (plus several spinning sidebands) when the sample is spun at the magic angle.

The spectra obtained for xenon trapped in the phenol clathrate (Fig. 5) shows three anisotropic shielding patterns which can be resolved by spinning the sample at the magic angle. The three signals can be assigned, in the direction of decreasing field, to xenon in the small phenol cages, near the ends of the elongated large cage, and in the middle of the large cage, respectively.

Isotropic nuclear shielding parameters as well as shielding anisotropies for xenon trapped in a number of clathrates are summarized in Table 3. The general trend of the shifts follows that for the pure phases of xenon, i.e. greater density causes a shift to lower fields. For clathrates this means that the tightest fitting cages should produce the largest shifts to low field.

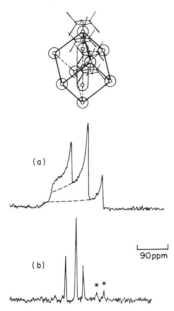

Fig. 5. ^{129}Xe proton-enhanced NMR spectra of Xe–β-phenol, (a) without and (b) with magic-angle spinning.

2.4. Dielectric behaviour

Dielectric relaxation studies have been used extensively since the early work of Dryden[21] on quinol clathrates to provide information about the generally rapid reorientational motion of encaged dipolar molecules.

A considerable number of new and previously known hydrates (Table 1) were identified as clathrates by the characteristic behaviour of the water relaxation region (Section 3.1.3) which occurs at kHz frequencies near 200 K.[29–33] Moreover, the permittivity $\varepsilon_{\infty 1}$ measured to the high-frequency side of the water dispersion range for clathrate hydrates includes a substantial contribution proportional to the square of the dipole moment of the reorienting guest molecule as well as to its concentration. This is not the case for ethanol hydrate which has an $\varepsilon_{\infty 1}$ value similar to ice[22] and is not a proper clathrate. The ethanol molecules do not change their orientation readily and are perhaps hydrogen bonded to water molecules in the hydrate.

Dielectric studies have shown that the hydrates of the dioxanes[23,24] and isoxazole,[25] despite the hydration numbers of 34 reported from some thermal analysis measurements,[26,27] are normal structure II clathrates with hydration numbers close to 17. The aldehyde hydrates must be formed rapidly from

Table 3. ^{129}Xe nuclear shielding in various environments

	σ_{av} (ppm)	$\Delta\sigma$ (ppm[a])
Xenon		
Gas, 0 density	0	
Gas, density 100 amagat	−52.8	
Liquid, 244 K	−161	
Liquid, 161 K	−229	
Solid, 161 K	−274	
Structure I Hydrate		
Small cage	−242	0
Large cage	−152	32
Structure II Hydrate		
Small cage	−229	15
β-Quinol	−210	−160
β-Phenol		
Small cage	−229	−171
Large cage I	−248	−105
Large cage II	−279	−53
p-Fluorophenol	−209	−164
α-Cyclodextrin	−192	23
Dianin's compound		
Site I	−133	−85
Site II	−152	−55

[a] Anisotropy of axially-symmetric shielding tensor. A positive value indicates greatest shielding along cage symmetry axis.

newly-prepared aqueous solutions and stored at low temperatures to prevent the hydrolysis reaction $RCHO + H_2O \rightarrow RCH(OH)_2$ from occurring.[28]

3. Clathrate hydrates

Unlike most clathrates and other inclusion compounds, the host water lattice of clathrate hydrates consists of molecules which are themselves reasonably mobile. NMR and dielectric techniques give useful information about the motion of the lattice water molecules, in addition to the kind of structural and guest-molecule motional information obtainable for inclusion compounds in general.

Of the more than 100 known simple clathrate hydrates,[1] NMR and/or dielectric information is available for about one-third. We choose the common hydrate of tetrahydrofuran (THF) to show the properties of a typical clathrate hydrate and then consider how these properties differ for other hydrates.

3.1. Illustration of the NMR and dielectric methods

3.1.1. NMR lineshapes and second moments

3.1.1.1. Proton resonance. At sufficiently low temperatures the shapes of NMR lines become characteristic of the structure of the rigid lattices, i.e. lattices in which, besides intramolecular vibration, only translational and librational oscillations occur.

For the THF–H_2O hydrate the rigid-lattice condition is achieved at temperatures of 10 K and lower.[34] More significant information about the guest and host molecules individually may be obtained by deuteration of either the host or guest molecules. The rigid H_2O spectrum of the THF–d_8–H_2O hydrate resembles the spectrum of H_2O ice.[35]

Although the proton lineshape of THF–D_2O hydrate may be calculated from the THF geometry, the lack of distinct spectral features for this 8-proton molecule makes it hardly worthwhile. More simply, the experimental second moment (mean square line width) may be compared with that calculated from the polycrystalline Van Vleck formula. The experimental rigid lattice second moments of 14.1 ± 0.3, 33 ± 2, and 29.5 ± 1.0 G^2 for THF–D_2O, THF–d_8–H_2O and THF–H_2O respectively agree well with the calculated values of 15.1, 31.5 and 30.8 G^2, respectively,[34] which include as well as intramolecular contributions, relatively small ($\sim 3\%$) contributions from lattice deuterons and the protons of neighbouring guest molecules.

Figure 6 shows how the second moments (SM) change as the temperature is raised. With the onset of reorientation of THF at rates comparable to the

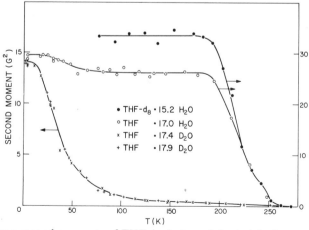

Fig. 6. Proton second moments of THF–hydrate and deuteriohydrate.

line width (~20 kHz) the intra- and inter-molecular proton interactions begin to be averaged, leading to a change in line shape and a fall in SM. It is characteristic of most molecules enclathrated in D_2O hydrates that this narrowing occurs over a great range of temperature, in contrast to most line narrowing "transitions" which occur in solids, and an increasingly gradual fall in SM extends to relatively high temperatures.

The THF reorientational motions affect the SM's of THF in H_2O at comparable temperatures to a lesser extent because most of the signal comes from protons in the H_2O lattice which is "rigid" below 200 K. Guest–molecule reorientational motion causes sufficient narrowing of the THF spectrum that it appears as a narrow line superimposed on a broad line from H_2O over a wide temperature range (60 to 200 K). This is a characteristic feature of clathrate hydrates.

The proton spectrum of THF–d_8 begins to narrow at 200 K because of motion of the water molecules. In fact, the narrowing occurs in two steps, of which the first is due to reorientation of water molecules and the second due to their translational diffusion. There is considerable evidence that water-molecule reorientation is faster than diffusion in clathrate hydrates, in contrast with the case in ice.[36] Reorientation alone leaves a substantial contribution (7 G^2) to the SM from inter-H_2O interactions; only diffusion can completely eliminate dipolar broadening.

At high enough temperatures such that the water molecules are undergoing rapid rotation on an NMR time scale, the THF line shape and SM finally achieve the characteristics of isotropic rotation. This is illustrated in Fig. 7 for the deuteriohydrates of THF and some other structure II hydrate-forming molecules.[1] The rapid reorientations of the THF molecules do not quite correspond to complete averaging of *intra*molecular proton interac-

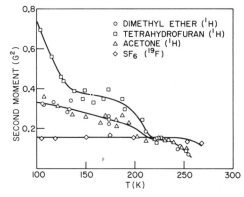

Fig. 7. Second moments of some structure II deuteriohydrates at high temperatures.

tions as long as the water lattice is rigid. With reorientation of the water molecules, however, the individual 16-hedral cages assume on time average the tetrahedral ($\bar{4}3m$) symmetry already present on space average[37] and all intramolecular contributions to the line width vanish.

3.1.1.2. *Deuterium resonance.* Deuterium has a spin 1 nucleus whose NMR behaviour in solids is normally dominated by the coupling between its quadrupole moment and the electric field gradient (efg) at the site of the nucleus. This interaction is larger than dipolar coupling by more than an order of magnitude and entirely dependent on molecular orientation. It is thus particularly useful for detecting small motional anisotropy.

Deuterium spectra[38] of THF–d_8 in H_2O and of THF in D_2O are shown in Figs. 8 and 9. The near identity of the quadrupole coupling constant e^2qQ/h and asymmetry parameter η for D_2O in ice[39] (213.4 kHz, 0.112) and THF hydrate (215 kHz, 0.11) shows that the hydrogen-bonded OD bond lengths are practically the same.

For a deuterated guest molecule one can follow the increasing rotational averaging effects as the temperature is raised by defining an average effective quadrupolar splitting

$$(a^2)^{1/2} = \langle 3e^2qQ/4h \rangle \tag{1}$$

obtained from the line width (or at high temperature from T_2 measurements).[38,40] This quantity is plotted against temperature in Fig. 10 for THF–d_8

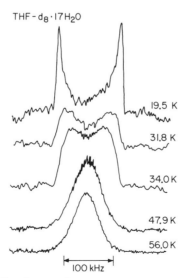

Fig. 8. Deuterium NMR lineshapes of THF–d_8 hydrate.

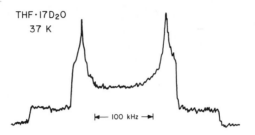

Fig. 9. Rigid lattice deuterium NMR lineshape of THF deuteriohydrate.

hydrate as well as for the structure I hydrate of ethylene oxide–d_4 (EO–d_4). For the former the average quadrupolar coupling drops to zero at temperatures where the water molecules rotate sufficiently rapidly; the finite residual width is then dipolar in origin. For the structure I hydrate a corresponding substantial fall occurs but only to an effective coupling constant of several kHz. In EO hydrate 90% of the guest molecules occupy structure I 14-hedral cages which, with H_2O molecules rotating, individually

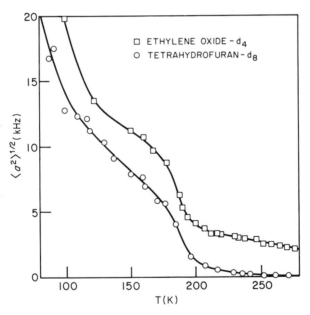

Fig. 10. Effective deuterium quadrupole splitting frequencies of deuterated guest molecules.

acquire $\bar{4}$2m symmetry.[41] The presence of a unique axis prevents "isotropic" motional averaging of the CD direction. These results complement the X-ray electron density maps which at 250 K are spherically symmetric for THF in structure II[37] but not for EO in structure I.[41]

3.1.2. Spin-lattice relaxation

Although considerable information about the rates of molecular motion is provided by line shape studies of the kind described above, more quantitative data may be obtained from measurements of spin-lattice relaxation times in the static (T_1) and rotating $(T_{1\rho})$ frame. These may be written in a form sufficient for present purposes as

$$\frac{1}{T_1} = A\left[\frac{\tau_c}{1+\omega_0^2\tau_0^2} + \frac{4\tau_c}{1+4\omega_0^2\tau_c^2}\right] \qquad (2)$$

and

$$\frac{1}{T_{1\rho}} = \frac{3A}{2}\left[\frac{\tau_c}{1+4\omega_1^2\tau_c^2}\right], \qquad (3)$$

where $A = (2/3)\gamma^2\Delta SM$, ω_0 is the Larmor frequency, τ_c is the second order reorientational correlation time which, when much smaller than ω_0^{-1}, reduces the dipolar second moment by ΔSM, and $\omega_1 = \gamma H_1$ is the frequency of the rf field H_1. Both equations assume the presence of a single correlation time. Equation 3 is approximately valid under conditions where $\omega_0\tau_c \gg 1$. For deuterium relaxation

$$A = 3\pi^2(e^2qQ)^2/(10h^2)$$

Figure 11 shows the dependence on temperature of the proton T_1 for THF hydrate and deuteriohydrate. At temperatures near 50 K the T_1 curves pass through a minimum which, although clearly associated with reorientation of THF, is much shallower and broader than predicted by Equation 2, which gives a minimum T_1 value of 6.2 ms for THF in D_2O at 10 MHz. The T_1 curves at low temperatures can be analysed[34] in terms of a broad distribution of correlation times, as can T_1 values measured[38] for D resonance in THF–d_8 in H_2O. The fall in T_1 at high temperatures is due to the motion of the water molecules. A minimum is not reached below the melting point since, for the water molecules, $\tau_c \gg \omega_0^{-1}$. At the relatively low H_1 fields, however, $T_{1\rho}$ passes through a minimum. Figure 12 in fact shows a double minimum, which may be attributed[34] to superposition of reorientation and the slower diffusional process already mentioned.

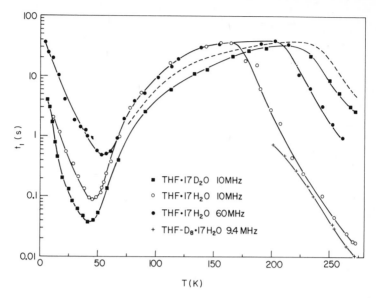

Fig. 11. Proton spin-lattice relaxation times of THF hydrates. The broken line represents 30 MHz data for THF · 17D$_2$O.[136]

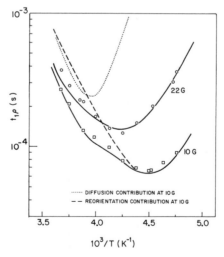

Fig. 12. Proton spin-lattice relaxation times in the rotating frame at fields of 10 and 22 G for THF–d$_8$ hydrate. Solid curves are calculated for the sum of the reorientation and diffusion processes, shown separately for $H_1 = 10$ G.

3.1.3. Dielectric relaxation

At frequencies low in comparison with the reorientation rates of the constituent molecules the static permittivities of hydrates of THF and other dipolar guest molecules include a major contribution from reorientation of water molecules and a smaller contribution from guest molecules.[42] With increase of frequency the permittivity decreases sharply and strong dielectric absorption occurs at frequencies comparable to the reorientation rates of the water molecules (~ 1 kHz at 200 K). At higher frequencies reorientation of water no longer contributes to the permittivity, which remains constant over a wide frequency range until (at ~ 10 GHz at 200 K) the onset of a dispersion–absorption region associated with reorientation of the guest molecules.

Figure 13 shows a dielectric loss (ε'') vs. permittivity (ε') plot for the water relaxation region in THF hydrate. This is typical of clathrate hydrates in showing a small departure from the semicircle which characterizes a

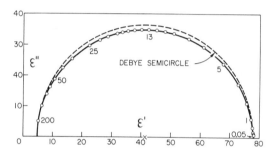

Fig. 13. Cole–Cole plot of water relaxation in THF hydrate at 211 K and 3.0 kbar. Numbers on the locus are frequencies in kHz.

single relaxation time. For structure II hydrates the shape may be adequately represented by two discrete relaxation times which differ by a factor of about two.[42] The temperature dependence of the frequency of maximum loss gives an activation energy of 30.9 kJ mol^{-1} for reorientation of water molecules in THF hydrate.

Although a few microwave dielectric measurements have been made in which dielectric absorption by the guest molecules in THF and acetone hydrates was detected at 90 K in the GHz range,[43] most measurements have been made on samples cooled sufficiently to bring the absorption frequencies down to the MHz range and below. At these low temperatures the absorption curves are very broad; Fig. 14 shows the case of THF hydrate[44] at 20.4 K where the half width extends over a factor of 10^4 in frequency compared with the factor of ~ 14 characteristic of a single relaxation time (broken

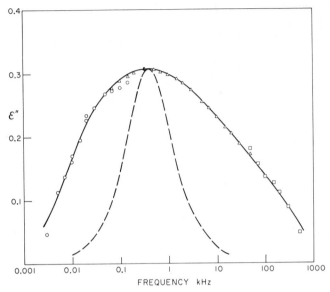

Fig. 14. Frequency dependence of dielectric loss from THF reorientation in its hydrate at 20.4 K.

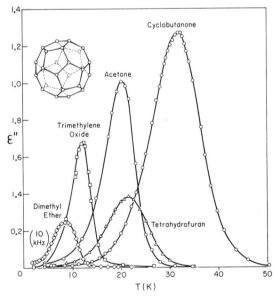

Fig. 15. Dielectric loss at 1 kHz from reorientation of guest molecules in a number of structure II hydrates.

curve). Broad dielectric absorption is a characteristic feature of the reorientation of guest molecules in clathrate hydrates.

Guest-molecule absorption may be conveniently shown by plots of ε'' vs temperature at fixed frequency as illustrated in Fig. 15. Effective activation energies for molecular reorientation (e.g. 3.80 kJ mol^{-1} for THF) may then be derived from the variation of the temperatures of maximum absorption measured at a number of fixed frequencies.

The permittivity measured on the high frequency side of the water dispersion region ($\varepsilon_{\infty 1} = \varepsilon_{02}$) rises as $1/T$ at relatively high temperatures but not at lower temperatures (Fig. 16). The departure becomes increasingly great for THF hydrate with fall of temperature below 100 K. This general behaviour of clathrate hydrates is significant in showing that the preferred orientations of guest molecules in individual cages have energies that become significantly different in comparison with kT. This anisotropy introduces a number of discrete relaxation times in each cage and a continuous distribution of relaxation times can follow from the variations from cage to cage as a result of disorder in the orientations of the water molecules.

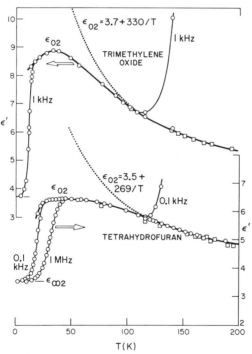

Fig. 16. Temperature dependence of ε_{02} of two structure II hydrates. Permittivities at some fixed frequencies are also shown.

We proceed now to a survey of the dielectric and NMR behaviour of the host and guest molecules in clathrate hydrates.

3.2. The host lattice

Over a range of temperature extending downward about 100 K from the melting points, the static permittivity of structure II hydrates, to within a few permittivity units, is given by [42,2]

$$\varepsilon_0 = \varepsilon_{\infty 1} + \frac{14900}{T} \tag{4}$$

The results of fewer measurements on hydrates of structure I suggest

$$\varepsilon_0 \approx \varepsilon_{\infty 1} + \frac{13300}{T} \tag{5}$$

for this structure. The relatively high values (e.g. ~58 and ~53 at 273 K from Equation 5 and 6 respectively) are due to a correlation between the directions of the water dipoles in the hydrogen-bonded lattice, as in ice itself ($\varepsilon_0 \approx 92$ at 273 K). Considerably higher values (73 and 70, respectively, at 273 K) have been calculated from a model of dipole–dipole correlation which assumed all the allowable orientations of water molecules to be equally likely.[45]

The complex permittivity curves of the water relaxation in Structure II clathrate hydrates all resemble Fig. 13. The presence of two or more relaxation times (as many as eight are predicted by the lattice structure[2]) clearly shows that not all orientations of the water molecules allowed by the Bernal–Fowler rules are equally stable, though the differences are minor at relatively high temperatures. The water dispersion region for structure I hydrates (where up to ten relaxation times are possible) is somewhat broader at comparable temperatures, as might be anticipated from the larger distribution of O···O···O angles in this structure. In terms of the relaxation model which attributes reorientation of water molecules in ice to the diffusion of orientational Bjerrum defects,[46] multiple relaxation times in both hydrate structures arise from the presence of small differences in the rate of defect diffusion along different pathways in the 4-connected networks.

Table 4 gives dielectric relaxation times and activation energies of the water molecules for hydrates of structures I and II. Trimethylene oxide forms hydrates of both structures,[29,47] the reorientation rate of water molecules being faster in structure I.

Table 4. Water molecule dielectric relaxation times and activation energies

Structure	Guest (ref.)	τ_0 at 233.2 K (μs)	$E_D^{(a)}$ (kJ mol^{-1})	$\mu^{(b)}$ (debye)
Ice Ih		1420	55.2	
I	Xe (54)	330	50	0
	cyclopropane (86)	310	47.6	0
	N$_2$ (87)	180		0
		(1.2 kbar)		
	Ar (87)	96		0
		(2.0 kbar)$^{(c)}$		
	CH$_3$Cl		(43.9)	1.87
	ethylene oxide (69)	0.33	32.2	1.90
			(34.07)	
	trimethylene oxide (29)	0.03$^{(d)}$	24.2	1.93
II	SF$_6$ (88)	780	51.4	0
	CCl$_3$F (50)	~450		0.46
	CCl$_2$F$_2$ (51)	~230		0.51
	CClF$_3$ (52)	~160		0.50
	CHCl$_2$F (50)	~130		1.30
	CBrClF$_2$ (53)	~60		—
	1,3-dioxolane (89)	5.4	36.4	1.47
	1,4-dioxane (23)	4.6	38.0	0
	propylene oxide (29)	2.0	33.4	2.00
	1,3-dioxane (24)	1.7	32.2	2.06
	2,5-dihydrofuran (29)	1.5	31.3	1.54
	tetrahydrofuran (29)	1.0	30.9	1.63
			(30.1)	
	isoxazole (25)	0.88	29.7	2.90
	dimethyl ether (49)	0.78	28.4	1.31
	acetone (31)	0.57	27.2	2.88
	cyclobutanone (31)	0.49	27.2	2.89
	trimethylene oxide (29)	0.48	29.3	1.93
Tetragonal	dimethyl ether (49)	0.087$^{(d)}$	24.7	1.31

(a) Bracketed values are from T_1 measurements.
(b) Dipole moment of guest molecule.
(c) Extrapolated from higher temperatures.
(d) Extrapolated from lower temperatures.

Dimethyl ether can form, in addition to a structure II hydrate, a second clathrate hydrate[48,49] which is stable below 238 K and in which the water molecules relax relatively rapidly.

Although neither the *amplitude* nor the *shape* of the water absorption-dispersion region is greatly dependent on the nature of the guest molecule, the reorientation *rates* vary by many orders of magnitude and the activation energies by a factor of almost two, for structure (54, 86–88). For SF_6 in structure II and non-dipolar guests in structure I, the water molecule reorientation rates approach those in ice Ih, while ethers and ketones promote much faster reorientation. The entries for the halogenated methanes[50–53] are approximate only, the complex permittivity plots not being accurately defined. Nevertheless these results suggest that the water reorientation in the presence of dipolar halogenated guests is generally more than 100 μs at 233 K, and therefore not nearly as rapid as when the guest molecules are ethers or ketones. The most obvious explanation of the difference is to be found in the ability of these latter molecules to inject Bjerrum defects into the lattice, perhaps by (very occasionally) forming hydrogen bonds with cage water molecules.

Figure 17 shows the temperature dependence of the proton second moments of four structure I hydrates.[54] The arrows, which indicate temperatures at which the dielectric relaxation times are 10^{-4} s, are seen to correlate well with the beginning of line narrowing due to water molecule reorientation. A similar correspondence occurs for structure II hydrates.[55]

For ethylene oxide–d_4 hydrate, Fig. 17 clearly shows two well-separated regions of line narrowing centered at 205 and 260 K. Again two corresponding minima occur in the $T_{1\rho}$ curve.[56] As for THF–d_8 hydrate, diffusion of water molecules is responsible for the higher temperature process. At 273 K, reorientation is two orders of magnitude faster than diffusion in EO hydrate and one order faster in THF hydrate. The diffusion rate of water molecules in xenon and cyclopropane hydrates is slower than in EO hydrate and

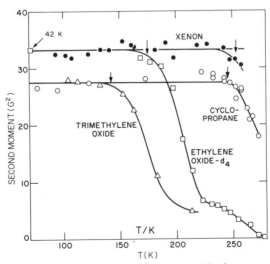

Fig. 17. Proton second moments of some structure I hydrates.

slower than in ice. The lattice-vacancy defect mechanism commonly invoked to account for self-diffusion in ice appears to predict faster diffusion in the hydrates. More extensive studies of diffusion are clearly required.

Measurements of the pressure dependence of water relaxation in THF hydrate at 211 K at pressures up to 3 kbar gave[42] a volume of activation of 4.4 ± 0.3 cm^3 mol^{-1}. Like the similar value of 4.1 ± 0.3 cm^3 mol^{-1} for ice,[57] this value is consistent with the Bjerrum model of reorientational relaxation but not with a local "melting" mechanism.

3.3. Guest molecules

The clathrate hydrates are almost unique among solid structures in the high rotational mobility of the guest molecules. This is a consequence of the nearly spherical shape of the cages and of the almost tetrahedral bonding structure of the water lattice which leads to resultant electrostatic fields due to water dipoles which are relatively small near the cage centres.[44] Since the potential barriers to molecular reorientation are normally small, even small variations resulting from the disordered orientations of the water molecules have a substantial effect on the reorientation rates. At the low temperatures where NMR and dielectric measurements are sensitive to guest molecule reorientation, these measurements have always given evidence of broad distributions of correlations times.

3.3.1. Spherical top molecules

The evidence for distributions of rotation rates of guest molecules of cubic point symmetry, i.e., tetrahedral and octahedral molecules, is particularly impressive. For these molecules the NMR spectrum changes as the temperature rises from a broad structured line characteristic of the rigid lattice to a narrow line of the kind expected for isotropic rotation. Over a considerable range of intermediate temperatures these two lines are superimposed, giving the appearance of the co-existence of two phases. Figure 18 shows this for SF$_6$[58] and neopentane.[59] Other examples are provided by the ^{19}F line shapes of CF$_4$ in structure I[58] and SeF$_6$ in structure II hydrates. The corresponding second moments (e.g., Fig. 19) fall regularly with rise in temperature as the narrow line acquires greater intensity and thereafter become independent of temperature at the values which correspond to isotropic rotation (cf. Fig. 7 for SF$_6$). Such behaviour is in marked contrast to that shown by THF and other guests of low symmetry.

The superposition of two lines is a good example of the Resing apparent-phase-change effect[60,61] arising from the presence of a very broad distribution

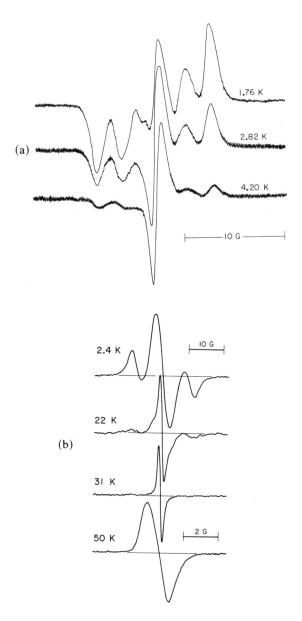

Fig. 18. Illustration of the Resing effect in (a) the ^{19}F lineshape of $SF_6 \cdot 17D_2O$ and (b) the proton lineshape of $(CH_3)_4C$ in the D_2O double hydrate with D_2S.

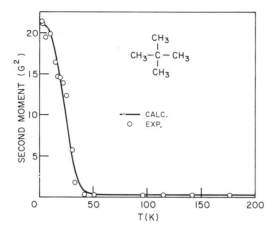

Fig. 19. Second moment of proton lineshape in $(CH_3)_4C-D_2S-D_2O$ hydrate. Solid line is the calculated value.

of correlation times. At temperatures within the transition range, practically all the correlation times lie either above or below the value $(\sim 10^{-5}\,s)$ which give lines of intermediate shape. That encaged spherical top molecules appear to rotate isotropically or not at all is the result of their high symmetry. Any reorientational process which rapidly exchanges the 12 (tetrahedral molecules) or 24 (octahedral molecules) energetically indistinguishable rotational configurations in an environment of any symmetry will give the effect of isotropic reorientation. It is immaterial whether there are one or more energetically distinct orientations in each cage. Since there is only a single correlation time within each cage, the wide distribution of correlation times must arise from differences between cages, that is, from the disorder in the orientations of the water molecules.

Included in Table 5 are the temperature $(T_{1/2})$ at which half the molecules are rotating fast enough to contribute to the narrow line, the mean activation energy E_m, and a parameter A specifying the width of a distribution of activation energies E according to a normal distribution

$$F(E) = \frac{\exp[-(E - E_m)^2/A^2]}{\sqrt{\pi A}} \tag{6}$$

Equation 6 was used[58,59] to calculate the distribution of correlation times, and hence the second moment, at different temperatures within the transition region from $\tau_c = 10^{-12}\exp(E/RT)$. Although this simple representation (cf. solid line of Fig. 19) accounts satisfactorily for the second moment behaviour, other distribution functions are also possible. However, the

Table 5. Reorientation correlation times and activation energies of encaged molecules

Structure	Guest (ref.)[a]	Temperature of ϵ'' max. at 100 kHz(K)	E_D (kJ mol⁻¹)	Temperature of min. T_1(K)	Larmor frequency (MHz)	E_{T_1}[b] (kJ mol⁻¹)	$T_{1/2}$ (K)	E_{SM}[c] (kJ mol⁻¹)
I	H₂S (59)	<1.8	<0.2		24	~0.071	<1.8	
	CH₄ (58)						<1.8	
	GeH₄						<3.5	
	CF₄ (58)						11.1	1.505[d], 0.468[e]
	C₂H₆ (59)			31.7	9.2	3.84	36	4.89
	CH₃Cl	20.3	3.34	27	9.4	1.75	45	6.10
	CH₃Br			40	9.4	2.72	53	7.19
	ClO₃F			28	9.2	2.67		
	SO₂F₂			38.5	10.0	2.51		
	ethylene oxide (69)	34.2	5.89				48	6.52
	ethylene sulphide						49	6.65
	trimethylene oxide (47)	66.4	8.8	63	9.2	7.1	78	10.57
		7.1	~0.58	15	9.2	~0.4		
II	SF₆ (58)			9.7	12.0	0.84	6.4	0.865[a], 0.422[e]
	SeF₆ (1)			15.5	10.0	1.25	7.7	1.057[d], 0.439[e]
	dimethyl ether (49)	9.4	0.88				16	2.17
	ethylene sulphide						20	2.72
	trimethylene oxide (42)	16.2	1.92	22.5	9.2	1.92	21	2.84
	C₃H₈ (59)	17.0	2.51				17	2.30
	CH₃I						10	1.38
	CF₃I	26.3	3.72	24	10.0	2.01	16	2.17
	CF₂Cl₂			23.8	10.0	1.71	18	2.42
	acetone (42)	24.3	4.26	36.5	9.2	3.26	24	3.26

II							
acetaldehyde (28)	36.1	5.60				33	4.47
propionaldehyde (28)	30.9	5.14					
isobutyraldehyde (33)	62	8.40				37	5.02
isobutane (59)	37.3	5.06					
1,3-dioxolane	23	3.80					
tetrahydrofuran (42, 34, 38, 65)	27.6	3.80	43	10.0	3.84	36	4.89
isoxazole (25)	26.2	5.22	55	9.2		32	4.35
cyclobutanone (42)	39.9	6.02	23	9.2	4.39	35	4.72
cyclopentane			29	9.2	2.67	22	2.97
1,4-dioxane (23)	44.4	7.19	63	9.2	6.48	47	6.35
1,3-dioxane (24)	58.3	8.28	84	9.2	8.4	56	7.56
Double hydrates with H$_2$S							
CH$_2$ClCH$_2$Cl (72)	26.2	3.64				23.5	3.106[d], 1.170[e]
n-butane (59)	39.6	5.89				43	5.81
neopentane (59)						51	6.90
(CH$_3$)$_3$COD							
(CH$_3$)$_2$CCl$_2$							
Double hydrates with							
THF–d$_8$ (42) H$_2$S (42)	<1.8		5.6	60	0.088	<1.8	
CH$_4$ (58)			4.85	9.2	0.247	<1.8	
Tetragonal							
dimethyl ether (49)	17.5	2.47					

[a] Results without reference are from unpublished work in this laboratory.
[b] From slope on high temperature side of T_1 minimum.
[c] From the approximation $E_{SM} = 135.4\,T_{1/2}$ J mol^{-1}.
[d] E_m and [e] A in Equation 6.

conclusion that there is a distribution of activation energies whose width
is an appreciable fraction of the most probable activation energy, is independent of the model chosen.

Additional information about correlation-time distributions is available
from low-temperature T_1 measurements of SF_6 and SeF_6 enclathrated in
D_2O. Each shows a broad T_1 minimum, the activation energy derived from
the log T_1 vs. $1/T$ slope above the minimum being similar to the mean
values E_m obtained from the SM analysis (Table 5). Figure 20 shows ω_0/T_1

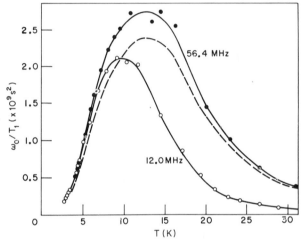

Fig. 20. ω_0/T_1 plot for ^{19}F resonance in $SF_6 \cdot 17D_2O$ at two frequencies.

plots for SF_6. This kind of plot is chosen for its similarity to ε'' vs. temperature
plots for dipolar guest molecules (see Section 3.3.3). At 56 MHz the ^{19}F T_1
values are affected by a chemical shift anisotropy of $\Delta\sigma = 310$ ppm.[62] This
contribution is

$$\left(\frac{\omega_0}{T_1}\right)_{CSA} = \frac{2}{15}(\omega_0\Delta\sigma)^2 \int \frac{(\omega_0\tau_c)F(\ln \tau_c)\,d\ln \tau_c}{1+\omega_0^2\tau_c^2} \tag{7}$$

where $F(\ln \tau_c)\,d\ln \tau_c$ specifies the (normalized) distribution of reorientational correlation times. The corresponding contribution from
intramolecular nuclear dipole–dipole interactions is

$$\left(\frac{\omega_0}{T_1}\right)_D = \frac{2}{5}\gamma^2\Delta SM\left[\int \frac{(\omega_0\tau_c)F(\ln \tau_c)\,d\ln \tau_c}{1+\omega_0^2\tau_c^2}\right.$$
$$\left. +2\int \frac{(2\omega\tau_c)F(\ln \tau_c)\,d\ln \tau_c}{1+4\omega_0^2\tau_c^2}\right] \tag{8}$$

where ΔSM is the rigid-lattice *intra*molecular dipolar second moment, $10.5 \, G^2$ for SF_6. Without a detailed knowledge of the distribution function one cannot quantitatively evaluate the relative contributions from Equations 7 and 8 except in the high and low temperature limits. For the former, $\omega_0 \tau_c \ll 1$ for all values of τ_c and

$$\frac{(\omega_0/T_1)_{CSA}}{(\omega_0/T_1)_D} = \frac{1}{25} \frac{(\omega_0 \Delta \sigma)^2}{\gamma^2 \Delta SM} \qquad (9)$$

At sufficiently low temperatures this ratio is $(\omega_0 \Delta \sigma)^2/(10 \gamma^2 \Delta SM)$.

The broken curve in Fig. 20 represents schematically the dipolar contribution to ω_0/T_1 at 56 MHz, obtained by subtracting the chemical shift anisotropy contribution. The latter is negligible at 12 MHz.

At relatively high temperatures $(\omega_0/T_1)_D$ should be proportional to ω_0, as it is, while at very low temperatures it should be proportional to ω_0^{-1}. This is not the case at the lowest temperatures of the measurements. These and comparable results for other hydrates have not been adequately explained.

In an early study of ^{19}F T_1 behaviour of SF_6 in D_2O it was observed[63] that T_1 passed through a flat maximum near 50 K and thereafter decreased monotonically with increasing temperature with values which were independent of Larmor frequency. This behaviour was attributed[63] to spin-rotation interaction, a process well known in liquids and gases to dominate the spin-lattice relaxation of nuclei which like ^{19}F exhibit large chemical shielding effects.[64]

Figure 21 shows this T_1 behaviour for three sets of measurements[63,65] of $SF_6 \cdot 17D_2O$. The curves labelled A and B represent means of measurements on two samples at a number of Larmor frequencies. The differences from sample C are attributed to the effects of small amounts of enclathrated

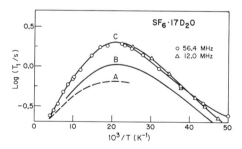

Fig. 21. ^{19}F spin-lattice relaxation times for different samples of $SF_6 \cdot 17D_2O$ above 20 K showing the effects of spin-rotation at the higher temperatures and of small amounts of encaged O_2 impurity near the maximum. (A) is from ref. 63, (B) and (C) from ref. 65.

molecular oxygen in samples A and B. The solubility of air in structure II hydrates is particularly great since N_2 and O_2 molecules can occupy the otherwise empty 12-hedral cages. Paramagnetic oxygen introduces an almost temperature-independent contribution to $1/T_1$ and in relatively large concentrations seriously changes, for example, the shape of the proton resonance in THF · 17D_2O.[65] The temperature dependence for "oxygen-free" sample C at 56.4 MHz can be represented between 20 and 250 K by the sum of

$$\left(\frac{1}{T_1}\right)_{SR} = 8.6 \exp\left(-\frac{185}{T}\right) \tag{10}$$

and

$$\left(\frac{1}{T_1}\right)_D + \left(\frac{1}{T_1}\right)_{CSA} = 0.40 \exp\left(\frac{101}{T}\right) \tag{11}$$

Since the latter is effectively proportional to the orientational dipolar correlation time τ_D, it is seen that $(1/T_1)_{SR}$ is far from being proportional to τ_D^{-1}, in contrast to the case for many liquids at low temperatures.[64] The T_1 behaviour for SeF_6 · 17D_2O is similar.

No adequate molecular theory of spin-rotation in solids appears to exist. Spin-rotation was given as the probable explanation for the fall in T_1 observed with increasing temperature just below the melting point of solid SF_6 itself,[66] but increasing rapid translational diffusion of SF_6 molecules is a more likely explanation.[67,68] Such a diffusional process cannot occur for SF_6 enclathrated in hydrate.

3.3.2. Symmetric top molecules

Among guest molecules of this symmetry, the methyl halides have rigid-lattice proton second moments of 6 G^2, a characteristic of rotation of the CH_3 group or of the whole molecule about its axis of 3-fold symmetry (see also Section 3.3.5 below). The corresponding line shape is shown in Fig. 22 for CH_3Br. The narrowing which takes place with increasing temperature then results from reorientation about other axes. Temperatures, $T_{1/2}$, at which the SM has fallen to half the rigid lattice value are included in Table 5.

For CF_3X molecules (X = Cl, Br, or I) the ^{19}F rigid lattice line shapes show the effect of anisotropy of chemical shielding (cf. Fig. 23). With increasing temperature behaviour reminiscent of the Resing effect already described for spherical top molecules at first occurs (Fig. 23a). Superimposed on the rigid lattice line is a triplet arising from rotation about the symmetry

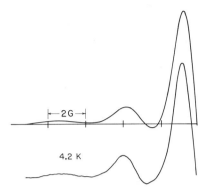

Fig. 22. Half of the "rigid-lattice" proton derivative lineshape of CH_3Br deuteriohydrate.

axis. The latter has become dominant at 24.9 K. At higher temperatures other reorientational processes occur.

Since the dipolar axis coincides with the symmetry axis, dielectric relaxation is only observed for reorientation of this axis. A useful comparison may be made between the dielectric loss

$$\varepsilon'' = (\varepsilon_{02} - \varepsilon_{\infty 2}) \int \frac{\omega \tau F(\ln \tau) \, d(\ln \tau)}{1 + \omega^2 \tau^2}, \tag{12}$$

where $F(\ln \tau)$ expresses the distribution of dielectric relaxation times, and is obtained from ω_0/T_1 as shown in Equation 8. Although neither the correlation times nor the distribution functions used in Equations 12 and 8 are the same—the dielectric time refers to decay of the autocorrelation function for electric dipole direction, the NMR time to averaging of the directions of the vectors connecting interacting magnetic dipoles—they should be quite similar when they apply to the same reorientational process. Moreover when the correlation times are widely distributed, as they are for guest molecules in clathrate hydrates, the presence of the second term in Equation 8 adds little to the width already implicit in $F(\ln \tau)$. However, the maximum in ω_0/T_1 occurs at a most probable correlation time $\tau_m \approx 0.615/\omega_0$ for all reasonable distribution functions whereas the corresponding ε'' maximum occurs at $\tau_m = 1/\omega$. Maximum ω_0/T_1 therefore occurs at a slightly higher temperature than maximum ε'' for measurements at a fixed frequency $\omega = \omega_0$ when the $F(\ln \tau)$'s are similar.

A comparison of this kind for $CF_3I \cdot 17D_2O$ is shown in Fig. 24. The dielectric data extend only to 1 MHz but simple extrapolation indicates that the dielectrically active process should give ω_0/T_1 maxima at the temperatures marked by arrows for the two Larmor frequencies. Substantial

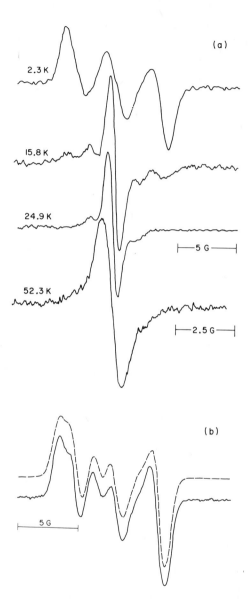

Fig. 23. ^{19}F derivative lineshapes of CF_3I deuteriohydrate at (a) 10 MHz and various temperatures, (b) 56 MHz and 2.5 K.

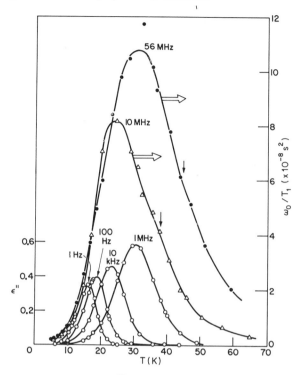

Fig. 24. Comparison of ω_0/T_1 for ^{19}F in $CF_3I \cdot 17D_2O$ and ε'' for $CF_3I \cdot 17H_2O$.

contributions occur near these temperatures but the actual peaks occur at much lower temperatures and must arise from reorientation about the CF_3I symmetry axis. Two overlapping broad distributions of correlation times are present, that centred at the lower temperature making the larger contribution to ω_0/T_1 as expected, since the contribution ΔSM removed by rotation about the 3-fold axis is about three times as large as the residual ΔSM removed by reorientation of this axis (cf. Equation 8).

3.3.3. *Asymmetric top molecules*
Most asymmetric guest molecules whose motion has been studied by both NMR and dielectric methods contain 3- to 6-membered rings with, as for THF, a dipolar axis which is, at least very nearly, a two-fold axis of symmetry.

Comparisons of ε'' and ω_0/T_1 low-temperature behaviour are made for a number of structure II hydrates in Figs. 25–27. In the case of 1,3-dioxane[24] (Fig. 25) a remarkable similarity exists between the ε'' and ω_0/T_1 curves.

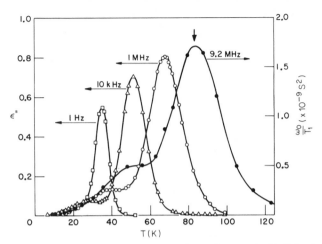

Fig. 25. Comparison of ω_0/T_1 for protons of 1,3-dioxane in D_2O hydrate with ε'' of 1,3-dioxane in H_2O hydrate.

There are two distinct reorientation processes, of which the faster reduces the proton second moment[24] by about 10%. The origin of this second process is not clear; it may be related to the geometry of the molecule which makes it possible for both oxygen atoms to be "stuck" to the cage near the centres of hexagonal rings. Only one general relaxation process, which makes comparable contributions to ε'' and ω_0/T_1, is observed for 1,4-dioxane (Fig. 26). The T_1 measurements for this guest provide confirmation of the interpre-

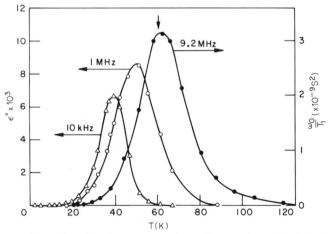

Fig. 26. Comparison of ω_0/T_1 for protons of 1,4-dioxane in D_2O with ε'' of 1,4-dioxane in H_2O.

tation of the dielectric measurements.[23] At low temperatures 1,4-dioxane exists in the non-dipolar chair configuration. The existence of very weak dielectric absorption (ε'' always less than 10^{-2}) may be accounted for by dipole moments induced in water molecules of the cage by the large quadrupole moment of the dioxane molecule. These relax at rates determined by the reorientation rates of the dioxane molecules. The presence of two distinct broad distributions of correlation times for cyclobutanone is shown by the ω_0/T_1 plot in Fig. 27. The early ε'' measurements[31] show only one such process. Rotation of the cyclobutanone molecule about its long dipolar axis is responsible for the ω_0/T_1 peak at 23 K, while much slower reorientation about the other axes gives rise to the peak at 55 K.

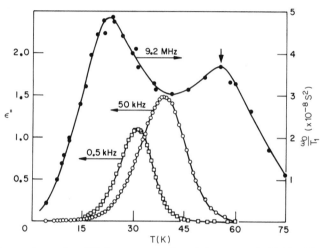

Fig. 27. Comparison of ω_0/T_1 for protons of cyclobutanone in D_2O with ε'' of this molecule in H_2O hydrate.

At relatively high temperatures where the differences in energies between preferred orientation in structure II become less important in comparison with kT, the guest dipoles make effectively the full isotropic contribution to the permittivity ε_{02}. Thus at 168 K, ε_{02} is very nearly proportional to the square of the dipole moment μ of the guest species (Fig. 28a). Within experimental error ε_{02} may be calculated explicitly from the Onsager model according to[41]

$$OP = \frac{(\varepsilon_{02}-1)(2\varepsilon_{02}+1)}{3\varepsilon_0} - \frac{(\varepsilon_{\infty2}-1)(2\varepsilon_{\infty1}+1)}{3\varepsilon_{\infty2}} = \frac{4\pi N\mu^2}{3kT(1-f\alpha)^2} \quad (13)$$

where N = the number density of guest molecules of polarizability α and

f is the reaction field factor. With $f = 2(\varepsilon_{02} - 1)/[(2\varepsilon_{02} + 1)a^3]$ and the cavity radius a equal to the free radius of the 16-hedral cage (3.2 Å), Equation 13 gives the line shown in Fig. 28b.

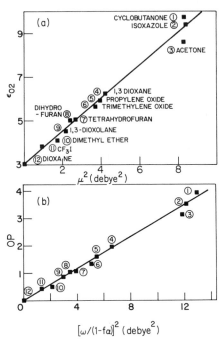

Fig. 28. (a) Dependence of the permittivity ε_{02} at 168 K on dipole moment μ of the guest molecule in structure II hydrates. (b) Dependence of the Onsager polarization (OP) (Equation 13) on the reaction-field enhanced dipole moment. The line is drawn with the slope $4\pi N/(3kT)$.

In principle, ordering of the guest dipole directions in structure I along the $\bar{4}$ axis of the 14-hedral cage is possible. The dielectric behaviour[69] of encaged ethylene oxide appears to show no sign of an ordering transition. For the structure I hydrate of trimethylene oxide, which is too large to occupy the 12-hedral cages, there is strong dielectric evidence[47] of the presence of such a transition near 105 K. As shown in Fig. 29 the frequency-independent permittivity $\varepsilon_{\infty 1} = \varepsilon_{02}$ shows an abrupt change with temperature in this region. The proton SM of TMO in D_2O (Fig. 30) has reached about half the rigid lattice value at the transition temperature and thereafter rapidly falls. A break occurs at the transition temperature in the ω_0/T_1 curves (Fig. 31). These show two peaks at temperatures below the transition. The major peak, which does not correspond to a region of appreciable dielectric

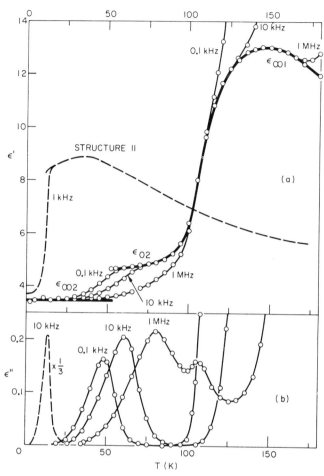

Fig. 29. Comparison of (a) permittivity and (b) loss behaviour of structure I (solid lines) and structure II (broken lines) hydrates of trimethylene oxide. The presence of a region of small dielectric loss below 10 K in structure I is not shown.

absorption, must be attributed to reorientation of TMO about its polar axis which is aligned along the $\bar{4}$ axis of the cage. The peak at lower temperatures, which corresponds to a region of weak dielectric absorption, is of less certain origin. It may arise from relatively rapid reorientation or partial reorientation of a fraction of the TMO molecules which are not aligned with the $\bar{4}$ axis. Although evidence of the existence of a *partial* ordering transition in structure I is thus provided by three kinds of experiment, the infrared spectrum appears to show no effects of such a transition.[70]

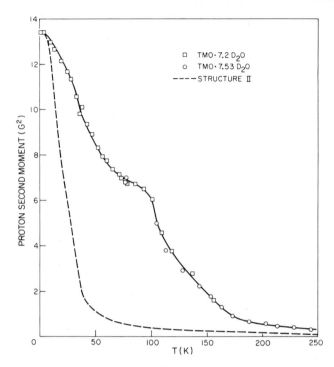

Fig. 30. Second moments of proton NMR lines of trimethylene oxide deuteriohydrates of structures I and II.

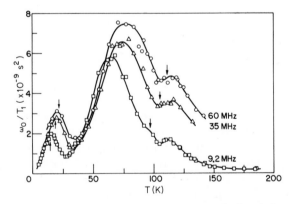

Fig. 31. ω_0/T_1 curves for structure I deuteriohydrate of trimethylene oxide.

Dielectric absorption in H_2S hydrate[59] at 1 MHz peaks below 2 K, as it does also for H_2S incorporated in the small cages of structure II.[42] In the proton NMR spectrum the H_2S signal is superimposed on the rigid H_2O spectrum as a narrow line which persists to very low temperature.[59] The barrier to H_2S reorientation is less than 210 J mol^{-1}.

Dimethyl ether has a relatively rapid reorientation rate in the 16-hedral cages of structure II. In the second, low-temperature hydrate, reorientation of DME is more complex, and like the composition,[48] is most easily interpreted in terms of the rare tetragonal hydrate structure. The unit cell of this

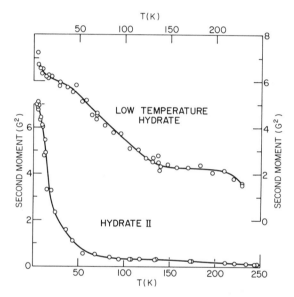

Fig. 32. Second moments of proton lines in the two dimethylether deuteriohydrates.

structure[71] contains 172 water molecules, ten 12-hedral cages too small to be occupied by DME molecules, sixteen 14-hedra similar to those which occur in structure I and four 15-hedra. The drop in *SM* below 12 K (Fig. 32) is attributed to reorientation of DME about its long axis in the 15-hedra, and the drop between 50 and 140 K to rotation of DME lying in the equatorial plane of the 14-hedra about the axis perpendicular to this plane. The plateau in *SM* at 2.2 G^2 between 140 and 210 K is unique among clathrate hydrates so far studied and illustrates a severe geometric restriction in the reorientational motions possible.[49]

3.3.4. Natural gas hydrates

The hydrocarbon hydrates have a special interest in view of their natural occurrence and the need for non-destructive methods of analysis. Despite their small dipole moments, low temperature dielectric absorption characteristic of the reorientation of enclathrated propane and isobutane has been measured.[59] The corresponding absorption by n-butane in its double hydrate with H_2S provides proof of its enclathration in a configuration different from the non-dipolar *trans* form which is normally the stable form of n-butane.

The proton NMR spectra[59] of ethane, propane, and isobutane hydrates all show a narrow guest line superimposed on a broad H_2O line at temperatures between 50 and 200 K. The narrow line of methane persists to below 2 K.[58] The NMR results for the neopentane–D_2S–D_2O system (cf. Figs. 18b and 19) showed for the first time the ability of neopentane to occupy the 16-hedral cages. The distortion of the geometry of encaged n-butane referred to above is likely to be similar to that found for encaged 1,2-dichloroethane.[72] The *gauche* form (dihedral angle = 60°) predominates in the encaged molecule, the *trans* form in solid 1,2-dichloroethane.

For analysis of the line narrowing from guest molecule reorientation in deuteriohydrates it may be assumed that the second moment has reached half its rigid lattice value at a temperature $T_{1/2}$ when the average reorientation rate is about 100 kHz. For the Waugh approximation[73] an estimate of the average barrier to reorientation is then given[59] by E_{SM} (J mol^{-1}) = 135.4 $T_{1/2}$. Comparisons of E_{SM} values with dielectric activation energies given in Table 5 show them to be similar for hydrocarbon guest species.

Proton-enhanced ^{13}C NMR at relatively high temperatures should provide identification of the hydrocarbons present and their relative abundances.

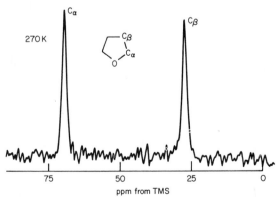

Fig. 33. Proton-decoupled ^{13}C spectrum of THF hydrate.

Once experimental difficulties related to hydrate instability are overcome, it is anticipated that ^{13}C analysis of mixed hydrocarbon hydrates will be possible. Figure 33 shows the ^{13}C spectrum of THF in H_2O with the two inequivalent carbon atoms easily resolved.

3.3.5. Magnetic isolation by enclathration in D_2O

Enclathration provides a method of matrix isolation in which the guest molecules, normally only one to a cage, are present in high concentration while at the same time are separated from one another in more or less well-defined environments. A particularly useful device is enclathration in D_2O hydrate where, because of the smallness of the magnetic dipole of deuterium, the NMR spectrum of the guest molecule is much less broadened by *inter*molecular dipolar interactions than is normally the case for solids.[74] This is illustrated by the low-temperature ^{19}F lineshapes of SF_6 shown in Fig. 34.

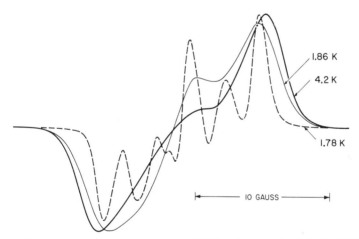

I.86 K

4.2 K

I.78 K

|← IO GAUSS →|

Fig. 34. Comparison of low-temperature ^{19}F lineshapes of solid SF_6 (thick line), $SF_6 \cdot 17H_2O$ (thin line), and $SF_6 \cdot 17D_2O$ (broken line).

The better definition of line shapes thus made possible has led to the development of methods of calculating polycrystalline rigid-lattice line shapes for molecules containing a considerable number of magnetic nuclei.[75] For protons these line shapes depend entirely on molecular geometry. Application to CH_2Cl_2, ethylene oxide, ethylene sulphide, cyclopropane and trimethylene oxide gave,[75] for example, interproton distances in the CH_2 groups of these molecules which were consistently almost 0.01 Å longer than given by microwave measurements of the molecules in the gas. Almost certainly, this effect results from the rotational oscillations of the encaged molecules.

For nuclei, like ^{19}F, with large chemical shielding, both geometry and chemical shielding tensors may be determined. An example[76] of the comparison of experimental and simulated ^{19}F spectra of $CF_4 \cdot 7\frac{2}{3}D_2O$ is given in Fig. 35. Values of ^{19}F chemical shielding anisotropy are important to the testing of molecular orbital theories. They also give fairly directly the molecular spin-rotation tensors necessary for deriving angular momentum correlation times from T_1 measurements of liquids and gases.[62,77–79]

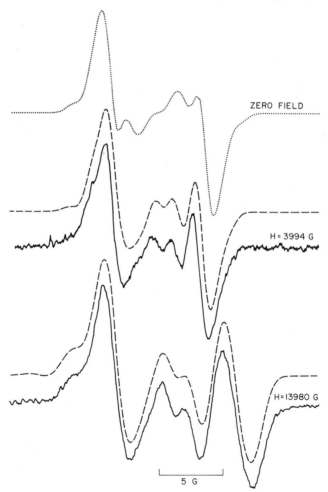

ZERO FIELD

H = 3994 G

H = 13980 G

5 G

Fig. 35. Comparison of experimental (solid curves) and calculated ^{19}F rigid lattice lineshapes of $CF_4 \cdot 7\frac{2}{3}D_2O$. The chemical shift anisotropy is 138 ± 4 ppm.

Another important application of enclathration in D_2O concerns CH_3 group rotation. The proton spectrum of CH_3 depends on the barrier to rotation and varies between the classical limits corresponding to very fast rotation of either the CH_3 group or the whole molecule about its C_3 axis (as for CH_3Br, Fig. 22) and no rotation at all (as for methylchloroform, Fig. 36a). At intermediate barriers to rotation (between about 15 and 21 kJ mol^{-1}) the line shapes are very sensitive to the tunnelling splitting of the torsional oscillation ground state of the methyl group. These may be calculated by use of the Apaydin and Clough treatment[80] of an isolated methyl group with three possible orientations separated by 120°. Fits comparable to that shown in Fig. 36 for CH_3CD_2Br were obtained[81,82] for the molecules given in Table 6, which gives the mean tunnelling frequencies, the Gaussian parameter β used to represent the broadening, the mean

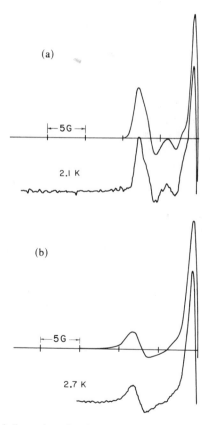

Fig. 36. Experimental (lower) and calculated derivative proton lineshapes for (a) CH_3CCl_3–D_2S–D_2O hydrate and (b) CH_3CD_2Br–D_2O hydrate.

Table 6. Methyl group tunnelling in enclathrated molecules. Barrier heights are in kJ mol^{-1}

Guest molecule	Clathrate	Tunnelling frequency (kHz)	β (G)	V_0	δV	V_0 in gas	V_0 in solid INS[a]	Far IR
CH_3Br	D_2O	>150	0.7					
CH_3CD_2Cl	D_2O	61	0.80	15.38	1.00	15.38		18.73
CH_3CD_2Br	D_2O	19	0.75	17.35	1.96	15.38		18.18
CH_3CD_2I	D_2O, D_2S	4.7	0.75	19.65	0.92	15.84		18.52
CH_3CCl_3	D_2O, D_2S	<2	0.6				21.82	
$(CH_3)_3CD$	D_2O	24	1.5	16.97	0.38	14.17		16.47
$(CH_3)_3CF$	D_2O	49	1.6	15.63	0.92	18.10	16.64	
$(CH_3)_3CCl$	D_2O, D_2S	9.4	1.3	18.47	0.42		19.06	18.85
$(CH_3)_3COD$	D_2O, D_2S	19	1.5	17.35	1.59		15.34	17.26
$(CH_3)_4C$	D_2O, D_2S	14	1.8	17.81	1.59		16.68	18.14
$(CH_3)_3N$	D_2O	18	1.5	17.43	0.42	15.17		

[a] Inelastic neutron scattering.

barrier to rotation V_0 as derived from the approximate Hecht–Dennison relationship[83] between tunnelling frequency and barrier heights, and barrier heights for the same molecules in the gas and solid states as estimated from far infrared studies of the torsional modes and inelastic neutron scattering.[84]

A good fit of the calculated to the experimental line shapes could not be obtained unless a distribution of tunnelling frequencies was assumed. The distribution function chosen[81,82] was of a form similar to that previously used to fit the Resing effect (cf. Equation 6). The width parameter for the barrier is given as δV in Table 6. Although the choice of distribution function is rather arbitrary, some distribution of barrier heights is required. It is very likely that this reflects a dependence of the contribution of $D_2O–CH_3$ group interactions to the torsional barrier on the disordered orientations of the water molecules. Line shapes of CH_3CD_2X molecules dissolved in perchloropropane glass[85] can equally well be analysed by a continuous distribution of barrier heights imposed by a disordered environment.[81]

Examination of the barriers to CH_3 rotation of enclathrated CH_3CD_2X molecules (Table 6) shows an increase with molecular size which is not seen in the gas or solid phases and which must be attributed to increasing tightness of fit of the molecule within the 16-hedral cages. This effect is less pronounced for $(CH_3)_3CX$ molecules which have approximately tetrahedral structures more easily accommodated in these cages. In general the mean barrier heights for encaged molecules are lower than for these molecules in the solid state, but the precise values depend on the type of torsional potential function employed.[84]

The proton second moment of CH_4 deuteriohydrate remains at the value (0.42 G^2) characteristic of isotropic reorientation until at temperatures below 4 K some increases takes place.[58] At 1.8 K the SM is still only 0.8 G^2 and tunnelling through a low torsional barrier is clearly present. The GeH_4 molecule has about the same van der Waals diameter as CF_4 and like it probably occupies mainly 14-hedral cages. Although encaged CF_4 is "rigid" below 6 K, GeH_4 at 3.5 K exhibits a *SM* (~0.82 G^2) only a sixth of its rigid lattice value. The long wings in the absorption at very low temperatures have not entirely disappeared at 20 K (Fig. 37). The line shape appears to be determined by tunnelling at the lowest temperatures and by more classical molecular reorientation at temperatures above roughly 10 K for CH_4 and 30 K for GeH_4.

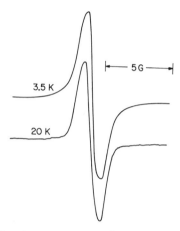

Fig. 37. Proton NMR lineshapes of $GeH_4 \cdot 7\frac{2}{3}D_2O$.

4. Quinol clathrates

The characterization of molecular motion in β-quinol clathrates is complicated by the dependence on the degree of cage occupancy, the occurrence of several β-quinol crystal structures of different symmetry, and by the recent confirmation[90,91] that α-quinol itself can enclathrate small molecules, to the extent of one guest to 18 hydroquinone molecules. The most significant dielectric and NMR information has been obtained for β-quinol crystallized from solution in the liquid form of the encageable species (CH_3OH, CH_3CN, HCN, HCOOH), which assures maximum cage occupancy and simplifies the growth of large single crystals.

114   D. W. Davidson and J. A. Ripmeester

4.1. Methanol

A transition at about 66 K for high occupancy samples is shown by heat
capacity,[92,93] dielectric[92,94–96] and NMR[96] measurements. The dielectric
measurements above the transition temperature (T_t) show [21,97,98] the presence
of two Debye processes which overlap at relatively high temperatures but
become well-separated at lower temperatures. The lower-frequency process
makes the dominant contribution to the static permittivity and from the
single-crystal measurements[92] corresponds to polarization along the c-axis
of the crystal. Relaxation rates for this process from six sources vary a
million-fold between 230 K and T_t (Fig. 38) and, except for some dielectric

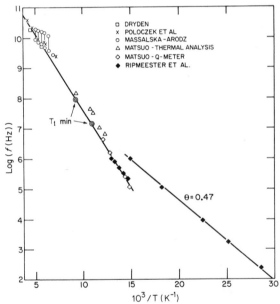

Fig. 38. Arrhenius plot of frequency of dominant relaxation process ($\mu_\parallel$ reorientation) in CH_3OH–β-quinol. All data refer to $\theta \approx 1$ except those labelled $\theta = 0.47$.

heating results[99] which show a systematic shift, fit together to give an
activation energy of 9.78 kJ mol^{-1}.[96] The static permittivity above T_t is much
higher than can be attributed to uncorrelated methanol dipole moments
and must be ascribed to a tendency for alignment of the component of
the methanol dipole moment ($\mu_\parallel$) along the c axis in the same direction
for molecules in adjacent cages. The higher-frequency relaxation can be
ascribed to reorientation of the dipole moment component perpendicular
to $c(\mu_\perp)$.

Below T_t the static permittivity becomes much smaller and, although the two relaxation regions are still present, the dielectric absorption associated with each is much smaller and broader than above the transition and shrinks rapidly with falling temperature. Thus for most methanol molecules both the $\mu_{\parallel}$ component and the $\mu_{\perp}$ component become frozen in below T_t, with the residual absorption arising from reorientation of a small fraction of molecules between preferred orientations of different energies, perhaps next to empty cages.

For a sample with low methanol content the static permittivity is, at all temperatures, much less, there is no sign of a transition, and the $\mu_{\parallel}$ relaxation process is faster ($\theta = 0.47$ in Figure 38) and the $\mu_{\perp}$ process slower at low temperatures than in the high occupancy ($\theta = 0.99$) sample. No transition was detected in the specific heat measurements for θ less than about 0.55.[92,93]

β-quinol containing small guest molecules like H_2S has R$\bar{3}$ symmetry.[100] For an unsymmetric guest molecule like methanol to conform to this symmetry, it must be disordered among three orientations with the same $\mu_{\parallel}$ component related by 3-fold rotation about the c-axis and a further three orientations with reversed $\mu_{\parallel}$ realized by inversion through the cage centre of the first three. The relaxation behaviour of methanol-β-quinol with $\theta = 0.47$ may be attributed to a very fast exchange within the groups of three orientations ($\mu_{\perp}$ reorientation at 1 MHz at 27 K)[96] and a slower exchange between the groups ($\mu_{\parallel}$ reorientation at 1 MHz at 66 K). For high occupancy samples of this clathrate, however, the early assignment[101] of the non-centrosymmetric space group R3 has recently been confirmed.[102] Thus the dielectric evidence for short-range order in the direction of $\mu_{\parallel}$ is complemented by X-ray evidence that even at room temperature the $\mu_{\parallel}$ direction is predominantly the same throughout a single crystal.

The NMR measurements provide additional information. Measurements[96,103] of the proton T_1 of enclathrated CH_3OH at 9.2 (Fig. 39) and 60 MHz each give a shallow minimum above T_t at temperatures corresponding to the $\mu_{\parallel}$ inversion process (see "T_1 min" points in Fig. 38) and an activation energy (Table 7) which agrees with the dielectric one. The minimum is shallow since spin-lattice relaxation is only effected by modulation of inter-molecular dipolar fields during inversion. This minimum is not seen at all in the deuterium T_1 curve for CD_3OH since quadrupolar coupling is unaffected by inversion. Above T_t the slope of the D T_1 curve is mainly determined by $\mu_{\perp}$ reorientation and agrees with other evidence (Table 7) that the activation energy is about 2.9 kJ mol^{-1} for this process. The corresponding T_1 minimum is not seen because rotation of most methanol molecules about the c axis does not occur below T_t. The deep minimum in the D T_1 curve at 24 K (Fig. 39) agrees with the model of classical rotation of the CD_3 group about its symmetry axis[96]; slopes to either side of the

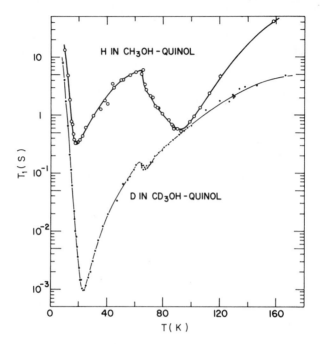

Fig. 39. Proton and deuterium spin-lattice relaxation times at 9.2 MHz.

minimum give $1.67\ \text{kJ mol}^{-1}$ for the barrier. The corresponding proton T_1 minimum for CH_3OH is much too shallow for classical rotation and the "activation energy" of $0.96\ \text{kJ mol}^{-1}$ may be identified with E_{01}, the difference between the ground and first excited CH_3 group torsional levels. In fact calculation of the torsional energy levels for a potential $\frac{1}{2}V\cos 3\theta$, where $V = 1.67\ \text{kJ mol}^{-1}$, gives $0.79\ \text{kJ mol}^{-1}$ for E_{01}.

Analysis of dielectric and several different kinds of NMR lineshape data shows that in its stable orientations the methanol molecule is oriented with respect to the c axis much as shown in Fig. 40, where the c axis is parallel to the COH plane. The available information about θ_{CO}, the inclination of the CO bond with the c axis, is summarized in Table 8. It has been assumed that the CH_3 symmetry axis coincides with the principal axis of the cylindrically symmetric ^{13}C shielding tensor and makes an angle of $3.3°$ with the CO axis.[104] The NMR method may be illustrated by reference to Fig. 41 which shows that the separation of the outer components of the derivative spectrum of CH_3 changes from 7.7 G at low temperature to 4.0 G above T_t. The narrowing from rotation of the CH_3 group about the c axis gives $(3\cos^2\gamma - 1)/2 = 4.0/7.7$ and $\gamma = 34.5°$ for the angle between the CH_3 symmetry axis and c. The other entries in Table 8 result from similar applications

Table 7. *Activation energies* $(kJ\ mol^{-1})$ *for reorientation of guest molecules in* β*-quinol*

Guest	Reorientation	Method	E_A	Temp. Range (K)
CH_3OH ($\theta \sim 1.0$)	Reversal of c component	Dielectric[96]	9.78 ± 0.29	66–230
		Proton T_1 at 9.2 MHz[96]	9.2	100–180
	About c axis	Proton T_1 at 60 MHz[103]	10.0 ± 0.4	90–200
		Dielectric	0.96 ± 0.21[96]	9–22
			2.5 ± 1.2[98]	153–193
		D T_1 of CD_3OH at 9.2 MHz[96]	3.05 ± 0.21	70–100
		Proton T_1 at 60 MHz[103]	2.9 ± 0.8	67–77
	About CH_3 axis	Proton T_1 at 9.2 MHz[2]	0.96	
		Proton T_1 at 60 MHz[109]	0.84	
		D T_1 of CD_3OH at 9.2 MHz[96]	1.67	
CH_3CN ($\theta \sim 1.0$)	Reversal of molecule along c	Dielectric[21]	75	196–293
		Dielectric[109]	67	112–133
HCN	Reversal of molecule along c	Dielectric ($\theta \sim 1.0$)[110]	16.3	133–213
		Dielectric ($\theta \sim 0.80$)[43]	5.27 ± 0.84	243–333
HCOOH	About c axis (?)	Dielectric ($\theta = 0.50$)[43]	13.96 ± 0.17	74–98
CH_3F	Reversal of molecule along c	Dielectric ($\theta = 0.81$)[115]	~ 15.8	42–75
		Dielectric ($\theta = 0.21$)[115]	11.08 ± 0.63	
SO_2	About c axis	Dielectric (various θ)[115]	~ 0.4	1.8–12
N_2	Reversal of molecule along c	^{14}N NQR($\theta = 0.50$)[119]	3.59	1.6–80
O_2	Reversal of molecule along c	Magnetic susceptibility[121]	0.54	0.25–10

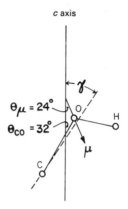

Fig. 40. Preferred orientation of methanol in β-quinol.

of the addition theorem of spherical harmonics to motional narrowing of ^{1}H, ^{2}H and ^{13}C resonances of methanol.

Recent single crystal studies[103] show that the single methanol ^{13}C line splits into three lines below T_t as most methanol molecules cease to rotate about the c axis. They become frozen in three-fold disordered orientations consistent with the (room temperature) unit cell[102] or the crystallographic

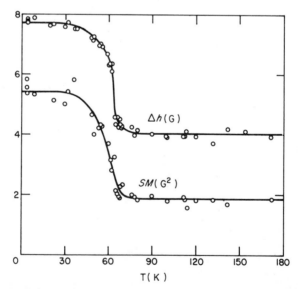

Fig. 41. Separation of outer components of proton CH_3 triplet (Δh) and second moment of CH_3OD in 93% deuterated β-quinol.

Table 8. *Angle between CO bond direction in methanol and c axis of β-quinol*

Method	Primary result (°)	θ_{CO}(°)
Below 100 K[a]		
Ratio of $\mu_\parallel^2$ to $\mu_\perp^2$ from dielectric measurements of low θ sample	$\theta_\mu = 23.1 \pm 3.4$	32
Proton NMR CH_3 triplet splitting of CH_3OD in 93% deuterated quinol	$\gamma = 34.5 \pm 1.2$	31
Deuterium quadrupole coupling constant of CD_3OH	$\gamma = 34.5 \pm 2.6$	31
^{13}C chemical shift anisotropy of CH_3OH (temperature dependent)[b]103	$\gamma = 31$ to 35	28–32
300 K[a]		
Proton CH_3 triplet splitting of CH_3OD in 86% deuterated quinol[b]105		
H_O parallel to c	$\gamma = 44.7$	41
H_O perpendicular to c	$\gamma = 42.3$	39
Deuterium quadrupole coupling constant		
CD_3OH	$\gamma = 44.3 \pm 1.2$	41
CH_3OD in $C_6H_4(OD)_2$	$\theta_{OD} = 71.4 \pm 1.0$	38
^{13}C chemical shift anisotropy of CH_3OH[b]		
	$\gamma = 42.1 \pm 1.3$	39
	$\gamma = 44$103	41
X-ray diffraction[b]102	θ_{CO}	35.5 ± 1.6

[a] Data from ref. 96 unless indicated.
[b] Single crystal.

structure changes somewhat below the transition to form a larger unit cell in which there is only one stable orientation in each cage. Some evidence for such ordering is provided by the observation[103] that at least some of the ^{13}C lines of hydroquinone split into triplets below T_t.

4.2. Acetonitrile

The crystal structure[106] is trigonal, space group P3, with the unit cell three times as large as for the R$\bar{3}$ and R3 structures. The three acetonitrile molecules are aligned with the c axis, two pointing in one direction, the third in the opposite direction. The proton-decoupled ^{13}C spectrum of the single crystal shows as many as 27 partly resolved hydroquinone lines,[107] in contrast with the 9 lines seen for general orientations of methanol-β-quinol crystals.[96,103,108] The ^{13}C spectra of hydroquinone in the acetonitrile clathrate for a variety of crystal orientations may be reasonably well calculated[107] from the same shielding tensors of the three inequivalent C atoms as those used to fit the methanol clathrate spectra,[108] when the structural differences are taken into account. The ^{13}C line of the acetonitrile CN group is asymmetrically split into three lines by coupling to the ^{14}N nuclei;[107] the magnitude of the splitting was used to provide evidence of a relatively large anisotropy of J-coupling between ^{13}C and ^{14}N.

Relatively small dielectric absorption[21,109] occurs at audio frequencies at room temperature and in single crystals only appears when the electric field is parallel to the c axis.[21] The absorption is broad and the dispersion amplitude decreases with decreasing temperature, much as for the $\mu_\parallel$ process below T_t in the methanol clathrate. The Arrhenius energy for reversal of the direction of occasional acetonitrile molecules is very high (Table 7) as expected for the snug fit within the cages.

4.3. Hydrogen cyanide

The large static permittivity rises as $1/T$ to a maximum near 170 K and thereafter falls rather slowly.[110] Excess heat capacity peaks at 178 K but apparently extends over a considerable range of temperature.[110] Dielectric absorption reported in the same study[110] at lower temperatures (e.g., maximum ε'' at 133 K at 1.5 MHz) is at much lower frequencies than given by a microwave study[43] (e.g., maximum ε'' at 133 K at 1.3 GHz) which also showed a maximum of static permittivity at 225 K. Further study is necessary to resolve these differences which may be related to different cage occupancy in the samples studied.

There seems little doubt that the HCN molecules are generally aligned with the c axis.

4.4. Formic acid

The only significant information is from the microwave dielectric study[43] which showed for $\theta = 0.5$ a single Debye process at GHz frequencies and a static permittivity said to scarcely vary between 213 and 333 K. The most likely orientations of HCOOH were considered to be those in which the line connecting the O atoms lies in the c axis with reorientation taking place about this axis. Attempts to define the orientations of HCOOH in single crystals of partly deuterated quinol by proton NMR were unsuccessful because of the inability to separate guest molecule lines from those of residual lattice protons.[105,111] No anomalous heat capacity effects were observed[93] for $\theta = 0.69$.

4.5. Hydrogen chloride and hydrogen bromide

The HCl-β-quinol structure has been found to be R3, with the HCl molecules apparently lying in disordered orientations on a cone whose surface is inclined by 33° to the c axis with Cl at the apex.[112] Early dielectric measurements[94] at 10 kHz showed the permittivity to rise almost as $1/T$ to pass through a flat maximum at temperatures which varied between 4 K for $\theta = 0.62$ and 12 K for $\theta = 0.77$. This evidence for very rapid HCl reorientation is supported by the microwave study[43] which showed maximum ε'' at 88 K to occur above 8.5 GHz. For HBr ($\theta = 0.51$) the 10 kHz permittivity was still rising with cooling at 4 K.[94]

4.6. CH$_3$F, CF$_3$H and NF$_3$

Measurement of the ^{19}F chemical shift anisotropy in single crystals at 1.3 K showed the molecular axis of 3-fold symmetry to be aligned with the c axis for CH$_3$F[113] and essentially perpendicular (80 ± 10°) to this axis for CHF$_3$ and NF$_3$.[114] A heat capacity anomaly occurred[93] for CH$_3$F ($\theta = 0.87$) at 110 K. Unpublished dielectric measurements[115] of this clathrate show a maximum static permittivity of 5.4 at 112 K for $\theta = 0.81$ and the absence of a maximum above 40 K for $\theta = 0.21$. Broad radio-frequency dielectric absorption occurs below 100 K, with a maximum ε'' at 2 kHz occurring at 86 K for $\theta = 0.81$ and 54 K for $\theta = 0.21$. Reversal of the direction of the CH$_3$F dipoles is thus accelerated by the presence of unoccupied cages.

4.7. Sulphur dioxide

SO$_2$-β-quinol was found by Palin and Powell[116] to have the R3 structure in which the unit cell becomes larger in the c-direction and smaller in a with increasing occupancy of the cages.[117] The ^{13}C NMR spectra with magic-angle spinning show increasing C$_2$, C$_3$ splitting (Table 2) with increasing θ. The SO$_2$ symmetry axis is 6-fold disordered in directions perpendicular to c or nearly so. The permittivity measured at 10 kHz for $\theta \sim 0.74$[94] passed through a maximum of 9 K, reached a plateau near 35 K and fell roughly as $1/T$ between 100 and 200 K. More recent measurements[115] at sub-MHz frequencies show the same general behaviour, but the details are found to depend sensitively on both average θ and on homogeneity of θ throughout the sample. The fall in permittivity below ~9 K results from dispersion, the corresponding broad sample-dependent dielectric absorption (e.g. Fig. 42) from reorientation of SO_2 formally giving activation energies of the order of 0.4 kJ mol^{-1}. A mixed β-quinol sample with $\theta(SO_2) = 0.05$ and $\theta(Xe) = 0.72$ showed 1 MHz dielectric loss which peaked below 2 K. Qualitatively similar behaviour was observed for a very low occupancy ($\theta \sim 0.04$) SO_2-quinol sample, pale yellow in colour, which was shown by its ^{13}C spectrum to be $\sim 90\%$ α-quinol. Thus SO_2 appears also to undergo very rapid reorientation when encaged in α-quinol. In this form there is one cage to every 18 hydroquinone molecules[91] and the reorientation behaviour is likely to be much less dependent on cage occupancy than for β-quinol.

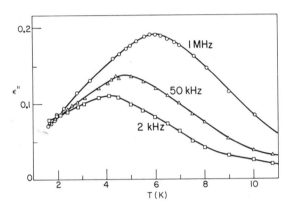

Fig. 42. Low-temperature dielectric loss from reorientation of SO_2 in β-quinol ($\theta = 0.57$). Measurements by J. E. D. Davies.

4.8. Nitrogen and oxygen

Early ^{14}N nuclear quadrupole measurements[118] of N_2-β-quinol showed that N_2 was aligned with the c axis below 8 K. A more recent NQR study[119] with $\theta \approx 0.5$ gave three well-resolved major components (L, M and H in Fig. 43), each with partly resolved fine structure. The assignment of L, M and H to N_2 molecules with two nearest cages empty, one nearest cage occupied, and both nearest cages occupied, respectively, was supported by analysis of the fine structure as due to *intra*molecular dipolar splitting (L and H) and, in addition, splitting from different field gradients at the two ^{14}N atoms in the asymmetric case (M). These results provide a clear indication of the effect of cage occupancy on the local geometry of the quinol lattice. The disappearance of splitting of M at 60 K and above is almost certainly due to reversal of the N_2 direction at a rate faster than the splitting (about 1 kHz) due to nonequivalence of the two N atoms.

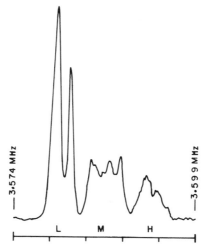

Fig. 43. ^{14}N NQR spectrum of N_2 in β-quinol ($\theta \sim 0.5$) at 1.6 K.[119]

Magnetic susceptibility measurements[120,121] of encaged O_2 show departures from the susceptibility expected for freely-rotating O_2 only at the lowest temperatures. Measurements to as low as 0.25 K were interpreted[121] in terms of a potential of the form $V_0(1-\cos 2\theta)$, where θ is the angle between the O_2 axis and c, with $V_0 = 267$ J mol^{-1}. The paramagnetic resonance (ESR) spectra[122] were complex but could be interpreted for dilute solutions of O_2 in N_2–quinol. The extent of alignment at different temperatures of N_2 and O_2 with the c-axis was calculated[123] from atom–atom Lennard-Jones and multipole potentials of guest–host interaction. Barriers to reorientation of N_2 and O_2 (taken as $2V_0$[124]) are included in Table 7.

5. Other clathrates

There is as yet little information about motional processes in the more complex clathrates.

Dianin's compound (4-p-hydroxyphenyl-2,2,4-trimethylchroman) can enclathrate more than 40 different molecular species[125] in a unit cell of R3̄ symmetry which contains hour-glass shaped cavities, one to every six chroman molecules.[126,127] (See Chapter 1, Volume 2). The most significant results of two dielectric studies[43,128] of molecules encaged in Dianin's compound are given in Table 9. The results for chloroform and ethanol were interpreted in terms of the disordered orientations of these guest

Table 9. Guest molecule reorientation in Dianin's compound[(a)]

Guest	$A(s^{-1})$	$E(kJ\,mol^{-1})$	Remarks
CHCl₃	$\sim 10^{13}$	12.1	A second absorption occurs at higher frequencies
C₂H₅OH	8×10^{13}	20.1	Inversion along c
	1×10^{14}	16.51	Rotation of OH about CO
	4×10^{12}	10.28	Rotation about c
CH₃CN	1.3×10^{12}	8.78 ± 0.63	Probably rotation about c of molecules aligned approximately perpendicular to c
CH₃CH₂CN			Absorption peaks at 0.6 and 3 GHz at 293 K
CH₃COCH₃			Maximum absorption at 1 GHz at 293 K
n-heptanol	1.4×10^{14}	32.6	Assignments uncertain
	4×10^{12}	26.7	
	1.8×10^{12}	17.1	
	1.2×10^{13}	14.2	

[(a)] Data from ref. 43 and 128 with $(1/\tau) = A \exp(-E/RT)$.

molecules revealed by the X-ray studies.[127] The relative contributions of the three relaxation processes to the static permittivity, if these are properly assigned, lead to an angle of about 70° between the CO bond of ethanol and the c axis. While most guest molecules can be accommodated two to a cage, the long n-heptanol molecule extends along the c axis through the waist of the cage to occupy both halves.[129] The multiple dielectric absorption regions must arise from reorientation about the c axis and internal rotation processes. Activation energies of a few kJ mol⁻¹ for reorientation of encaged o-dichlorobenzene, benzonitrile and benzyl cyanide were suggested[43] by low permittivity values at 8.5 GHz and 88 K.

A proton NMR study[130] of Dianin's compounds with a number of large guest molecules showed a very narrow line superimposed on the broad intense line from lattice protons. Although the narrow line was attributed to the guest species,[130] it is likely that it arose from liquid inclusions, as already discussed for β-quinol.

[129]Xe spectra with proton decoupling show two axially symmetric shielding tensors in Dianin's compound, identified with sites I and II in Table 3. At both "sites" the down-field shift is larger parallel to c than perpendicular to c. Site II contained about 80% of the xenon atoms and may correspond to positions at the "bulges" of the cage while site I may coincide with the "waist". Electron paramagnetic resonance spectra[131] of di-t-butylnitroxide

incorporated in Dianin's compound showed the encaged free radical to undergo reorientation about the c axis with a barrier of about 8 kJ mol^{-1}. A number of proton NMR studies have been made on clathrates of the Hofmann type $M(NH_3)_2M'(CN)_4 \cdot 2G$ where M and M' are transition metals (Ni, Cd, Mn, ...) and G is a guest molecule which normally contains an aromatic ring. The prototype structure was established for $M = M' = Ni$, and benzene as guest by Rayner and Powell,[132] who showed that the benzene molecules occupied sites half way between planes formed by the Ni^{2+} and CN^- ions in a tetragonal cell of P4/m symmetry. The normal to the benzene ring is perpendicular to the c axis. (See Chapter 2, Volume 1). This structure is paramagnetic with an NMR spectrum at 4 K which is an asymmetric doublet with the NH_3 resonance shifted by about 20 G with respect to the C_6H_6 resonance by the field of an adjacent paramagnetic Ni^{2+} ion.[133] Narrowing of the benzene resonance occurs at about 150 K due to reorientation of the benzene ring about its normal, the barrier height being 16.7 kJ mol^{-1}. At temperatures below a specific heat anomaly at 2.4 K NMR and other studies of single crystals showed[134] antiferromagnetic domains of magnetic unit cells four times as large as in the paramagnetic form. For $M = Cd$, $M' = Ni$ the clathrate is diamagnetic and, with NH_3 replaced by ND_3, well-defined line narrowing at ~ 140 K again shows reorientation of benzene in its plane.[135] Line narrowing is gradual when G is aniline or pyrrole, while thiophene is apparently "rigid" to 300 K.

References

1. D. W. Davidson and J. A. Ripmeester, *J. Glaciol.*, 1979, **21**, 33.
2. D. W. Davidson, in *Water: A Comprehensive Treatise*, (ed. F. Franks) Plenum Press, New York, Vol. 2, 1973, p. 115.
3. P. Sixou and P. Dansas, *Ber. Bunsenges. Phys. Chem.*, 1976, **80**, 364.
4. F. A. Rushworth and D. P. Tunstall, *Nuclear Magnetic Resonance*, Gordon and Breach Science Publishers Inc., New York, 1973.
5. J. A. Ripmeester and D. W. Davidson, *Mol. Cryst. Liq. Cryst.*, 1977, **43**, 189.
6. J. P. McTague, *J. Chem. Phys.*, 1969, **50**, 47.
7. J. D. Brown and J. E. D. Davies, *J. Magn. Reson.*, 1978, **29**, 167.
8. P. Gregoire, J. Gallier and J. Meinnel, *J. Chim. Phys.*, 1973, **70**, 1247.
9. A. Pines, M. C. Gibby and J. S. Waugh, *J. Chem. Phys.*, 1973, **59**, 509.
10. J. Schaefer, S. H. Chin and S. I. Weissman, *Macromol.*, 1972, **5**, 789.
11. J. A. Ripmeester, *Chem. Phys. Lett.*, 1980, **74**, 536.
12. J. A. Ripmeester, J. S. Tse and D. W. Davidson, *Chem. Phys. Lett.*, 1982, **86**, 428.
13. S. C. Wallwork and H. M. Powell, *J. Chem. Soc., Perkin Trans. 2*, 1980, 641.
14. D. D. MacNicol, J. J. McKendrick and D. R. Wilson, *Chem. Soc. Rev.*, 1978, **7**, 65.
15. S. J. Opella and M. H. Frey, *J. Am. Chem. Soc.*, 1979, **101**, 5854.

16. D. Brinkman and H. Y. Carr, *Phys. Rev.*, 1966, **150**, 174.
17. C. J. Jameson, A. K. Jameson and S. M. Cohen, *J. Chem. Phys.*, 1975, **62**, 4224.
18. J. A. Ripmeester and D. W. Davidson, *J. Mol. Struct.*, 1981, **75**, 67.
19. J. A. Ripmeester, *J. Am. Chem. Soc.*, 1982, **104**, 289.
20. J. S. Tse and D. W. Davidson, in *Proc. 4th Can. Permafrost Conf.*, (ed. H. M. French), Nat. Res. Council, Ottawa, 1982, p. 329.
21. J. S. Dryden, *Trans. Faraday Soc.*, 1953, **49**, 1333.
22. A. D. Potts and D. W. Davidson, *J. Phys. Chem.*, 1965, **69**, 996.
23. S. R. Gough, J. A. Ripmeester and D. W. Davidson, *Can. J. Chem.*, 1975, **53**, 2215.
24. D. W. Davidson, S. R. Gough, F. Lee and J. A. Ripmeester, *Rev. Chim. Minérale*, 1977, **14**, 447.
25. S. R. Gough, S. K. Garg, J. A. Ripmeester and D. W. Davidson, *Can. J. Chem.*, 1974, **52**, 3193.
26. L. Carbonnel and J. C. Rosso, *J. Solid State Chem.*, 1973, **8**, 304.
27. J. Kaloustian, J. C. Rosso and L. Carbonnel, *C.R. Hebd. Seances Acad. Sci., Paris, Ser. C*, 1972, **275**, 249.
28. D. W. Davidson, S. R. Gough and J. A. Ripmeester, *Can. J. Chem.*, 1976, **54**, 3085.
29. R. E. Hawkins and D. W. Davidson, *J. Phys. Chem.*, 1966, **70**, 1889.
30. G. J. Wilson and D. W. Davidson, *Can. J. Chem.*, 1963, **41**, 264.
31. B. Morris and D. W. Davidson, *Can. J. Chem.*, 1971, **49**, 1243.
32. A. Venkateswaran, J. R. Easterfield and D. W. Davidson, *Can. J. Chem.*, 1967, **45**, 884.
33. S. R. Gough, *Can. J. Chem.*, 1978, **56**, 2025.
34. S. K. Garg, D. W. Davidson and J. A. Ripmeester, *J. Magn. Reson.*, 1974, **15**, 295.
35. K. Kume, *J. Phys. Soc. Jpn.*, 1960, **15**, 1493.
36. L. Onsager and L. K. Runnels, *J. Chem. Phys.*, 1969, **50**, 1089.
37. T. C. W. Mak and R. K. McMullan, *J. Chem. Phys.*, 1965, **42**, 2732.
38. D. W. Davidson, S. K. Garg and J. A. Ripmeester, *J. Magn. Reson.*, 1978, **31**, 399.
39. D. T. Edmonds and A. L. MacKay, *J. Magn. Reson.*, 1975, **20**, 515.
40. J. A. Ripmeester, *Can. J. Chem.*, 1977, **55**, 78.
41. R. K. McMullan and G. A. Jeffrey, *J. Chem. Phys.*, 1965, **42**, 2725.
42. S. R. Gough, R. E. Hawkins, B. Morris and D. W. Davidson, *J. Phys. Chem.*, 1973, **77**, 2969.
43. M. Davies and K. Williams, *Trans. Faraday Soc.*, 1968, **64**, 529.
44. D. W. Davidson, *Can. J. Chem.*, 1971, **49**, 1224.
45. G. P. Johari, *J. Chem. Phys.*, 1981, **74**, 1326.
46. N. Bjerrum, *Kgl. Dan. Vidensk. Selsk., Met.-Fys. Medd.*, 1951, **27**, 3; *Science*, 1952, **115**, 385.
47. S. R. Gough, S. K. Garg and D. W. Davidson, *Chem. Phys.*, 1974, **3**, 239.
48. S. L. Miller, S. R. Gough and D. W. Davidson, *J. Phys. Chem.*, 1977, **81**, 2154.
49. S. R. Gough, S. K. Garg, J. A. Ripmeester and D. W. Davidson, *J. Phys. Chem.*, 1977, **81**, 2158.
50. W. S. Brey, Jr. and H. P. Williams, *J. Phys. Chem.*, 1968, **72**, 49.
51. D. Yu. Stupin, *Izv. Vyssh. Uchebn. Zaved., Khim. Tekhnol.*, 1979, **22**, 170.
52. D. Yu. Stupin, *Zh. Strukt. Khim.*, 1979, **20**, 744.
53. D. Yu. Stupin, *Teor. Eksp. Khim.*, 1979, **15**, 751.
54. S. K. Garg, Y. A. Majid, J. A. Ripmeester and D. W. Davidson, *Mol. Phys.*, 1977, **33**, 729.
55. S. K. Garg and D. W. Davidson, in *Physics and Chemistry of Ice*, (ed. E. Whalley, S. J. Jones and L. W. Gold), *Roy. Soc. Canada, Ottawa*, 1973, p. 56.

56. J. A. Ripmeester, *Can. J. Chem.*, 1976, **54**, 3677.
57. R. Taubenberger, M. Hubmann and H. Granicher, in *Physics and Chemistry of Ice* (ed. E. Whalley, S. J. Jones and L. W. Gold) *Roy. Soc. Canada, Ottawa*, 1973, p. 194.
58. S. K. Garg, S. R. Gough and D. W. Davidson, *J. Chem. Phys.*, 1975, **63**, 1646.
59. D. W. Davidson, S. K. Garg, S. R. Gough, R. E. Hawkins and J. A. Ripmeester, *Can. J. Chem.*, 1977, **55**, 3641.
60. H. A. Resing, *J. Chem. Phys.*, 1965, **43**, 669.
61. H. A. Resing and D. W. Davidson, *Can. J. Phys.*, 1976, **54**, 295.
62. S. K. Garg, J. A. Ripmeester and D. W. Davidson, *J. Magn. Reson.*, 1980, **39**, 317.
63. M. B. Dunn and C. A. McDowell, *Chem. Phys. Lett.*, 1972, **15**, 508.
64. R. E. D. McClung, *Adv. Mol. Relax. Interact. Processes*, 1977, **10**, 83 and references cited there.
65. J. A. Ripmeester, S. K. Garg and D. W. Davidson, *J. Magn. Reson.*, 1980, **38**, 537.
66. R. Blinc and G. Lahajnar, *Phys. Rev. Lett.*, 1967, **19**, 685.
67. J. Virlet and P. Rigny, *Chem. Phys. Lett.*, 1970, **6**, 377.
68. S. K. Garg, *J. Chem. Phys.*, 1977, **66**, 2517.
69. S. K. Garg, B. Morris and D. W. Davidson, *J. Chem. Soc., Faraday Trans. 2*, 1972, **68**, 481.
70. J. E. Bertie and S. M. Jacobs, *Can. J. Chem.*, 1977, **55**, 1777.
71. K. W. Allen and G. A. Jeffrey, *J. Chem. Phys.*, 1963, **38**, 2304.
72. S. K. Garg, D. W. Davidson, S. R. Gough and J. A. Ripmeester, *Can. J. Chem.*, 1979, **57**, 635.
73. J. S. Waugh and E. I. Fedin, *Fiz. Tverd. Tela Leningrad*, 1962, **4**, 2233.
74. S. K. Garg and D. W. Davidson, *Chem. Phys. Lett.*, 1972, **13**, 73.
75. S. K. Garg, J. A. Ripmeester and D. W. Davidson, *J. Magn. Reson.*, 1979, **35**, 145.
76. S. K. Garg, D. W. Davidson and J. A. Ripmeester, *J. Magn. Reson.*, 1979, **36**, 325.
77. S. K. Garg, J. A. Ripmeester and D. W. Davidson, *J. Chem. Phys.*, 1980, **72**, 567.
78. S. K. Garg, J. A. Ripmeester and D. W. Davidson, *J. Chem. Phys.*, 1982, **77**, 2847.
79. S. K. Garg, J. A. Ripmeester and D. W. Davidson, *J. Chem. Phys.*, 1980, **73**, 2005.
80. F. Apaydin and S. Clough, *J. Phys. C, Ser. 2*, 1968, **1**, 932.
81. J. A. Ripmeester, *J. Chem. Phys.*, 1978, **68**, 1835.
82. J. A. Ripmeester, *Can. J. Chem.*, 1982, **60**, 1702.
83. K. T. Hecht and D. M. Dennison, *J. Chem. Phys.*, 1957, **26**, 31.
84. C. I. Ratcliffe and T. C. Waddington, *J. Chem. Soc., Faraday Trans. 2*, 1976, **72**, 1821 and references cited there.
85. C. Mottley and C. S. Johnson, Jr., *J. Chem. Phys.*, 1974, **61**, 1078.
86. Y. A. Majid, S. K. Garg and D. W. Davidson, *Can. J. Chem.*, 1969, **47**, 4697.
87. S. R. Gough, E. Whalley and D. W. Davidson, *Can. J. Chem.*, 1968, **46**, 1673.
88. Y. A. Majid, S. K. Garg and D. W. Davidson, *Can. J. Chem.*, 1968, **46**, 1683.
89. A. Venkateswaran, J. R. Easterfield and D. W. Davidson, *Can. J. Chem.*, 1967, **45**, 884.
90. G. N. Chekhova, T. M. Polyanskaya, Yu. A. Dyadin and V. I. Alekseev, *Zh. Strukt. Khim.*, 1975, **16**, 1054.
91. S. C. Wallwork and H. M. Powell, *J. Chem. Soc., Perkin Trans. 2*, 1980, **4**, 641.
92. T. Matsuo, *J. Phys. Soc. Jpn.*, 1971, **30**, 794.
93. J. F. Belliveau, Ph.D. Dissertation, Boston University, 1970, University Microfilms, Ann Arbor, 70-22418.
94. D. Buss, M.Sc. Thesis, Mass. Inst. Technol., 1960.

95. M. Jaffrain, J. L. Siemens and A. Lebreton, C.R. Hebd. Seances Acad. Sci., Paris, Ser. C, 1969, **268**, 2240.
96. J. A. Ripmeester, R. E. Hawkins and D. W. Davidson, J. Chem. Phys., 1979, **71**, 1889.
97. P. Poloczek, M. Massalska, J. M. Janik and J. A. Janik, Acta Phys. Polonica, 1976, **A50**, 803.
98. M. Massalska-Arodz, Acta Phys. Polonica, 1977, **A52**, 555.
99. T. Matsuo, H. Suga and S. Seki, J. Phys. Soc. Jpn., 1967, **22**, 677.
100. T. C. W. Mak, J. S. Tse, C. Tse, K. Lee and Y. Chong, J. Chem. Soc., Perkin Trans 2, 1976, 1169.
101. D. E. Palin and H. M. Powell, J. Chem. Soc., 1948, 571.
102. T. C. W. Mak, J. Chem. Soc., Perkin Trans. 2, 1982, 1435.
103. S. Matsui, T. Terao and A. Saika, J. Chem. Phys., 1982, **77**, 1788.
104. P. Venkateswarlu and W. Gordy, J. Chem. Phys., 1955, **23**, 1200.
105. J. Gallier and J. Meinnel, J. Chim. Phys., 1974, **71**, 920.
106. T. C. W. Mak and K. Lee, Acta Crystallogr., 1978, **B34**, 3631.
107. J. A. Ripmeester, J. S. Tse and D. W. Davidson, Chem. Phys. Lett., 1982, **86**, 428.
108. T. Terao, S. Matsui and A. Saika, Chem. Phys. Lett., 1979, **64**, 582.
109. P. Dansas and P. Sixou, Mol. Phys., 1976, **31**, 1319.
110. T. Matsuo, H. Suga and S. Seki, J. Phys. Soc. Jpn., 1971, **30**, 785.
111. J. Gallier, Chem. Phys. Lett., 1975, **30**, 306.
112. J. C. A. Boeyens and J. A. Pretorius, Acta Crystallogr., 1977, **B33**, 2120.
113. E. Hunt and H. Meyer, J. Chem. Phys., 1964, **41**, 353.
114. A. B. Harris, E. Hunt and H. Meyer, J. Chem. Phys., 1965, **42**, 2851.
115. Unpublished data by J. E. D. Davies, S. R. Gough and D. G. Leaist.
116. D. E. Palin and H. M. Powell, J. Chem. Soc., 1947, 208; 1948, 571.
117. D. E. Palin and H. M. Powell, J. Chem. Soc., 1948, 815.
118. H. Meyer and T. A. Scott, J. Phys. Chem. Solids, 1959, **11**, 215.
119. A. A. V. Gibson, R. Goc and T. A. Scott, J. Magn. Reson., 1976, **24**, 103.
120. A. H. Cooke, H. Meyer, W. P. Wolf, D. F. Evans and R. E. Richards, Proc. R. Soc. London, 1954, **225A**, 112.
121. H. Meyer, M. C. M. O'Brien and J. H. Van Vleck, Proc. R. Soc. London, 1957, **243A**, 414.
122. S. Foner, H. Meyer and W. H. Kleiner, J. Phys. Chem. Solids, 1961, **18**, 273.
123. S. Hirokawa, Mol. Phys., 1978, **36**, 29.
124. J. H. van der Waals, J. Phys. Chem. Solids, 1961, **18**, 82.
125. W. Baker and J. F. W. McOmie, Chem. Ind. (London), 1955, 256.
126. H. M. Powell and B. D. P. Wetters, Chem. Ind. (London), 1955, 256.
127. J. L. Flippen, J. Karle and I. L. Karle, J. Am. Chem. Soc., 1970, **92**, 3749.
128. J. S. Cook, R. G. Heydon and H. K. Welsh, J. Chem. Soc., Faraday Trans. 2, 1974, **70**, 1591.
129. J. L. Flippen and J. Karle, J. Phys. Chem., 1971, **23**, 3566.
130. P. Grégoire and J. Meinnel, C.R. Hebd. Seances Acad. Sci., Ser. C, 1971, **272**, 347.
131. A. A. McConnell, D. D. MacNicol and A. L. Porte, J. Chem. Soc. A, 1971, 3516.
132. J. H. Rayner and H. M. Powell, J. Chem. Soc., 1952, 319.
133. H. Nakajima, J. Phys. Soc. Jpn., 1965, **20**, 555.
134. S. Takayanagi and T. Watanabe, J. Phys. Soc. Jpn., 1971, **31**, 109 and references given there.
135. T. Mikamoto, T. Iwamoto and Y. Sasaki, J. Mol. Spectrosc., 1970, **35**, 244.
136. R. J. Hayward and K. J. Packer, Mol. Phys., 1973, **25**, 1443.
137. J. A. Ripmeester, J. Incl. Phenom., 1983, **1**, 87.

4 · CRYSTALLOGRAPHIC STUDIES OF INCLUSION COMPOUNDS

G. D. ANDREETTI
Università di Pàrma, Parma, Italy

1. Introduction

A statement which often appears in papers published before the 70's is "Clathrates or molecular inclusion adducts are *crystalline* solids in which *guest* molecules occupy cavities, channels or tunnels in the host lattices". This can be considered to be a general definition of the systems we are dealing with and it is now commonly accepted. Thus, all the techniques operating on *crystalline* samples have played a crucial role in the growth of knowledge of the nature of the compounds that were first discussed twenty years ago in the treatise *Non-Stoichiometric Compounds.*[1] A substantial contribution has been made by X-ray crystallography. Papers[2,3] by H. M. Powell can be considered as the starting points of the quantitative description of the physicochemical properties of clathrates.

A few years later six basic types of inclusion compounds were described in the paper by J. F. Brown, Jr.[4] In retrospect we have to agree that the results from crystallographic studies form the basis of our knowledge of the structural features of clathrate compounds. In the future an improvement of all the structural aspects, both static and dynamic will be needed.

INCLUSION COMPOUNDS III
ISBN 0-12-067103-4

Several monographs have been published recently on some aspects of the structural chemistry of inclusion compounds: zeolite framework sites,[5] coordination structural chemistry of macrocyclic compounds,[6] concept, structure and binding in complexation,[7] the structural chemistry of natural and synthetic ionophores and their complexes with cations,[8] and ethers, crown ethers, hydroxyl groups and their sulphur analogues.[9] In the majority of these papers the geometrical aspects of host–guest interactions are considered with diagrams of the complexes and with the relevant specific bond distances and angles or torsion angles tabulated. However little data are available to the reader who wants to test ideas resulting from reading the papers. For that purpose he needs the primary crystallographic data, and this is available in the form of the Cambridge Crystallographic Data Base.[10] A summary of references on inclusion compounds is given in Table 1. In Table 1 the compound name, molecular formula, authors' names and bibliographic information are given. Also accessible are the cell dimensions, space group, and the atomic coordinates of the atoms, if available. Ordering information for the Cambridge Data Base is obtainable from the Editors. In this way the reader can recall the structural information for a given compound on his own computer. It is surprising to note that about 40% of the total information is lost mainly because it was published in journals for which the presentation of preliminary crystallographic data is not compulsory. In some cases only the drawings are available while in others only secondary data such as some bond lengths or angles are found. Certainly, one of the future goals of the inclusion chemistry community is the recovery of these data.

Some general remarks can be extracted from the analysis of the papers connected with crystallographic studies of inclusion compounds and will be discussed in the following sections.

2. Experimental techniques

The first experimental required for a crystallographic study is the preparation of the crystalline sample. The problem is not always simple since the stability of inclusion compounds at ambient conditions is not yet predictable. Lifetime periods range from a few seconds to years. For example, crystals of calix[8]arene[11] are stable in chloroform solution, but decompose in air in less than one second. Therefore, the stability of an inclusion compound is an essential parameter which must be known in order to devise the crystallization strategy. If the final product looses the guest in air in a few minutes the crystallization has to be stopped at a certain stage and then the crystal

Table 1. *Bibliography of structural analysis of inclusion compounds*

α-Cyclodextrin *N,N*-dimethylformamide clathrate pentahydrate
$C_{36}H_{60}O_{30}$, C_3H_7NO, $5(H_2O)$
K. Harata, *Bull. Chem. Soc. Jpn.*, 1979, **52**, 2451.

α-Cyclodextrin 2-pyrrolidone clathrate pentahydrate
$C_{36}H_{60}O_{30}$, C_4H_7NO, $5(H_2O)$
K. Harata, *Bull. Chem. Soc. Jpn.*, 1979, **52**, 2451.

Diammine-nickel(II) tetracyanonickelate(II) bis(biphenyl)
$C_4H_6N_6Ni_2$, $2(C_{12}H_{10})$
T. Iwamoto, T. Miyoshi and Y. Sasaki, *Acta Crystallogr.*, *Sect. B*, 1974, **30**, 292.

1,6,20,25-Tetra-aza[6.1.6.1]paracyclophane hydrochloride durene tetrahydrate
$C_{34}H_{44}N_4^{4+}$, $C_{10}H_{14}$, $4(Cl^-)$, $4(H_2O)$
K. Odashima, A. Itai, Y. Iitaka and K. Koga, *J. Am. Chem. Soc.*, 1980, **102**, 2504.

(Allylidene-triphenylphosphorane) molybdenum benzene
$C_{25}H_{19}MoO_4P$, C_6H_6
I. W. Bassi and R. Scordamaglia, *J. Organomet. Chem.*, 1973, **51**, 273.

(9,10-Bis(methyl-piperazin-1,4-diyl-ethylidene)-anthracene)-(2,12-dimethyl-1,5,9,13-tetra-azacyclohexadeca-1,4,9,12-tetraene-3,11-diylidene-N,N',N'',N''')-nickel hexafluorophosphate methyl cyanide clathrate (at 16 °C)
$C_{42}H_{54}N_8Ni^{2+}$, $2(F_6P)^-$, $2(C_2H_3N)$
K. J. Takeuchi, D. H. Busch and N. Alcock, *J. Am. Chem. Soc.*, 1981, **103**, 2421.

Hexa-aziridino-cyclotriphosphazene benzene clathrate
$2(C_{12}H_{24}N_9P_3)$, C_6H_6
T. S. Cameron, C. Chan, J.-F. Labarre and M. Graffeuil, *Z. Naturforsch.*, *Teil B*, 1980, **35**, 784.

Cyclo(tetrakis(5-*t*-butyl-2-hydroxy-1,3-phenylene)methylene) toluene clathrate
$C_{44}H_{56}O_4$, C_7H_8
G. D. Andreetti, R. Ungaro and A. Pochini, *J. Chem. Soc.*, *Chem. Commun.*, 1979, 1005.

Tetra-*n*-butyl ammonium benzoate hydrate
$C_{16}H_{36}N^+$, $C_7H_5O_2^-$, $39.5(H_2O)$
M. Bonamico, G. A. Jeffrey and R. K. McMullan, *J. Chem. Phys.*, 1962, **37**, 2219.

Tetra-*n*-Propylammonium bromide tri(urea) monohydrate
$C_{12}H_{28}N^+$, $3(CH_4N_2O)$, Br^-, H_2O
R. D. Rosenstein, R. K. McMullan, D. Schwarzenbach and G. A. Jeffrey, *A.C.A.* (Summer), 1973, 152.

t-Butylamine hydrate
$C_4H_{11}N$, $9.75(H_2O)$
R. K. McMullan, G. A. Jeffrey and T. H. Jordan, *J. Chem. Phys.* 1967, **47**, 1229.

β-Cyclodextrin *p*-ethylaniline hydrate
$2(C_{42}H_{70}O_{35})$, $2(C_8H_{11}N)$, $32(H_2O)$
R. Tokuoka, T. Fujiwara and K.-I. Tomita, *Acta Crystallogr.*, *Sect. B*, 1981, **37**, 1158.

Table 1. Continued

α-Cyclodextrin–n-propanol
$C_{36}H_{60}O_{30}$, C_3H_8O
W. Saenger and P. C. Manor, *Eur. Cryst. Meeting*, 1973.

α-Cyclodextrin dihydrate
$C_{36}H_{60}O_{30}$, $2(H_2O)$
W. Saenger and P. C. Manor, *Eur. Cryst. Meeting*, 1973.

α-Cyclodextrin potassium γ-aminobutyrate decahydrate clathrate
$C_{36}H_{60}O_{30}$, $C_4H_8NO_2^-$, K^+, $10(H_2O)$
R. Tokuoka, M. Abe, K. Matsumoto, K. Shirakawa, T. Fujiwara and K.-I. Tomita,
Acta Crystallogr., Sect. B, 1981, **37**, 445.

α-Cyclodextrin m-nitroaniline hexahydrate
$C_{36}H_{60}O_{30}$, $C_6H_6N_2O_2$, $6(H_2O)$
K. Harata, *Bull. Chem. Soc. Jpn*, 1980, **53**, 2782.

Cyclohexa-amylose 1-propanol hydrate
α-Cyclodextrin 1-propanol hydrate
$C_{36}H_{60}O_{30}$, C_3H_8O, $4.8(H_2O)$
W. Saenger, R. K. McMullan, J. Fayos and D. Mootz, *Acta Crystallogr., Sect. B*,
1974, **30**, 2019.

Copper(II) bis-l-ephedrine benzene clathrate
$C_{20}H_{28}CuN_2O_2$, $0.67(C_6H_6)$
Y. Amano, K. Osaki and T. Watanabe, *Bull. Chem. Soc. Jpn.*, 1964, **37**, 1363.

Diammine-cadmium tetracyanomercury(II) dibenzene
$C_4H_6CdHgN_6$, $2(C_6H_6)$
R. Kuroda, *Inorg. Nucl. Chem. Lett.*, 1973, **9**, 13.

Diammine-cadmium(II) tetracyanonickelate(II) dibenzene
$C_4H_6CdN_6Ni$, $2(C_6H_6)$
Y. Sasaki, *Bull. Chem. Soc. Jpn.*, 1969, **42**, 2412.

1,4,7,10,13,16-Hexaoxacyclo-octadecane bis(2,4-dinitrophenylhydrazine)
18-Crown-6 bis(2,4-dinitrophenylhydrazine)
$C_{12}H_{24}O_6$, $2(C_6H_6N_4O_4)$
R. Hilgenfeld and W. Saenger, *Z. Naturforsch., Teil B*, 1981, **36**, 242.

2-Phenyl-3-p-(2,2,4-trimethylchroman-4-yl)phenyl-quinazolin-4(3H)-one
methylcyclohexane clathrate
$2(C_{32}H_{28}N_2O_2)$, C_7H_{14}
C. J. Gilmore, A. D. U. Hardy, D. D. MacNicol and D. R. Wilson, *J. Chem. Soc.,
Perkin Trans. 2*, 1977, 1427.

1,4,7,10,13,16-Hexaoxacyclo-octadecane pentakis(urea) complex (at 148 K)
18-Crown-6 pentakis(urea)complex
$C_{12}H_{24}O_6$, $5(CH_4N_2O)$
S. Harkema, G. J. van Hummel, K. Daasvatn and D. N. Reinhoudt, *J. Chem. Soc.,
Chem. Commun.*, 1981, 368.

Table 1.   Continued

10,15-Dihydro-2,3,7,8,12,13-hexahydroxy-5H-tribenzo[*a,d,g*]cyclononene di-2-propanolate clathrate
Cyclotricatechylene di-2-propanolate
$C_{21}H_{18}O_6$, $2(C_3H_8O)$
J. A. Hyatt, E. N. Duesler, D. Y. Curtin and I. C. Paul, *J. Org. Chem.*, 1980, **45**, 5074.

Diammine-copper(II) tetracyano-nickel(II) dibenzene
$C_4H_6CuN_6Ni$, $2(C_6H_6)$
T. Miyoshi, T. Iwamoto and Y. Sasaki, *Inorg. Chim. Acta*, 1973, **7**, 97.

Cycloveratril benzene clathrate monohydrate
$C_{27}H_{30}O_6$, $0.5(C_6H_6)$, H_2O
S. Cerrini, E. Giglio, F. Mazza and N. V. Pavel, *Acta Crystallogr., Sect. B*, 1979, **35**, 2605.

γ-Cyclodextrin *n*-propanol clathrate hydrate
$C_{48}H_{80}O_{40}$, C_3H_8O, $x(H_2O)$
K. Lindner and W. Saenger, *Biochem. Biophys. Res. Commun.*, 1980, **92**, 933.

3α,12α-Dihydroxy-5β-cholan-24-oic acid bicyclo[2.2.1] hepta-2,5-diene (2:1) clathrate
Deoxycholic acid norbornadiene clathrate
$2(C_{24}H_{40}O_4)$, C_7H_8
A. D'Andrea, W. Fedeli, E. Giglio, F. Mazza and N. V. Pavel, *Acta Crystallogr., Sect. B*, 1981, **37**, 368.

N,N-Diethyl-β-alanine benzene clathrate (at 160 K)
$C_7H_{15}NO_2$, C_6H_6
M. A. Peterson, H. Hope and C. P. Nash, *J. Am. Chem. Soc.*, 1979, **101**, 946.

Diethylamine hydrate
$C_4H_{11}N$, $8.67(H_2O)$
T. H. Jordan and T. C. W. Mak, *J. Chem. Phys.*, 1967, **47**, 1222.

4-*p*-Hydroxyphenyl-2,2,4-trimethylchroman–chloroform
$C_{18}H_{20}O_2$, $0.167(CHCl_3)$
J. L. Flippen, J. Karle and I. L. Karle, *J. Am. Chem. Soc.*, 1970, **92**, 3749.

4-*p*-Hydroxyphenyl-2,2,4-trimethylchroman–ethanol
$C_{18}H_{20}O_2$, $0.33(C_2H_6O)$
J. L. Flippen, J. Karle and I. L. Karle, *J. Am. Chem. Soc.*, 1970, **92**, 3749.

4-*p*-Hydroxyphenyl-2,2,4-trimethylchroman–*n*–heptanol
Dianin's compound–*n*-heptanol
$6(C_{18}H_{20}O_2)$, $C_7H_{16}O$
J. L. Flippen and J. Karle, *J. Phys. Chem.*, 1971, **75**, 3566.

Dioxane diammine-cadmium-tetracyano-nickel(II) clathrate
$C_4H_6CdN_6Ni$, $2(C_4H_8O_2)$
E. Kendi and D. Ulku, *Z. Kristallogr.*, 1976, **144**, 91.

Table 1. Continued

Deoxycholic acid—di-t-butyl-diperoxycarbonate
$C_{24}H_{40}O_4$, $0.25(C_9H_{18}O_5)$
N. Friedman, M. Lahav, L. Leiserowitz, R. Popovitz-Biro, C. P. Tang and Z. Zaretskii,
J. Chem. Soc., Chem. Commun., 1975, 864.

μ-Ethylenediamine cadmium(II) tetra-μ-cyano-nickel(II) pyrrole
$C_6H_8CdN_6Ni$, $2(C_4H_5N)$
T. Iwamoto and M. Kiyoki, Bull. Chem. Soc. Jpn., 1975, 48, 2414.

Catena-μ-ethylenediamine-cadmium(II) tetracyanocadmate(II) dibenzene
$C_6H_8Cd_2N_6$, $2(C_6H_6)$
T. Iwamoto, Chem. Lett., 1973, 723.

Catena-μ-ethylenediamine-cadmium(II) tetracyanonickelate(II) dibenzene
$C_6H_8CdN_6Ni$, $2(C_6H_6)$
T. Miyoshi, T. Iwamoto and Y. Sasaki, Inorg. Chim. Acta 1972, 6, 59.

Ethylene oxide hydrate
C_2H_4O, $7.2(H_2O)$
R. K. McMullan and G. A. Jeffrey, J. Chem. Phys., 1965, 42, 2725.

Ethylene oxide deuterohydrate (neutron study, at 80 K)
C_2H_4O, $7.6(D_2O)$
S. Takagi and G. A. Jeffrey, Am. Crystallogr. Assoc., Abst. Papers (Summer Meeting),
1976, 50.

Ethylene oxide deuterohydrate clathrate (neutron study, at 80 K)
C_2H_4O, $7.67(D_2O)$
F. Hollander and G. A. Jeffrey, J. Chem. Phys., 1977, 66, 4699.

Hydrogen tetrakis (benzoylacetonato) europate diethylamine complex
$C_{40}H_{37}EuO_8$, $x(C_4H_{11}N)$
A. L. Il'Inskii, M. A. Porai-Koshits, L. A. Aslanov and P. I. Lazarev,
Zh. Strukt. Khim. 1972, 13, 277.

Exo-2,exo-6-dihydroxy-2,6-dimethyl-bicyclo[3.3.1]nonane ethylacetate clathrate
$3(C_{11}H_{20}O_2)$, $C_4H_8O_2$
R. Bishop and I. Dance, J. Chem. Soc., Chem. Commun., 1979, 992.

Thiourea-ferrocene clathrate
$C_{10}H_{10}Fe$, $3(CH_4N_2S)$
E. Hough and D. G. Nicholson, J. Chem. Soc., Dalton Trans., 1978, 15.

Furaltadone hydrochloride acetic acid clathrate
$C_{13}H_{17}N_4O_6^+$, $C_2H_4O_2$, Cl^-
I. Goldberg, Eur. Cryst. Meeting, 1980, 6, 49.

Furaltadone hydrochloride propionic acid clathrate
$C_{13}H_{17}N_4O_6^+$, $C_3H_6O_2$, Cl^-
I. Goldberg, Eur. Cryst. Meeting, 1980, 6, 49.

Cyclohepta-amylose 2-bromo-5-t-butylphenol clathrate octahydrate
$C_{42}H_{70}O_{35}$, $C_{10}H_{13}BrO$, $8(H_2O)$
J. A. Hamilton, M. N. Sabesan and L. K. Steinrauf, Carbohydr. Res., 1981, 89, 33.

Table 1. Continued

2,2,4,4,6,6-Hexa(1-aziridinyl)-cyclotriphosphazene carbon tetrachloride anti-clathrate
$C_{12}H_{24}N_9P_3$, 3(CCl_4)
J. Galy, R. Enjalbert and J.-F. Labarre, *Acta Crystallogr., Sect B*, 1980, **36**, 392.

Hexakis(benzylthiomethyl)benzene 1,4-dioxane clathrate (monoclinic form)
$C_{54}H_{54}S_6$, $C_4H_8O_2$
A. D. U. Hardy, D. D. MacNicol, S. Swanson and D. R. Wilson, *Tetrahedron Lett.*, 1978, 3579.

Hexakis(benzylthiomethyl)benzene 1,4-dioxane clathrate (monoclinic form)
$C_{54}H_{54}S_6$, $C_4H_8O_2$
A. D. U. Hardy, D. D. MacNicol, S. Swanson and D. R. Wilson, *J. Chem. Soc., Perkin Trans. 2*, 1980, 999.

Cyclohepta-amylose 2,5-di-iodobenzoic acid clathrate heptahydrate
$C_{42}H_{70}O_{35}$, $C_7H_4I_2O_2$, 7(H_2O)
J. A. Hamilton, M. N. Sabesan and L. K. Steinrauf, *Carbohydr. Res.*, 1981, **89**, 33.

Hexakis(phenylthio)benzene carbon tetrachloride clathrate
$C_{42}H_{30}S_6$, 2(CCl_4)
A. D. U. Hardy, D. D. MacNicol and D. R. Wilson, *J. Chem. Soc., Perkin Trans. 2*, 1979, 1011.

4-*p*-Hydroxyphenyl-2,2,4-trimethylthiachroman–ethanol
3($C_{18}H_{20}OS$), C_2H_6O
D. D. MacNicol, H. H. Mills and F. B. Wilson, *J. Chem. Soc. D*, 1969, 1332.

Hydroquinone acetonitrile clathrate
3($C_6H_6O_2$), C_2H_3N
T. C. W. Mak and K. Lee, *Acta Crystallogr., Sect. B*, 1978, **34**, 3631.

Hydroquinone–hydrogen sulphide clathrate
$C_6H_6O_2$, 0.256(H_2S)
T. C. W. Mak, J. S. Tse, C. Tse, K. Lee and Y. Chong, *J. Chem. Soc., Perkin Trans. 2*, 1976, 1169.

Hexamethylenetetramine hexahydrate
$C_6H_{12}N_4$, 6(H_2O)
T. C. W. Mak, *J. Chem. Phys.* 1965, **43**, 2799.

Bis(10,22-dimethyl-1,4,7,13,16,19-hexaoxa-10,22-diazacyclotetracosane) benzylammonium thiocyanate clathrate
2($C_{18}H_{38}N_2O_6$), $C_7H_{10}N^+$, CNS^-
M. J. Bovill, D. J. Chadwick, M. R. Johnson, N. F. Jones, I. O. Sutherland and R. F. Newton, *J. Chem. Soc., Chem. Commun.*, 1979, 1065.

5,18-Dimethyl-5,11,12,18-tetrahydrotribenzo[*b,f,j*][1,4]diazacyclododecine-6,17-dione *o*-xylene
$C_{24}H_{22}N_2O_2$, C_8H_{10}
W. D. Ollis, J. S. Stephanatou, J. F. Stoddart, G. G. Unal and D. J. Williams, *Tetrahedron Lett.*, 1981, **22**, 2225.

Table 1.   Continued

Hydroquinone hydrogen chloride clathrate
$C_6H_6O_2$, 0.33(HCl)
J. C. A. Boeyens and J. A. Pretorius, *Acta Crystallogr. Sect. B*, 1977, **33**, 2120.

Hydroquinone hydrogen chloride clathrate (neutron study)
$C_6H_6O_2$, 0.33(HCl)
J. C. A. Boeyens and J. A. Pretorius, *Acta Crystallogr., Sect. B*, 1977, **33**, 2120.

Bis(isothiocyanato)-tetrakis(4-methylpyridine)-nickel(II) 1-methylnaphthalene
clathrate
$C_{26}H_{28}N_6NiS_2$, $2(C_{11}H_{10})$
J. Lipkowski, G. D. Andreetti and P. Sgarabotto, *Acta Crystallogr., Sect. A* 1978,
34, S145.

Bis(isothiocyanato)-tetrakis(4-methylpyridine)-nickel(II) 2-bromonaphthalene
clathrate
$C_{26}H_{28}N_6NiS_2$, $2(C_{10}H_7Br)$
J. Lipkowski, P. Sgarabotto and G. D. Andreetti, *Acta Crystallogr., Sect. B*, 1980,
36, 51.

Bis(isothiocyanato)-tetrakis(4-methylpyridine)-nickel(II) 2-methylnaphthalene
clathrate
$C_{26}H_{28}N_6NiS_2$, $2(C_{11}H_{10})$
J. Lipkowski, P. Sgarabotto and G. D. Andreetti, *Acta Crystallogr., Sect. B*, 1980,
36, 51.

Isopropylamine octahydrate (at −160 °C)
C_3H_9N, $8(H_2O)$
R. K. McMullan, G. A. Jeffrey and D. Panke, *J. Chem. Phys.*, 1970, **53**, 3568.

Tetraisoamylphosphonium bromide hydrate (at −120 °C)
$C_{20}H_{44}P^+$, Br^-, $32.6(H_2O)$
S. F. Solodovnikov, T. M. Polyanskaya, V. I. Alexseev, Yu. A. Dyadin and V. V.
Bakakin, *Dokl. Akad. Nauk SSSR*, 1979, **247**, 357.

Lithium triphenylphosphine–oxide iodide complex
$5(C_{18}H_{15}OP)$, I^-, Li^+
Y. M. G. Yasin, O. J. R. Hodder and H. M. Powell, *J. Chem. Soc., Chem. Commun.*, 1966,
705.

N,N'-Dimethyl-*N''*-benzyltri-3-methyltrianthranilide toluene clathrate
$C_{33}H_{31}N_3O_3$, C_7H_8
S. J. Edge, W. D. Ollis, J. S. Stephanatou, D. J. Williams, and K. A. Woode,
Tetrahedron Lett., 1981, **22**, 2229.

5-Methylbenzene-1,3-dicarbaldehyde-bis(*p*-tolylsulphonylhydrazone) benzene
clathrate
$C_{23}H_{24}N_4O_4S_2$, $2(C_6H_6)$
T. Chan, T. C. W. Mak and J. Trotter, *J. Chem. Soc., Perkin Trans. 2*, 1980, 672.

Diammine-manganese(II) tetracyano-nickel(II) dibenzene
$C_4H_6MnN_6Ni$, $2(C_6H_6)$
R. Kuroda and Y. Sasaki, *Acta Crystallogr., Sect. B*, 1974, **30**, 687.

Table 1. Continued

2,3,4,5-Bis(1,2-(3-methylnaphtho))-1,6,9,12,15,18-hexaoxacycloeicosa-2,4-diene
t-butylammonium perchlorate clathrate benzene solvate (at 113 K)
$C_{32}H_{36}O_6$, $C_4H_{12}N^+$, ClO_4^-, C_6H_6
I. Goldberg, *J. Am. Chem. Soc.* 1980, **102**, 4106.

4-*p*-Mercaptophenyl-2,2,4-trimethyl-chroman carbon tetrachloride clathrate
$3(C_{18}H_{20}OS)$, CCl_4
A. D. U. Hardy, J. J. McKendrick, D. D. MacNicol and D. R. Wilson, *J. Chem. Soc., Perkin Trans. 2*, 1979, 729.

Tetra(4-methylpyridine) nickel(II) thiocyanate *p*-terphenyl
$C_{26}H_{28}N_6NiS_2$, $C_{18}H_{14}$
G. D. Andreetti, L. Cavalca and P. Sgarabotto, *Gazz. Chim. Ital.*, 1970, **100**, 697.

4-*p*-Hydroxyphenyl-2,2,4,8-tetramethyl-thia-chroman cyclooctan clathrate
$4.5(C_{19}H_{22}OS)$, C_8H_{16}
A. D. U. Hardy, J. J. McKendrick and D. D. MacNicol, *J. Chem. Soc., Perkin Trans. 2*, 1979, 1072.

Tris(2,3-naphthalenedioxy)cyclotriphosphazene benzene
$C_{30}H_{18}N_3O_6P_3$, $3(C_6H_6)$
H. R. Allcock and M. T. Stein, *J. Am. Chem. Soc.*, 1974, **96**, 49.

Nickel iodide-urea
$C_6H_{24}N_{12}NiO_6^{2+}$, $4(CH_4N_2O)$, $2(I^-)$
X. Suleyman, M. A. Porai-Koshits, A. S. Antsyshkina and K. Sulaimankulov, *Zh. Neorg. Khim.*, 1971, **16**, 3394.

Tris(1,8-naphthalenedioxy)cyclotriphosphazene *p*-xylene
$C_{30}H_{18}N_3O_6P_3$, $0.5(C_8H_{10})$
H. R. Allcock, M. T. Stein and E. C. Bissell, *J. Am. Chem. Soc.*, 1974, **96**, 4795.

Bis(tri-*o*-thymotide) (R)-2-butanol clathrate (at 123 K)
$C_{33}H_{36}O_6$, $0.5(C_4H_{10}O)$
J. Allemand and R. Gerdil, *Cryst. Struct. Commun.*, 1981, **10**, 33.

Bis(triphenylbenzylphosphonium) tetrachloro-cadmium dichloroethane clathrate
$2(C_{25}H_{22}P^+)$, $CdCl_4^{2-}$, $2(C_2H_4Cl_2)$
J. C. J. Bart, I. W. Bassi and M. Calcaterra, *J. Organomet. Chem.*, 1980, **193**, 1.

Hexakis(2-phenylethylthiomethyl)benzene 1,4-dioxane clathrate
$C_{60}H_{66}S_6$, $C_4H_8O_2$
K. Burns, C. J. Gilmore, P. R. Mallinson, D. D. MacNicol and S. Swanson, *Eur. Cryst. Meeting*, 1980, **6**, 23.

Hexakis(2-phenylethylthiomethyl)benzene 1,4-dioxane clathrate
$C_{60}H_{66}S_6$, $C_4H_8O_2$
K. Burns, C. J. Gilmore, P. R. Mallinson, D. D. MacNicol and S. Swanson, *J. Chem. Research* (S), 1981, 30; (M), 1981, 501.

Hexakis(R-α-phenylethylsulfonyl-methyl)benzene acetic acid clathrate
$C_{60}H_{66}O_{12}S_6$, $4(C_2H_4O_2)$
A. Freer, C. J. Gilmore, D. D. MacNicol and S. Swanson, *Tetrahedron Lett.*, 1980, **21**, 205.

138

G. D. Andreetti

Table 1. Continued

Tris(o-phenylenedioxy)-phosphonitrile trimer bromobenzene inclusion compound
$C_{18}H_{12}N_3O_6P_3$, $0.5(C_6H_5Br)$
L. A. Siegel and J. H. van den Hende, *J. Chem. Soc. A*, 1967, 817.

Tris(o-phenylenedioxy)-phosphonitrile trimer benzene inclusion compound
$C_{18}H_{12}N_3O_6P_3$, $0.5(C_6H_6)$
L. A. Siegel and J. H. van den Hende, *J. Chem. Soc. A*, 1967, 817.

Trans, anti, trans, anti, trans-perhydrotriphenylene–chloroform
$C_{18}H_{30}$, $0.5(CHCl_3)$
G. Allegra, M. Farina, A. Immirzi, A. Colombo, U. Rossi, R. Broggi and G. Natta,
J. Chem. Soc. B, 1967, 1020.

Trans,anti,trans,anti,trans-perhydrotriphenylene-*n*-heptane
$C_{18}H_{30}$, $0.225(C_7H_{16})$
G. Allegra, M. Farina, A. Immirzi, A. Colombo, U. Rossi, R. Broggi and G. Natta,
J. Chem. Soc. B, 1967, 1020.

Pinacol hexahydrate
2,3-Dimethyl-2,3-butane diol hexahydrate
$C_6H_{14}O_2$, $6(H_2O)$
H. S. Kim and G. A. Jeffrey, *J. Chem. Phys.*, 1970, **53**, 3610.

4-*p*-Hydroxyphenyl-*cis*-2,4-dimethyl-chroman carbon tetrachloride clathrate
$C_{17}H_{18}O_2$, $0.17(CCl_4)$
J. H. Gall, A. D. U. Hardy, J. J. McKendrick and D. D. MacNicol, *J. Chem. Soc.,
Perkin Trans. 2*, 1979, 376.

1-Phenylpropyl-9-anthroate *n*-hexane clathrate
$6(C_{24}H_{20}O_2)$, C_6H_{14}
M. Lahav, L. Leiserowitz, L. Roitman and C. P. Tang, *J. Chem. Soc., Chem. Commun.*,
1977, 928.

Tetracyclo[14.2.2.2^{4,7}.2^{10,13}]tetracosa-4,6,10,12,16,18,19,21,23-nonaene   silver
trifluoromethane-sulphonate
[2.2.2]paracyclophane silver triflate
$C_{24}H_{24}$, $CF_3O_3S^-$, Ag^+
J.-L. Pierre, P. Baret, P. Chautemps and M. Armand, *J. Am. Chem. Soc.*, 1981,
103, 2986.

n-Propylamine hydrate (Model 1, at $-100\,°C$)
C_3H_9N, $6.5(H_2O)$
C. S. Brickenkamp and D. Panke, *J. Chem. Phys.*, 1973, **58**, 5284.

n-Propylamine hydrate (Model 2, at $-100\,°C$)
C_3H_9N, $x(H_2O)$
C. S. Brickenkamp and D. Panke, *J. Chem. Phys.*, 1973, **58**, 5284.

Tris(1,2-bis(diphenylphosphino-selenoyl)ethane) *p*-xylene clathrate
$3(C_{26}H_{24}P_2Se_2)$, C_8H_{10}
D. H. Brown, R. J. Cross, P. R. Mallinson and D. D. MacNicol, *J. Chem. Soc.,
Perkin Trans. 2*, 1980, 993.

Table 1. Continued

Tris(o-phenylenedioxy)cyclotriphosphazene benzene
$C_{18}H_{12}N_3O_6P_3$, $x(C_6H_6)$
H. R. Allcock, R. W. Allen, E. C. Bissell, L. A. Smeltz and M. Teeter, *J. Am. Chem. Soc.*, 1976, **98**, 5120.

Tris(o-phenylenedioxy)cyclotriphosphazene o-xylene
$C_{18}H_{12}N_3O_6P_3$, $x(C_8H_{10})$
H. R. Allcock, R. W. Allen, E. C. Bissell, L. A. Smeltz and M. Teeter, *J. Am. Chem. Soc.*, 1976, **98**, 5120.

Quinol–sulphur dioxide
$C_6H_6O_2$, O_2S
D. E. Palin and H. M. Powell, *J. Chem. Soc.*, 1947, 208.

Hexakis(p-t-butylphenylthiomethyl)benzene squalene clathrate
$C_{72}H_{90}S_6$, $0.5(C_{30}H_{50})$
A. Freer, C. J. Gilmore, D. D. MacNicol and D. R. Wilson, *Tetrahedron Lett.*, 1980, **21**, 159.

Tetra isoamyl ammonium fluoride hydrate
$C_{20}H_{44}N^+$, F^-, $38(H_2O)$
D. Feil and G. A. Jeffrey, *J. Chem. Phys.*, 1961, **35**, 1863.

Tetra-n-butyl ammonium fluoride hydrate
$C_{16}H_{36}N^+$, F^-, $32.8(H_2O)$
R. K. McMullan, M. Bonamico and G. A. Jeffrey, *J. Chem. Phys.*, 1963, **39**, 3295.

Tri-n-butyl sulphonium fluoride hydrate (cubic form)
$C_{12}H_{27}S^+$, F^-, $20(H_2O)$
G. A. Jeffrey and R. K. McMullan, *J. Chem. Phys.*, 1962, **37**, 2231.

Tri-n-butyl sulphonium fluoride hydrate (monoclinic form)
$C_{12}H_{27}S^+$, F^-, $23(H_2O)$
P. T. Beurskens and G. A. Jeffrey, *J. Chem. Phys.*, 1964, **40**, 2800.

4-p-Hydroxyphenyl-2,2,4-trimethylthiachroman 2,5,5-trimethylhex-3-yn-2-ol
$6(C_{18}H_{20}OS)$, $C_9H_{16}O$
D. D. MacNicol and F. B. Wilson, *J. Chem. Soc. D.*, 1971, 786.

Tetrahydrofuran hydrogen sulphide hydrate
$8(C_4H_8O)$, $7.33(H_2S)$, $136(H_2O)$
T. C. W. Mak and R. K. McMullan, *J. Chem. Phys.*, 1965, **42**, 2732.

(+)-Tri-o-thymotide-S-(+)-2-bromo-butane clathrate (at $-50\,^\circ C$)
$2(C_{33}H_{36}O_6)$, C_4H_9Br
R. Arad-Yellin, B. S. Green, M. Knossow and G. Tsoucaris, *Tetrahedron Lett.*, 1980, **21**, 387.

(−)-Tri-o-thymotide-RR-(+)-2,3-Dimethyl-thiirane clathrate (at $-50\,^\circ C$)
$2(C_{33}H_{36}O_6)$, C_4H_8S
R. Arad-Yellin, B. S. Green, M. Knossow and G. Tsoucaris, *Tetrahedron Lett.*, 1980, **21**, 387.

Table 1. Continued

Trimesic acid dimethylsulphoxide clathrate
$C_9H_6O_6$, C_2H_6OS
F. H. Herbstein, M. Kapon and S. Wasserman, *Acta Crystallogr., Sect. B*, 1978, **34**, 1613.

Tetramethylammonium fluoride tetrahydrate
$C_4H_{12}N^+$, F^-, $4(H_2O)$
W. J. McLean and G. A. Jeffrey, *J. Chem. Phys.*, 1967, **47**, 414.

Tetramethylammonium hydroxide pentahydrate
$C_4H_{12}N^+$, HO^-, $5(H_2O)$
R. K. McMullan, T. C. W. Mak and G. A. Jeffrey, *J. Chem. Phys.*, 1966, **44**, 2338.

Trimethylamine hydrate
C_3H_9N, $10.25(H_2O)$
D. Panke, *J. Chem. Phys.* 1968, **48**, 2990.

Tri-*o*-thymotide cetyl alcohol
$0.2(C_{33}H_{36}O_6)$, $C_{16}H_{34}O$
D. J. Williams and D. Lawton, *Tetrahedron Lett.*, 1975, 111.

Tri-*o*-thymotide ethanol
$C_{33}H_{36}O_6$, C_2H_6O
D. J. Williams and D. Lawton, *Tetrahedron Lett.*, 1975, 111.

Tri-*o*-thymotide pyridine clathrate
$2(C_{33}H_{36}O_6)$, C_5H_5N
S. Brunie, A. Navaza, G. Tsoucaris, J. P. Declercq and G. Germain, *Acta Crystallogr., Sect. B*, 1977, **33**, 2645.

Trans-perhydrotriphenylene cyclohexane solvate
$C_{18}H_{30}$, $0.39(C_6H_{12})$
A. Immirzi and G. Allegra, *Atti Accad. Naz. Lincei*, 1967, **43**, 181.

Tris(thiourea) 2,3-dimethylbutadiene (at $-130\,°C$)
$3(CH_4N_2S)$, C_6H_{10}
Y. Chatani and S. Nakatani, *Z. Kristallogr.*, 1976, **144**, 175.

Urea-1,4-dichlorobutane
CH_4N_2O, $C_4H_8Cl_2$
J. Otto, *Acta Crystallogr., Sect. B*, 1972, **28**, 543.

Tri-*o*-thymotide *trans*-stilbene clathrate
$C_{33}H_{36}O_6$, $C_{14}H_{12}$
R. Arad-Yellin, S. Brunie, B. S. Green, M. Knossow and G. Tsoucaris, *J. Am. Chem. Soc.*, 1979, **101**, 7529.

---

has to be transferred to a capillary directly from the solution. The sealing procedure has to be quite quick.

A general procedure for crystallization used in our laboratory (provided there are no phase transitions of the inclusion compound which imposes a rigorous control of the temperature) is to dissolve the host material in an appropriate solvent not forming a stable inclusion compound at room temperature. The solution, saturated at about 40° C, is filtered and poured into a flask with a large neck previously cleaned very carefully with a buckskin. To the flask is also added a solution in the same solvent of the guest component in some tentative stoichiometric ratio to the amount of host (e.g., 1:1 or 1:large excess). The flask is closed with a soft plastic cover (parafilm) and left to stand for one day. Usually a limited number of crystal seeds start to grow. When too many crystals appear, the operation is repeated but with the addition of more solvent (usually up to 10% more). If after one day, no seeds are observed then a small hole is made in the plastic cover with a sharp pin, and the flask left to stand. When the crystals have reached the right size a few drops of the solvent are added and the crystals can be transferred to the capillaries for the X-ray study. In the capillary a small drop of the solution has to be present in order to avoid rapid decomposition.

The sample prepared in this way could also be used for X-ray analysis at low temperature, but there have been very few such studies. The high stability of the crystal against the escape of the guest, which is usually more volatile than the host, is certainly an advantage. On the other hand, some experiments carried out with inclusion compounds of Werner complexes at different temperatures have shown that the mosaicity of the crystals greatly increases upon lowering the temperature.[12] This is caused by the different thermal compressibility of the host and guest molecules. More work is needed in this area to give a quantitative answer to the question. If the X-ray data are good enough to explain the thermal motion of guest molecules inside the channels or cavities of inclusion crystals, then quantitative answers to important questions can be obtained.

3. Crystal packing

The problem of a theoretical basis for the description of crystal packing has been considered in some detail by A. I. Kitaigorodsky,[13,14] but it seems that up to now no general rules have been found. Obviously, from a thermodynamic point of view, the free energy of a crystal composed of two molecular species, as in an inclusion compound, is lower than the average energy of a crystal composed of only one species. It can therefore be shown that the packing of a host–guest compound could have a better density (corresponding to a lower interaction potential energy) than the packing of its components. For this reason, in several cases, the crystalline form of the

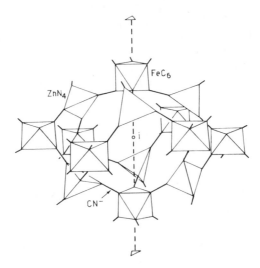

Fig. 1. Illustration of the structure of the cavity found in $Na_2Zn_3[Fe(CN)_6]_2 \cdot 9H_2O$. (Reproduced by permission from ref. 19.)

empty phases of some inclusion compounds are not known. Furthermore, it has been shown, by calculating the packing coefficients for host–guest crystals and for components, that the molecular packing principle is no different than in crystals of a single species and that the main energy component, the sum of the atom–atom potentials, appears to remain practically unchanged. For the formation of host–guest crystals, Kitaigorodsky assumes the cause could be the fact that mixing of dipolar molecules with molecules with an induced dipole moment results in a decrease of the electrostatic part of the interaction energy. It is considered doubtful that a transfer of the charge would significantly affect the lattice energy. Since the

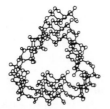 

Fig. 2. Stereoview illustrating the helical host packing in the ethyl acetate inclusion compound of *exo*-2,*exo*-6-dihydroxy-2,6-dimethylbicyclo[3.3.1]nonane. Intermolecular hydrogen bonds are denoted solid. (Drawn from data of ref. 20.)

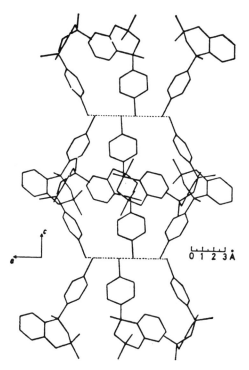

Fig. 3. Illustration of the host packing in the CCl$_4$ clathrate of 4-*p*-mercaptophenyl-2,2,4-trimethylchroman.[21]

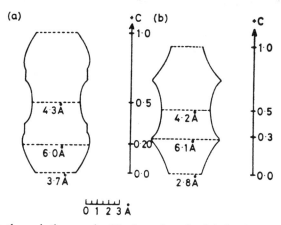

Fig. 4. Section through the van der Waals surface for (a) the CCl$_4$ clathrate of the thiol host in Fig. 3, compared with (b) the corresponding section for Dianin's compound as its CHCl$_3$ clathrate. (Reproduced by permission from ref. 21.)

publication of the Kitaigorodsky book very little has been done to verify experimentally that conclusion.

As far as the symmetries observed for the crystal packings, it is too early to try to deduce general rules. In several cases high symmetry space groups have been observed: P4, P4/m, $P4_2/m$, P4/n, $I4_1/a$, I422, $P4_2/mnm$, P3, $R\overline{3}$, P321, P3, 2, 1, R3c, P65, $P6_3$, $P6_3/m$, $I\overline{4}3d$, Fd3m. Analysis of possible arrays of molecules on the basis of close packing principles considers the above space groups rather improbable in the sense that the crystals exhibit a lowering of packing density. So the retention of higher symmetry of the crystals causes a significant lowering in the packing coefficient. The geometrical model of close packing cannot explain this preference for symmetrical arrangements. Probably the decrease in the packing density due to the high symmetry is partially compensated by host–guest interactions.

4. Description of cavities, channels, and layers

Having at our disposal the atomic coordinates of the atoms of a crystal structure, all types of calculations and drawings can be performed. Several styles of molecular drawings have been used to describe cavities, channels or layers in inclusion compounds. Some of them are represented in Figs. 1–4, and the reader can deduce from a qualitative point of view the space available in the host structure for the arrangement of the guest molecules. Three-dimensional drawings are now available through graphics displays and good representations can be obtained with them, but their reproduction in journals becomes very expensive. Moreover, a quantitative description of the size of the empty space in host structures is, in principle, possible both semiquantitatively from the knowledge of the volume increments of the groups of the host molecules[15] or, more accurately, by direct integration with Monte Carlo methods of the volume occupied by the host molecule from the knowledge of van der Waals radii.[16] The evaluation of an "average volume of cavities" has been suggested[17] assuming that the average shape of a cavity could be an ellipsoid, whose boundaries are the van der Waals radii of the atoms in the cavity and whose semi-axes and volume are calculated by least-squares methods. The results obtained with this procedure for the cavities housing the acetic acid dimer are quite encouraging.[18]

5. Conclusions

From the previous considerations, crystallography has contributed greatly to the description of the geometrical properties of clathrates and of inclusion compounds. All new compounds showing host properties will need a structural description of the nature of host–guest interactions. The availability of NMR, infrared and Raman data both in the solid state and in solutions now also allows some reliable extrapolations to the structure of the complex in solution. Morever, a large number of studies and more experimental work are needed to try to elucidate a quantitative theory to explain the static behaviour of clathrate crystals in view of the development of crystal engineering for the preparation of synthetic hosts tailored for a given guest molecule (see Chapter 5, Volume 2).

References

1. L. Mandelcorn (ed.) *Non-Stoichiometric Compounds*, Academic Press, New York, 1964.
2. D. E. Palin and H. M. Powell, *J. Chem. Soc.*, 1947, 208.
3. H. M. Powell, *J. Chem. Soc.*, 1948, 61.
4. J. F. Brown, *Sci. Amer.*, 1962, **207**, 82.
5. W. J. Mortier, *Compilation of Extra-Framework Sites in Zeolites*, IPC Science and Technology Press, Guildford, 1982.
6. N. F. Curtis, in *Coordination Chemistry of Macrocyclic Compounds*, (ed. G. A. Melson) Plenum Press, New York, 1979.
7. D. J. Cram and K. N. Trueblood, in *Topics in Current Chemistry No. 98. Host Guest Complex Chemistry I*, (ed. F. Vögtle) Springer–Verlag, Berlin, Heidelberg, New York, 1981.
8. M. Dobler, *Ionophores and their Structures*, John Wiley and Sons, New York, 1981.
9. I. Goldberg, in *The Chemistry of Ethers, Crown Ethers, Hydroxyl Groups and their Sulphur Analogues*, (ed. S. Patai) J. Wiley & Sons, New York, 1980.
10. F. H. Allen, S. Bellard, M. D. Brice, B. A. Cartwright, A. Doubleday, H. Higgs, R. Hummelink, B. G. Hummelink-Peters, O. Kennard, W. D. S. Motherwell, J. R. Rodgers and D. Watson, The Cambridge Crystallographic Data centre: Computer-Based Search, Retrieval Analysis and Display of Information, *Acta Crystallogr.*, 1979, **B35**, 2331.
11. G. D. Andreetti and R. Ungaro, results to be published.
12. J. Lipkowski and G. D. Andreetti, results to be published.
13. A. I. Kitaigorodsky, *Molecular Crystals and Molecules*, Academic Press, New York and London, 1973.
14. A. I. Kitaigorodsky and A. A. Frolova, *Izv. Sekt. Fiz. Khim. Anal. Inst. Obshch. Neorg. Khim. Akad. Nauk USSR*, 1949, **19**, 306.

146

G. D. Andreetti

15. A. Immirizi and B. Perini, *Acta Crystallogr.*, 1977, **A33**, 216.
16. G. Zanotti and A. Del Pra, *I Quaderni dell'Elaborazione automatica*, 1978, **2**, 17.
17. V. Sangermano, Internal Report, Institute of Structural Chemistry of the University of Parma, 1981.
18. C. Rizzoli, G. D. Andreetti, R. Ungaro and A. Pochini, *J. Mol. Struct.*, 1982, **82**, 133.
19. E. Garnier, P. Gravereau and A. Hardy, *Acta Crystallogr.*, 1982, **B38**, 1401.
20. R. Bishop and I. Dance, *J. Chem. Soc., Chem. Commun.*, 1979, 992.
21. A. D. U. Hardy, J. J. McKendrick, D. D. MacNicol and D. R. Wilson, *J. Chem. Soc., Perkin Trans. 2*, 1979, 729.

5 · HOST LATTICE–GUEST MOLECULE ENERGY TRANSFER PROCESSES

A. GUARINO

Istituto di Chimica Nucleare, Rome, Italy

1. Introduction

At first glance, this chapter may seem rather out of place amongst the other chapters in this volume. However, it simply stresses the importance of excited state interactions between a host lattice and the guest molecules in an inclusion compound. In fact, most of the papers devoted to inclusion compounds concern ground state effects and interactions: very few refer to UV and γ-ray radiation effects. The main purpose of examining excited state effects is the possibility of controlling reaction pathways inside the host lattice cavities, under irradiation. To fulfil this goal it is absolutely essential to have a quantitative knowledge of the processes of diffusion of electronic energy for a specific lattice of an inclusion compound. Consequently this chapter should supply a basis to link the crystal structures and well established electronic energy transfer processes in the solid state.

INCLUSION COMPOUNDS III
ISBN 0-12-067103-4

Very many theories and experiments have been applied to this subject, particularly on lightly doped substitutional mixed crystals: their application to host–guest inclusion compounds implies a loss of rigorousness and should be taken as estimates of the energy diffusion in an inclusion compound.

In any case, the busy reader anxious to move on to the next chapter needs only to observe the conclusions of the first Section (2.6) and then to read Sections 3 and 4 which are devoted to the experimental aspects of the problem.

It seemed useful throughout the text of the chapter to shorten "electronic energy transfer" to EET and denote a clathrate with host–guest pairs in an excited state as an "excithrate": it will be a personal contribution to the pollution of chemical nomenclature.

2. Theoretical aspects of EET processes in mixed crystals and inclusion compounds

A central problem in photochemistry and radiation chemistry is the study of the overall deactivation channels of any excited metastable state created by the absorption of UV or gamma photons by molecules in the gaseous, liquid or solid state; an impressive number of original papers and reviews has been devoted to the theoretical and experimental aspects of this subject.[1,2,3] Unfortunately, as a consequence of the many different mathematical formalisms employed in dealing with these processes, much confusion exists in the literature. Limiting ourselves to the solid state, the usual "model" studied, both theoretically and experimentally, is a mixed crystal consisting of an inert host (H), lightly doped with donor molecules (D) *randomly* distributed, and acceptor molecules (A), again *randomly* distributed. The relative order of their concentrations is $c_H \gg c_A \gg c_D$.

2.1. Energy transfer laws: $w(R_i)$

Starting from a single excited donor–acceptor pair, let us denote by $w(R_i)$ the microscopic energy transfer law from one molecule to another; $w(R_i)$ may be of the multipolar type:

$$w(R_i) = \frac{1}{\tau_0}\left(\frac{d}{R_i}\right)^s \tag{1}$$

or of the exchange type:

$$w(R_i) = \frac{1}{\tau_0}\exp\left[\gamma(d - R_i)\right] \tag{2}$$

where d is the next-neighbour distance between the molecules within the crystal lattice, $1/\tau_0$ is the transfer rate from an excited donor to an acceptor at a distance d, and R_i is the position of any acceptor occupying a site i of the lattice; γ is a constant relating to donor–acceptor distance; s is the multipolar interaction, (6 for dipole–dipole, 8 for dipole–quadrupole, etc.).

It is worth observing that, for the moment, we are considering these interactions as being isotropic; we will eventually relax this rather unrealistic condition. If we now follow the time evolution of the excitation $E_i(t)$ for this single donor–acceptor pair,[5] then

$$\frac{dE_i(t)}{dt} = -w(R_i)E_i(t) \tag{3}$$

and, consequently

$$E_i(t) = \exp[-tw(R_i)] \tag{4}$$

2.2. Deactivation functions: $\bar{\rho}_K(t), \bar{\rho}_{ET}(t)$

We may define a deactivation function $\rho_K(t)$ to show any deactivation process of an excited donor in the presence of acceptors, A, present at sites i of the crystal lattice and for a particular configuration K

$$\rho_K(t) = \exp\left[-\frac{t}{\tau_D} - t \sum_A \sum_i w(R_i) \right] \tag{5}$$

where $\rho_K(t)$ represents the overall deactivation process for a pulsed photonic irradiation of the crystal: the first term of the exponential shows the rate of any channel of *intramolecular* deactivation of the excited donor, t/τ_D both non-radiative and radiative, $1/\tau_D$: the second term shows any *intermolecular* energy transfer process between the excited donor and any acceptor A present in any site i of the crystal lattice, in a particular configuration K.

The deactivation function, in the following discussion, will be restricted only to the second term of the exponential, i.e., to the EET processes, $\rho_{ET}(t)$; the overall deactivation $\rho_K(t)$, is obtained by multiplying $\rho_{ET}(t)$ by $\exp(-t/\tau_D)$; i.e.

$$\rho_K(t) = \exp\left(-\frac{t}{\tau_D} \right) \rho_{ET}(t) \tag{6}$$

However, it is experimentally more useful to deal with deactivation functions *averaged* over all the possible configurations of the donor–acceptor pairs, i.e.:

$$\bar{\rho}_{ET}(t) = \sum_K \rho_{ET}(t) P(K) \tag{6a}$$

where K is a particular configuration and $P(K)$ is its probability of occurrence. The ensemble average $\bar{\rho}(t)_{ET}$ can be shown[4,5] to be

$$\bar{\rho}(t)_{ET} = \prod_i \sum_A p_A E(t)_{iA} \tag{7}$$

where p_A is the probability that molecules of acceptors A are distributed on the different sites i. In the case of a single acceptor species Equation 7 reduces to

$$\bar{\rho}(t)_{ET} = \prod_i \{1 - p_A[1 - E_i(t)]\} \tag{8}$$

Then, taking into account the two different $E_i(t)$ functions related to multipolar or exchange donor–acceptor interactions, Equation 8 takes the form

$$\bar{\rho}(t)_{ET} = \prod_i \left\{ 1 - p_A \left[1 - \exp\left(-\frac{t}{\tau_0} \left[\frac{d}{R_i} \right]^s \right) \right] \right\} \tag{9}$$

and, respectively

$$\bar{\rho}(t)_{ET} = \prod_i \left\{ 1 - p_A \left[1 - \exp\left(-\frac{t}{\tau_0} \exp[\gamma(d - R_i)] \right) \right] \right\} \tag{10}$$

The previous deactivation functions may be rearranged in a more physical meaningful way[4] if we denote by μ_i the number of sites of the crystal lattice at an equal distance R_i from the excited donor: there are then i shells. Consequently, all the unexcited molecules will be distributed in "*shells*" of increasing radius, centred at the site of the excited donor. Then Equations 9 and 10 take the form

$$\bar{\rho}(t)_{ET} = \prod_i \left\{ 1 - p_A \left[1 - \exp\left(-\frac{t}{\tau_0} \left[\frac{d}{R_i} \right]^s \right) \right] \right\}^{\mu_i} \tag{11}$$

and, respectively

$$\bar{\rho}(t)_{ET} = \prod_i \left\{ 1 - p_A \left[1 - \exp\left(-\frac{t}{\tau_0} \exp[\gamma(d - R_i)] \right) \right] \right\}^{\mu_i} \tag{12}$$

In conclusion, these two equations represent EET processes between excited donors and acceptors, under the condition of low donor concentration with respect to the acceptors, so that donor–donor EET processes may be taken as negligible. The products of Equation 8 refer to all the i sites except the origin.

Equation 8 is particularly useful for *numerical* calculations; it is possible to obtain plots of $\bar{\rho}(t)_{ET}$ vs. t/τ_0 for various values of p_A, which show *characteristic behaviours* according to the *specific crystal lattices* taken into account. For $c = c_H + c_A + c_D$, where c is the total concentration of the lattice molecules, c_H is the inert host, c_A and c_D are the acceptor and donor concentrations, then the value of p_A is c_A/c.

In the case where the acceptor concentration is very low, i.e. $c_A \ll c$, then Equations 9 and 10 may be approximated to the following analytical functions: the first one, valid for multipolar interactions, is

$$\bar{\rho}(t) \simeq \exp[-V_\Delta d^\Delta c_A \Gamma(1 - \Delta/s)(t/\tau_A)^{\Delta/s}] \tag{13}$$

The second one, valid for exchange interactions,[8] is

$$\bar{\rho}(t) \simeq \exp[-V_\Delta c_A \gamma^{-\Delta} g(u)] \tag{14}$$

where V_Δ is the volume of a unit sphere in a space of Δ dimensions; $\Gamma(1 - \Delta/s)$ is a gamma function; s is the multipolar interaction, $(6, 8, \ldots)$; $u = t/\tau_A \exp(\gamma d)$; $\Gamma^j(1)$ are the j-derivatives of the gamma function; and

$$g(u) \simeq \sum_{j=0}^{\Delta} \binom{\Delta}{j} (-1)^j \Gamma^j(1)(\ln u)^{\Delta-j}$$

2.3. Energy transfer functions, $H(t)$

It is possible to represent the same EET processes in a simplified way by means of a function, $H(t)$, which can be called the "*energy transfer function*".[6] In fact $\bar{\rho}(t)$, the overall deactivation function of Equation 6, can also take the form

$$\bar{\rho}(t) = \exp\left[-\frac{t}{\tau_D} - c_A H(t)\right] \tag{15}$$

where, as usual, the first term of the exponential shows any *unimolecular* deactivation process of the excited donor and the second term shows any *bimolecular* EET process in the presence of acceptor A; c_A is their number per unit volume; $H(t)$ is a time dependent function, expressed in volume units, which can be thought of as an "*interaction volume*" expanding with time around an excited *randomly distributed* donor; this "*interaction volume*" will contain other acceptor and unexcited donor molecules (also *randomly distributed*).

Under the same approximation of Equation 6, i.e. $c_D \ll c_A \ll c$, this energy transfer function $H(t)$ can be expressed, for multipolar interactions

$$H(t)_{mp} = V_\Delta d^\Delta \Gamma(1 - \Delta/s) \left(\frac{t}{\tau_0}\right)^{\Delta/s} \tag{16}$$

where the parameters are the same as those discussed in Equation 15 and

14. If $s = 6$ (dipolar interaction), $H(t)_{d-d}$ is, for $\Delta = 3$

$$H(t)_{d-d} = 1.77 \frac{4\pi R_0^3}{3} \left(\frac{t}{\tau_0}\right)^{1/2} \text{cm}^3 \tag{17}$$

where $\Gamma(1 - 3/6) = 1.77$; if Equation 17 is substituted in Equation 15, the theoretical equation of Förster for $\bar{\rho}(t)$ is obtained.[7]

The $H(t)$, for exchange interactions, takes the form

$$H(t)_{exc} = V_\Delta d^\Delta \gamma^{-\Delta} \sum_{j=0}^{\Delta} \binom{\Delta}{j} (-1)^j \Gamma^j(1) \left(\ln A \frac{t}{\tau_0}\right)^{\Delta - j} \tag{18}$$

where $A = \exp(\gamma)$; $\gamma = 2R_0/L$; R_0 is a "critical" distance at which unimolecular and bimolecular processes occur with the same probability; i.e. when the first and the second term of Equation 15 become equal; L is the average Bohr radius of either the excited donor or acceptor.

Equation 18, for $\Delta = 3$ takes the form:

$$H(t)_{exc} = \frac{\pi L^3}{6} \left[\left(\ln A \frac{t}{\tau_0}\right)^3 + 1.73 \left(\ln A \frac{t}{\tau_0}\right)^2\right.$$
$$\left. + 5.9 \left(\ln A \frac{t}{\tau_0}\right) + 5.4\right] \text{cm}^3 \tag{19}$$

If this $H(t)_{exc}$ energy transfer function is substituted in Equation 15 the theoretical equation of Inokuti–Hirayama for $\bar{\rho}(t)$ is obtained.[8]

2.4. EET processes in inclusion compounds

The lattice of an inclusion compound differs significantly from that of a substitutional mixed crystal, as shown in Fig. 1. In fact, in an inclusion compound only two kinds of molecules are present: host and guests, according to their relative concentrations. However, the filling of the host cavities, (i.e. the sites i where the acceptor or guests are present), is *not* complete in any "*real*" inclusion compound. Hence, the guest molecules are *randomly* distributed over the available host cavity sites i, in about one out of many empty cavities.[9]

The model of EET for an inclusion compound then becomes the following: depending on the relative excitation energy levels of the host and guest molecules, the impinging photons excite the host molecules, which consequently act as excited donors; it will be assumed that the concentration of excited host is $c_{H^*} \ll c_H$, where c_H is the unexcited host concentration in the lattice. However a significant difference exists between an inclusion compound and the previous mixed crystal model: in fact, for the latter[10–12]

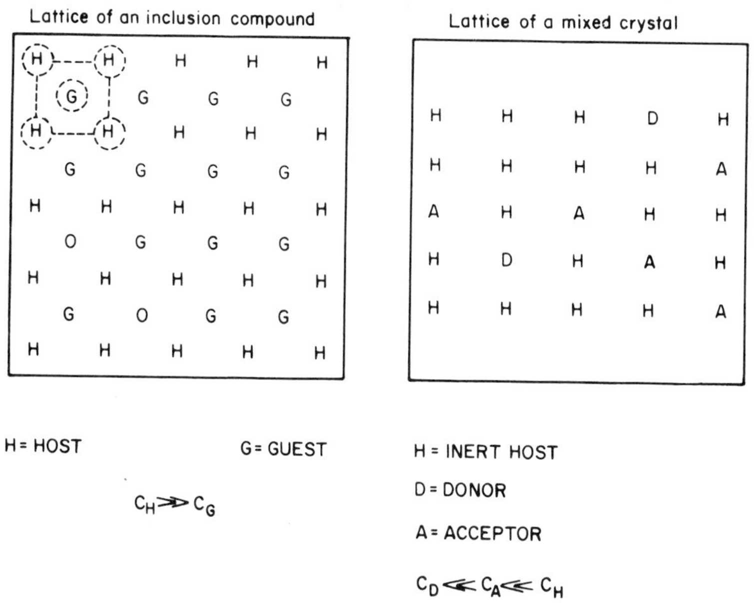

Lattice of an inclusion compound   Lattice of a mixed crystal

H = HOST   G = GUEST   H = INERT HOST

$$c_H \gg c_G$$

D = DONOR

A = ACCEPTOR

$$c_D \lll c_A \lll c_H$$

Fig. 1. Comparisons of the lattice of a mixed crystal and an inclusion compound.

the condition $c_D \ll c_A$ has been assumed in deriving all the equations for EET processes. But, in the case of an inclusion compound the condition $c_D \gg c_A$ holds in any case. Hence, following a pulsed photon absorption, the excited host molecule will either deactivate intramolecularly or transfer nonradiatively its electronic excitation to other hosts or to guests. This fact implies either the occurrence of a *single step* mechanism (i.e. excited host → guest), or of a *multiple step* mechanism, (i.e. excited host–guest migration of excitation energy followed by transfer of excitation to a guest as the final trap). Employing the same formalism of the donor–acceptor pair in mixed crystals, in the case of an inclusion compound, for a *fixed configuration* K of the host and guest components, the deactivation function takes the form

$$\rho(t)_{(\text{clathrate})_K} = \exp\left[-t \sum_A \sum_i w_{iA}(R_{iA})\right]$$
$$= \prod_A \prod_i \exp[-tw_{iA}(R_{iA})] = \prod_A \prod_i E_{iA}(t) \qquad (20)$$

In this *ideal* inclusion compound, the acceptors A are either unexcited host molecules or guests. The sites i correspond to all the sites containing acceptors (hosts and guests) with the exception of the site where the excited host is present. However, a *real* inclusion compound is far removed from a fixed configuration model: many host cavities are empty, a rather *random* distribution of guests through the lattice exists and consequently many host*–guest configurations K have to be taken into account; if their probability of occurrence is P(K) then the *average* deactivation function takes the form

$$\bar{\rho}(t)_{\text{(clathrate)}} = \sum_{\text{(clathrate)}} \rho(t)P(K) = \prod_i \sum_A p_A E_{iA}(t) \tag{21}$$

where, as in the previous equation, A refers to all the possible acceptors, (hosts, H, and guests, G,) and i refers to all the sites where these acceptors are present in the lattice, except the excited host; p_A gives the probability of finding the acceptors A in sites i.

Then, if the total concentration of the hosts and guests is $c = c_H + c_G$, Equation 21 becomes

$$\bar{\rho}(t)_{\text{(clathrate)}} = \prod_i \left[\frac{c_H}{c} E_{Hi}(t) + \frac{c_G}{c} E_{Gi}(t) \right] \tag{22}$$

This function, in the case of multipolar interactions, takes the form

$$\bar{\rho}(t)_{\text{(clathrate)}} = \prod_i \left\{ \frac{c_H}{c} \exp\left[-\frac{t}{\tau_{EM}} \left(\frac{d_H}{R_{Hi}} \right)^s \right] + \frac{c_G}{c} \exp\left[-\frac{t}{\tau_{ET}} \left(\frac{d_G}{R_{Gi}} \right)^s \right] \right\} \tag{23}$$

and, in the case of exchange interactions

$$\bar{\rho}(t)_{\text{(clathrate)}} = \prod_i \left\{ \frac{c_H}{c} \exp\left[-\frac{t}{\tau_{EM}} \exp\left\{ \gamma[d_H - R_{Hi}] \right\} \right] \right.$$
$$\left. + \frac{c_G}{c} \exp\left[-\frac{t}{\tau_{ET}} \exp(\gamma[d_G - R_{Gi}]) \right] \right\} \tag{24}$$

where $1/\tau_{EM}$ and $1/\tau_{ET}$ are the EEM (electronic energy migration) and EET rates, respectively; d_H and d_G are the next-neighbour distance between host and host–guests, respectively; R_{Hi} and R_{Gi} are the hosts and guests present in sites i of the clathrate lattice. Simple analytical approximations are obtained in two cases.[13]

(a) when $1/\tau_{ET} \gg 1/\tau_{EM}$, i.e. when the rate of EET is larger than the EEM, *single steps* H* → G are more favourable and the deactivation function may be approximated, for a multipolar interaction, to

$$\bar{\rho}(t)_{\text{(clathrate)}} = \exp\left[-V_\Delta d^\Delta c_G \Gamma(1 - \Delta/s) \left(\frac{t}{\tau_{ET}} \right)^{\Delta/s} \right] \tag{25}$$

(b) when $1/\tau_{ET} \ll 1/\tau_{EM}$, i.e. when the rate of EET is less 'han the EEM, *multiple steps* H* → H → G are to be accounted for and, analogously, the deactivation function takes the form

$$\bar{\rho}(t)_{(\text{clathrate})} = \exp\left[-V_\Delta d^\Delta c_H \Gamma(1 - \Delta/s) \left(\frac{t}{\tau_{EM}}\right)^{\Delta/s} \right]$$ (26)

(c) when $1/\tau_{ET} \approx 1/\tau_{EM}$ the analytical approximation takes a rather complex formulation.[14,15]

2.5. Anisotropic EET processes in inclusion compounds

In Section 2.1. the energy transfer laws were considered to be independent of the orientations of the donor–acceptor pairs; this model is highly unrealistic in the case of an inclusion compound. For instance, in the case of multipolar interactions

$$w(R_i) = f(R_i) \frac{1}{\tau_0} \left(\frac{d}{R_i}\right)^s$$ (27)

where $f(R_i)$ is a function of the directions of the vectors of the transition multipolar moments of the excited host and the unexcited hosts or guests. In the specific case of a dipolar interaction, $w(R_i)$ takes the form[6]

$$w(R_i) = k_{d-d} = \frac{1}{\tau_0} \left(\frac{d}{R_i}\right)^s = \frac{\alpha}{\tau_{H_0}} \frac{\chi^2}{R_i^6}$$ (28)

where α is a constant related to spectroscopic characteristics of the donor–acceptor pair and χ is a dimensionless geometric parameter determined by the steric distribution of the donor and acceptor transition dipole moments (μ_D and μ_A, respectively). The parameter χ is a function of the angle θ between the vectors μ_D and μ_A, of the angle θ_A between the vector μ_A and R_i and of the angle θ_D between μ_D and R_i, i.e.

$$\chi^2 = (\cos \theta - 3 \cos \theta_D \cos \theta_A)^2$$ (29)

Obviously, the donor D is, in this case, the excited host, and the acceptors A are either unexcited host or guests. In the next section some examples of this anisotropic behaviour of the energy transfer law will be reported.

2.6. Conclusions

It seems useful to summarize in two simple statements the conclusions obtained from the above treatments.
1. The electronic energy transfer processes in an inclusion compound are strongly dependent on the structure of its crystal lattice. This fact implies,

experimentally, that, for instance, photoisomerizations or radical-pair formation inside the cavity of an inclusion compound are strongly dependent on the host lattice crystallographic characteristics. Even if this result is *qualitatively* intuitive, the equations of this section give some *quantitative* estimate of the processes.

2. The electronic energy transfer processes in an inclusion compound are strongly anisotropic. This fact is a direct consequence of the first statement, and stresses the point that the host and guest interactions depend on the relative steric distributions of the two components of an inclusion compound.

3. Experimental aspects of EET processes in mixed crystals and inclusion compounds

The previous section was devoted to the derivation of some equations describing EET processes either in mixed crystals or inclusion compounds. It now seems useful to illustrate briefly some experiments carried out by various authors on EET processes for mixed crystals and inclusion compounds irradiated by gamma or UV photons. Quite obviously, this section is by no means a review of the subject and only a few examples will be reported.

The first example refers to a mixed crystal[16] where fluorene is the principal component and acridine is a random substitutional guest. By UV photon irradiation, a triplet state complex is formed, which consists of a radical-pair involving hydrogen abstraction from a fluorene and hydrogen addition to an acridine molecule, as shown in Fig. 2. The existence of this particular radical-pair, called by the authors "*hetero-excimer*" has been demonstrated by EPR techniques.

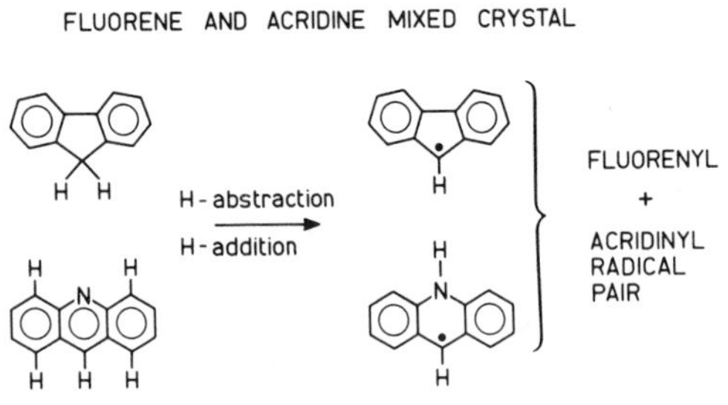

Fig. 2. Formation of the fluorene-acridine radical pair.

In particular it is worth observing that:

(a) light absorption by acridine leads to the formation of a radical-pair (heteroexcimer) in the excited state.

(b) this heteroexcimer is formed among host and guest molecules in fixed lattice sites; by means of the observed hyperfine tensor the authors were able to assess which, out of the three fluorene neighbours to a particular acridine molecule, forms the *heteroexcimer* complex.

(c) complete dissociation occurs of the heteroexcimer into the monomer molecules in the ground state.

In conclusion, the major evidence obtained in this experiment is the fact that the reaction takes place under solid state conditions, with the host and guest molecules in precisely known sites of the crystal lattice.

The next example refers to the adamantane adducts.[17] Actually, photonic irradiation of adamantane containing organic molecules is a routine method to observe radicals by means of the EPR technique. It is thought that the energy of the photons is absorbed by the adamantane and transferred to the guests by an excitonic mechanism: Fig. 3 shows the kind of aminoalkyl

Adamantane adducts with aliphatic amines

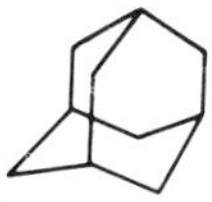

$+ R - NH_2 \rightarrow$ Adducts

Adamantane

Guest amine	Radical structure
Ethylamine	$CH_3 - \overset{\bullet}{C}H - NH_2$
n-Propylamine	$CH_3 - CH_2 - \overset{\bullet}{C}H - NH_2$
Isopropylamine	$(CH_3)_2 - \overset{\bullet}{C} - NH_2$
n-Butylamine	$CH_3 - CH_2 - CH_2 - \overset{\bullet}{C}H - NH_2$
n-Pentylamine	$CH_3 - (CH_2)_3 - \overset{\bullet}{C}H - NH_2$

Fig. 3. Radicals formed in adamantane adducts.

radicals formed in adamantane adducts containing aliphatic amines: quite interestingly they are formed in the *anti*-conformation. In other adamantane adducts containing furan and thiophene derivatives[18] it was shown by EPR techniques that the initially formed adamantyl radicals, (host radicals), at 77 K, were quantitatively replaced by guest radicals on warming: the guest radicals were formed by hydrogen abstraction from carbon atoms adjacent to the heteroatoms.

With inclusion compounds various authors have demonstrated the formation of guest radicals by EPR techniques; in particular, the clathrate formed by Dianin's compound (host) and acetone (guest) was γ-irradiated and studied by EPR: the guest molecules undergo radiolysis with the formation of secondary radicals $CH_3-\dot{C}OH-CH_3$; the same host (Dianin's compound) and nitroparaffins as guests lead to the formation of secondary radicals of the type $R-\dot{N}O_2H$, again with the addition of H atoms.[19] These guest radicals are particularly stable: $C_2H_5\dot{N}O_2H$ remains stable for about 10 min at 300 K.[20]

In the case of the inclusion compound formed by urea and palmitic acid it was observed by EPR that under γ-irradiation alkyl radicals $-CH_2-\dot{C}H-CH_2-$ and the anion $-CH_2-\dot{C}(O^-)OH$ were formed; the latter decomposed above 78 K yielding $-CH_2-\dot{C}H-C=O$, which decayed at room temperature to the radical $-CH_2-\dot{C}H-COOH$.[21]

In conclusion, evidence can be obtained from these examples of the existence of host–guest radical pairs, with hydrogen abstraction from the host moiety. The next two examples demonstrate a particular reaction process, i.e. isomerization of the guests, under photonic irradiation of some inclusion compounds. A β-cyclodextrin-tropolone inclusion compound was irradiated, either as an aqueous dispersion or as a powder, by UV photons; the guest isomerized giving a 64% yield of optically active 1-hydroxy-bicyclo[3.2.0]hepta-3,6-dien-2-one. Analogously, when the guest was 2-methoxytropone the optically active photoisomer 1-methoxybicyclo-[3.2.0]hepta-3,6-dien-2-one was obtained.[22]

Another isomerization process occurring in an inclusion compound under UV irradiation has been recently reported in the inclusion compound of tri-o-thymotide (TOT) containing cis-stilbene or trans-stilbene.[23]

Both inclusion compounds crystallize in the triclinic space group p $\bar{1}$ and have very similar cell constants. Each unit cell of the trans-stilbene-TOT inclusion compound contains four TOT molecules and two stilbene molecules; the stilbenes lie on crystallographic centres of symmetry within two crystallographically independent sausage-like channels. The cis-stilbene-TOT inclusion compound contains partially empty channels. On irradiation by UV photons the cis-stilbene-TOT inclusion compound yields trans-stilbene and some phenanthrene. This result is consistent with a pathway involving photoisomerization of the cis-stilbene within the host lattice. On the contrary the trans-stilbene-TOT inclusion compound is rather stable under UV irradiation. These observations have been rationalized stressing the importance of the coincidence or noncoincidence of molecular symmetry and cavity symmetry. In other words, a centrosymmetric cavity appears to stabilize centrosymmetric guest molecules and favour pathways from non-centrosymmetric reactants to centrosymmetric products.

The next example of EET processes which occur in inclusion compounds refers to the γ-radiolysis and photolysis of the compounds formed by the complex $Ni(4\text{-picoline})_4(NCS)_2$ acting as the host and 1-Br-naphthalene or 2-Br-naphthalene acting as the guests.[24,25] These two halonaphthalenes, which possess very similar energetic properties, luminescence lifetimes, carbon–halogen bond strengths, etc., were chosen as acceptors to emphasize the role of the *relative spatial distributions* of the host and guests in *determining* the kind of EET processes which occur during radiolysis or photolysis. The structures of both inclusion compounds are fully reported in Chapter 3, Volume 1; they were kindly supplied privately by Dr. J. Lipkowsky of the Polish Academy of Sciences, Warsaw.[26]

The only radiolytic or photolytic products observed, for both inclusion compounds, were naphthalene and traces of Br_2 and HBr; no higher halogenated naphthalenes or binaphthyl were detected. However, the naphthalene yield C dehalogenation product) was significantly higher in the case of the 2-Br-naphthalene compound. Figures 4 and 5 show the structures of the two compounds whilst Figs. 6 and 7 show the relative spatial orientations of the host and guest. Table 1 contains the structural parameters of the lattices and the computed distances of the host–host, host–guest and guest–guest interactions for the two inclusion compounds.

This experiment was carried out with the specific purpose of observing the influence of the lattice structure on the EET processes between hosts (donors) and guests (acceptors).

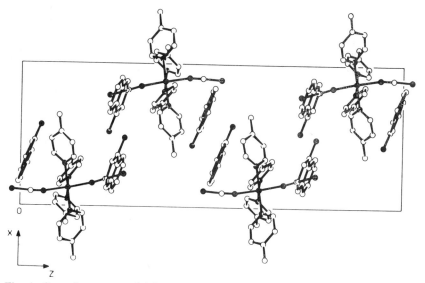

Fig. 4. Crystal structure of 1-Br-naphthalene-Ni(II)complex inclusion compound.

A. Guarino

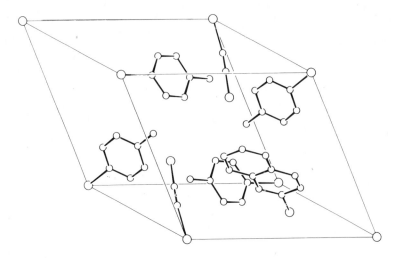

Fig. 5. Crystal structure of 2-Br-naphthalene-Ni(II)complex inclusion compound.

It is worth considering separately the ground state and excited state host–host and host–guest interactions.

(a) Ground state host–host and host–guest interactions. These interactions have been correlated with the crystal structures of the two inclusion compounds, the site symmetries of the host and guest molecules, and their relative spatial orientations: in particular, shifts were observed in their infrared and visible spectra, as reported in Table 1.[27,28]

(b) Excited state host–host and host–guest interactions. Taking into account the energy levels of the excited host and the guest molecules, shown in Fig. 8, the following reactions may rationalize the observed photolytic products. Reactions **7**, **8** and **11** refer to processes that are highly improbable inside the host cavities for steric reasons.

Ni(II)complex + Bromonaphthalene $(S_0) \rightarrow$

$$\text{Ni(II)complex} + \text{Bromonaphthalene } (S_2) \tag{1}$$

$$\text{Bromonaphthalene } (S_2) \rightarrow \text{Bromonaphthalene } (S_1) \tag{2}$$

$$\text{Bromonaphthalene } (S_1) \xrightarrow{\text{ISC}} \text{Bromonaphthalene } (T_{\text{diss}}) \tag{3}$$

$$\text{Bromonaphthalene } (T_{\text{diss}}) \rightarrow \text{Br} \cdot + \text{Naphthyl} \cdot \tag{4}$$

$$\text{Naphthyl} \cdot + \text{R–H} \rightarrow \textit{Naphthalene} + \textit{R} \cdot \tag{5}$$

$$\text{Naphthyl} \cdot + \text{Br} \cdot \rightarrow \textit{Bromonaphthalene} \tag{6}$$

$$\text{Naphthyl} \cdot + \text{Naphthyl} \cdot \rightarrow \textit{Binaphthyl} \tag{7}$$

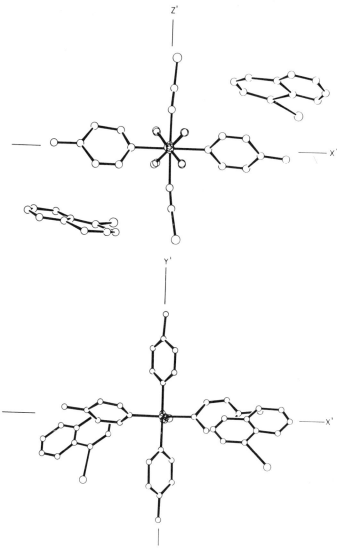

Fig. 6. Spatial orientations of host vs. guest molecules in 1-Br-naphthalene-Ni(II)complex inclusion compound.

Naphthyl· + Bromonaphthalene → *Bromobinaphthyl*   **(8)**

Br· + Br· → *Br₂*   **(9)**

Br· + R–H → *HBr* + R·   **(10)**

Br· − Bromonaphthalene → *Dibromonaphthalene*   **(11)**

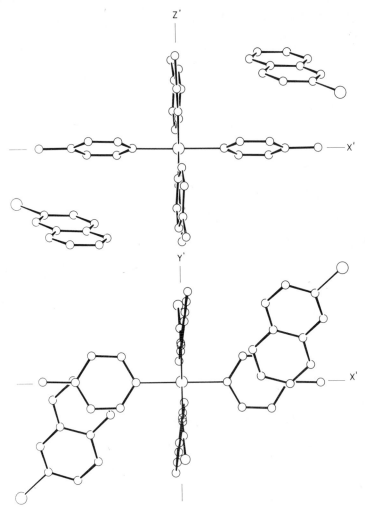

Fig. 7. Spatial orientations of host vs. guest molecules in 2-Br-naphthalene-Ni(II)complex inclusion compound.

In particular, triplet naphthyl and bromine radicals are formed through intersystem crossing (ISC) from S_1 (π, π^*) states (reaction **3**) to dissociative triplet states $T_{diss(\pi, \sigma^*)}$ localized on the C–Br bonds.[29] The bromine radicals may either diffuse up to the crystal surface, yielding Br_2 and/or HBr, or recombine with naphthyl according to reaction **6**.

Some experiments were carried out at 77 K, where almost no bromine diffusion occurs: under these conditions the *same yield* of naphthalene was

Table 1a. The structural parameters of the lattices for, (I) Ni(II)-complex + 1-Br-naphthalene inclusion compound (Space group: Monoclinic $P2_1/c$) and (II) Ni(II)-complex + 2-Br-naphthalene inclusion compound (Space group: triclinic, $P\bar{1}$)

	Molecules per unit cell		Lattice parameters $(\text{Å}, °)$					
	Host	Guest	a	b	c	α	β	γ
I	4	8	11.88	11.82	32.79		102.0	
II	1	1	11.30	9.54	11.76	115.9	81.7	109.6

Table 1b. Nearest-neighbour interactions for, (I) Ni(II)-complex + 1-Br-naphthalene inclusion compound (Space group: Monoclinic $P2_1/c$) and (II) Ni(II)-complex + 2-Br-naphthalene inclusion compound, (Space group: triclinic, $P\bar{1}$)

	Host–host		Host–guest[a]			Guest–guest[a]		
	$R(\text{Å})$	χ	$R(\text{Å})$	χ_L	χ_S	$R(\text{Å})$	χ_L	χ_S
I	6.05	0.044	4.29	0.04	0.31	7.64	5.16	4.67
II	5.06	0.084	4.05	0.59	2.20	5.80	0.81	1.48

Guest I.R. band shifts $\gamma(C–H)$, $\Delta\bar{\nu}$ (cm^{-1})	Host visible band shift $Ni^{2+}(^3A_{2g} \rightarrow {}^3T_{1g})$, $\Delta\bar{\nu}$ (cm^{-1})
12	128
20	60

[a] χ_L calculated for the transition dipole moment through the long axis of Br-naphthalene; χ_S calculated through the short axis of the same molecules.

observed for *both* inclusion compounds: i.e. reaction **6** dominates over reaction **5**. Exactly the opposite occurs at room temperatute, with a significant difference in the naphthalene yield for the two compounds, i.e. a ratio of 16:1 of naphthalene yield for the 2-Br-naphthalene compound. The irradiation by UV photons of 1-bromo or 2-bromonaphthalene, *in solution*, where orientational and steric distribution plays no role, gave the *same* yield of naphthalene.

Hence, taking also into account the fact that these two guests were specifically chosen because of their similar energetic characteristics as shown in Table 1, in order to explain the rather large yield difference in dehalogenation at room temperature, the density and spatial distribution of the host and guest moieties should strongly influence the EET processes for the two inclusion compounds.

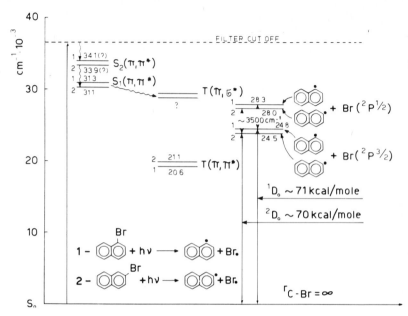

Fig. 8. Energy diagram for the halonaphthalenes-Ni(II)complex inclusion compounds.

The number N_{g1} of 1-Br-naphthalene molecules per cm^3 of inclusion compound is about 1.7×10^{21}; analogously N_{g2} is about 0.8×10^{21} for 2-Br-naphthalene; i.e. $N_{g1} > N_{g2}$. If the experimental packing of these inclusion compounds is taken into account then N_{g2} is even smaller: i.e. $N_{g1} \gg N_{g2}$. Hence, if the same steady state population of excited guest molecules is formed by EET processes for both compounds, i.e. $N_{g1}^* \approx N_{g2}^*$, then for the large density of 1-Br-naphthalene molecules in its inclusion compound, $(N_{g1} \gg N_{g2})$, the probability of naphthyl and bromine radical recombination should be larger in the case of the 1-bromo-naphthalene compound, with a consequent lower yield of the dehalogenation product, naphthalene, as was observed experimentally. However, since $N_{g2} \ll N_{g1}$, in order to fulfil the condition $N_{g2}^* \approx N_{g1}^*$ it is necessary to assume that the EET rate from the excited hosts to the guests is *much faster* in the case of the 2-Br-naphthalene than is the case of the 1-Br-naphthalene inclusion compound, or $k_{ET2} \gg k_{ET1}$. With the help of the equations reported in Section 2.5, and employing the crystallographic data of Table 1 it is possible to calculate the ratio between the EET rate constants k_{ET1}/k_{ET2} for the two inclusion compounds. Using H* to indicate the excited host (Ni(II)complex), H the unexcited host, and G1 and G2 the guests 1-Br-naphthalene and

2-Br-naphthalene, respectively, then:

$$\frac{k_{ET_1}(H^* \to G2)}{k_{ET_2}(H^* \to G1)} = \frac{R^6_{H-G1}\chi^2_{H-G2}}{R^6_{H-G2}\chi^2_{H-G1}} \simeq 10-20$$

where the first value of the ratio holds when χ_L is used in the calculation (i.e. in the case when the S_1 states of excited Br-naphthalenes are taken into account), and the second value holds when χ_s is employed, (S_2 states of the excited Br-naphthalene). It is worth observing that for the reaction mechanism previously reported, (see reactions 1–11) it has been estimated that excitation occurs to the S_2 singlet states of the halonaphthalenes; then $k_{ET2} \simeq 20\ k_{ET1}$. This ratio correlates with the ratios between electronic energy migration between hosts, EEM, and EET to the guests for the two clathrates. In fact, for the 1-Br-naphthalene inclusion compound:

$$q_1 = \frac{k_{ET}(H^* \to H)}{k_{ET}(H^* \to H)} = \frac{R^6_{H-G}\chi^2_{H-H}}{R^6_{H-H}\chi^2_{H-G}} \simeq 0.16$$

In the case of the 2-Br-naphthalene inclusion compound:

$$q_2 = \frac{k_{ET}(H^* \to H)}{k_{ET}(H^* \to G)} = \frac{R^6_{H-G}\chi^2_{H-H}}{R^6_{H-H}\chi^2_{H-G}} \simeq 0.005$$

All these ratios show that:

(a) the rate of EET from the excited host to a 2-Br-naphthalene guest is approximately 10–20 times *faster* than that from the same excited host to a 1-Br-naphthalene guest.

(b) for the 1-Br-naphthalene inclusion compound a negligible amount of *multistep* host* → host EEM occurs before EET occurs to the guest; $q_1 = 0.16$.

(c) for the 2-Br-naphthalene inclusion compound practically only *single step* host* → guest EET processes occur; $q_2 = 0.005$.

These calculations confirm the experimental results, i.e. a larger yield, at room temperature, of naphthalene for the photolysis of the 2-Br-naphthalene inclusion compound.

A rather unusual pathway of EET processes will be described as a last example in this section: in this case the excited donors are the guests.[30] Many inclusion compounds were prepared using the hosts: TOT, β-cyclo-dextrin, cycloveratryl, deoxycholic acid, and guest methyl-stearate-9,10-^{3}H. As a consequence of the tritium decay, the tritiated guest generated a small amount of parent cations and electrons of about 18 keV of max. energy, which irradiated the whole lattice:

$$(R - T)_{guest} \to R-He^+ + \beta \to R^+ + He$$

The radiation intensity depends in such a case on the specific activity of the guest, and the total dose depends on the time of "storage" of the tritiated inclusion compound. The main reason for preparing the compounds was to afford "protection" of the labelled molecules, acting as guests, from self-destruction or self-radiolysis, by means of absorption of the beta radiation on the inert host molecules. Consequently, with the exception of the unavoidable primary radiolysis, (the formation of R^+ ions), the secondary radiolysis effects (i.e. the reactions of the reactive intermediates, ions and radicals, with other tritiated molecules) can be decreased in two ways:

(a) in an inclusion compound, each labelled guest is imbedded in a "cage" formed by inert hosts: hence the reactive intermediates are prevented from reaching other labelled guest molecules. Compared to the same labelled molecule stored in solution, the diffusion of radicals, ions, etc. is substantially reduced.

(b) secondary radiolysis effects are also reduced by the transfer of the excitation to the inert host moiety: an additional advantage is that the radiolysis labelled fragments cross-link with the surrounding cage molecules. Once the inclusion compound is destroyed in order to recover the stored labelled compound, the radioactive impurities that usually contaminate a labelled molecule stored under vacuum in the solid state or in solution with a solvent, are decreased.

In conclusion, inclusion compound formation may prove of practical interest for storing labelled molecules of high specific activity, provided that good techniques are available to prepare the adducts and simple methods can be used for a quantitative recovery after storage. In the case

Table 2. Self-radiolysis of Me-stearate-9,10-^{3}H in different systems.

System	Time of storage (days)	Absorbed dose (eV g^{-1})	Decomposition (%)	Percent decrease of decomposition in inclusion compound.
Me-stearate-9-10-^{3}H stored under vacuum	71	$1.37 \cdot 10^{21}$	24.90	—
Deoxycholic acid inclusion compound	72	$1.37 \cdot 10^{21}$	19.75	16.95
Tri-o-thymotide inclusion compound	72	$1.37 \cdot 10^{21}$	18.04	27.10
Cycloveratryl inclusion compound	71	$1.37 \cdot 10^{21}$	20.10	19.35
β-Cyclodextrin inclusion compound	101	$1.91 \cdot 10^{21}$	21.82	12.10

of cyclodextrins there is less necessity to free the labelled molecules, when employed for biological studies, owing to the possibility of an enzymatic hydrolysis of the adducts. The experiments described in the work by Guarino *et al.*[30] gave the results shown in Table 2: in particular, the TOT inclusion compound gave the maximum protection of self-radiolysis for the storage of methyl-stearate, 9,10-^{3}H in the solid state under vacuum.

It seems useful to summarize this section: depending on the crystallographic properties of the irradiated clathrates, EET processes lead to the formation of radical pairs between host and guests; these excited state complexes or "excithrates" decay to ground state final products, analogous to the situation which occurs in mixed crystal "heteroexcimers".

4. Host lattice stereocontrol of reaction pathways under EET processes

Many reviews have been devoted to reactions in the solid state, which are useful for synthetic purposes.[31] Syntheses with high regio- and stereo-selectivity by means of processes occurring within a host lattice have been carried out. A systematic examination of the overall potentials of this approach must include ground state host–guest interactions and excited state host–guest interactions. Let us first examine the ground state effects:

(a) Host and achiral guest. In this case the host cavities may determine the reaction pathway provided that a *matching* exists between the guest molecular symmetry and the cavity symmetry. The main effect of inclusion is consequently a *steric regioselectivity*. An example of this effect is shown in Fig. 9: anisole, when chlorinated in solution, gives an *ortho/para* mixture of chloroanisole; if the same molecule is "blocked" within the cage of a α-cyclodextrin, only one isomer is formed.[32] (See Chapter 14, Volume 3.)

(b) Host and prochiral guest. In an achiral host a guest processing a prochiral site, i.e. a single improper rotational symmetry element (for instance, a ketone with non-equivalent groups bonded to the carbonyl group), may react at this prochiral site,[33] by an addition reaction from above or from below its reference plane: the two possible products are chiral but unfortunately they are generated in enantiomerically equivalent quantities: no net stereoselectivity. But, if the same prochiral molecule is present as a guest within a chiral host lattice then, for steric reasons, compared with the reaction for the first case, the products will be generated in enantiomerically unequal quantities: a net *steric stereoselectivity* results. See Fig. 10. Recently Tanaka *et al.*[36] have demonstrated the formation of a 100% optically pure product during the solid phase chlorination of methacrylic acid as a guest in α-cyclodextrin.

Ground-state steric regioselectivity in a clathrate
(Achiral guest)

CH3
O
← α-Cyclodextrin
O
Cl

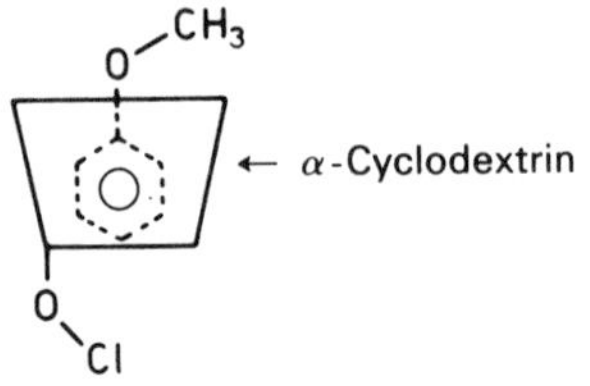

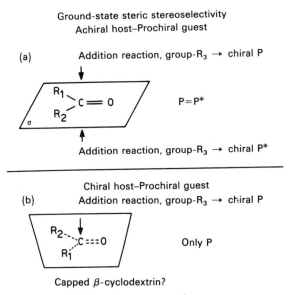

% Anisole guest	Chloroanisole ratio p:o
0	1.48
20	3.43
43	7.42
72	21.6

Fig. 9. Stereocontrol in inclusion compounds.

(c) Chiral host, chiral guest. It is known that some hosts crystallize in chiral forms (for instance, tri-*o*-thymotide). It has been shown that, when chiral guest molecules are included within these chiral cavities a certain guest enantiomeric excess occurs.

Ground-state steric stereoselectivity
Achiral host–Prochiral guest
(a)
Addition reaction, group-R3 → chiral P
R1
C = O
R2
σ
P = P*
Addition reaction, group-R3 → chiral P*
Chiral host–Prochiral guest
(b)
Addition reaction, group-R3 → chiral P
R2
C = O
R1
Only P
Capped β-cyclodextrin?

Fig. 10. Stereoselectivity in inclusion compounds.

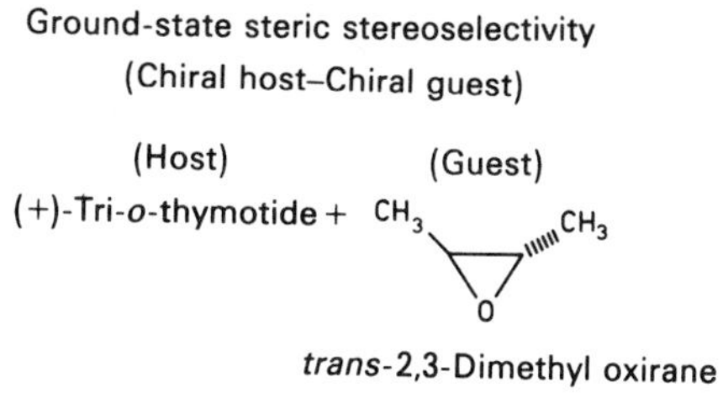

Ground-state steric stereoselectivity
(Chiral host–Chiral guest)

(Host)   (Guest)

(+)-Tri-o-thymotide +   CH$_3$\ ╱CH$_3$
                              ╲╱ ''''
                              │
                              O

trans-2,3-Dimethyl oxirane

Fig. 11. Stereoselectivity in inclusion compounds. 1. Host/guest = 2 : 1; 2. Enantiomeric excess 47%; 3. Guest configuration (S,S)-(−)-.

For instance, trans-2,3-dimethyloxirane is included by (+)-TOT with a 47% excess of one of the two enantiomers; it has also been shown that the excess guest enantiomeric configuration is (S,S)-(−).[34] (See Fig. 11 and Chapter 9, Volume 3.) Analogously, in the case of the intercalate formed by (R)-N-lauroyl-α-(1-naphthyl)ethylamine, (host), and N-trifluoroacetyl-α-phenylethylamine, (chiral guest with R or S configuration) an enantiomeric guest excess is obtained; it has been employed in GLC for the resolution of optical isomers.[35] Other examples are the inclusion compounds between TOT in the M configuration and 2-bromobutane or ethyl-methylsulphoxide (chiral guests) when a significant enantiomeric excess was obtained experimentally.[9]

In all the examples previously reported a direct correlation exists between the host cavity symmetry and the guest molecular symmetry. However, the main application of stereocontrol inside a host lattice in the ground state has been restricted to the separation of optically active molecules. If we now examine excited state effects coming from the irradiation of inclusion compounds, more potentialities of stereocontrol can be envisaged, starting from simple isomerizations up to true regio and/or stereoselective syntheses. The isomerization of cis-stilbene to trans-stilbene inside centrosymmetric TOT cavities, described in Section 3, is an example of this type of excited state effect; also the isomerization of 2-methoxytropone to an optically active photoisomer inside β-cyclodextrin is an example of reaction stereocontrol at the excited state.

The existence of an "heteroexcimer", formed by the radical-pair fluorene-acridine has been reported in Section 3. Analogously, in clathrates irradiated by UV and gamma photons, the existence of radical-pairs between the host and the guest, with hydrogen abstraction from the host molecule, has been shown using EPR techniques for the case of Dianin's compound (host) and nitroparaffin (guest). These short lived complexes, existing only in the excited state, could be given the name of "excithrate", from "excited clathrates", according to the similar nomenclature for "excimers" and "exciplexes".

Consequently, "excithrates" should be the *determining intermediates* of the final products: this fact implies that the "excithrate" symmetry determines the reaction pathway inside an irradiated clathrate, and not simply the cavity symmetry and guest molecular symmetry. However, how the symmetry of these "excithrates" affects the chirality of the final products remains to be studied either theoretically or experimentally; see Figure 12.[37]

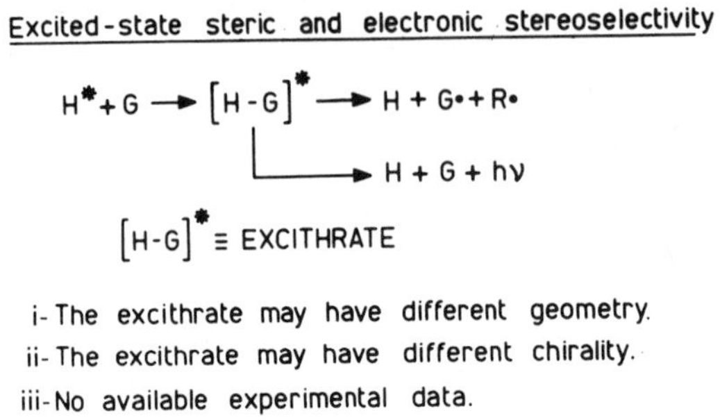

Excited-state steric and electronic stereoselectivity

$$H^*+G \longrightarrow \left[H\text{-}G\right]^* \longrightarrow H + G\bullet + R\bullet$$
$$\longrightarrow H + G + h\nu$$

$$\left[H\text{-}G\right]^* \equiv \text{EXCITHRATE}$$

i- The excithrate may have different geometry.
ii- The excithrate may have different chirality.
iii- No available experimental data.

Fig. 12. Excithrate formation.

Acknowledgments

The author wishes to thank his co-workers for their many contributions to the work carried out in the Institute of Nuclear Chemistry in Rome: Drs. G. Occhiucci, R. Pizzella, E. Possagno, Mr. R. Bassanelli and Mr. A. Patacchiola. Special thanks are due to Dr. J. Lipkowski for the structures supplied privately; to Dr. Cerrini for some computer work on these structures and to Miss M. L. Pompili for her scrupulous typewriting and proofreading. Finally, the author is deeply grateful to Dr. J. E. D. Davies for his editorial help.

References

1. G. O. Phillips, *Energy Transfer in Radiation Processes*, Elsevier, London, 1966.
2. J. B. Birks, *Photophysics of Aromatic Molecules*, Wiley, London, 1970.
3. R. G. Bennet and R. E. Kellog, *Prog. React. Kinet.*, 1967, **4**, 217.
4. A. Blumen and J. Manz, *J. Chem. Phys.*, 1979, **71**, 4694.
5. A. Blumen, *J. Chem. Phys.*, 1970, **72**, 2632.
6. A. Guarino, *J. Photochem.*, 1980, **12**, 147.
7. F. Förster, *Z. Naturforsch., Teil A*, 1949, **4**, 321.
8. M. Inokuti and F. Hirayama, *J. Chem. Phys.*, 1965, **43**, 1978.

9. R. Gerdil and J. Allemand, *Helv. Chim. Acta*, 1980, **63**, 1750.
10. A. Blumen and R. Silbey, *J. Chem. Phys.*, 1979, **70**, 3707.
11. (a) S. I. Golubov and Yu. V. Konobeev, *Phys. Stat. Sol.*, 1975, **70**, 373; (b) A. I. Burshtein, *Sov. JEPT Phys.*, 1972, **35**, 882.
12. K. Godzik and J. Jortner, *J. Chem. Phys.*, 1980, **72**, 4471.
13. A. Blumen and G. Zumofen, *Chem. Phys. Lett.*, 1980, **70**, 387.
14. (a) A. Blumen and G. Zumofen, *J. Chem. Phys.*, 1981, **75**, 893; (b) G. Zumofen and A. Blumen, *Chem. Phys. Lett.*, 1981, **83**, 372.
15. R. Twardowski, J. Kusba and C. Bojarski, *Chem. Phys.*, 1982, **64**, 239.
16. R. Furrer, M. Heinrich, D. Stehlik and H. Zimmermann, *Chem. Phys.*, 1979, **36**, 27.
17. D. E. Wood and R. Lloyd, *J. Chem. Phys.*, 1970, **53**, 3932.
18. G. C. Dissunkes and J. E. Willard, *J. Phys. Chem.*, 1976, **80**, 1435.
19. A. P. Kuleshov and V. I. Trofimov, *Khim. Vys. Energ.*, 1972, **6**, 79.
20. A. P. Kuleshov, V. I. Trofimov and I. I. Chkheidze, *Khim. Vys. Energ.*, 1975, **7**, 82.
21. A. Faucitano, A. Perotti and G. Allara, *J. Phys. Chem.*, 1972, **76**, 801.
22. H. Takeshita, M. Kumamoto and I. Kouno, *Bull. Chem. Soc. Jpn.*, 1980, **53**, 1006.
23. R. Arad-Yellin, S. Brunie, B. S. Green, M. Knossow and G. Tsoucaris, *J. Am. Chem. Soc.*, 1979, **101**, 7529.
24. A. Guarino, G. Occhiucci, E. Possagno and R. Bassanelli, *J. Chem. Soc., Faraday Trans. 1*, 1976, **72**, 1848.
25. A. Guarino, G. Occhiucci and A. Patacchiola, *J. Photochem.*, 1980, **12**, 147.
26. J. Lipkowski, P. Sgarabotto and G. D. Andreetti, *Acta Crystallogr.*, 1980, **B36**, 51.
27. A. Guarino, G. Occhiucci, E. Possagno and R. Bassanelli, *Spectrochim. Acta*, 1977, **33A**, 199.
28. A. Guarino, E. Possagno and R. Bassanelli, *J. Chem. Soc., Faraday Trans. 1*, 1980, **76**, 2003.
29. M. Dvzonik, S. Yang and R. Bersohn, *J. Chem. Phys.*, 1974, **61**, 4408.
30. A. Guarino, R. Pizzella and E. Possagno, *J. Labelled Comp.*, 1965, **1**, 1.
31. M. D. Cohen, *Angew. Chem. Int. Ed. Engl.*, 1975, **14**, 386.
32. (a) R. Breslow and P. Campbell, *J. Am. Chem. Soc.*, 1969, **91**, 3085; (b) R. Breslow, M. F. Czarniecki, J. Emert and H. Hamaguchi, *J. Am. Chem. Soc.*, 1980, **102**, 762.
33. P. Schipper, *Chem. Phys.*, 1980, **45**, 315.
34. R. Arad-Yellin and B. S. Green, *J. Am. Chem. Soc.*, 1980, **102**, 1157.
35. S. Weinstein, L. Leiserowitz and E. Gil-Av, *J. Am. Chem. Soc.*, 1980, **102**, 2768.
36. Y. Tanaka, H. Sakuraba and H. Nakanishi, *J. Chem. Soc., Chem. Commun.*, 1983, 947.
37. A. Guarino, *Chem. Phys.*, in press.

6 · APPLICATIONS OF INCLUSION COMPOUNDS IN CHROMATOGRAPHY*

D. SYBILSKA
Polish Academy of Sciences, Warsaw, Poland

E. SMOLKOVÁ-KEULEMANSOVÁ
Charles University, Prague, Czechoslovakia

1. Introduction

The selective separation of substances, made possible by the formation of inclusion compounds, has also been used in multi-stage chromatographic processes (LLC, GLC, GSC, GPC and TLC). The work published so far indicates that the chromatographic application of inclusion processes allows

* This chapter is dedicated to our great friend, teacher, well known researcher and advocate of clathrate chromatography Prof. Dr Wiktor Kemula on his 80th birthday.

INCLUSION COMPOUNDS III
ISBN 0-12-067103-4

the solution of specific analytical problems and that chromatography also frequently becomes a very effective and important method for the study of inclusion compounds.[1]

Of the many types of inclusion compounds, molecular sieves have found the widest use in chromatographic methods; here the inclusion structure of the host is permanent and independent of the content of guests molecules. The study and use of these compounds have therefore been the subject of many publications, which have not always been connected with the inclusion process. Consequently, in this chapter attention will be paid to other types of inclusion compounds, whose use in chromatography has not yet been systematically treated i.e. to the application of inclusion compounds of some Werner type complexes as well as to the inclusion compounds of urea, thiourea and the cyclodextrins.

The separation of the compounds in a chromatographic column depends on three main factors:

the selectivity of the chromatographic system towards the separated compounds,

the capacity of the stationary phase,

the efficiency of the column.

The proportion of these three factors is given by the basic equation, describing the so called resolution (R) of two compounds:

$$R = \frac{1}{4} \cdot \frac{\alpha - 1}{\alpha} \cdot \frac{k_2'}{k_2' + 1} \cdot N^{1/2} \tag{1}$$

where the relative retention (selectivity factor) $\alpha = K_2/K_1$, where K_2 and K_1 are the equilibrium distribution constants of two substances; the subscripts "2" and "1" represent their elution order; the capacity factor $k_2' = K_2 \cdot V_S/V_M$, where V_S and V_M are the volumes of the stationary and mobile phases in the chromatographic column; the number of the theoretical plates of the column $N = (t_R/\sigma_t)^2$, where t_R is the retention time of the given component and σ_t equals the standard deviation of its elution curve (Gaussian function) in units of time. The separation is usually satisfactory, when $R = 1$; if better separation is required R should be at least 1.5; at $R < 0.8$ the separation is unsatisfactory.

In the early years of modern liquid chromatography it was assumed that it was the efficiency which determined the separation even for systems of relatively poor selectivity. It has turned out however, that the efficiency of a chromatographic column cannot be increased indefinitely, mainly because of practical reasons.

Because of this, more and more attention is being paid to the problems of the selectivity of the chromatographic systems and to possible ways of increasing it, with particular reference to groups of compounds which are difficult to separate. This can be achieved in different ways—either by new, more selective stationary phases, or by making the mobile phase more selective for a given system.

In addition, even in gas chromatography where very complex mixtures can be separated, the problem of getting more informative chromatograms, based on the selective interactions in the chromatographic system has arisen.

One of the most important problems in the case of organic compounds is to find chromatographic systems which can separate isomers, since there are many analytical problems which remain unsolved, or only partly solved.

In this respect inclusion compounds are a very promising group of compounds due to the nature of the interaction of their components. Steric factors are of crucial importance for inclusion compound formation and stability. The most important consequence of this phenomenon is the ability of a host to include in its crystal lattice molecules of a given shape and size. This property results in preparative separations of chemically similar compounds, differing in their shape.

However the selectivity and capacity of a sorbent, which determine the α and k' values, are not sufficient to make a suitable chromatographic system. Any process or a system to be adapted in chromatography should also meet other requirements, connected with the dynamic nature of the method, in particular reversibility and correct kinetics.

Moreover a sorbent, when applied in liquid-chromatography, LC, conditions should be stable and insoluble in the mobile phase solutions and should be thermally stable in gas chromatography, GC, conditions.

2. Inclusion compounds of Werner type MX_2A_4 hosts as sorbents for LC separations

2.1. Historical outline

The separation of aromatic compounds by means of coordination complexes has been long known. Towards the end of the 19th century Hofmann and Küspert[2] obtained the first compound of the complex $NiNH_3(CN)_2$ with benzene; its clathrate nature was found by two crystallographers, Rayner and Powell,[3] as late as 1952. High purity benzene was then obtained in this way.[4] See Chapter 2, Volume 1.

Remarkable progress in this field was made by Schaeffer and his co-workers[5] and by de Radzitzky and Hanotier[6] who discovered two families of complexes, able to selectively include various organic molecules. These two families are of two different formulae:

1. $M(anion)_2(pyridine\ base)_4$ [Ref. 5]

where M is a bivalent metal cation: Ni^{2+}, Co^{2+}, Fe^{2+}, Mn^{2+} or Cu^{2+}, the anion being NCS^-, NCO^-, Cl^-, Br^- or NO_2^-. See Chapter 3, Volume 1.

2. $Ni(NCS)_2\left[NH_2CHR_1\left\langle\!\!\!\!\bigcirc\!\!\!\!\right\rangle\!\!-_{R_2}\right]_4$ [Ref. 6]

where R_1 is a hydrogen atom or a first-order alkyl group (up to C_{10}) and R_2 is a hydrogen atom or an alkyl group, or another group of reactivity low enough not to interfere with the formation of the Werner-complex. See Chapter 4, Volume 1.

Very fine preparative separations of mixtures of various organic isomers have been performed by means of these host lattices. Some of these mixtures are practically inseparable by other methods e.g. *ortho-*, *meta-*, and *para*-disubstituted benzene derivatives and α- and β-substituted naphthalenes.

The scope of the application of these complexes has not been fully determined, though there have been several attempts to use them on a pilot-plant scale.

Kemula and Sybilska found in 1960,[7] that some of these complexes can be used as stationary phases for the selective separation of various compounds under dynamic chromatographic conditions.

2.2. Preparation of the clathrate chromatographic systems

Three complex compounds, which can be used as chromatographic sorbents are now known, or more exactly—for these three complexes the conditions which are necessary to make them "chromatographically active" have been found. They are

$Ni(NCS)_2(4\text{-MePy})_4{}^*$

$Co(NCS)_2(4\text{-MePy})_4$

$Fe(NCS)_2(4\text{-MePy})_4$

The column fillings made of these compounds are of similar character with regard to their capacity and selectivity towards the same groups of compounds. However they differ in solubility, liability to ligand exchange

* 4-MePy = 4-Methylpyridine (γ-Picoline).

and oxidation. The complexes of Co and Fe are less stable than that of Ni and this is why the latter has been chosen as the model compound to study the selectivity, capacity and the nature of the chromatographic separation process.

In particular it should be explained that to make a sorbent chromatographically active it is necessary to use an inclusion compound of the β-polymorph of formula:

$$\beta\text{-}Ni(NCS)_2(4\text{-}MePy)_4 \cdot G$$

where G is a mixture of a pyridine base, mainly 4-MePy and of an aliphatic solvent. The composition, structure and the chromatographic properties of such sorbents depend on the composition of the corresponding mobile phase. To obtain a chromatographic sorbent it is necessary to let the α-polymorph of the empty host lattice $\alpha\text{-}Ni(NCS)_2(4\text{-}MePy)_4$ or the inclusion compound $\beta\text{-}Ni(NCS)_2(4\text{-}MePy)_4 \cdot (4\text{-}MePy)_{0.7}$ equilibrate with the mobile phase of a selected composition, at the constant ratio of the sorbent mass to the mobile phase volume. Some rules of selection of the right composition of the mobile phase solution which would be appropriate for a given mixture of compounds will be discussed later. The detailed directions for obtaining chromatographic sorbents of the above type have already been published.[8-10] One of the most widely used is the following:

2.2.1. *Preparation of the basic sorbent*[10]

$50\ cm^3$ of an aqueous 40% solution of 4-MePy (0.21 mole) is added slowly, i.e. over a period of 3 h, into 500 ml of an aqueous solution of composition: 5.94 g (0.025 mole) $NiCl_2.6H_2O$ and 10.85 g (0.11 mole) KSCN, with continuous stirring. The total volume is then reduced to 150 ml by decantation. 10 ml of pure 4-MePy is added and the mixture is allowed to stand for 48 h. The precipitate is then filtered and dried in air at room temperature.

In this way a very permeable sorbent of formula $\beta\text{-}Ni(NCS)_2(4\text{-}MePy)_4.(4\text{-}MePy)_{0.7}$ is obtained. Its composition can be checked by means of titration.

2.2.2. *Mobile phase solutions*

A mobile phase in clathrate systems is composed of an aqueous solution of an organic aliphatic solvent miscible with water, such as methyl alcohol, acetone, formamide, some 4-MePy, NCS^- ions, and if necessary other aromatic bases or other aromatic compounds. The role of these components in controlling the chromatographic properties will be discussed in a later section.

It should be noted here that the solubility of $Ni(NCS)_2\ (4\text{-}MePy)_4$ in water is rather high for chromatographic requirements and even higher in the organic solvents used. For example the solubility (in g per 100 g of solution) of the complex is 0.093 in water; 2.165 in methanol and 7.125 in acetone.[11,12] To stop the dissociation of the complex and to reduce its solubility at least tenfold it is necessary to add NCS^- ions and 4-MePy to the mobile phase at concentrations of 0.1 and 1.0 mol dm^{-3} respectively.

Before being packed into a chromatographic column both the mobile and stationary phases should be equilibrated with each other for 24 hours and the ratio of the sorbent mass to the mobile phase volume should be constant within the entire set of experiments.

2.2.3. Columns
Owing to the great selectivity of the inclusion compounds the dimensions of the chromatographic columns can be reduced. Glass columns of inner diameter 4–5 mm and length 20–60 mm are filled with the suspension of the sorbent (0.2–1.0 g; particle size 0.2–10 μm) in the mobile phase, at a pressure of 2 kg m^{-2}, being slightly higher than the working pressure of the column.

2.3. Examples of analytical separations

Examples of chromatographic separations have been selected which illustrate the most characteristic features, such as selectivity and the capacity of the sorbents in order to evaluate their applicability in analytical liquid chromatography.

All the examples given below use the basic sorbent β-Ni(NCS)$_2$(4-MePy)$_4$.(4-MePy)$_{0.7}$, unless stated otherwise.

The samples injected onto the columns were 0.2–10 μ dm^3; with a concentration of 0.001–0.02 mol dm^{-3} in the mobile phase solutions.

2.3.1. Benzene derivatives
Figures 1, 2 and 3 show three examples of the separation of di-substituted benzene derivatives: the dinitrobenzenes, nitrotoluenes and nitroanisoles.[13] The values of the dipole moments for these compounds display the following relationships:

ortho > para for dinitrobenzenes

ortho < para for nitrotoluenes

ortho ≈ para for nitroanisoles

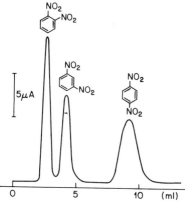

Fig. 1. Elution curve of a mixture of dinitrobenzenes. Working conditions; column, clathrate 0.3 g, dimensions 26×5 mm i.d.; mobile phase, 0.1 mol dm^{-3} NH$_4$SCN, 0.2 mol dm^{-3} 4-MePy, 30% (v/v) methanol in water; flow rate, 8.0 ml h^{-1}; d.c. polarographic detection at constant potential -1.2 V (vs. NCE).

The elution order of the isomers of all these groups of compounds is the same: namely *ortho*-, *meta*- and *para*-. About 30 combinations have been examined and none has deviated from this rule. This order is found regardless of the nature of the functional groups and the dipole moment of the molecules and is absolutely specific for inclusion chromatography.

Figure 4 shows the separation of ^{82}Br labelled bromobenzenes formed by the neutron irradiation of bromobenzene. In this case Siekierski and

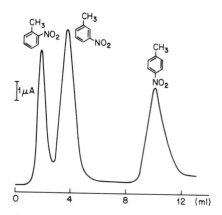

Fig. 2. Elution curve of a mixture of nitrotoluenes. Working conditions; column, clathrate 0.3 g, dimensions 26×5 mm i.d.; mobile phase, 0.1 mol dm^{-3} NH$_4$SCN, 1.0 mol dm^{-3} 4-MePy, 40% (v/v) dimethylformamide in water; flow rate, 8.0 ml h^{-1}; d.c. polarographic detection at constant potential -1.2 V (vs. NCE).

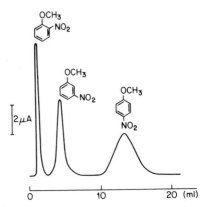

Fig. 3. Elution curve of a mixture of nitroanisoles. Working conditions; column, clathrate 0.2 g, dimensions 20×5 mm i.d.; Mobile phase, $0.2 \, \text{mol dm}^{-3}$ NH$_4$SCN, $0.3 \, \text{mol dm}^{-3}$ 4-MePy, 50% (v/v) methanol in water; flow rate, $10 \, \text{ml h}^{-1}$; d.c. polarographic detection at constant potential -1.2 V (vs. NCE).

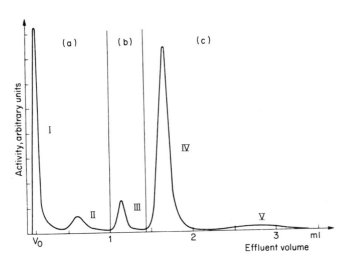

Fig. 4. The course of the elution of the active ^{82}Br products formed during the neutron irradiation of bromobenzene. Peaks: I inorganic bromine; II o-dibromobenzene; III m-dibromobenzene; IV bromobenzene; V p-dibromobenzene. Working conditions; column, clathrate 0.1 g, dimensions 18×3 mm i.d.; mobile phase, $0.2 \, \text{mol dm}^{-3}$ NH$_4$SCN and a, 3%; b, 4% and c, 60% of 4-MePy in 70% aqueous ethanol; flow rate, $0.5 \, \text{ml cm}^{-2} \, \text{s}^{-1}$. Detection: the 5 drop fractions of the effluent were collected by an automatic collector and their ^{82}Br gamma activity was measured using a well type scintillation counter. Reproduced with permission.

Narbutt[14] used inclusion compounds to identify the products of the neutron irradiation of bromo-derivatives of benzene. Very good separation of bromobenzene and *o*-, *m*-, and *p*-dibromobenzenes was obtained. The elution order is the same as previously, but the affinity of bromobenzenes towards the sorbent is much higher than those of nitro-, methoxy- or methyl derivatives, thus it was necessary to raise the eluting power of the mobile phase by increasing the 4-MePy concentration. This method was then used in the investigation of the chemical effects of neutron capture in bromobenzoic acids[15] and also for the investigation of thermal annealing of ^{82}Br in neutron-irradiated *p*-dibromobenzene.[16]

Figure 5 shows an example of the elution analysis of a mixture of dinitrotoluenes.[17] The elution order in the case of tri-substituted benzene

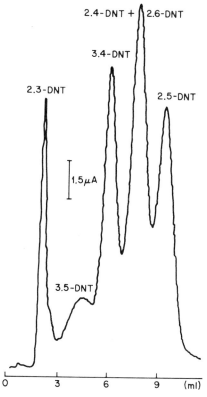

Fig. 5. Elution curve of a mixture of dinitrotoluenes. Working conditions; column, clathrate 1.0 g, dimensions 110 × 4 mm i.d.; mobile phase, 0.2 mol dm^{-3} NH$_4$SCN, 0.1 mol dm^{-3} 4-MePy, 20% (v/v) dimethylformamide in water; flow rate, 8.0 ml h^{-1}; d.c. polarographic detection at constant potential −1.1 V (vs. NCE).

derivatives is not as simple as that of the disubstituted isomers, since there are twice as many possible isomers. On the other hand trisubstituted benzene derivatives, especially dinitro- and trinitro-, are susceptible to various interactions with the components of the mobile phase solutions. These interactions vary from weak dispersion forces to the formation of stable molecular complexes, influencing the values of the distribution coefficients of the investigated molecules. This can be seen by comparing the separations of 2,4- and 2,6-dinitrotoluenes in Figs. 5 and 6. In the example given in Fig. 5 2,4- and 2,6-dinitrotoluenes could not be separated. They can be separated if the solvent in the mobile phase solution is changed e.g. replacing dimethyl-formamide by methyl alcohol, as shown in Fig. 6. In a similar way the separation and determination of the components of crude dinitrotoluene can be carried out.[18]

Under the conditions given in Fig. 5 very small quantities (0.006%) of 2,4,6-trinitrotoluene in the mixture of dinitrotoluenes, were determined with a relative error of ±10% after removing oxygen from the solution reaching the detector.[17]

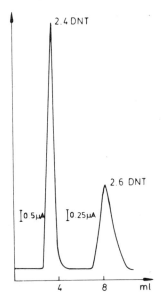

Fig. 6. Elution analysis of a crude dinitrotoluene. Working conditions, column, clathrate 0.35 g, dimensions 30×5 mm i.d.; mobile phase, 0.2 mol dm^{-3} NH$_4$SCN, 0.15 mol dm^{-3} 4-MePy, 20% (v/v) methanol in water; flow rate, 4 ml h^{-1}; d.c. polarographic detection at constant potential -1.1 V (vs. NCE).

2.3.2. Polycyclic aromatic derivatives

Figure 7 shows the elution curve of a mixture of α- and β-methylnaphthalenes and naphthalene.[13] The sorbent itself is extremely β-selective; the column filling affinity towards β-methylnaphthalene is very pronounced, while the α-isomer is hardly retained in the column, which results in a very high value of the selectivity factor $\alpha = K_\beta / K_\alpha$. It should be noted that, contrary to the classical reversed phase RP systems, unsubstituted naphthalene is much more strongly sorbed than substituted naphthalene on this Werner-clathrate column.

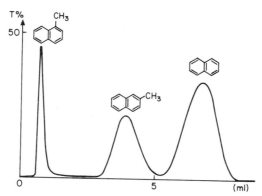

Fig. 7. Elution curve of a mixture of α- and β-methylnaphthalenes and naphthalene. Working conditions; column, clathrate 0.3 g, dimensions 30×5 mm i.d.; mobile phase, 0.4 mol dm^{-3} NH$_4$SCN, 0.43 mol dm^{-3} 4-MePy, 60% (v/v) methanol in water; flow rate, 7.0 ml h^{-1}; spectrophotometric detection at 278 nm.

The elution order given here: α-, then β- is also valid in the case of nitronaphthalenes[19] (Fig. 8), acetylnaphthalenes, naphthols (Fig. 19),[13] naphthylamines, ethylnaphthalenes, chloronaphthalenes, bromonaphthalenes, naphthoic acids (Fig. 18) and others.

The two chromatograms shown in Fig. 8 of a mixture of α- and β-nitronaphthalenes were obtained using two columns of the same dimensions, but filled with different sorbents.[19] The first one [a] was filled with the basic sorbent, i.e. β-Ni(NCS)$_2$(4-MePy)$_4$.(4-MePy)$_{0.7}$ and called "clathrate" in all the examples here; the second one [b] was filled with the sorbent Co(NCS)$_2$(4-MePy)$_4$(4-MePy)$_{0.8}$. These two chromatograms not only prove the high β-selectivity of the sorbents in the separation of mono-substituted naphthalenes but also support the previously mentioned fact (Section 2.2), that both Ni(NCS)$_2$(4-MePy)$_4$ and Co(NCS)$_2$(4-MePy)$_4$-based sorbents have similar chromatographic properties of selectivity, capacity and efficiency.

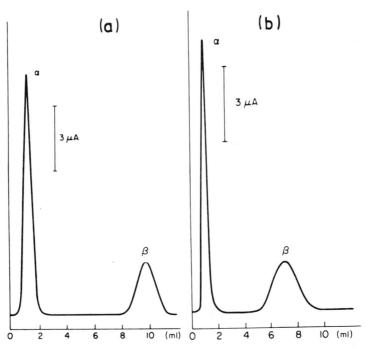

Fig. 8. Elution curves of a mixture of α- and β-nitronaphthalenes. Working conditions; columns, (a) β-Ni(NCS)$_2$(4-MePy)$_4$(4-MePy)$_{0.7}$ − 0.3 g, (b) β-Co(NCS)$_2$(4-MePy)$_4$(4-MePy)$_{0.8}$ − 0.3 g; dimensions, 28 × 5 mm i.d.; mobile phase, 0.2 mol dm^{-3} NH$_4$SCN, 0.3 mol dm^{-3} 4-MePy, 40% (v/v) ethanol in water; flow rate, 6.0 ml h^{-1}; d.c. polarographic detection at constant potential − 1.0 V (vs. mercury pool anode).

It should be noted here that in the case of disubstituted benzene derivatives the high *para*-selectivity takes place both in dynamic chromatographic conditions and in static preparative experiments. In the case of α- and β-substituted naphthalenes in the chromatographic conditions there is β-selectivity, as opposed to the α-selectivity in static preparative experiments, first described by Shaeffer and his co-workers.[5]

This apparent contradiction has not been previously explained because of the lack of detailed investigations of the composition, structure and physicochemical properties of various inclusion compounds formed by Ni(NCS)$_2$(4-MePy)$_4$. It has however now been shown[20–32] that, depending on the guest to be included, its concentration and the procedure used, structures other than β can be formed; the so called γ-modification showing an entirely different selectivity. The detailed physicochemical data concerning these phenomena can be found in Chapter 3, Volume 1.

Figure 9 shows the separation of phenanthrene, carbazole and anthracene[13] and Fig. 20 shows the separation of fluorene and acenaphthene. These two examples demonstrate the potential applications of clathrate sorbents in the separation of some aromatic polycyclic hydrocarbons.

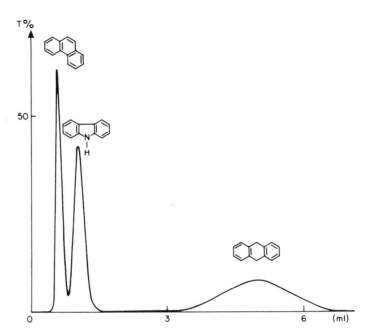

Fig. 9. Elution curve of a mixture of anthracene, carbazole and phenanthrene. Working conditions; column, clathrate 0.35 g, dimensions 34×5 mm i.d.; mobile phase, 0.4 mol dm^{-3} NH_4SCN, 0.45 mol dm^{-3} 4-MePy, 75% (v/v) methanol in water; flow rate; 6.0 ml/h^{-1}; spectrophotometric detection at $\lambda = 292$ nm for phenanthrene and carbazole and 356 nm for anthracene.

It is now possible to formulate some general rules concerning the chromatographic sorption of polycyclic hydrocarbons and their derivatives:
(a) All compounds derived from β-substituted naphthalene (including anthracene and tetracene) are selectively sorbed in a clathrate chromatographic column. An additional aromatic ring weakens the sorption i.e. there is an order of distribution coefficient values: tetracene < anthracene < naphthalene. Benzene molecules are likely to be sorbed most strongly, but it has not yet been proved, since no adequate detection method has been found. This order is the complete opposite to that of other known RP chromatographic systems.

(b) Any α substitution destroys the naphthalene molecule affinity towards the clathrate sorbent. This phenomenon is independent of the nature of the substituent. As a consequence neither fluorene nor phenanthrene, being α-derivatives of naphthalene, are sorbed under chromatographic conditions. Further confirmation of such specific dependence between the chromato-graphic retention data and the geometric features of the molecules to be sorbed is provided by the behaviour of quinoline, i.e. a compound of similar shape to naphthalene, but of different chemical character. Figure 10 shows the chromatogram of a mixture of several nitroquinolines. The derivatives substituted in the 5 and 8 positions, i.e. analogous to α-substituted naph-thalene are not sorbed on the clathrate sorbent, but the isomer substituted in the 6 position, analogous to β-substituted naphthalene, is sorbed.

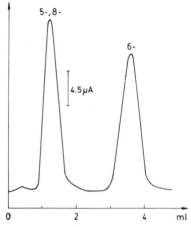

Fig. 10. Elution curve of a mixture of 5-, 6- and 8-nitroquinolines. Working condi-tions; column, clathrate 0.4 g, dimensions 40×5 mm i.d.; mobile phase, 0.2 mol dm^{-3} KSCN. 0.2 mol dm^{-3} 4-MePy, 30% (v/v) methanol in water; flow rate, 4.0 ml h^{-1}; d.c. polarographic detection at constant potential -0.8 V (vs. mercury pool anode).

2.3.3. Cis-trans isomers

Figure 11 shows the chromatogram of a mixture of two geometric azoben-zene isomers.[33] The difference in the affinity of *trans*-azobenzene and *cis*-azobenzene towards the sorbent is so large that it was possible to separate them in columns as short as 13 mm. It is even possible to obtain a satisfactory separation using a 5 mm column, but it is difficult to pack uniformly such short columns and to inject extremely small samples.

Cis-trans methylazobenzenes, methoxyazobenzenes (Fig. 16), and *para* nitrocinnamic acids (Fig. 12) can be also separated in clathrate-filled columns.

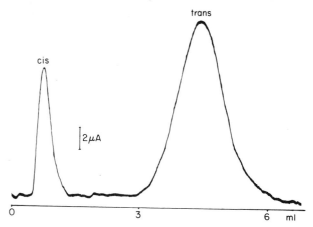

Fig. 11. Elution curve of a mixture of *cis* and *trans* azobenzenes. Working conditions; column, clathrate 0.14 g, dimensions 13 × 5 mm i.d.; mobile phase, 0.4 mol dm^{-3} NH$_4$SCN, 0.45 mol dm^{-3} 4-MePy, 60% (v/v) dimethylformamide in water; d.c. polarographic detection at constant potential −0.8 V (vs. NCE).

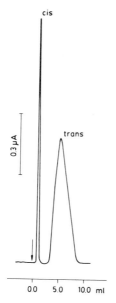

Fig. 12. Elution curve of a mixture of *cis* and *trans* p-nitrocinnamic acids. Working conditions; column, clathrate 0.3 g, dimensions 32 × 5 mm i.d.; mobile phase, 0.1 mol dm^{-3} NH$_4$SCN, 0.84 mol dm^{-3} 4-MePy, 0.56 mol dm^{-3} CH$_3$COOH, 30% (v/v) methanol in water (pH 5.5); flow rate, 8.0 ml h^{-1}; d.c. polarographic detection at constant potential −0.9 V (vs. mercury pool anode).

An example of the separation of *syn-anti* isomers is shown in Fig. 13 which illustrates the chromatogram of a mixture of 5-nitrofurfuraldoximes as well as furazolidone on a very short column.[35] Investigations of the stability and isomerization of 5-nitrofuraldoximes in solution vs. pH, temperature and light irradiation have been carried out using this separation.[36]

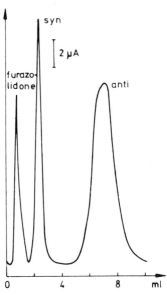

Fig. 13. Elution curve of a mixture of furazolidone [3-(5-nitro-2-furfurylidene-amino)-2-oxazolidynon] and *syn* and *anti* 5-nitrofurfuraldoximes. Working conditions; column, clathrate 0.25 g, dimensions 25×5 mm i.d.; mobile phase, 2.0 mol dm^{-3} NH$_4$SCN, 0.3 mol dm^{-3} 4-MePy, 50% (v/v) dimethylformamide in water; flow rate, 5.0 ml h^{-1}; DC polarographic detection at constant potential −1.1 V (vs. mercury pool anode).

These separations show that the non-planar *syn* and *cis* isomers are not included by the host lattice to the same extent as the planar *anti* and *trans* isomers.

2.3.4. Aliphatic derivatives

Figure 14 shows the separation obtained for four isomers of nitrobutane[37] in 1961. At that time it was thought surprising that a sorbent which could separate aromatic molecules so efficiently was also efficient in separating isomers of aliphatic molecules. Many authors had maintained that only aromatic molecules could be included in Werner-type complexes. On the basis of further investigations one may now conclude that this phenomenon results from the existence of two types of sorption centres in the clathrate β-phase.[38] The detailed explanation for this is given by Lipkowski. (See Chapter 3, Volume 1.)

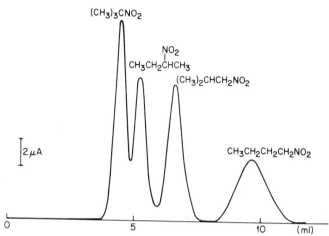

Fig. 14. Elution curve of a mixture of nitrobutanes. Working conditions; column, clathrate 0.8 g, dimensions 60×6 mm i.d.; mobile phase, 0.1 mol dm^{-3} NH$_4$SCN, 0.2 mol dm^{-3} 4-MePy in water; flow rate, 6 ml h^{-1}; d.c. polarographic detection at constant potential -1.2 V (vs. NCE).

With aliphatic compounds such as nitroalkanes, the selectivity factors do not attain values as high as those obtained with aromatic compounds, but they are still sufficiently high for separation. There is a general rule of sequence of elution for aliphatic compounds: the more branched a molecule, the weaker it is sorbed in a clathrate-filled column.

The selectivity of the sorbent towards homologous series is however remarkably lower than that of conventional RP chromatographic systems. Only by combining these two systems, i.e. producing a mixed clathrate-partition (RP) column, was it possible to separate a mixture of eight C_1–C_4 mono-nitroalkanes.[39]

All the examples quoted above lead to the general conclusion that the selectivity of the investigated sorbents is based more on the shape of the molecules, than on their volume. The rules of selectivity which have been found are of general applicability and they justify the application of clathrate sorbents for the chromatographic separation of isomers.

The compounds to be separated can have various functional groups, provided that they do not destroy the host complex, e.g. strongly acidic or alkaline media decompose the $Ni(NCS)_2(4\text{-}MePy)_4$ complex. However aromatic weak acids e.g. o-, m- and p-nitrobenzoic acids, o-, m- and p-nitrocinnamic acids and α-, and β-naphthoic acids have been separated on clathrate columns using a mobile phase solution whose pH value is about 1.5–2.0 units higher than their pK_a values.[34]

2.4. Applicability of the known methods of detection

In modern liquid chromatography the problem of detection has become as important as the methods of separation. Numerous physicochemical methods have been applied to the detection of the separated components at the exit of a chromatographic column. Their usefulness varies with the dependence of the measured signal on the concentration of the substance to be determined, and with the sensitivity, accuracy and scope of the method. It should be noted that the scope of a method is limited not only by the nature of the compounds to be separated, but also by the nature of the chromatographic separating system.

In spite of many achievements in the field of chromatographic detection— this problem has not been satisfactorily solved and improvements are still being sought, both by improving the method of detection and/or by transforming the eluted substances into derivatives which can be easily detected.

The main limitations of the methods of detection used in clathrate chromatography result from the composition of the mobile phase. As mentioned in Section 2.2.2 the mobile phase must contain small amounts of thiocyanate ions and 4-MePy molecules in order to suppress the solubility and to prevent the complex dissociating, due to the relatively low chemical stability of the complex in water and in aqueous solutions of solvents, such as methanol, ethanol, acetone, formamide etc. These free ligands, present in rather high concentration are the main obstacle in detection.

2.4.1. d.c. polarographic detection

In the case of compounds which are reducible or oxidizable on a dropping mercury electrode, d.c. polarography can be used in chromatographic "reversed phase" systems i.e. with a polar mobile phase. Direct current polarographic detection can be used only with polar solvents, when the inert "background" electrolyte is present in the percolating solution, to provide diffusion mass transport control of the redox substance to the electrode surface, usually at a constant pH value. It was introduced to chromatography almost thirty years ago and the author called this method chromato-polarography.[40] This method is characterized by high sensitivity (about 10^{-8} g ml^{-1}), good accuracy, a linear relationship between the current and concentration and the simplicity of automatic recording.

This method of polarographic detection has been used by many authors studying the separation of nitrocompounds and other compounds with polarographically active groups. Ammonium thiocyanate, usually present in the mobile phase solutions of clathrate systems not only stabilizes the complex, but also accomplishes the function of the inert "background" electrolyte and buffering component.

The disadvantage of classical d.c. polarographic detection, coupled with clathrate columns, is its limitation to compounds reducible at potentials no more negative than -1.4 to -1.8 V (vs. NCE), depending on pH. Above this limit, i.e. at higher negative potentials, there is a catalytic wave of hydrogen reduction, originated by protonated 4-MePy ions. Compounds which are either oxidizable or which react with mercury, cannot be detected because of the anodic oxidation current of mercury, which takes place in the presence of thiocyanates.

The examples of chromatograms recorded with d.c. polarographic detection given in Section 2.3 were obtained using classical glass cells of rather large volume (about 250 μ dm^3). Only recently, in the course of development of high performance liquid chromatography, a new type of detector-cell, the FTPD-101, has been designed with a fixed, precisely defined geometry, small volume (10 μ dm^3), and giving reproducible results.[41] It is manufactured in the workshop of the Institute of Physical Chemistry in Warsaw. The examples given in Figs. 15 and 16 illustrate the separation of nitropropanes and p-methoxyazobenzenes recorded using such a detector.[42]

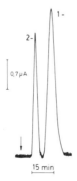

Fig. 15. Elution curve of (1) 2-nitropropane (0.02 mol dm^{-3}) and (2) 1-nitropropane (0.02 mol dm^{-3}). Working conditions; column, clathrate 0.45 g, dimensions 60 × 4 mm i.d.; mobile phase, 0.1 mol dm^{-3} NH$_4$SCN, 0.1 mol dm^{-3} 4-MePy in water; flow rate 10 ml h^{-1}; d.c. polarographic detection with the flow-through FTPD-101 cell at -1.0 V (vs. mercury pool anode).

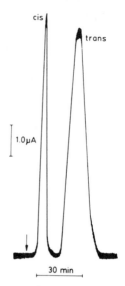

Fig. 16. Elution curve of the *cis* and *trans* p-methoxyazobenzenes mixture. Initial concentration of *trans* p-methoxyazobenzene was 10 mmol dm^{-3}. The *cis*-isomer was generated by photoirradiation of the solution. Working conditions; column, clathrate 0.45 g, dimensions 60×4 mm i.d.; mobile phase, 0.2 mol dm^{-3} NH$_4$SCN, 2.2 mol dm^{-3} 4-MePy, 58% (v/v) methanol in water; d.c. polarographic detection with flow-through FTPD-101 cell at −0.7 V (vs. mercury pool anode).

2.4.2. a.c. polarographic detection

Difficulties in detecting polarographically inactive compounds in clathrate chromatographic systems highlighted the need for a new electrochemical method of detection. It is based upon the changes of the capacity of the electrochemical double-layer on the mercury surface caused by the adsorption of the solute molecules. Therefore a.c. polarographic measurements can be used as a means of detection and determination of eluting substances, which are adsorbed on the surface of a mercury dropping electrode.[43]

Figures 17 and 18 show two examples of chromatographic separation of methylnaphthalenes[44] and naphthoic acids,[34] obtained with the use of a.c. polarography. It should be noted that it would not be possible to use spectrophotometric detection for the methylnaphthalenes mixture used in Fig. 17,[44] because of the presence of acetone in the mobile phase solution.

The fact that in such complex multicomponent solutions it was possible to record detectable effects associated with the adsorption of the eluted molecules on a dropping mercury electrode, initiated further investigation of the basic mechanism of the adsorption process on the mercury surface from the multicomponent solutions.[45]

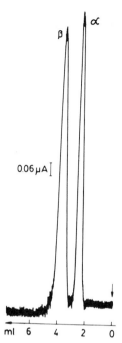

Fig. 17. Elution curve of a mixture of α- and β-methylnaphthalenes. Working conditions; column, clathrate 0.6 g, dimensions 43 × 6 mm i.d.; mobile phase, 1.0 mol dm^{-3} NH$_4$SCN, 0.2 mol dm^{-3} 4-MePy, 45% (v/v) acetone in water; flow rate, 6.0 ml h^{-1}; a.c. polarographic detection at constant potential −1.02 V (vs. NCE).

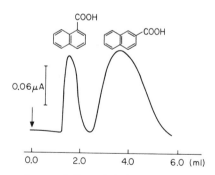

Fig. 18. Elution curve of α- and β-naphthoic acids. Working conditions; column, clathrate 0.6 g, dimensions 57 × 5 mm i.d.; mobile phase, 0.1 mol dm^{-3} NH$_4$SCN, 0.84 mol dm^{-3} 4-MePy, 0.56 mol dm^{-3} CH$_3$COOH in 30% (v/v) aqueous methanol (pH 5.5); flow rate, 8.0 ml h^{-1}; a.c. polarographic detection at constant potential −0.8 V (vs. NCE).

2.4.3. Spectrophotometric detection

The range of application of spectrophotometry, which is a common method in modern liquid chromatography is limited for clathrate systems to wavelengths higher than 270–300 nm, depending on the concentration of 4-MePy and of the other bases in the solution. Thus the currently used single wavelength detector, measuring at the 254 nm mercury line cannot be used with clathrate systems. The accessible range of UV radiation is then small; polycyclic aromatic hydrocarbons and their derivatives are examples of compounds which can be selectively sorbed in clathrate columns and which absorb radiation at wavelengths higher than 270 nm.

An example of the application of spectrophotometric detection in the analysis of α- and β-naphthols[13] is shown in Fig. 19.

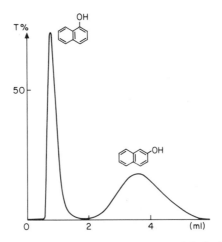

Fig. 19. Elution curve of a mixture of α- and β-naphthols. Working conditions; column, clathrate 0.28 g, dimensions 26 × 5 mm i.d.; mobile phase, 0.4 mol dm⁻³ NH₄SCN, 0.43 mol dm⁻³ 4-MePy, 70% (v/v) methanol in water; flow rate, 5.0 ml h⁻¹; spectrophotometric detection at λ = 279 nm.

Rapid-scanning spectrophotometry[46] was first applied in clathrate chromatography for the analysis of the eluate composition.[47] This new detection system enables the absorption spectra of the components at the exit of the column to be displayed continuously and instantaneously at several fixed wavelengths. This procedure permits the identification of the separated compounds during the actual process of separation. In Figs. 20 and 21 the method is illustrated by the separation of a mixture of fluorene and acenaphthene.

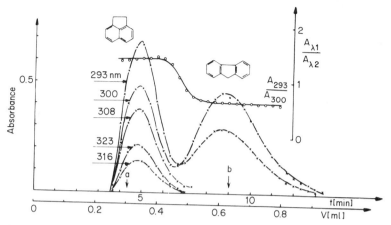

Fig. 20. Elution curves of a mixture of fluorene and acenaphthene recorded by the use of the rapid scanning spectrophotometer (×−×) curves measured as absorbance at constant wavelength; (○−○) ratio $A_{293\,nm}/A_{300\,nm}$. Working conditions; column, clathrate 0.3 g, dimensions 30×5 mm i.d.; mobile phase, 0.4 mol dm^{-3} NH$_4$SCN, 0.43 mol dm^{-3} 4-MePy, 60% (v/v) methanol in water. The significance of the vertical arrows "a" and "b" is explained in Fig. 21.

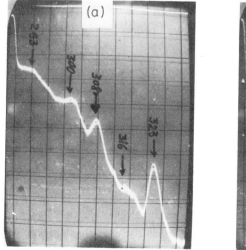
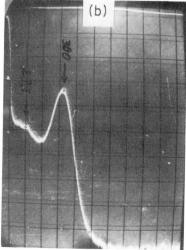

Fig. 21. Spectral transmittance curves in the near UV region, as photographed from the spectromonitor screen, at the moments indicated by the vertical arrows "a" and "b" in Fig. 20. 0% transmittance at the top, 100% at the bottom of the screen.

2.4.4. Radiometric method

This survey of the use of various methods of detection for clathrate chromatography ends with the case of the detection of active ^{82}Br products (Fig. 4) with a scintillation counter.[14-16] This means of detection allowed the monitoring of the separation process in solutions consisting of 60% 4-MePy and 40% ethanol, concentrations which could not be examined by any other detection method.

2.5. Physicochemical conditions for chromatographic activity

The ability to form inclusion compounds with various organic molecules seems to be a common feature of octahedral complexes of the MX_2A_4 type (see Section 2.1 formulae 1 and 2). Thus the number of possible complexes and their inclusion compounds is very large and gives a wide range of selective or specific chromatographic sorbents.

The question which then arises is which compounds are chromatographically active and which ones will provide the desired static and dynamic aspects of the sorption–desorption process?

Experimental investigations seemed to be the proper course of action as a first step towards deducing the properties of chromatographic clathrate sorbents thus leading to the explanation for the mechanism of the chromatographic activity of clathrates.

This section attempts to summarize briefly the latest knowledge on the topic, based upon the investigations made with the model sorbent $Ni(NCS)_2(4\text{-}MePy)_4$.

The determination of the physicochemical conditions for the chromatographic activity of clathrate systems derived from $Ni(NCS)_2(4\text{-}MePy)_4$ requires the solution of a few problems. The most important one is the unequivocal identification of the chromatographically active solid phase i.e. its *in situ* composition and structure. The composition of the mobile phase which is in equilibrium with the solid phase, the forms of the sorbed species and the general features of sorption–desorption processes are the remaining problems.

2.5.1. Solid phase identification

It has been found that chromatographic activity is shown by the tetragonal $I4_1/a$ inclusion compounds of general formula

$$\beta\text{-}Ni(NCS)_2(4\text{-}MePy)_4 \cdot G,$$

where G is a mixture of a pyridine base (4-MePy in most of the cases) and an aliphatic solvent such as: methanol, acetone, formamide, ethylacetate[10] etc. The β-phase exhibits zeolitic properties, permitting guest desorption without the collapse of the host lattice.[25] The properties of the sorbent are determined by the composition of G, which in turn depends on the composition of the liquid mobile phase, being in equilibrium with the clathrate.

The above statements are based upon the results of a detailed examination of the relationship between the chromatographic properties of the sorbents, their composition and structure and the composition of the mobile phase.[10,38,48–51] Figures 22 and 23 show the data, supporting the above statement.

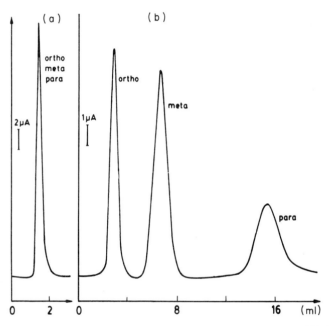

Fig. 22. Elution curves of *o*-, *m*- and *p*-dinitrobenzenes on columns filled with the same quantity of the initial sorbent (0.3 g) with the two mobile phases of different 4-MePy concentration, (a) 0.3% 4-MePy, (b) 1.0% 4-MePy, 0.1 mol dm^{-3} NH$_4$SCN in the aqueous 30% methanol. Column dimensions 26 × 5 mm i.d.; d.c. polarographic detection at the constant potential −1.1 V (vs. mercury pool anode).

In Fig. 22 the effect of the concentration of 4-MePy in the eluent on the separation is illustrated by presenting two limiting cases: (a) no separation and (b) the optimum separation.[10] In case (a) the 4-MePy concentration was 0.3%, the three isomers of dinitrobenzene, when injected onto the column separately, were eluted in the same elution time at the volume corresponding to the free volume of the column. Curve (b) is a chromatogram of a mixture of *o*-, *m*- and *p*-dinitrobenzenes at a 4-MePy concentration in the mobile phase solution of 1.0%. Under these conditions all three isomers were eluted separately. In both experiments the column was filled with 0.3 g of the basic clathrate of the tetragonal form $I4_1/a$ and formula found analytically to be β-$Ni(NCS)_2(4\text{-}MePy)_4.(4\text{-}MePy)_{0.7}$. At 0.3% or lower concentrations of 4-MePy the clathrate appeared to be unstable and it recrystallized into the non-clathrate "α"-monoclinic form $P2_1/c$ of the complex[52] which has no chromatographic activity.

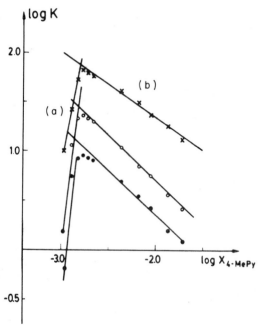

Fig. 23. The log K values of *ortho* (●), *meta* (○) and *para* (×) dinitrobenzenes plotted against log $X_{4\text{-}MePy}$, where $X_{4\text{-}MePy}$ is the molar fraction of 4-MePy in the solution of 0.1 mol dm^{-3} NH_4SCN in aqueous 30% (v/v) methanol.

Figure 23 shows the distribution coefficient (K) of the separated compounds plotted against the 4-MePy concentration in the mobile phase solution, having been equilibrated with the basic sorbent.

The results of the chemical and phase analyses show, that in the concentration range from 0.0 to 0.3% of 4-MePy in the solution, i.e. in the range of no chromatographic activity, the solid phases contain no clathrate. Its main component is the non-clathrate modification α-$Ni(NCS)_2(4\text{-}MePy)_4$.

For solutions with a 4-MePy concentration higher than 0.6%, i.e. in the range of chromatographic activity, the only component of the solid phase is β-Ni(NCS)$_2$(4-MePy)$_4$.G, where G is a mixture of 4-MePy and methanol.[10]

It was possible to evaluate the equilibrium concentration C_R, of 4-MePy in the solution of 0.1 mol dm^{-3} NH$_4$SCN in 30% aqueous methanol at the point of phase transformation as lying within the limits: 0.051 mol dm^{-3} < C_R < 0.061 mol dm^{-3}. Application of the simple considerations of Hanotier, Brandli and de Radzitzky[53] using the relationship $(1/C_R) = K_R$ (where K_R is an equilibrium constant of clathration: complex$\rightleftarrows$clathrate) results in the following limits for K_R: 19.6 > K_R > 16.4.[10]

All these data show that the phase transition $\alpha \to \beta$ accompanies the generation of chromatographic activity.

2.5.2. Molecular model of the sorption–desorption process
The elution force of the mobile phase solution is determined mainly by its 4-MePy content. Within the range of 0.8–8.0% 4-MePy (range b in Fig. 23), corresponding to the two phase equilibrium, the values of log K decrease almost linearly as log $X_{4\text{-MePy}}$ increases (where $X_{4\text{-MePy}}$ is the molar fraction of 4-MePy in the mobile phase solution).

Similar plots had formerly been found in ion exchange chromatography and in the processes of a competitive adsorption of polar molecules on polar adsorbents.[54–55]

The straight line dependence

$$\log K = a - \log X_{4\text{-MePy}} \qquad (2)$$

agrees with the molecular exchange model, which can be described as follows:

$$\beta\text{-Ni(NCS)}_2\text{(4-MePy)}_4.\text{(4-MePy)} + p\text{-dinitrobenzene} \rightleftarrows$$

$$\rightleftarrows \beta\text{-Ni(NCS)}_2\text{(4-MePy)}_4.(p\text{-dinitrobenzene}) + 4\text{-MePy}.$$

2.5.3. pH limits for the mobile phase solution
The fact that the concentration of 4-MePy in the mobile phase must exceed the equilibrium value, C_R, of the phase transition $\alpha \rightleftarrows \beta$ also determines the lower limit of pH in the solutions.

The experimental k' values indicate that the practical lower pH limit for chromatography using β-Ni(NCS)$_2$(4-MePy)$_4$.G, is 5.2.[35]

This result was obtained by studying the retention of the nondissociating dinitrobenzenes. From the data given in Table 1 it can be seen that the lower pH limit of chromatographic activity of the system is related to the

Table 1. The dependence of the capacity factors (k') of dinitrobenzenes (DNB) on the total 4-methylpyridine concentration in the mobile phase, the pH being kept constant at 5.2[(a)]

Mobile phase	Total concentration of 4-MePy in the mobile phase mol dm^{-3}	k'			Solid stationary phase
		o-DNB	m-DNB	p-DNB	
a	⩽0.2!	0.0	0.0	0.0	Inactive α-Ni(NCS)$_2$(4-MePy)$_4$
b	0.84	1.4	2.7	10.5	Active sorbent β-Ni(NCS)$_2$(4-MePy)$_4$.G
c	⩾0.84	0 < k' < 1.4	0 < k' < 2.7	0 < k' < 10.5	Active sorbent β-Ni(NCS)$_2$(4-MePy)$_4$.G

[(a)] Mobile phase: 0.1 mol dm^{-3} NH$_4$SCN/4-MePy in 30% aqueous methanol; the pH was kept constant at 5.2 by addition of a suitable amount of acetic acid. Column dimensions: 35 × 5 mm i.d.

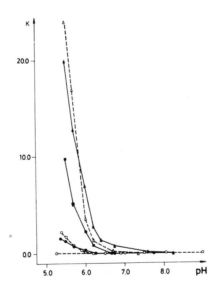

Fig. 24. Plot of distribution coefficients K of o- (○), m- (□) and p- (△) nitrobenzoic acid and o- (●), m- (■) and p- (▲) nitrocinnamic acid (in the *trans* configuration) versus pH of the mobile phase. Mobile phase: 0.1 mol dm^{-3} NH$_4$SCN, 0.2 mol dm^{-3} 4-MePy in 30% (v/v) aqueous methanol. The pH of the solutions were adjusted by adding acetic acid.

stability of the clathrate crystalline phase. Simple calculations of the molar concentration of non protonated 4-MePy molecules in the mobile phase (a), (b) and (c) in Table 1 (pK_a of 4-MePy = 6.02) give values of *ca.* 0.035 mol dm^{-3} in (a) and *ca.* 0.14 mol dm^{-3} in (b). The concentration of 4-MePy molecules in the mobile phase necessary to make Ni(NCS)$_2$(4-MePy) chromatographically active (C_R) has been evaluated as *ca.* 0.06 mol dm^{-3}.

These results as well as those shown in Fig. 24 enable a clearcut decision to be made as to whether the β-form clathrate absorbs molecules only or also ionic species. It seems clear that protonated 4-methylpyridine cannot be exchanged or included in the form of ion-pairs (e.g. with CH$_3$COO$^-$). Figure 24 illustrates the dependence of the equilibrium distribution coefficients (K) of *o*-, *m*- and *p*-nitrobenzoic and *trans o*-, *m*- and *p*-nitrocinnamic acids on the pH of the mobile phase.[34] One can see that at pH > pK_a +2 (corresponding to practically complete dissociation) the distribution coefficients of the acids are equal to zero.

Therefore the clathrate sorbents are able to absorb molecular rather than ionic species. As in partition (or adsorption) reversed phase chromatography, an increase in the distribution coefficients of the acids is observed as the pH decreases. By using the known equation[56,57]

$$K^0 = K(1 + 10^{pH\text{-}pK_a}) \tag{3}$$

the distribution coefficients of the undissociated acid molecules (K^0) can be calculated (Table 2). These values are unobtainable by any other means because the solid phase is unstable at pH < 5.2.

Table 2. *Distribution coefficients (K^0) calculated for undissociated acid molecules, experimental distribution coefficients ($K^{5.5}$) taken at pH 5.5 and dissociation constants (given as pK_a) of nitrobenzoic (NB) and nitrocinnamic (NC) acids and $K^{5.5}$ of dinitrobenzenes (DNB)*

Compound	pK_a	$K^{5.5}$	K^0
o-NB	2.17	0.0	—
m-NB	3.49	1.7	175
p-NB	3.43	24.0	2845
o-NC	4.15	1.3	30
m-NC	4.12	9.8	245
p-NC	4.05	20.0	585
o-DNB	—	6.0	—
m-DNB	—	11.0	—
p-DNB	—	42.0	—

Mobile phase as in Fig. 24.

The range of acidic pH values for mobile phase solutions of the investigated systems is very limited, while alkaline media are more accessible, as the upper pH limit is equal to about 12.

As a consequence of the fact that β-Ni(NCS)$_2$(4-Mepy)$_4$.G is able to absorb molecules but not to exchange ions, the selectivity observed for acids follows the same rules as reported for neutral molecules in Section 2.3. Thus o-, m- and p-nitrobenzoic and o-, m- and p-nitrocinnamic acids show $K_p \geqslant K_m > K_o$, $K_\beta > K_\alpha$ for naphthoic acids, and $K_{trans} > K_{cis}$ for p-nitrocinnamic acids.

2.5.4. General features of the process

Another interesting aspect of these systems is whether the separation process takes place on the surface only or also inside the crystals.

It has been found that the distribution coefficient, given as the retention volume per gram of sorbent is independent of the particle size of the clathrate sorbent (Fig. 25).[9] In other words it is independent of the surface area of the crystals. This suggests that the sorption involves the whole volume of the solid phase, and not only its surface. This conclusion also explains the observed great capacity of clathrate columns and their relatively low efficiency, which must be limited by the mass-transfer effects in the stationary phase.

It has been found that the dilation effects in the crystal lattice of the β-structure caused by clathration are also essential.[25] The so called free volume, that is the volume available for sorption varies between 47 and 90 cm^3 mol^{-1} of clathrate depending on the nature of the guest. These changes influence not only the values of the capacity factor (k') but also to some extent the selectivity and efficiency.[50] One can take advantage of this phenomenon in the design of the chromatographic conditions for separation.

The sorption isotherms, found in both static and chromatographic conditions are linear over a wide range up to concentrations of the order of 10^{-2} mol dm^{-3}, if the same composition of the mobile phase is strictly maintained and if the concentration of 4-MePy is at least 10 times higher than the highest local concentration of the substance to be separated.

3. Inclusion compounds of Werner type MX$_2$A$_4$ hosts as sorbents for GC and TLC separations

The enormous selectivity towards various isomers of the (Fe, Co,)Ni-(NCS)$_2$(4-MePy)$_4$ complexes under dynamic chromatographic conditions

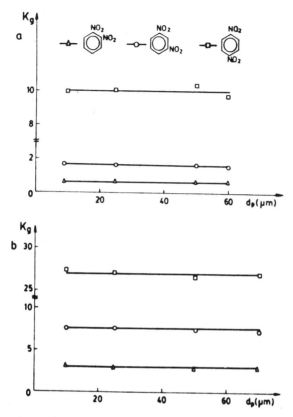

Fig. 25. Dependence of the distribution coefficient (K_g) on particle size (d_p). Column dimensions: 90×5 mm i.d.; flow rate 10.0 ml h^{-1}. The clathrate sorbents were equilibrated with the mobile phases: (a) 7.6% 4-MePy, 27% acetone in water, (b) 3.4% 4-MePy, 27% acetone in water.

has stimulated efforts to utilize them in other chromatographic techniques, i.e. in gas chromatography (GC) and in thin layer chromatography (TLC).

3.1. Gas chromatography separations

The first attempts to apply (Fe, Co)Ni(NCS)$_2$(4-MePy)$_4$ and Ni(NCS)$_2$(α-phenylethylamine)$_4$ were made by Bhattacharyya.[58,59] The author found that (Fe, Co)Ni(NCS)$_2$(4-MePy)$_4$ were the only complexes of those examined to be selective under gas chromatography conditions. Some aromatic and aliphatic hydrocarbons and other compounds were separated. The author, on the assumption that the columns had been made of just stoichiometric complexes, put forward a hypothesis of the presence of some permanent clathration holes in the crystal structure of their non clathrate modifications and believed them to have sorbing properties.

Sybilska and her co-workers, knowing of Lipkowski's results concerning the structure and composition of "active phases"[52] and their properties in LC systems,[13] made attempts to use inclusion compounds of the $Ni(NCS)_2(4-MePy)_4$ complex in gas-chromatography.[60]

The physicochemical conditions necessary for the chromatographic activity of clathrates in LC systems were discussed in Section 2.5. It is thus possible that the β-form clathrates, which are responsible for the chromatographic activity in LC systems, were probably present in the sorbents used by Bhattacharyya.[58,59] These inclusion compounds containing 4-MePy molecules as the guest component are formed easily, while it is difficult to obtain the pure α non-clathrate modification of $Ni(NCS)_2(4-MePy)_4$.[52]

3.1.1. Preparation of the sorbent[60]

The inclusion compound of formula β-$Ni(NCS)_2(4-MePy)_4(4-MePy)_{0.7}$ used frequently in column liquid chromatography is unstable when dry, and in a stream of an inert gas it slowly loses the included 4-MePy molecules,[12] finally forming the chromatographically inactive form even at room temperature. This reaction is rapid at the relatively low temperature of ~70° C, which limits the possibility of its application in GC. Some improvement was obtained by substituting some of the included 4-MePy molecules, by other molecules such as hydroquinone and benzoic acid. Such inclusion compounds were stable and active in gas chromatography conditions up to a temperature of 90° C. The physicochemical reasons for this phenomenon remain unexplained.

The solid support was impregnated with the inclusion compound using the following method: a chloroform solution of the basic inclusion compound β-$Ni(NCS)_2(4-MePy)_4.(4-MePy)_{0.7}$ (0.8 g in 30 ml of $CHCl_3$) was mixed with a hydroquinone solution (0.07 g in 30 ml $CHCl_3$ and 6 ml CH_3OH), and the resulting solution was poured onto 14 g of dried Celite (BDH, 80–120 mesh). The solvents were then evaporated in a rotary vacuum dryer at a temperature not higher than 60° C. The column was conditioned for 4 h at 60° C using a nitrogen flow rate of 50 ml min^{-1}.

3.1.2. Examples of applications

In Fig. 26 the chromatogram of a mixture of *o*-, *m*- and *p*-xylene is presented. The separation of *m*- and *p*-xylene is better than with any other

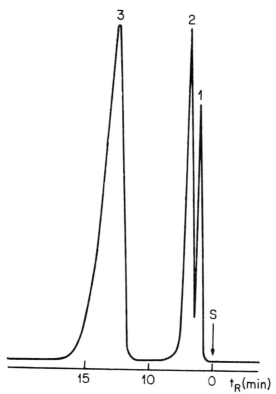

Fig. 26. Chromatogram of a mixture of (1) *o*-, (2) *m*- and (3) *p*-xylenes on a 1 m long column containing β-Ni(NCS)$_2$(4-MePy)$_4$ (benzoic acid) impregnated on Celite with a ratio of 1/14. Sample size 0.8 µl, nitrogen flow 50 ml min^{-1}, column temperature 60° C.

kind of filled columns. Especially valuable is the fact that the sequence of elution of isomers is opposite to that obtained in all other known systems, including columns filled with heavy metal salts,[61] but the same as the sequence obtained using clathrate liquid chromatography (see Section 2.3).

Figure 27 shows the separation of a mixture of *o*- and *p*-diethylbenzene. The sequence in which the isomers are eluted is the same as in the case of the xylenes. In Fig. 28 the chromatogram of a mixture of *m*-xylene and ethylbenzene is presented. Quantitative analysis of ethylbenzene and all the xylene isomer mixtures can be made using this separation.

The mixture of benzene and thiophene is known to be very difficult to separate. Because of the sequence of elution, seen in Fig. 29, the above column is particularly convenient for the determination of the purity of thiophene. It was possible to determine 0.1% of benzene in thiophene.

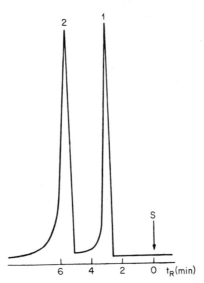

Fig. 27. Chromatogram of a mixture of (1) o- and (2) p-diethylbenzenes on a 1 m long column containing β-Ni(NCS)$_2$(4-MePy)$_4$ (hydroquinone) impregnated on Celite with a ratio of 1/14. Sample size, 0.2 μl, nitrogen flow 45 ml min^{-1}, column temperature 85° C.

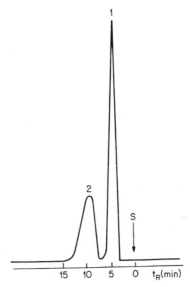

Fig. 28. Chromatogram of a mixture of (1) m-xylene and (2) ethylbenzene on the same column as in Fig. 27. Sample size 0.5 μl, nitrogen flow 20 ml min^{-1}, column temperature 85° C.

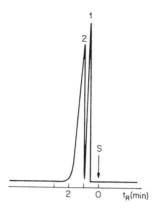

Fig. 29. Chromatogram of a mixture of (1) benzene and (2) thiophene on the same column as in Fig. 27. Sample size 0.2 μl, nitrogen flow 60 ml⁻¹, column temperature 80° C.

The alkanes and cycloalkanes are sorbed less strongly than aromatic hydrocarbons. The sorbent seems to be useful for the group separation of the cycloalkanes from *n*-alkanes.

All the above separations were made using thermal conductivity or flame ionization detectors.

These encouraging results have not been extensively used in practice and the comparatively low thermal stability of the columns seems to be the main reason for this with the columns losing their chromatographic activity within about 50 hours.

A possible solution to this problem is to look for new guests which would increase the thermal stability of the inclusion compounds or for new hosts. In this respect the application of tetracyanocomplexes in gas chromatography has recently been presented by Sopková and her co-workers.[62,63] (See Chapter 7, Volume 3.)

3.2. Thin layer chromatography separation

An attempt to apply inclusion compounds in thin layer chromatography for the separation of *o*-, *m*- and *p*-cresol has been made.[64,65]

3.2.1. Preparation of the sorbent

The sorbent of formula β-Ni(NCS)$_2$(4-MePy)$_4$(4-MePy)$_{0.7}$ and particle size range 0.9–1.8 μm was prepared according to the former procedure (Section 2.2) but with two modifications: the order of pouring of the solutions was reversed and the precipitation process was much faster (about 15 min.).

3.2.2. Separation

The best separation of the cresols was attained on support-plates coated with a 0.2 mm layer of the mixture, containing 25% of the clathrate and 75% gypsum, using a solution of composition: 0.2 mol dm^{-3} NH_4SCN, 0.4 mol dm^{-3} 4-MePy in 50% aqueous methanol for the development of the plates.

The cresols were detected with a 10% aqueous solution of $NaNO_2$ and a 10% aqueous solution of NaOH.

The R_F values found under these conditions were:

$$R_{F_{para}} = 0.2, \qquad R_{F_{ortho}} = 0.38 \quad \text{and} \quad R_{F_{meta}} = 0.53$$

The method described resulted in the separation and identification of all the isomers of cresol in the mixture.[65]

It should be noted that when using TLC to separate cresols some deviation from the rule, concerning the elution order of disubstituted benzene derivatives (see Section 2.3) was found for the first time. The preferential sorption of the *para*-isomer observed in column liquid chromatography is also observed in thin layer chromatography. However $R_{F_{meta}} > R_{F_{ortho}}$ which does not agree with the usual LC order:

$$K_{para} > K_{meta} > K_{ortho}$$

Cresols have not however been separated using column liquid chromatography because of the difficulty in detecting cresols. Nitroanilines however show the usual *para* > *meta* > *ortho* order in both liquid and thin layer chromatography.

4. A summary of the applications of Werner type complexes in chromatography

Because of the specific kind of selectivity and of the nature of the molecular exchange process, clathrate chromatography is an individual kind of chromatography and cannot be classified as any other presently known branch of chromatography.

It is the method of choice for the chromatographic separations of isomers. The selectivity factors (α) reach very high values depending on the differences in the shape of the molecules to be separated. In some cases the method becomes almost specific (e.g. with α- and β-substituted naphthalenes) and this is a disadvantage from the chromatographic point of view.

The very high selectivity is obtained at the expense of the efficiency of the clathrate columns, which is mediocre, because of the very nature of the process of separation on clathrates. But taking into account both of these effects complete separation i.e. $R_s \geqslant 1$ (see equation (1)) can be obtained using extremely short columns.

In contrast to the conventional sorbents or adsorbents the chromatographic properties of inclusion compounds can be extensively controlled by:

(a) modifying the dilation of the crystal lattice, caused by inclusion.

(b) producing various sorption centres, by selecting the composition of the guests (G).

It has not yet been found whether these differences are of a kinetic, or of a structural nature. But the possibility of the fine control of the properties of the sorbents look extremely promising.

A disadvantage of the clathrate-hosts discussed here is their comparatively low chemical and thermal stability, and this probably explains why they have not been extensively used. The search for more stable inclusion compounds showing similar chromatographic properties is an area worthy of more research. The conditions necessary for the existence of chromatographic activity outlined in this chapter, could be used as a guide in the study of other clathrate systems.

5. Application of urea and thiourea in chromatography

Urea and thiourea inclusion compounds belong to that class of compounds in which the host structure is stable only in the presence of the guest molecule. In contrast to the tetragonal lattice of urea alone, urea as the host in inclusion compounds crystallizes hexagonally with channels *ca.* 5 Å in diameter, whereas the thiourea channels are approx. 6 Å in diameter, owing to the presence of the sulphur atom instead of oxygen (see Fig. 30). [See Chapter 2, Volume 2.]

The arrangement and the size of the channels determine the type of possible guests. Urea forms inclusion compounds preferentially with straight chain hydrocarbons, and compounds with side chains or larger substituents are not included; on the other hand, thiourea preferably forms inclusion compounds with branched substances.

Under liquid chromatography conditions, branched fatty acids were separated from unbranched ones,[66] on the basis of the formation of stable inclusion compounds with linear fatty acids, the branched acids being eluted from the column. When thiourea was used as the stationary phase for the separation of a mixture of fluoroderivatives of pentene,[67] the unbranched compounds were eluted after the branched ones.

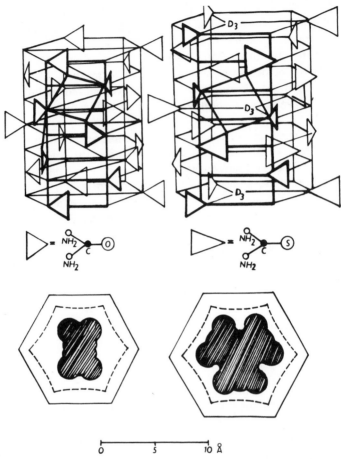

Fig. 30. The hexagonal structures of the urea and thiourea inclusion compounds.

The basic properties of urea inclusion compounds with various types of organic guests have been studied in detail by Schlenk.[68] They have been successfully applied to laboratory and industrial separations of linear and branched compounds in solution, e.g. for the separation of n-alkanes from branched alkanes.

To date, little is known about urea inclusion compound formation with guest molecules in the gaseous phase. To answer this question, gas chromatography has been used as a method that permits a modelling of experimental conditions over a wide range and thus enables the formation and properties of inclusion compounds to be studied.

5.1. Inclusion compounds of urea as stationary phases in gas chromatography

Urea inclusion compounds have been studied as stationary phases,[69,70] to obtain information on the behaviour and properties of these compounds under gas chromatographic conditions, to draw conclusions concerning their stability and selectivity and finally to use them as selective stationary phases for analytical purposes.

5.1.1. *Stability and selectivity of urea inclusion compounds with n-alkanes*

Inclusion compounds of urea with *n*-alkanes were used as model substances, as these materials are readily obtainable, their structure and their chemical properties are known.

The urea inclusion compounds with *n*-octane to *n*-hexadecane were prepared by shaking urea (grain size, 50–80 mesh) with excess alkane mixed with benzene for 24 h. The inclusion compound obtained was checked by the X-ray method and packed into a glass column 1.4 m long, 4 mm i.d.

The sorbate–inclusion compound interaction was studied with inclusion compounds of varying stability. The results are summarized in Tables 3 and 4, giving the specific retention volumes per m² on the adduct, V_{sa}^0, and on urea, V_{su}^0. The ratio of the two values and % V_{sa}^0, corresponding to the effect of the inclusion process, indicate that *n*-alkanes are most influenced; e.g. *n*-butane is eluted after 3-methyl pentane (see Table 3) in spite of the large difference in their boiling points and the specific retention volume of *n*-butane is 26 times larger than on pure urea.

The measurements carried out with different types of alkanes as sorbates have shown that selective adsorption takes place even with relatively stable inclusion compounds such as the ones formed between urea and *n*-tetradecane and *n*-hexadecane. It follows from the data that even a very small amount of dissociation is sufficient to bestow selective properties on an inclusion compound.

Table 3. Retention data for the urea inclusion compounds with n-octane. Column temperature 23°

Sorbate	B.p. (° C)	V_{sa}^0	V_{su}^0	V_{sa}^0/V_{su}^0	% V_{sa}^0
2,2-Dimethylbutane	49.7	0.427	0.437	0.98	—
2,3-Dimethylbutane	58.0	0.541	0.548	0.99	—
3-Methylpentane	63.2	1.11	0.635	1.75	42
n-Butane	−0.5	2.42	0.092	26.3	96
2-Methylpentane	60.2	2.64	0.632	4.18	76

Table 4. Retention data for the urea inclusion compounds with n-decane, n-dodecane, n-tetradecane and n-hexadecane. Column temperature 26°

Sorbate	n-C$_{10}$		n-C$_{12}$		n-C$_{14}$		n-C$_{16}$	
	$\frac{V^0_{sa}}{V^0_{su}}$	% V^0_{sa}	$\frac{V^0_{sa}}{V^0_{su}}$	% V^0_{sa}	$\frac{V^0_{sa}}{V^0_{su}}$	% V^0_{sa}	$\frac{V^0_{sa}}{V^0_{su}}$	% V^0_{sa}
n-Butane	3.85	74	1.71	42	1.10	9	1.02	
n-Pentane	10.00	90	2.62	62	1.42	30	1.08	7
2,2-Dimethylbutane	0.94		0.95		0.97		0.97	
2,3-Dimethylbutane	0.97		0.98		1.00		0.97	
3-Methylpentane	1.11	10	1.04		1.02		1.01	
2-Methylpentane	1.38	28	1.12	11	1.06	6	1.03	
n-Hexane	—	—	5.93	83	2.62	62	1.46	31
2,2,3-Trimethylbutane	0.98		0.98		1.00		0.96	
2,3-Dimethylpentane	1.05	5	1.03		1.05		1.02	
2,4-Dimethylpentane	1.08	7	1.05	5	1.02		1.03	
3-Methylhexane	1.34	25	1.14	12	1.05	5	1.02	
2-Methylhexane	2.67	63	1.46	31	1.13	11	1.06	6
n-Heptane	—	—	—	—	3.52	72	1.93	48
2,2,4-Trimethylpentane	1.01		1.03		0.99		1.02	
2,2,3-Trimethylpentane	1.00		1.04		1.01		1.01	
2,3,3-Trimethylpentane	0.98		1.00		1.01		0.98	
2,3,4-Trimethylpentane	0.99		1.01		1.00		0.99	
2,2-Dimethylhexane	1.13	12	1.07	7	1.02		1.02	
2,5-Dimethylhexane	1.11	10	1.09	9	1.01		1.00	
2,4-Dimethylhexane	1.09	8	1.04		1.02		1.01	
2,3-Dimethylhexane	1.08	7	1.07	6	1.03		1.01	
3,4-Dimethylhexane	1.04		1.02		1.01		1.01	
3,3-Dimethylhexane	1.04		1.04		1.02		1.01	
3-Ethylhexane	1.04		1.02		1.01		1.00	
4-Methylheptane	1.44	30	1.19	16	1.09	8	1.04	
3-Methylheptane	2.26	56	1.46	32	1.19	17	1.08	8
2-Methylheptane	—	—	2.24	55	1.53	35	1.16	14

5.1.2. Factors affecting the selective sorption from the gaseous phase

Partial dissociation of the inclusion compound is a necessary condition for selective sorption from the gaseous phase. Therefore, all factors that affect the dissociation of the inclusion compound, i.e. the amount of guest present, its chain length and vapour pressure as well as the temperature of the inclusion compound, must all be considered.

It follows from the dependence on the guest molecule content that the inclusion compound selectivity decreases with the loss of guest molecule. The effect of the chain length of the guest molecule demonstrates that the inclusion compound stability increases with increasing chain length, leading to a decrease in the retention of all sorbates capable of inclusion.

The influence of the temperature on the selectivity is illustrated by the changes in the elution order of an active sorbate, e.g. n-hexane, with

Table 5. *The influence of temperature on the elution of an active sorbate (n-hexane). Column: urea inclusion compound with n-dodecane*

Elution order	26° C	30° C
1	2,5-Dimethylhexane	2,5-Dimethylhexane
2	**n-Hexane**	3-Ethylhexane
3	3-Ethylhexane	**n-Hexane**
4	4-Methylheptane	4-Methylheptane
5	3-Methylheptane	3-Methylheptane
6	2-Methylheptane	2-Methylheptane

Elution order	35° C	40° C
1	2,5-Dimethylhexane	2,5-Dimethylhexane
2	3-Ethylhexane	3-Ethylhexane
3	4-Methylheptane	4-Methylheptane
4	**n-Hexane**	3-Methylheptane
5	3-Methylheptane	**n-Hexane**
6	2-Methylheptane	2-Methylheptane

variations in the temperature (see Table 5), confirming that the selectivity of the inclusion compound increases with increasing temperature.

An increase in the vapour pressure of the guest molecule leads to a decrease in the dissociation of the inclusion compound and consequently to a decrease in its selectivity. On the other hand, control of the vapour pressure of the appropriate guest molecule, when using unstable inclusion compounds as stationary phases, makes the system more convenient and reproducible for analytical applications, as is shown below (Section 5.1.4).

On the basis of these results, the sorption character can be described. On dissociation, the hexagonal cavities are partially uncovered, i.e. active sites for the guest molecules from the gaseous phase are formed. An increase in the degree of dissociation causes an increase in the inclusion compound selectivity. The selectivity of unstable inclusion compounds is dependent on the guest molecule content and is always greater than that of stable inclusion compounds.

5.1.3. Interaction of urea n-alkane inclusion compounds with different types of sorbate

Gas chromatographic studies with different types of sorbate allow a more detailed evaluation of the effect of the sorbate structure to be made. Studies of the formation of inclusion compounds by interaction from the liquid phase have shown that n-alkanes react selectively from n-hexane upwards and branched alkanes with a single methyl group in the side chain from 2-methyl-decane upwards. The multiple repetition of the equilibrium process under chromatographic column conditions, using a relatively unstable inclusion compound as the stationary phase, permits the determination of the inclusion structure selectivity towards n-alkanes from n-butane upwards and towards branched alkanes from 3-methylpentane upwards. Moreover, it is also possible to differentiate the degree of selectivity for individual sorbates (see Table 4). n-Alkanes are most affected by the inclusion structure among all hydrocarbons with the same number of carbon atoms. They are followed by alkanes with one methyl group in position 2-, 3- and 4-, whose selectivities decrease in the given order. Alkanes with three methyl groups in the side chain do not form inclusion compounds with urea.

Gas chromatographic measurements on the urea n-hexadecane inclusion compound,[71] which is stable up to 120° C, enabled a study of the adduct formation and selectivity with other sorbate types, such as alkenes, alcohols, ketones and aromatic compounds to be carried out. The retention data plotted against the corresponding boiling points of the test substances (Fig. 31) show that sorbates such as n-alkanes, alkenes and alcohols interact

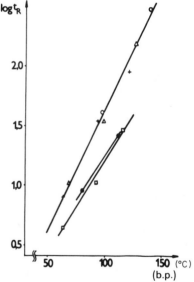

Fig. 31. The relationship between the retention data for sorbates with different structures and their boiling points; △, alkanes; +, alkenes; ○, n-alcohols; ●, aromatics; □, monomethylalkanes.

using the same mechanism as that involved in the formation of their inclusion compounds, while branched alkanes and aromatics exhibit lower retention times than those expected on a non-polar stationary phase, such as *n*-hexadecane.

5.1.4. Analytical applications

From the analytical point of view it is possible to readily separate active sorbates from inactive ones, as well as active sorbates with different selectivities towards the inclusion structure.

The chromatogram of a mixture of 2,3-dimethylbutane and *n*-pentane shown in Fig. 32 is an example of the separation of an inactive and an active sorbate and illustrates the reversed elution order with respect to the boiling points (58 and 36.1° C). The peak of the inactive sorbate is sharp while that of the *n*-alkane is affected by the diffusion character of the sorption (inclusion) process.

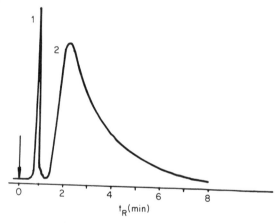

Fig. 32. Chromatogram of a mixture of 2,3-dimethylbutane (1) and *n*-pentane (2). Stationary phase, urea inclusion compound with *n*-decane; column temperature, 26° C.

The broadening and tailing of the peaks caused by the diffusion character of the inclusion process and the dependence of the retention time on the guest molecule content with unstable inclusion compounds are serious drawbacks for analytical applications. However, they can be removed by saturating the carrier gas with the vapours of the appropriate *n*-alkane. Loss of the guest molecule is then prevented and the diffusion character of the active sorbate peaks is suppressed to a certain degree and the peak symmetry is much improved.

Figure 33 depicts an example of the separation of the C_8 isomers, 4-methylheptane and 2-methylheptane, which have identical boiling points but different selectivities towards the inclusion structure, on a column packed with the urea n-octane adduct at 51° C and using the carrier gas saturated with n-octane vapour. It should be pointed out that these isomers can only be separated on highly efficient capillary columns.

0   1   2 t_R(min)

Fig. 33. Chromatogram of a mixture of 4-methylheptane (1) and 2-methylheptane (2). Stationary phase, urea inclusion compound with n-octane; column temperature, 51° C. Mobile phase, nitrogen with n-octane, vapour pressure 15.7 T.

A separation of 2,2,3-trimethylbutane, 2,2,5-trimethylhexane and n-hexane is shown in Fig. 34 where the branched hydrocarbons are eluted before the active n-alkane. The experimental conditions were the same as in the previous separation. An application of this separation would be the checking of the purity of n-alkanes because even very small amounts of impurities would be shown by sharp peaks preceding the main component.

5.2. Urea and thiourea as stationary phases in gas chromatography

Interactions occurring in gas-solid chromatographic systems involving a common adsorbent or support coated with the host molecules have also been studied,[72-75] using a wide-pore silica as a strongly absorbing substance and an inert chromatographic support.

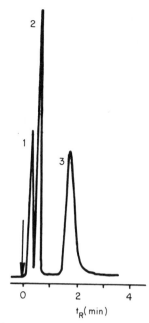

Fig. 34. Chromatogram of a mixture of 2,2,3-trimethylbutane (1), 2,2,5-trimethyl-hexane (2) and *n*-hexane (3). The working parameters are the same as in Fig. 33.

5.2.1. Urea deposited on an adsorbent

A system involving wide-pore silica (Silochrom S-80) coated with urea has been investigated.[72,73] The experiments utilized the properties of sorbates formed by the deposition of a monomolecular layer of a strongly adsorbed substance on a strongly adsorbing surface. The adsorption on a solid sorbent was replaced by adsorption on the monolayer. The specific properties of the adsorbed substances can then be utilized, retaining the main advantage of GSC, namely, rapid partition kinetics permitting rapid analysis.

The measurements were carried out over the temperature range of 40 to 120° C with a great variety of substances (aliphatic and aromatic hydrocarbons, their halogenoderivatives, alcohols, and ketones) that were selected on the basis of their ability to form inclusion compounds. The results obtained on pure silica and on silica coated with urea were compared.[72]

The temperature dependences of the specific retention volume (V_g) for substances capable of forming urea inclusion compounds (e.g. *n*-alkanes, chlorinated hydrocarbons) and for inactive substances (e.g. benzene, toluene) show that these two groups behave differently. While the decrease in the retention of inactive substances is caused by their inability to interact with hydroxyl groups over the whole temperature range, hydrocarbons and chlorinated derivatives exhibit higher retention on the urea-coated silica than on pure silica. This phenomenon can be explained by assuming that the total energy of interaction is increased by a contribution from the specific interaction, due to the formation of inclusion compounds with urea. This contribution increases with the increasing number of carbon atoms in the n-alkane molecules (Fig. 35) and with the increasing number of chlorine atoms in chloroethanes (Fig. 36) and chloroethylenes.

These findings again confirm the effect of the chain length on the formation and stability of the inclusion compounds. In agreement with the behaviour in the liquid phase, the elution order of the tetrachloro derivatives changes.

However, an assumed increase in the thermal stability of urea due to a decrease in the volatility on adsorption on a strongly adsorbing surface has

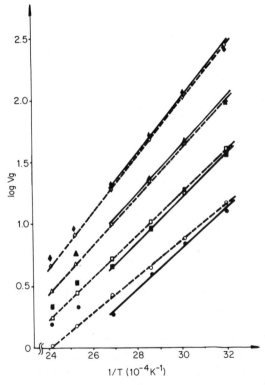

Fig. 35. Temperature dependences of V_g for n-alkanes on Silochrom S-80 (dotted line) and on a monomolecular layer of urea on Silochrom S-80 (solid line) ○, n-pentane; □, n-hexane; △, n-heptane; ◇, n-octane.

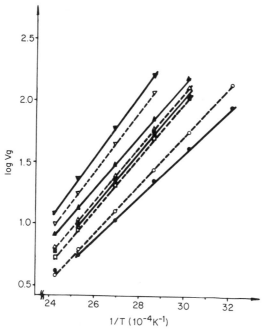

Fig. 36. Temperature dependences of V_g for chlorinated ethanes using the same column as in Fig. 35. ○, 1,2-dichloro-; □, 1,1,2-trichloro-; △, 1,1,1,2-tetrachloro; ▽, 1,1,2,2-tetrachloroethane.

not been observed. Changes in the properties of the column due to the decomposition of urea occur even below the melting point of urea, apparently due to a catalytic effect of the silica acidic sites. Ammonia was detected in the carrier gas at 120° C and for this reason the retention data obtained at higher temperatures deviated from the linear dependence of log V_g vs. $1/T$.

5.2.2. Urea and thiourea deposited on an inert support[72,74,75]

Sorbate-urea and sorbate-thiourea interactions have been examined using chromatographic packings containing 10 to 20 wt.% urea (or thiourea) deposited from a methanolic (ethanolic) solution on Chromosorb W 60/80.

In contrast to pure urea as a stationary phase[69,70] where there was no selectivity toward different sorbates, the phase with urea coated on an inert support permitted a detailed study[74] with various substances including aliphatic and aromatic hydrocarbons, their halogeno derivatives, alcohols, ethers, esters, carbonyl compounds, amines and organic acids, over a temperature range from 40 to 120° C. It was found that urea was lost from the column at temperatures above 120° C, but it was sufficiently thermally stable to be used at column temperatures up to 100° C.

Interesting conclusions can be drawn from the measurements with urea and thiourea deposited on an inert support, using the differences in the retention. However, so far it is impossible to decide whether the inclusion process or the sorbent polarity plays the predominant role. To solve this problem, temperature dependences were obtained for V_g of various sorbates, from which thermodynamic data were calculated[76] using the equation,

$$\Delta H = 2.3 R \ln \alpha \qquad (4)$$

where α is the slope of the log V_g vs. $1/T$ dependence.

The ΔH dependence on the number of carbon atoms for alkanes and alcohols on urea suggest that adsorption plays the predominant role. However, the comparison of the ΔH values published by Schlenk[68] for n-octane and n-decane with the gas chromatographic data suggest that inclusion is the predominant process. (See Table 6.)

Table 6. Comparison of ΔH values (kJ mol^{-1}) for n-alkane urea inclusion compounds

Adduct	$\Delta H_{theor.}$	ΔH_{exp} [68]	ΔH_{exp} [76]
n-octane	39.04	29.98	34.6
n-decane	48.92	38.10	44.38

It follows from the above data that the agreement between the theoretical and experimental ΔH values is better for the values obtained from the chromatographic data in the system of the host and a guest in the gaseous phase than for inclusion from solution. This supports the assumption that inclusion also plays a role in the urea–n-alkane interaction in the gaseous phase and that the interaction mechanism is similar to that of inclusion compound formation from solution.

The ΔH values found for aromatic hydrocarbons demonstrate an energy contribution from the π-electron system of the arene to the interaction with urea, which is e.g. 11.97 kJ mol^{-1} for the cyclohexane–benzene pair whereas the value for less polar thiourea is only 2.6 kJ mol^{-1}. These data suggest that the polar character of the stationary phases plays a role in the interaction.

The selectivity of urea and thiourea towards alcohols is very pronounced. For example methanol (b.p. 64.7° C) exhibits a higher retention on both urea and thiourea than n-octane (b.p. 125.7° C). Among substances with similar boiling points, urea and thiourea differentiate sorbates with active hydrogen from non-polar substances, e.g. *n*-octane, chlorobenzene, 2,4-dimethylheptene and isopentanol which all have boiling points near 130° C exhibit orders of magnitude differences in their V_g values. These data support the assumption that the sorbate-urea (thiourea) interaction is controlled by adsorption.

The type of interaction cannot be found even from the calculated entropy changes. The values obtained are less than zero in all cases, which indicates that the ordering of the system increases. The solution of this question will require further accurate measurements both in dynamic systems, e.g. chromatographic, and in a static arrangement.

5.2.3. Analytical applications

Although it is impossible to draw unambiguous conclusions about the interaction mechanism in the sorbate–urea and thiourea systems, both substances can be used to advantage as stationary phases exhibiting selective properties in certain cases, which suggest that inclusion processes contribute to the overall retention.

An example is the separation of cyclohexane and *n*-hexane on urea deposited on Chromosorb W at 40° C (Fig. 37), where the substances are eluted in the order opposite to the order of their boiling points; this behaviour cannot be attributed to any type of interaction other than inclusion. Another example is the analysis of light petroleum performed under identical conditions (Fig. 38). The separation is poor on urea, (a) but the components can readily be differentiated on the thiourea-coated column (b) because of the formation of inclusion compounds with the branched alkanes that constitute light petroleum. The resolution of a mixture of benzene, chlorobenzene and *o*-dichlorobenzene (Fig. 39) also differs significantly, on urea and on thiourea especially for *o*-dichlorobenzene, which can be explained by a steric hindrance on urea.

In many other analytically useful applications, the separation and retention of individual substances cannot be unambiguously attributed to the formation of inclusion compounds, rather it can be assumed that the separation mechanism depends on the polarity. The rapid kinetics in the GSC system leading to fast analyses and the high symmetry of the peaks obtained, however, demonstrate the advantages of both column packings.

The practical importance can be illustrated by several examples. Figure 40 depicts a difficult separation of isopropanol (b.p. 82.3° C) from *t*-butanol (b.p. 82.4° C) on urea at 70° C. Mixtures of butanols (Fig. 41) can be separated on urea at 80° C within 4 min.

Ketones and ethers are also readily and rapidly separated. A mixture of ketones (Fig. 42) is separated at 70° C within 90 s. The chromatogram in

Fig. 37. Chromatogram of a mixture of cyclohexane (1) and *n*-hexane (2). Stationary phase: urea deposited on Chromosorb W, column temperature 40° C.

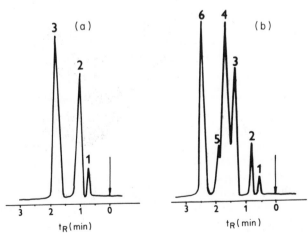

Fig. 38. Analysis of light petroleum on urea (a) and thiourea (b) deposited on Chromosorb W under identical conditions.

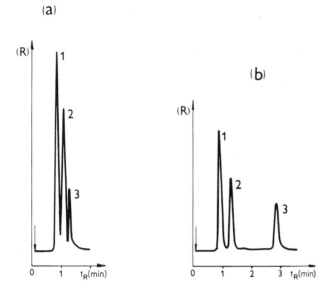

Fig. 39. Chromatograms of a mixture of benzene (1), chlorobenzene (2) and *o*-dichlorobenzene (3) on urea (a) and thiourea (b) deposited on Chromosorb W at 70° C.

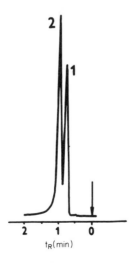

Fig. 40. Separation of isopropanol (1) and *t*-butanol (2) on urea deposited on Chromosorb W at 70° C.

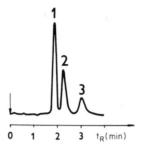

Fig. 41. Chromatogram of a mixture of butanols on urea deposited on Chromosorb W at 80° C; *sec*-butanol (1), isobutanol (2), *n*-butanol (3).

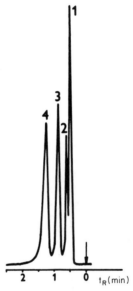

Fig. 42. Chromatogram of a mixture of ketones separated on urea deposited on Chromosorb W at 70° C; acetone (1), methylethylketone (2), methyl-*n*-propylketone (3), methylisobutylketone (4).

Fig. 43 illustrates the rapid analysis of a mixture of halogenobenzenes with different boiling points. Fluorobenzene (b.p. 84.7° C), chlorobenzene (131.7° C) and bromobenzene (156.2° C) are eluted on urea at 70° C in 60 s.

Benzene, toluene and *p*-xylene, and a mixture of benzene, chlorobenzene and *o*-dichlorobenzene were separated on thiourea within 3 minutes at 70° C. A very good separation of a mixture of C_1–C_5 *n*-alcohols, C_3–C_5 isoalcohols, and *sec*-butanol at 80° C is depicted in Fig. 44.

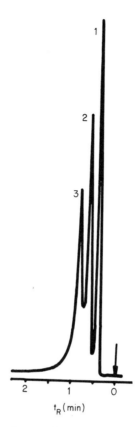

Fig. 43. Chromatogram of a mixture of halogenobenzenes separated using the same conditions as in Fig. 42; fluorobenzene (1), chlorobenzene (2), bromobenzene (3).

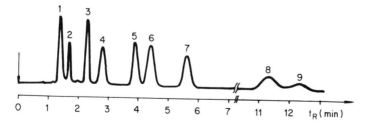

Fig. 44. Chromatogram of a mixture of alcohols separated on thiourea deposited on Chromosorb W at 80° C; methanol (1), ethanol (2), isopropanol (3), *n*-propanol (4), *sec*-butanol (5), isobutanol (6), *n*-butanol (7), isopentanol (8), *n*-pentanol (9).

The chromatogram in Fig. 45 of a pure isoamylalcohol preparation (a commercial product) is also analytically interesting. The substance behaved as an individual chemical species on common chromatographic packings. In the analysis on urea at 80° C, two comparable peaks were obtained, the first probably corresponding to optically active amylalcohol (2-methylbutanol-1) and the second to isoamylalcohol.

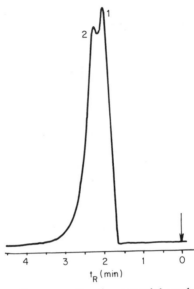

Fig. 45. Chromatogram of a preparation (commercial product) of "pure" isoamylalcohol using urea deposited on Chromosorb W at 80° C (1) probably optically active amylalcohol, (2) probably isoamylalcohol.

It is also possible to successfully perform other analyses of interest in environmental and human health protection. A separation of non-polar substances from alcohols can e.g. be used in environmental monitoring (atmosphere, water), as can the separations of halogenoderivatives, aliphatic and aromatic substances and other pollutants in industrial atmospheres.

5.3. Physicochemical studies of urea inclusion compounds by GC

Gas chromatography can also be used for the study of the physico-chemical properties of inclusion compounds during temperature variations and for determining their phase transitions or decomposition. When the test substance functions as a stationary phase, the retention characteristics are followed under isothermal conditions over a wide temperature range. The measurement is based on the relationship between the log of a retention characteristic and the absolute temperature of the column,

$$\log V_g = A + B/T, \tag{5}$$

where A and B are experimental constants. The linear dependence of log V_g vs. $1/T$ is disturbed on a phase change in the test substance.

This method has been used for the study of the stationary phases in GLC and GSC systems and also for the study of liquid crystals. It has been used by Knight[77] and later by McAddie[78] for the study of urea inclusion compounds. The measurements were carried out using the urea inclusion compound with n-hexadecane dispersed on Chromosorb W,[79] with toluene as the sorbate. The reason for the selection of toluene, which cannot form another inclusion compound with urea, is not quite clear and is not fully justified from the point of view of the model experiment. However, these studies have shown that this procedure can be used to advantage in the study of inclusion compounds.

5.3.1 Urea–n-hexadecane inclusion compound

The urea inclusion compound with n-hexadecane, which is sufficiently stable for measurements at temperatures up to 120° C, was used for monitoring the formation, transition state and decomposition process.[71] The chromatographic column was directly packed with the urea inclusion compound with n-hexadecane (without support), which was prepared by shaking urea (50–80 mesh) with n-hexadecane in benzene for 24 h. The formation of the inclusion compound was verified by X-ray analysis.

Measurements carried out with substances of various structural types and varying abilities to act as guest molecules at temperatures of up to 75° C demonstrated specific sorption properties which can be attributed to the inclusion compound formation mechanism (see Section 5.1). It can be seen (Fig. 46) from the plot of the temperature dependence of the retention data measured over a wide temperature range that benzene and toluene, which do not form inclusion compounds, have very low retention data in the region up to 90° C, where the urea inclusion compound acts as a solid stationary phase, while in the region above 90° C an increase in the retention data reflects decomposition of the n-hexadecane inclusion compound. Above 100° C, the retention data are comparable with those observed for n-hexadecane as a stationary liquid phase. For sorbates such as n-octane, 1-octene and acetone, which can act as guest molecules, the retention data below 90° C are higher than for benzene and toluene. The data above 100° C correspond to the values obtained on n-hexadecane as a stationary phase,

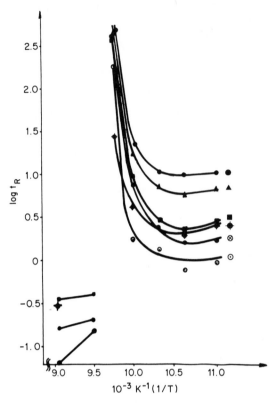

Fig. 46. Temperature dependence of the retention data for various sorbates on the urea inclusion compound with n-hexadecane; ●, octane; ▲ 1-octene; ■, 3 methyl-heptane; ◆, acetone; ⊗, toluene; ⊙ benzene.

obeying the rules of the retention behaviour of solutes in a non-polar solvent. With increasing temperature, the drift and noise increased, due to the decomposition of the inclusion compound and the bleeding of n-hexadecane. Pure urea is finally obtained, which is reflected in the extremely low retention data at 105–110° C, corresponding to the values obtained on a column packed with pure urea. The rapid increase in the retention data in the region from 90° C to 102.5° C indicates the decomposition of the inclusion compound to give urea coated with liquid n-hexadecane. Gas chromatography has also shown that the urea inclusion compounds with n-alkanes are thermally decomposed below the melting point of urea.[71,78,79]

5.3.2. Urea n-hexadecanol inclusion compound

An important part of the study of inclusion compounds, in addition to the study of their formation and decomposition, is a determination of the contents of the guest and host components and determination of the presence of free components. The results can be used both for identification, i.e. for finding whether an inclusion compound was formed, and for the preparation. Gas chromatography enables a complex solution of these problems, as demonstrated e.g. by the results of the study of phase changes of variously prepared urea inclusion compounds with *n*-hexadecanol. X-ray analysis verified in all cases the change of tetragonal urea into the hexagonal inclusion compound. The chromatographic measurements were performed with the inclusion compound alone (without support), over a temperature interval of 40 to 120° C. The behaviour and properties of the inclusion compounds by gas chromatography is shown by the results obtained with an atypical procedure for the inclusion compound preparation. X-ray analysis revealed pronounced inclusion structure; the cetylalcohol content was also similar to the values in other inclusion compounds. However, the log V_g vs. $1/T$ plot (Fig. 47) exhibited a maximum at 50° C, corresponding to the melting point of cetylalcohol. This maximum and the identical temperature dependence for cetylalcohol indicate that free cetylalcohol is present, forming a

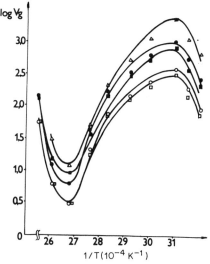

Fig. 47. Temperature dependence of the retention data for various sorbates on the urea inclusion compound with *n*-hexadecanol; ○, heptane; ●, octane; □, isooctane; ■, toluene; △ butanol.

surface layer. The change in the plot slope at 100° C is caused by decomposition of the inclusion compound. It can be concluded from the shape of the temperature dependence that free hexadecanol was present in the sample, in addition to the inclusion compound confirmed by the X-ray analysis. These results show that chromatographic data can be used to describe the properties and behaviour of inclusion compounds, which can then be used to advantage in their preparation, characterization and identification.

5.4. Concluding remarks

It follows from the study of urea, thiourea and their inclusion compounds as stationary phases that gas chromatography can be used to study weak interactions such as the formation of inclusion compounds from the gaseous phase and to use this process for selective separations and rapid analysis. Simple chromatographic experiments yield data on the preparation, behaviour and properties of inclusion compounds and thus suitably supplement spectral methods.

6. Application of cyclodextrins in chromatography

Cyclodextrins (CD's) are cyclic oligosaccharides composed of D(+)-glucopyranose units interconnected by α-(1,4) bonds and are interesting chiefly because of their inclusion properties, both in the solid state and in aqueous solutions. Inclusion complexes are formed inside the CD cavity whose geometry and chemical composition determine the selectivity of the inclusion process. By enzymatic degradation of starch, CD's with various numbers of glucose units are obtained; α, β, and γ-CD, formed by 6, 7 and 8 glucose units, respectively, have cavities with a size of 5–8 Å that enable the inclusion of molecules (or their parts) with similar dimensions (see Chapter 8, Volume 2).

The specific properties of CD's are retained even if they are used in the form of polymers which are less soluble in water than the original CD. This is an important fact, because the ability of CD to form water-soluble inclusion compounds was to a certain degree a limiting factor, especially when CD is to be used as a stationary phase in liquid chromatography.

The exceptional properties of CD's have been described in many papers, interesting monographs and reviews.[80-85] Increased attention has recently been paid to the possibilities of the use of chromatographic methods for the study of CD's and of the use of the selective formation of inclusion compounds for separations and analytical purposes using various chromatographic procedures.[86,87]

6.1. Cyclodextrins as stationary phases in chromatography

The fact that CD's can form inclusion complexes preferentially with certain types of compound, depending primarily on the shape of the molecule, has led to considerations about their use as stationary phases in chromatography.

6.1.1. Liquid chromatography

The mechanism of the inclusion process, which has been described in most papers as a process occurring in the aqueous phase, is in agreement with the considerations about the use of the selective character of inclusion in liquid chromatography. For use as a stationary phase it is necessary for the host to be insoluble in aqueous solutions; this requirement is satisfied by cyclodextrin polymers (CDP). These substances have become suitable chromatographic materials and have been used chiefly in gel inclusion chromatography.

Water-insoluble CDP in the form of gels can interact with various compounds according to various mechanisms,[88] involving: (a) interactions in the cavities, i.e. the formation of inclusion compounds, (b) interactions in the internal pores of polymeric particles, (c) interactions on the surface. These mechanisms, compared with those on Sephadex gels as non-inclusion analogues, have become the basis of gel inclusion chromatography.[89]

A comparison of the interaction isotherms on CDP and Sephadex gels, of aniline, pyridine, benzaldehyde, butyric acid, o- and p-nitrophenol, has indicated large differences that confirm the inclusion character of interaction with CDP. From the shape of the isotherms for o- and p-nitrophenol (Fig. 48) it follows that these substances cannot be separated on Sephadex, whereas the different inclusion interactions on CDP can be used to advantage for their chromatographic separation.

The increased affinity of CD towards aromatic molecules and steric specificity towards their isomers have been examined in detail with amino acids as model substances and using CDP prepared for chromatographic purposes by polymerization with ethyleneglycol-di-(epoxypropyl)ether in the presence of polyvinyl acetate.[90,91] Comparison experiments performed on α-, β- and γ-CD polymers in weakly acid solutions (pH 5–6) showed the greatest differences in the retention data on the β-CD gel. This polymer has been found to be extremely suitable for separations of aromatic amino

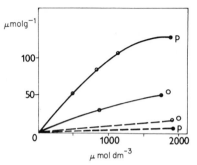

Fig. 48. Interaction isotherms of o- and p-nitrophenol (o, p) on CD-resin (solid line) and epichlorhydrin-dextran resin (broken line).

acids (phenylalanine, tyrosine, tryptophan). A high separation efficiency ($H = 0.7$–0.8 mm) was attained for these substances at laboratory temperature and a flow-rate of 10–20 ml h^{-1}. Under these conditions aromatic acids could be separated from non-aromatic ones (lysine, alanine), as well as from one another. The behaviour of these substances on α-, β- and γ-CD polymers can be illustrated by chromatograms obtained under the same experimental conditions (Fig. 49). The results obtained for the aromatic acids, especially on β-CDP, illustrate that the separation depends primarily on inclusion, while the mechanism of the separation of lysine and alanine depends primarily on adsorption. This conclusion is also confirmed by

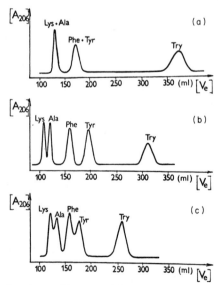

Fig. 49. Separation of amino acids on α-CDP (a), β-CDP (b) and γ-CDP (c).

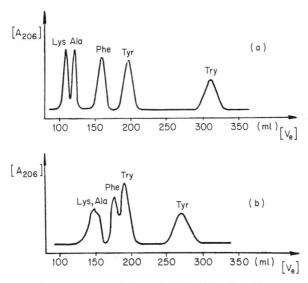

Fig. 50. Separation of amino acids on β-CDP (a) and on Sephadex G-25 gel (b).

comparing the chromatograms obtained with β-CDP and Sephadex under identical experimental conditions (Fig. 50).

The ability of some components of nucleic acids, especially those with an adenine base, to form compounds with β-CD, can also be readily used for chromatographic separations of various nucleotides and nucleosides.[92]

The guest–host interaction is weaker with oligonucleotides containing adenine. Therefore, anion-exchanging diethylaminoethyl groups were built into a CD gel. The gel thus obtained exhibits both ion-exchange and inclusion properties and has also been used to separate t-RNA.[93]

CD polymers can also be used for the separation of racemic mandelic acid.[94] The β-CDP preferentially forms inclusion compounds with L-(+) isomers, making it possible to separate DL methylesters of mandelic acid; the first fraction contains quite pure D-(−) isomer. The opposite effect is exhibited by α-CDP that preferentially binds D-(−) isomers; however, the separation of the racemate is then not complete.

A competitive effect of α-CD on the activity of β-amylase has been utilized for a solution of a special analytical problem.[95] If α-CD is bound to epoxy-Sepharose 6B, a gel is formed that retains β-amylase, whereas α-amylase is not retained. After elution of β-amylase by a highly specific eluent with a competitive counter-ligand, α-CD, highly pure enzyme with a high activity is obtained.

Alternatively, β-CD has been used as an immobilized ligand on epoxy-Sepharose 6B in biospecific affinity chromatography.[96] Here α-amylase was selectively retained and separated from proteins and then selectively eluted by a buffer containing β-CD. This chromatographic procedure yielded a recovery of 90% with a degree of purification of up to 180 fold compared with crude extracts.

β-CD with other oligosaccharides has also been used for selective retardation of phosphorylases by affinity electrophoresis on polyacryl gel. From the changes in the mobility of phosphorylase and its dependence on the concentration of the oligosaccharides in the gel, the dissociation constants of the complex have been calculated.[97]

6.1.2. Gas chromatography

The results obtained when using CD's in liquid chromatography have confirmed many assumptions about the inclusion mechanism and their use as stationary phases has resulted in many important analytical applications. However, the present knowledge has not made it possible to describe in detail the inclusion of guests present in the gaseous phase.

CD, like urea, exhibits selectivity towards e.g. linear and branched alkanes. In contrast to urea, the CD cavity is presented before the inclusion process. In common with other inclusion compounds, molecular dimensions play a predominant role; very small guests, e.g. rare gases (He, Ne, Ar) are not accepted by the relatively large β-CD cavity, because of the short-range character of the operative forces, whereas Kr and Xe do form inclusion compounds.[98] Similarly, α-CD forms very stable inclusion compounds with O_2, CO_2 and Cl_2 under high pressure.[99]

Acylated cyclodextrins (α, β, γ-CD acetate, β-CD propionate, butyrate and valerate) have been used[100,101] as polar stationary phases in gas-liquid chromatography; however, the separation was not based on an inclusion mechanism. Their use was based on the finding that polyesters used for the chromatographic separation of fatty acids have a C/O ratio similar to that in saccharide esters, whose applicability is, however, limited by temperature. Acylated cyclodextrins have a relatively high thermal stability (220–236° C) and a good separation efficiency for various polar compounds (α-alkenes, aldehydes, alcohols, esters, aldehyde-esters and diesters).

The methylated CD's have also been used as stationary phases,[102] either coated on silanized Chromosorb W or as a mixed phase containing 10% methylated CD in silicone oil. Hydrocarbons were separated above 100° C. The elution order suggests the participation of inclusion. Isooctane has a larger retention time on β-MeCD than on α-MeCD, as the larger cavity accommodates branched hydrocarbons better than the smaller cavity of α-CD.

A detailed study[103] deals with a comparison of the behaviour and properties of CD and MeCD used as stationary phases in gas chromatography.

It has been found that inclusion predominates in the gas-so-lid interaction of substances with suitable geometric dimensions, while the retention of polar substances is primarily determined by the interaction with cyclodextrin hydroxyl groups. On a decrease in the cyclodextrin polarity by methylation, stereoselectivity appeared even in the separation of alcohols. On the other hand, methyl groups cause certain steric hindrance to the guest penetration into the cyclodextrin cavity. The selectivity of the inclusion process is especially important for separations of positional isomers.

Thermal stability of the stationary phase and the presence of the inclusion process even at high temperatures are the necessary conditions for gas chromatographic use. Recent results have shown that these requirements can be satisfied by macroporous polymers with inbuilt CD molecules and by CD deposited on a chromatographic support.

Cyclodextrin-polyurethane resins (CDPU) have been used as stationary phases in GSC[104] and their sorption properties have been studied, indicating that a specific interaction governed by the dimensions and configuration of the host molecules takes place. The retention data for many organic substances were correlated with the inclusion phenomenon, i.e. with the size of the α- and β-CD cavities, and with the effect of π-electrons and heteroatoms in the guest molecule. Compared with the common organic polymers (Amberlite, Porapak Q, Tenax GC), an advantage of polymers containing CD should be their selectivity, which could also be utilized in trace analysis for preconcentration of the test substances.

The formation and properties of the inclusion compounds of α- and β-CD can also be studied in the GSC system if CD (10 wt.%) is deposited on Chromosorb W from a dimethylformamide solution.[105] The sorbates were chosen to include organic molecules of various structural types and geometries (hydrocarbons, hydrocarbon halogeno derivatives, alcohols, ethers and aromatic substances).

In the evaluation and comparison of the experimental results with CDPU resins and CD deposited on an inert support as stationary phases it is suitable to quote some data which can help in explaining the interaction character. With the CDPU resin it has been found that the preparative conditions affect the physicochemical properties of these materials (Table 7). The differences in the specific surface area can affect, as shown in the Table, the magnitude of the retention data and make unambiguous interpretation more difficult. On the other hand, the retention on the material with the lowest content of the residual hydroxyl groups demonstrates that with polar substances the interaction is not markedly affected by these processes and the results can be interpreted as being based on the inclusion process.

Although the measurements were carried out at various temperatures, i.e. 150–170° C with CDPU resin and 50–80° C with CD on an inert support,

Table 7. Physical properties and retention data on CD-polyurethane resins[104]

Property	Resin[a]		
	β-HDI-DMF-5.5-A	α-HDI-DMF-5.9-A	α-HDI-DMF-13.3-A
Temperature limit (° C)	200	230	230
Specific surface area ($m^2 g^{-1}$)	170	180	280
OH residues per CD molecule	13.5	10.1	0.9
Retention times of various sorbates[b]			
n-Hexane	0.04	0.16	0.27
n-Heptane	0.06	0.24	0.33
Cyclohexane	0.05	0.07	0.12
Benzene	1.00 (16.68)	1.00 (6.59)	1.00 (4.16)
Toluene	1.20	2.73	1.96
Methanol	0.36	0.53	1.06
Ethanol	0.78	0.83	1.53
Propanol	1.96	2.03	2.80

Column temperature 150° C; column length 120 or 80 cm; i.d. 3 mm; carrier gas, nitrogen (30 ml min^{-1}); detector FID.
[a] Resins obtained by polymerization of CD with hexamethylenediisocyanate (HDI) in N,N-dimethylformamide (DMF). Precipitant: acetone (A) (feed composition: β-CD 10 g + 5.5 g HDI; α-CD 8.6 g + 5.9 g HDI; α-CD 8.6 g + 13.3 g HDI).
[b] Relative to benzene = 1.00.
[c] Actual retention time (min).

certain correlations can be made and some general conclusions can be drawn (Tables 7 and 8), which can be supported by the results of measurements in aqueous solutions of α- or β-CD.

The differences of several orders of magnitude in the retention of aliphatic hydrocarbons between α- and β-CD confirm the inclusion character of the interaction. With benzene, toluene and cyclohexane, an increased retention of benzene was found on α-CD, whereas with the CD-polyurethane resin the stronger interaction with β-CDPU can be explained by the existence of π-bonds. The greater interaction with α-CD, however, fully corresponds to the model in which the dimensions of the benzene molecule better matches those of the cavity of α-CD as the host.

The great differences in the retention of the hydrocarbon halogeno derivatives indicate that with α-CD the bonding forces can differ from those encountered with β-CD. The difference in the retention data of trichloroethylene and 1,2-dichloroethane, compared with the tetrachloro derivatives, is especially marked in view of the boiling points. The experi-

Table 8. *Retention data of various sorbates on α-CD and β-CD*

Sorbate	B.p. (°C)	t'_R (s) β-CD	t'_R (s) α-CD
n-Pentane	36.07	5	496
n-Hexane	68.7	25	(a)
n-Heptane	98.42	75	(a)
Cyclohexane	80.7	81	1588
Benzene	80.1	133	1628
Toluene	110.8	191	1379
1,2-Dichloroethane	83.7	154	1429
Trichloroethylene	87.0	111	1730
1,1,2,2-Tetrachloroethane	146.0	228	850
Tetrachloromethane	76.8	105	254
Chlorobenzene	132.1	501	3328
Bromobenzene	156.2	1008	(a)
Methanol	64.7	686	437
Ethanol	78.4	348	790
Propanol	97.8	1119	2169
Isopropanol	82.4	487	1577
Diethyl ether	34.51	914	723
Diisopropyl ether	67.8	107	232

Column temperature 80°C; column length 120 cm; i.d. 3 mm; carrier gas, nitrogen (30 ml/min); detector FID.
[a] Does not leave the column.

mental results for benzene halogeno derivatives, chloro- and bromobenzene, on α- and β-CD agree with the measurements in aqueous solutions. The effect of the bulky bromo derivative is reflected in an increased retention.

The increased stability of the inclusion compounds of α-CD with unbranched aliphatic alcohols compared with the interaction with β-CD agrees with the association constants calculated for aqueous solutions. The interaction of unbranched and branched compounds (alcohols, ethers) shows a lower retention for branched compounds, whose geometry does not correspond to the dimensions of the CD cavity. In the GSC system with β-CD deposited on Chromosorb W, interesting results were also obtained with positional isomers of aromatic hydrocarbons.[103] It is evident from the dependence of the retention data on the polarizability (which is used as a criterion of the intensity of dispersion forces inside the CD ring) (Fig. 51) that their retention behaviour is different. Among the xylene isomers, the strongest interaction with β-CD is exhibited by p-xylene. Group positions at 1,2 and 1,3 somewhat suppress the inclusion process, which appears as a decrease in the V_g values for these isomers. A similar conclusion can be

D. Sybilska and E. Smolková-Keulemansová

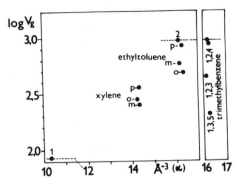

Fig. 51. Dependence of the specific retention volume on the polarizability of the positional isomers of some aromatic compounds. Benzene (1), propylbenzene (2).

drawn with ethyltoluene. Substituents in positions 1,4 do not increase the molecular volume significantly and thus permit inclusion into the CD cavity, whereas substituents in positions 1,2 and 1,3 lead to a more bulky grouping suppressing inclusion. This effect is even more pronounced with trimethylbenzene isomers that are eluted in the order, 1,3,5-, 1,2,3- and 1,2,4-trimethylbenzene.

Although the results obtained are rather qualitative, it is clear that both with a macroporous polymer with inbuilt CD molecules and with CD deposited on a chromatographic support, the forces operating inside the CD ring also play a role under the GSC conditions, i.e. the inclusion compounds are formed even when cyclodextrin as the host is in contact with sorbates in the gaseous state.

Further interesting results have also been obtained in GLC systems by dissolving the cyclodextrins in a suitable matrix.[106] A comparative study of the chromatographic properties of α- and β-CD in GSC and GLC systems[106a] has shown that in both cases CD acts as a selective agent, though to a different extent.

6.2. Cyclodextrins as Mobile Phase Components in TLC and HPLC

Recently the ability of CD's to selectively bind molecules whose shape and dimensions correspond to those of the cavity were utilized under conditions when a discrete amount of CD was added to an aqueous mobile phase.[87] It has been shown that a variety of substituted aromatic compounds, e.g. o-, m- and p-disubstituted and some trisubstituted phenols, nitriles, anilines and benzoic acids can be separated on polyamide TLC sheets using an aqueous solution of α-CD.[107,108] The R_F values depend mainly on the concentration of the CD in the mobile phase. In general, the p-substituted isomers have larger R_F values than m-isomers, which in turn have larger R_F values than o-substituted isomers. This order corresponds to the affinity of these isomers to CD.

Preliminary results indicate that isomeric bromobenzoic acids can be separated[87] using a 0.10 mol dm^{-3} aqueous solution of α-CD as the mobile phase and a C$_{18}$ reversed stationary phase in HPLC. The HPLC separation of some prostaglandins using a 0.5% aqueous solution of α- or β-CD as the mobile phase and a column with AV-02-500 anion-exchanger has been reported.[109] The use of β-CD permitted a better separation than α-CD, due to differences in the binding constants of the prostaglandins. Thus a very sensitive and rapid method for the separation and determination of prostaglandins was obtained.

Recently, β-CD was used in HPLC as an optically active component of the mobile phase for the resolution of racemates.[110] The separation of mandelic acid into its enantiomers was attained on a column packed with a C$_{18}$ reversed phase system and with 5×10^{-3} mol dm^{-3} CD in aqueous phosphate buffer (pH 2.1) as the mobile phase.

In all TLC and HPLC studies, the retention data are dependent on the CD concentration in the mobile phase. With increasing CD concentration, the R_F values in TLC increase and the retention times in HPLC decrease. This can be attributed to an increase in the solubility and partitioning of the solutes by complexation with the CD molecules in the mobile phase.

6.3. Concluding remarks

The exceptional properties of CD's have been known since the end of the last century and have been extensively utilized since the 1950s in various fields of research and practice. However, it is only recently that they have been used in chromatography. Therefore, it is too early to draw general conclusions, as new aspects of their use in chromatography are continually appearing. However, it follows from the research carried out so far that the stability of inclusion compounds and the selectivity of the inclusion process can be successfully studied by chromatographic methods, both using CD as the stationary phase in liquid or gas chromatography and as an active component of mobile phases in TLC and HPLC. Selectivity of the inclusion process is advantageous in analytical applications which would be difficult to carry out by other techniques. Many advantages following from their use as stationary or mobile phases in the separation or in improving the detection sensitivity are being studied and critically evaluated.

7. Conclusions

The range of applications of inclusion compounds in chromatography depends on their stability and their selectivity properties. A primary condition for chromatographic separation is the existence of an interaction between the substance forming the stationary phase and the eluent, the interaction in inclusion compounds being van der Waals forces, donor–acceptor interactions, electrostatic interaction and hydrogen bonds. The interaction often has a mixed character.

The spatial character of all types of interaction often leads to a high stabilization of the guest–host lattice, which contributes to the marked stereoselectivity.

In chromatography, only zeolites have been studied to a greater extent, i.e. inclusion compounds of a permanent host structure, independent of the kind of guest and its content. The knowledge and practical use of numerous other inclusion compounds, such as those described in this chapter, compounds of Werner type complexes, urea, thiourea and cyclodextrins, are still rather limited. Compounds, whose structure and properties depend on the kind and content of the guest molecules constitute an interesting and promising group of sorbents for chromatography. There is a great potential for the rational control of the properties of chromatographic systems to attain, first of all, a high selectivity. This control is only possible on the basis of a thorough knowledge of the inclusion processes occurring in the chromatographic column.

The authors hope that this chapter will increase the interest in the study and application of inclusion phenomena in chromatography. The great diversity of substances capable of forming inclusion compounds supports the conclusion that they can be employed to advantage for large scale selective chromatographic separations and that chromatographic methods can in turn be used to study their unusual properties.

References

1. E. Smolková-Keulemansová and S. Krýsl, *J. Chromatogr.*, 1980, **184**, 347.
2. K. A. Hofmann and F. Küspert, *Z. Anorg. Allg. Chem.*, 1897, **15**, 204.
3. J. H. Rayner and H. M. Powell, *J. Chem. Soc.*, 1952, 312.
4. A. L. Jones and P. S. Fay (Esso Research and Engineering Co.) U.S. Patent 2,732,413, 1956.
5. W. D. Schaeffer, W. S. Dorsey, D. A. Skinner and C. G. Christian, *J. Am. Chem. Soc.*, 1957, **79**, 5870.

6. P. de Radzitzky and J. Hanotier, *Ind. Eng. Chem. Process Des. Dev.*, 1962, **1**, 10.
7. W. Kemula and D. Sybilska, *Nature*, 1960, **185**, 237.
8. W. Kemula, D. Sybilska and A. Kwiecińska, *Roczniki Chem.*, 1965, **39**, 1101.
9. M. Pawłowska, D. Sybilska and J. Lipkowski, *J. Chromatogr.*, 1979, **177**, 1.
10. W. Kemula, D. Sybilska, J. Lipkowski and K. Duszczyk, *Pol. J. Chem.*, 1980, **54**, 317.
11. S. Brzozowski and M. Broniarek, *Roczniki Chem.*, 1974, **48**, 1213.
12. J. Czarnecki, Ph.D. Thesis, Polish Academy of Sciences, Warsaw, 1977.
13. D. Sybilska, *Proc. VI Intern. Symp. on Chromatography and Electrophoresis*, Presses Académique Européennes, Brussels, 1971, pp. 212–221.
14. S. Siekierski and J. Narbutt, *Nukleonika*, 1967, **12**, 487.
15. K. Sawlewicz and A. Halpern, *Radiochim. Acta*, 1969, **11**, 86.
16. J. Narbutt, D. Dancewicz and A. Halpern, *Radiochim. Acta*, 1967, **7**, 55.
17. W. Kemula, A. Kwiecińska and D. Sybilska, in *Proc. III Czechoslovak-Polen Symp. on Petro and Carbochemistry*, Nováky, IChO Warsaw, 1967, pp. 95–103.
18. W. Kemula, A. Kurjan and A. Kwiecińska, *Chem. Anal.* (*Warsaw*), 1967, **12**, 869.
19. W. Kemula, D. Sybilska and K. Duszczyk, *Microchem. J.*, 1966, **11**, 296.
20. M. J. Hart and N. O. Smith, *J. Am. Chem. Soc.*, 1962, **84**, 1816.
21. D. Belitskus, G. A. Jeffrey, R. K. McMullan, and N. C. Stephenson, *Inorg. Chem.*, 1963, **2**, 873.
22. F. Caselato and B. Casu, *Erdol Kohle Erdgas Petrochem.*, 1969, **22**, 71.
23. J. Lipkowski, A. Bylina, K. Duszczyk, K. Leśniak and D. Sybilska, *Chem. Anal.* (*Warsaw*), 1974, **19**, 1051.
24. W. Kemula, J. Lipkowski and D. Sybilska, *Roczniki Chem.*, 1974, **48**, 3.
25. J. Lipkowski and S. Majchrzak, *Roczniki Chem.*, 1975, **49**, 1655.
26. E. R. de Gil and I. S. Kerr, *J. Appl. Crystallogr.*, 1974, **10**, 315.
27. J. Lipkowski and G. D. Andreetti, *Transition Met. Chem.*, 1978, **3**, 117.
28. J. Lipkowski, *J. Mol. Struct.*, 1981, **75**, 13.
29. J. Lipkowski, K. Suwińska, G. D. Andreetti and K. Stadnicka, *J. Mol. Struct.*, 1981, **75**, 101.
30. J. Lipkowski and G. D. Andreetti, *Acta Crystallogr.*, 1982, **B38**, 607.
31. J. Lipkowski, P. Sgarabotto and G. D. Andreetti, *Acta Crystallogr.*, 1982, **B38**, 416.
32. J. Lipkowski, *Acta Crystallogr.*, 1982, **B38**, 1745.
33. W. Kemula, Z. Borkowska and D. Sybilska, *Monatsh. Chem.*, 1972, **103**, 860.
34. W. Kemula, D. Sybilska and J. Lipkowski, *J. Chromatogr.*, 1981, **218**, 465.
35. W. Kemula, D. Sybilska and K. Chlebicka, *Roczniki Chem.*, 1965, **39**, 1499.
36. W. Kemula, D. Sybilska and K. Chlebicka, *Rev. Chim.* (*Bucharest*), 1962, **7**, 1003.
37. W. Kemula and D. Sybilska, *Acta Chim. Acad. Sci. Hung.*, 1961, **27**, 137.
38. W. Kemula, D. Sybilska, J. Lipkowski and K. Duszczyk, *J. Chromatogr.*, 1981, **204**, 23.
39. W. Kemula and D. Sybilska, *Roczniki Chem.*, 1964, **38**, 861.
40. W. Kemula in (eds. L. Zuman and I. M. Kolthoff) *Progesss in Polarography*, Wiley-Interscience, New York, 1962, p. 397.
41. W. Kutner, J. Dębowski and W. Kemula, *J. Chromatogr.*, 1980, **191**, 47.
42. J. Dębowski, K. Duszczyk, W. Kutner, D. Sybilska and W. Kemula, *J. Chromatogr.*, 1982, **241**, 141.
43. W. Kemula, B. Behr, K. Chlebicka and D. Sybilska, *Roczniki Chem.*, 1965, **39**, 1315.
44. Z. Borkowska, D. Sybilska and B. Behr, *Roczniki Chem.*, 1971, **45**, 269.

45. J. E. B. Randles, B. Behr and Z. Borkowska, *J. Electroanal. Chem.*, 1975, **65**, 775.
46. Z. R. Grabowski and J. Koszewski, *Acta IMEKO* 1967, **3**, Hungarian Scientific Society for Measurements and Automation, Budapest 1968, p. 125.
47. A. Bylina, D. Sybilska, Z. R. Grabowski and J. Koszewski, *J. Chromatogr.*, 1973, **83**, 357.
48. S. Brzozowski, Z. Dobkowska and W. Kemula, *Roczniki Chem.*, 1972, **46**, 2305.
49. S. Brzozowski and A. Lewartowska, *Roczniki Chem.*, 1975, **49**, 1283.
50. J. Lipkowski, M. Pawłowska and D. Sybilska, *J. Chromatogr.*, 1979, **176**, 43.
51. W. Kemula, D. Sybilska, J. Czarnecki and J. Lipkowski, *Pol. J. Chem.*, 1980, **54**, 2319.
52. J. Lipkowski, Ph.D. Thesis, Institute of Physical Chemistry, Polish Academy of Sciences, Warsaw, 1972.
53. J. Hanotier, J. Brandli and P. de Radzitzky, *Bull. Soc. Chim. Belg.*, 1966, **75**, 265.
54. E. Soczewiński, *Anal. Chem.*, 1969, **41**, 179.
55. E. Soczewiński, *Ann. Univ. Marie Curie Skłodowska Lublin, Polonia Sec. D.* 1969, **24**, 21.
56. W. Kemula and H. Buchowski, *J. Phys. Chem.*, 1959, **63**, 155.
57. E. Soczewiński, *Adv. Chromatogr.*, 1968, **5**, 3.
58. A. C. Bhattacharyya, *J. Chromatogr.*, 1969, **41**, 446.
59. A. C. Bhattacharyya and A. Bhattacharjee, *Anal. Chem.*, 1969, **41**, 2055.
60. D. Sybilska, K. Malinowska, M. Siekierska and J. Bylina, *Chem. Anal.* (*Warsaw*), 1972, **17**, 1031.
61. L. B. Rogers and A. G. Altenau, *Anal. Chem.*, 1964, **36**, 1727.
62. A. Sopková, *J. Mol. Struct.*, 1981, **75**, 81.
63. M. Šingliar, A. Sopková and M. Dzurillova, International Micro-Symposium on Clathrate and Molecular Inclusion Phenomena, Czechoslovakia, 1981, Summaries of papers, p. L-VI/12, 13.
64. D. Sybilska, F. Werner-Zamojska and A. Kinowski, *Chem. Anal.* (*Warsaw*), 1973, **18**, 157.
65. F. Werner-Zamojska and A. Kinowski, *Chem. Anal.* (*Warsaw*), 1973, **18**, 299.
66. J. Cason, G. Sumrell, C. F. Állen and G. A. Gilles, *J. Biol. Chem.*, 1953, **205**, 435.
67. C. H. Mailen, T. M. Reed and J. A. Yong, *Anal. Chem.*, 1964, **36**, 1883.
68. W. Schlenk jr., *Ann. Chem.*, 1949, **565**, 204–240.
69. K. Mařík and E. Smolková, *Chromatographia*, 1973, **6**, 420.
70. K. Mařík and E. Smolková, *J. Chromatogr.*, 1974, **91**, 303.
71. E. Smolková-Keulemansová, *Chromatographia*, 1978, **11**, 70.
72. E. Smolková, L. Feltl, and J. Všetečka, *J. Chromatogr.*, 1978, **148**, 3.
73. A. A. Choudhury, *UNESCO Postgraduate Course Report*, Charles University, Prague, 1976.
74. E. Smolková, L. Feltl, and J. Všetečka, *Chromatographia*, 1979, **12**, 5.
75. I. Kolísková, Thesis, Charles University, Prague, 1982.
76. I. Kolísková and E. Smolková, unpublished results.
77. H. B. Knight, L. P. Witnauer, J. E. Coleman, W. R. Noble and D. Swern, *Anal. Chem.*, 1952, **24**, 1331.
78. H. G. McAddie, *Can. J. Chem.*, 1962, **40**, 2195.
79. P. F. McCrea in *New Developments in Chromatography*, (ed. H. Purnell) J. Wiley, New York, 1973, p. 107.
80. F. Cramer, *Einschlussverbindungen*, Springer Verlag, Berlin: Göttingen, 1954.
81. L. Mandelcorn, *Non-Stoichiometric Compounds*, Academic Press, New York, 1964.

82. J. A. Thoma and L. Stewart, in *Starch, Chemistry and Technology* (eds. R. L. Whistler and E. E. Paschall) Academic Press, New York, 1965.
83. M. L. Bender and M. Komiyama, *Cyclodextrin Chemistry*, Springer Verlag, Berlin, Heidelberg, New York, 1978.
84. W. Saenger, *Angew. Chem.*, 1980, **92**, 343.
85. J. Szejtli, *Cyclodextrins and their Inclusion Compounds*, Akademia Kiado, Budapest, 1982.
86. E. Smolková-Keulemansová, *J. Chromatogr.*, 1982, **251**, 17.
87. W. L. Hinze, *Sep. Purif. Methods*, 1981, **10**, 159.
88. J. Szejtli, E. Fenyvesi and B. Zsadon; *Stärke*, 1978, **30**, 127.
89. J. Solms and R. H. Egli; *Helv. Chim. Acta*, 1965, **48**, 1225.
90. B. Zsadon, M. Szilasi, K. H. Otta, F. Tüdös, E. Fenyvesi and J. Szejtli, *Acta Chim. Acad. Sci. Hung.* 1979, **100**, 265.
91. B. Zsadon, M. Szilasi, F. Tüdös, E. Fenyvesi and J. Szejtli, *Stärke*, 1979, **31**, 11.
92. J. L. Hoffmann, *Anal. Chem.*, 1970, **33**, 209.
93. J. L. Hoffmann, *J. Macromol. Sci., Chem.*, 1973, **7**, 1147.
94. A. Harade, M. Furuke and S. I. Nozakura, *J. Polym. Sci.*, 1978, **16**, 189.
95. J. A. Thoma, D. E. Koshland, J. Ruscica and R. Baldwin, *Biochem. Biophys. Res. Commun.*, 1963, **12**, 184.
96. M. P. Silvanovich and R. D. Hill, *Anal. Biochem.*, 1976, **73**, 430.
97. K. Takeo and S. Nakamura, in *Affinity Chromatography* (ed. Hoffmann-Ostenhof et al.), Pergamon Press, Oxford, New York, 1978, p. 67.
98. F. Cramer and F. M. Henglein, *Chem. Ber.*, 1957, **90**, 2572.
99. J. N. J. J. Lammers, PhD Thesis, TH-Eindhoven, 1970.
100. D. M. Sand and H. Schlenk, *Anal. Chem.*, 1961, **33**, 1624.
101. H. Schlenk and D. M. Sand, *Anal. Chem.*, 1962, **34**, 1529, 1676.
102. B. Casu, M. Reggiani and G. R. Sanderson, *Carbohydr. Res.*, 1979, **76**, 59.
103. J. Mráz, L. Feltl and E. Smolková-Keulemansová, *J. Chromatogr.*, 1984, **286**, 17.
104. Y. Mizobuchi, M. Tanaka and T. Shono, *J. Chromatogr.*, 1980, **194**, 153.
105. E. Smolková, H. Králová, S. Krýsl and L. Feltl, *J. Chromatogr.*, 1982, **241**, 3.
106. (a) T. Kościelski, D. Sybilska, L. Feltl and E. Smolková-Keulemansová, *J. Chromatogr.*, 1984, **286**, 23; (b) D. Sybilska and T. Kościelski, *J. Chromatogr.*, 1983, **261**, 357.
107. W. G. Buckert, C. N. Owensby and W. L. Hinze; *J. Liq. Chromatogr.*, 1981, **4**, 1065.
108. W. L. Hinze and D. W. Armstrong, *Anal. Lett*, 1980, **13**, 1093.
109. K. Uekama, F. Hirayama, K. Ikeda and K. Inaba, *J. Pharm. Sci.*, 1977, **66**, 706.
110. J. Debowski, D. Sybilska and J. Jurczak, *J. Chromatogr.*, 1982, **237**, 303.

7 · THE SORPTIVE ABILITIES OF TETRACYANOCOMPLEXES

A. SOPKOVÁ

P. J. Šafarik's University, Košice, Czechoslovakia

M. ŠINGLIAR

Research Institute for Petrochemistry, Nováky, Czechoslovakia

1. Introduction

The preceding chapter discussed the use of host lattices such as $Ni(4MePy)_4(NCS)_2$, urea, thiourea and the cyclodextrins as stationary phases in gas and liquid chromatography. Although these host lattices can give very effective separations, they do suffer from the disadvantage of having quite low upper temperature limits of 80–85° C for all of the above compounds.

The host lattices of Hofmann type inclusion compounds, such as $Ni(NH_3)_2Ni(CN)_4.2C_6H_6$, are much more stable and thus offer the possibility of working at higher temperatures.

Sorptive abilities[1,2] have been found not only with inclusion compounds such as $Ni(NH_3)_2Pt(CN)_4.2C_6H_6$; $Ni(NH_3)_2Pt(CN)_4.2C_6H_5OH$, and

INCLUSION COMPOUNDS III
ISBN 0-12-067103-4

$Zn(NH_3)_2Ni(CN)_4.0.2C_6H_6$, but also with tetracyanocomplexes such as $Ni[Pt(CN)_4].6H_2O$, $Zn[Ni(CN)_4].H_2O$, $Zn(en)_3Ni(CN)_4.H_2O$, $Cu(NH_3)_4$-$Ni(CN)_4.3H_2O$, $Cu(en)_2Ni(CN)_4.0.14C_6H_6$, and $Zn(en)_2Ni(CN)_4$-$0.14C_6H_6$. The anhydrous forms of the hydrated complexes listed above show no sorptive abilities.[1]

Structural details of the Hofmann, Hofmann–en, and the Hofmann–en–Td inclusion compounds and of the $Ni(H_2O)_2Ni(CN)_4.nH_2O$ complexes have been described in detail by Iwamoto in Chapter 2, Volume 1.

Oswald and his coworkers found[3] $Ni(NH_3)_2Pt(CN)_4.2C_6H_6$ to be topotactic with $Ni(NH_3)_mM'(CN)_4.nH_2O$. By following the thermal decomposition using the Weissenberg X-ray method and scanning electron microscopy it was found that the structure of the inclusion compound was retained until the guest molecule was completely lost (i.e. until $n \to 0$). The host lattice has a layered structure and during the deammination reaction:

$$Ni(NH_3)_2Pt(CN)_4(s) \to NiPt(CN)_4(s) + 2NH_3(g)$$

the ammonia molecules diffuse parallel to the 001 planes which are retained in the structure.

The structure of $Cu(NH_3)_4Ni(CN)_4.nH_2O$ has not been reported in the literature, but may be similar to those of $Ni(NH_3)_2Ni(CN)_4.nH_2O$[4] and $Ni(NH_3)_4Ni(CN)_4.nH_2O$.[5]

The crystal of $Zn(en)_3Ni(CN)_4.H_2O$ is built up from $[Zn(en)_3]^{2+}$ and $[Ni(CN)_4]^{2-}$ ions and molecules of H_2O which do not enter into the coordination sphere of the metal atoms. (Fig. 1). Each water forms two hydrogen bonds with nitrogen atoms of two $[Ni(CN)_4]^{2-}$ anions forming $[Ni(CN)_4]^{2-}\cdots H_2O\cdots[Ni(CN)_4]^{2-}\cdots H_2O$ chains in the z axis direction. The hydrogen bond lengths are 2.89(3) Å and 2.91(3) Å. The shortest water–cation distance, $H_2O\cdots N[Zn(en)_3]^{2+}$, is 3.07(3) Å. All other contacts are greater than 3.15 Å, in agreement with its sorptive abilities.[6]

The crystal of the anhydrous form $Zn(en)_3Ni(CN)_4$ is built up only from $[Zn(en)_3]^{2+}$ and $[Ni(CN)_4]^{2-}$ ions and the structure is of the NiAs type. On dehydration of $Zn(en)_3Ni(CN)_4.H_2O$ the change[7] from the CsCl to the NiAs structure can be observed.

The tetracyanocomplex[7] $Cd(NH_3)_2Ni(CN)_4.0.5H_2O$ is isostructural with $Ni(NH_3)_2Ni(CN)_4.0.5H_2O$.[4] It is built up from slightly distorted $CdNi(CN)_4$ layers shifted with respect to one another. The H_2O molecules are situated in the layers. The shortest distance between a H_2O molecule and another atom is 3.02 Å ($Ni\cdots H_2O$).

The preparation and identification of the tetracyanocomplexes $Cu(en)_2Ni(CN)_4.0.14C_6H_6$ and $Zn(en)_2Ni(CN)_4.0.14C_6H_6$ have been described[1,8] but their structures are not yet known.

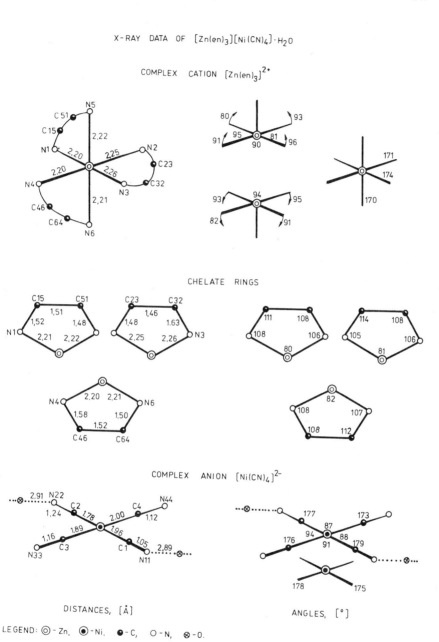

Fig. 1. X-ray data[7] for Zn(en)$_3$Ni(CN)$_4$·H$_2$O.

2. The thermal decomposition of tetracyanocomplexes

The thermal stabilities of the inclusion compounds are dependent on their structure, their composition and on the nature of the guest molecule.

$Ni(NH_3)_2Pt(CN)_4.2C_6H_6$ loses the guest in the temperature range[1] 68–190° C, whereas with phenol as the guest molecule the comparable temperature range[10] is 55–244° C. The zinc containing inclusion compound $Zn(NH_3)_2Ni(CN)_4.2C_6H_6$ is rather unstable losing the guest in the temperature range 26–138° C.

The stabilities of the tetracyanocomplexes also vary:[11,12] $Zn[Ni(CN)_4].H_2O$ loses the water[1] in the 34–150° C range.

Tetracyanocomplexes with a higher water content e.g. $Zn[Ni(CN)_4]2.5H_2O$, lose the water in the 25–175° C range.[14] Sorption experiments with β-picoline indicate two different kinds of water: two molecules being coordinated and the remaining molecules are in the inter-layer space where they can be exchanged for other organic molecules.

$Ni[Pt(CN)_4].6H_2O$ loses the water content in a stepwise fashion:[13] three molecules are lost in the 30–100° C range, the fourth is lost in the 100–135° C range, and the two remaining molecules in the 135–200° C range.

The $Cu(NH_3)_4Ni(CN)_4.3H_2O$ complex loses[15] two molecules of ammonia in the 20–80 °C range, the water content is lost in the 80–150° C range, and the third and fourth molecules of ammonia are lost in the 150–210° C and 210–293° C ranges respectively, leaving $Cu[Ni(CN)_4]$ as the final product.

The two complexes containing non-stoichiometric amounts of benzene, $Cu(en)_2Ni(CN)_4.0.14C_6H_6$ and $Zn(en)_2Ni(CN)_4.0.14C_6H_6$ have been reported as decomposing in similar ways.[1,8,9] Some en is lost in the 210–238 °C range in the former, and in the 167–266° C range in the latter. Both complexes have lost the benzene content near 300° C.

3. The non-stoichiometry of tetracyanocomplexes

Many of the above compounds exist in non-stoichiometric forms which can be obtained[1] in one of three ways:

(i) The preparation of complexes such as $M(NH_3)_2M'(CN)_4.nG$ and $M(en)_mM'(CN)_4.nG$ can result in a non-stoichiometric product i.e. n is non integer.

(ii) Thermal decomposition of a fully stoichiometric compound such as $M(NH_3)_2M'(CH)_4.2G$ can result in the loss of some of the guest molecules.

(iii) Heating a stoichiometric or a non-stoichiometric form of a tetracyanocomplex, and allowing it to remain in contact with the products of decomposition. In such cases, it is possible for the original guest molecule to be reabsorbed or for another species to be absorbed as the guest molecule.

It has been found that non stoichiometry is important for a tetracyanocomplex to show sorptive ability.

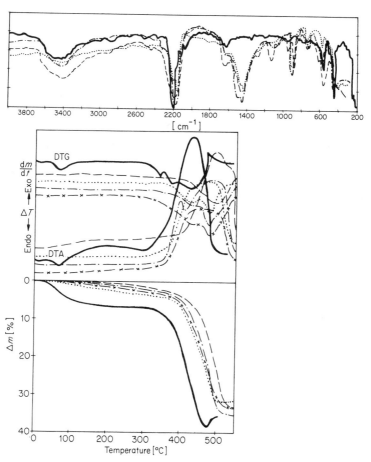

Fig. 2. The infrared spectra and the TG, DTG and DTA curves[14] for: —— $Zn[Ni(CN)_4].H_2O$; $---$ Zn complex treated with liquid C_6H_6 at 25° C for 1 h. Measurements made immediately after preparation. $\cdots\cdots$ As above, but measurements made two months after preparation. $-\cdot-\cdot-$ Zn complex treated with refluxing C_6H_6 for 1 h. Measurements made two months after preparation. $-\times-\times-$ Zn complex treate᠃ with refluxing CH_3OH for 1 h. Measurements made two months after preparation.

4. Sorptive abilities

Heating the inclusion compound $Zn(NH_3)_2Ni(CN)_4.2C_6H_6$ results in the loss of the benzene. In the compound $Zn(NH_3)_2Ni(CN)_4.0.2C_6H_6$, the structure will consist of the typical layered structure characteristic of Hofmann host lattices but with the majority of the voids being empty. If this non-stoichiometric compound is now exposed to organic vapours it has been found to be capable of absorbing compounds such as[1] β-picoline and alcohols.

It has also been found that the benzene guest molecules in $Ni(NH_3)_2Pt(CN)_4.2C_6H_6$ can be wholly or partially replaced by solvent molecules when suspended in solvents such as $CHCl_3$, dimethylformamide, nitrobenzene, and ethyleneglycol monoethyl ether.[1,16] When the inclusion compound $Ni(NH_3)_2Pt(CN)_4.2C_6H_5OH$ is heated in vacuo, the phenol guest molecules are liberated, and the resulting voids can be reoccupied by phenol or by another guest molecule such as benzene.[10]

It has also been reported that tetracyanocomplexes such as $Zn[Ni(CN)_4].H_2O$ will absorb organic guest molecules, such as benzene or methanol either by suspension of the complex in the liquid guest for 24 h, or by a continuous method using a Soxhlet apparatus.[1,14,16] Fig. 2 illustrates the infrared spectra, and the TG, DTG and DTA curves of the starting material and different products. Since the infrared spectra recorded two months after preparation still indicate the presence of benzene and methanol in the samples, this would seem to eliminate the possibility of surface absorption. The infrared spectra also point to the presence of both water and the organic compound present in the same sample. The thermal analysis curves also indicate the loss of two guests, the presence of the benzene giving added thermal stability.

5. Tetracyanocomplexes as stationary phases in gas chromatography

Since several tetracyanocomplexes have been shown to have sorptive properties an obvious next step was to try them as stationary phases in gas chromatography to investigate their separation ability.[1,2,9]

The complexes were deposited (10% w/w) on silanized Chromaton N-AW-HMDS, particle size 0.14–0.17 mm using a cyclohexane suspension, or they were mixed with the inert support and then packed into the column, of length 2.5 m and 3 mm i.d. The chromatograph was a Chrom 4 (LP-Prague) with a flame ionization detector.

The log t_R vs. $1/T$ plots for a homologous series of n-paraffins when $Cu(NH_3)_4Ni(CN)_4.3H_2O$ was used as the stationary phase,[15] are shown in Fig. 3.

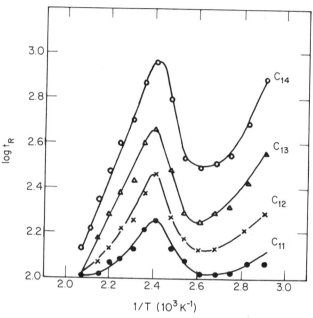

Fig. 3. Log t'_R vs. $1/T$ dependence for n-paraffins.[15] Stationary phase: 10% $Cu(NH_3)_4Ni(CN)_4.3H_2O$ on Chromaton N-AW-HMDS.

The shape of the plot can be correlated with the thermal decomposition of the complex as outlined in Section 2. In the temperature range 20–80 °C, two molecules of ammonia are lost, and the complex behaves as a surface adsorbent. It is only in the range 80–150° C that all the water is lost and the space occupied by the water is now accessible to the paraffin molecules. The maximum retention times are achieved at the temperature corresponding to the almost complete loss of water. A further increase in the column temperature leads to a decrease in the retention time as further ammonia molecules are lost and the complex once again behaves as a surface adsorbent.

Similar plots are obtained with aliphatic alcohols with the maximum and minimum retention times occurring at the same temperature as for the n-paraffins.

A. Sopková and M. Šingliar

The separating ability of the column is also found to be temperature dependent (Fig. 4), with the best resolution being obtained at the temperature (142 °C) corresponding to the almost complete loss of the water content. The high thermal stability of the complex is obvious from Fig. 5 which illustrates some chromatograms obtained after heating the column to 210° C for 1 h i.e. the temperature at which a third ammonia molecule is lost. There is a close similarity between Figs. 4c and 5c, the main effect of losing the third ammonia molecule being a reduction in the retention times which is similar to the behaviour of $Zn(NH_3)_2Ni(CN)_4.2H_2O$.

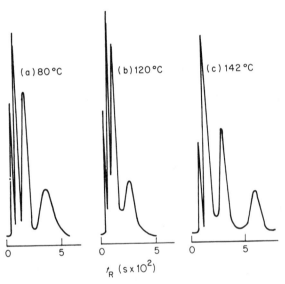

Fig. 4. Chromatograms of the $C_{11}-C_{14}$ *n*-paraffins[15] at (a) 80° C; (b) 120° C; (c) 142° C. Stationary phase as in Fig. 3.

The non-stoichiometric complexes $Cu(en)_2Ni(CN)_4.0.14C_6H_6$ and $Zn(en)_2Ni(CN)_4.0.14C_6H_6$ have also been used as stationary phases.[1,9] A mixture of the former and the inert support was found to give an acceptable separation of a mixture of aniline, methyl aniline and dimethyl aniline. It also showed no evidence of irreversible sorption of pyridines. This is very different from the behaviour of $Cu(NH_3)_4Ni(CN)_4.3H_2O$ which shows irreversible sorption of various pyridines at 150° C, (Fig. 6) probably due to their coordination to the metal atoms.

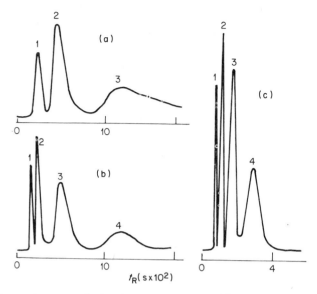

Fig. 5. Chromatograms of the C_{11}–C_{14} *n*-paraffins[15] at (a) 80° C; (b) 100° C; (c) 150° C after heating the column to 210° C. Stationary phase as in Fig. 3.

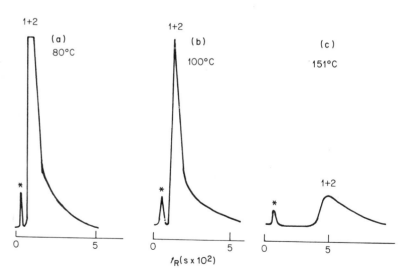

Fig. 6. Chromatograms of a mixture of pyridine (1) and 3-methyl pyridine (2) at (a) 80° C (b) 100° C (c) 151° C. Stationary phase as in Fig. 3. *Residual acetone.

When $Cu(en)_2Ni(CN)_4.0.14C_6H_6$ is deposited on the support material,[17] its sorption activity is found to increase. An example of the separation of aromatic hydrocarbons is illustrated in Fig. 7.

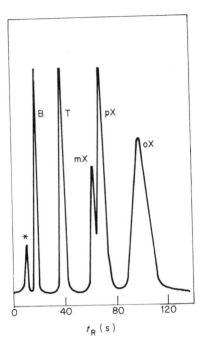

Fig. 7. Chromatograms of aromatic hydrocarbons.[17] Stationary phase: 10% $Cu(en)_2Ni(CN)_4.0.14C_6H_6$ on Chromaton N-AW-HMDS, column temperature 80° C. B, Benzene; T, Toluene; oX, mX, pX, *ortho, meta* and *para* xylene. *Residual acetone.

The separation of a mixture of aromatic amines using $Zn(en)_2Ni(CN)_4.0.14C_6H_6$ is illustrated in Fig. 8. The elution order is in the reverse order of their boiling points indicative of hydrogen bonding interactions. Such an inverse relationship between the elution order and boiling point is also observed when using other host lattices as stationary phases (see preceding chapter).

The thermal stability of tetracyanocomplexes is also illustrated in Fig. 8. Figures 8b and 8c were obtained after heating the column to 160° and 192° C respectively, temperatures at which some ethylenediamine is lost. The column still retains its separating ability but the retention time is reduced.

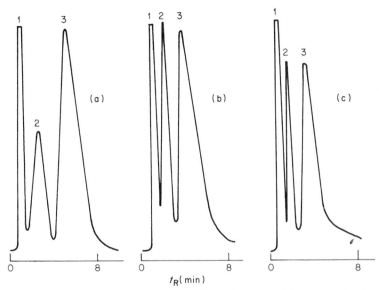

Fig. 8. Chromatograms of aromatic amines[17] at (a) 100° C; (b) 100° C after heating the column to 160° C; (c) 100° C after heating the column to 192° C. Stationary phase: 10% $Zn(en)_2Ni(CN)_4.0.14C_6H_6$ on Chromaton N-AW-HMDS. 1, Dimethyl aniline; 2, Methylaniline; 3, Aniline.

6. Conclusions

The study of their compounds tetracyanocomplexes and even containing non-stoichiometric amounts of guest (including water) has shown that they can be used as stationary phases in gas chromatography. The study of the temperature dependence of the retention times of the inclusion compounds and of hydrated complexes suggests an optimum separating ability during the temperature range when the guest is being liberated. Guest free compounds show no sorptive abilities.

Due to their high thermal stability tetracyanocomplexes can be used at much higher temperatures than other host lattices (see preceding chapter), and may thus be useful for separations which cannot otherwise be achieved.

References

1. (a) A. Sopková, *J. Mol. Struct.*, 1981, **75**, 81. (b) A. Sopková and M. Šingliar, *Collect. Czech. Chem. Commun.*, 1984 in the press. (c) A. Sopková, M. Šingliar, J. Bubanec, T. Görnerová and P. Králik, *Czech. Patent*, **222**, 610, 9 March 1983.

2. M. Šingliar, A. Sopková, J. Chomič, J. Skoršepa and E. Matejčíková, *Czech. Patent* 185986, 6 June 1978.
3. (a) H. R. Oswald, *Proc. 6 ICTA, Thermal Analysis*, Vol. 1, Bayreuth, 1980, 1. (b) V. Bachmann, I. R. Günther and H. R. Oswald, *Experienta* (*Suppl.*) 1979, **37**, 36.
4. J. H. Rayner and H. M. Powell, *J. Chem. Soc.*, 1958, 3412.
5. (a) A. G. Sharpe, *The Chemistry of Cyanocomplexes of the Transition Metals*, Academic Press, London, 1976. (b) E. E. Aynsley and W. A. Campbell, *J. Chem. Soc.*, 1958, 1723.
6. A. Sopková, J. Černák, M. Šingliar, J. Bubanec and P. Králik, *Abstracts of First Int. Symp. Clathrates and Molecular Inclusion Phenomena*, Jachranka, Warsaw 1980, 46.
7. C. Kappenstein and J. Černák, *Acta Crystallogr.*, 1984, in the press.
8. A. Sopková, J. Chomič and E. Matejčíková, *Monatsh. Chem.*, 1976, **102**, 961.
9. A. Sopková, M. Šingliar, J. Chomič, J. Skoršepa, and E. Matejčíková, *Czech. Patent* 186492, 7 July 1978.
10. A. Sopková and J. Bubanec, *J. Mol. Struct.*, 1981, **75**, 73.
11. I. Feigl, P. Démant and E. Oliviers, *Ann. Assoc. Quim.*, *Brazil*, 1945, **3**, 72.
12. P. Pascal, *Nouveau Traité de Chimie Mineral*, Vols. 17 and 18, Masson, Paris, 1963.
13. A. Sopková and J. Bubanec, *J. Therm. Anal.*, 1977, **12**, 97.
14. A. Sopková, M. Šingliar and P. Králik, *J. Incl. Phenom.*, 1983/4, **1**, in the press.
15. M. Šingliar and A. Sopková, *Petrochémia*, 1977, **17**, 18.
16. A. Sopková and P. Králik. *Proc. Intern. Microsymp. Clathrate and Molecular Inclusion Compounds*, 1981, Stará Lesná (Czechoslovakia) 87.
17. M. Šingliar and A. Sopková. *Proc. Intern. Microsymp. Clathrate and Molecular Inclusion Compounds*, 1981, Stará Lesná (Czechoslovakia), 223.

8 · ISOTOPIC FRACTIONATION USING INCLUSION COMPOUNDS

N. O. SMITH

Fordham University, New York, U.S.A.

1. Coordination compounds as hosts

Coordination compounds, with their arm-like ligands, crystallize to give structures with voids into which guest molecules can fit. They are versatile as hosts and can include within their cavities or channels a wide variety of guests. A given host, which forms an inclusion compound with both guest A and guest B when the guests are made available in separate experiments, will commonly show a preference for A over B when both are made available simultaneously. Thus the phenomenon of inclusion of two or more guests at a time shows a selectivity for one guest over another, and this property forms the basis for the separation, or at least fractionation, of mixtures of guests. Even structural isomers can be so fractionated. This subject has been covered elsewhere in this work. (See Chapters 3 and 4, Volume 1).

The remarkable selectivity towards structural isomers led the author to determine whether the selectivity could be extended to guests differing only isotopically, although it was anticipated that the effect, if any, would be a small one. Schaeffer *et al.*[1] had shown that $Ni(4\text{-MePy})_4(NCS)_2$ (**1**) would fractionate *ortho-*, *meta-* and *para-*xylenes, so attempts were made to fractionate mixtures of $p\text{-}(CH_3)_2C_6H_4(A)$ and $p\text{-}(CD_3)_2C_6D_4(B)$ and mixtures of A and $p\text{-}(CD_3)_2C_6H_4(C)$, using **1** as host.

INCLUSION COMPOUNDS III
ISBN 0-12-067103-4

Two approaches were used. In the first approach excess of mixtures of A and B and of A and C, dissolved in n-pentane, were stirred with 1 at room temperature for several days, during which period the two guests distributed themselves between the solid and liquid phases. The two phases were then separated and the relative amounts of A and B and of A and C in them determined by mass spectrometry. The details are given by Ofodile and Smith.[2] If the mole ratios B/A or C/A in the solid and liquid phases are denoted by R_S and R_L, respectively, it was found that the separation factor, $\alpha = R_S/R_L$, for the B/A system averaged $1.10 \pm .03$ for seventeen different mixtures in which R varied from 0.2 to 1.7. For the C/A system the value of α averaged $1.13 \pm .04$ for sixteen mixtures in which R varied from 0.3 to 5.5. Thus the values of α in the two systems are the same within experimental error. Nevertheless, the fact that R_S is uniformly about 10% larger than R_L is significant, and shows that the solid phase has a higher concentration of the more deuterated isotopomer than has the coexisting liquid phase. A similar result was obtained in a single experiment with the guest pair naphthalene–naphthalene-d_8.

In the second approach Ofodile and Smith[3] attempted to improve the fractionation by using liquid chromatography. Kemula and Sybilska[4] and Lipkowski et al.[5] had shown that 1, properly pre-treated, could be used as a stationary phase in the chromatographic separation of certain aromatic compounds, (see Chapter 6, Volume 3) and their technique was modified and used for the chromatographic fractionation of mixtures of A and B, of A and C, and of B and C. With R representing the mole ratios B/A, C/A and B/C in the three systems, the R values in the original mixture (R_0) and in the material included on the column (R_S) were measured. In one set of experiments R_S/R_0 was 1.4, 1.4 and 1.01 for the pairs B/A, C/A and B/C, respectively. Thus, again, the solid phases retained more of the more deuterated isotopomer, although the selectivity in the pair B/C was barely more than the experimental error. These results are consistent with those found in the first approach. They also show that deuteration of the methyl groups of the p-xylene is more influential in the fractionation than deuteration of the aromatic ring.

The explanation for the selectivities described above is a matter of interest and importance. In the fractionation of *ortho, meta,* and *para* isomers, one cannot overlook the differences in the shapes of the guest molecules. The marked preference of 1 for the *para* isomer is undoubtedly related to its elongated shape. The detailed crystal structure of the p-xylene inclusion compound with 1 has been published by Lipkowski et al.[6] following earlier studies of the host structure by Andreetti et al.[7] and by deGil and Kerr.[8] Such studies show interconnecting, zeolite-like channels spiralling through the host into which it would be "easier" for an elongated molecule to fit. The work of Lipkowski et al.[6] shows that the thermal motion of the guest molecules is more pronounced than for the host, indicating a comparatively loose fit of the guest in the channels. This, in turn, suggests that any difference shown by **1** towards guests differing only isotopically may be attributable not to differences in guest–host forces but to differences in the fugacities of the two guests in the liquid phase. These would be related to the vapour pressures of the pure liquid guests. Although the vapour pressures of the deuterated *p*-xylenes appear not to have been measured, one can estimate from the boiling points ($A = 138.3°\,C$, $B = 135°\,C$, $C = 135°\,C$) and the analogy to the isotopic benzenes that the vapour pressures of B and C are roughly 10% greater than that of A at room temperature. If this is so the fugacity of the more deuterated isomer would be 10% greater than that of the non-deuterated for a given concentration, and this could account for the observed separation factors. It would also indicate why mixtures of B and C could barely be fractionated.

Very recent, and as yet unpublished work by the author, with hosts in which there is a tight fit of the guest in the host cavities, has shown in two different systems that there is a noticeable preference for the protiated over the deuterated isotopomer even though, in these instances also, the deuterated compound has the greater vapour pressure.

2. Zeolites as Hosts

Attempts have been made for a number of years to fractionate mixtures of molecules such as D_2 and H_2 by using differences in the sorption qualities of such materials by zeolites. See Chapter 6, Volume 1 for a discussion of the zeolite structures. As early as 1958 Ohkoshi *et al.*[9] reported the separation of hydrogen isotopes by gas chromatography using molecular sieves, and since then a number of publications have appeared on the subject in journals and in the patent literature.

In 1960, for example, Basmadjian[10,11] made detailed measurements of the sorption of H_2 and D_2 on molecular sieves of the type 4A, 5A and 13X, in the neighbourhood of liquid air temperatures, and found a clear preference for D_2 sorption over H_2. In 1967, Parbuzin *et al.*[12] reported the selectivity shown by NaA, Linde 5A and Linde AW-500 toward the pairs D_2–HD, HD–H_2 and D_2–H_2, again at liquid air temperatures. They found that the separation factors (the ratio of the more to the less deuterated species in the sorbed gas divided by that in the gas phase) were all greater than unity and as large as four under some conditions. The values also depended on the degree of saturation of the zeolite. The differences in the behaviour of the isotopic forms are attributed to their different interaction energies with the host. Work has also been performed on the differential sorption behaviour of the isotopes of nitrogen, carbon, oxygen and the inert gases.[13-15] In all these cases the heavier isotopic form is held by the sorbent to a slightly greater extent than the lighter one.

The guests discussed in the preceding two paragraphs are all neutral molecules, but the use of zeolites as ion exchange media for the fractionation of isotopes has also been investigated. In functioning in this way, ions of the original host zeolite are *replaced* by ions from the fluid phase to a greater or lesser extent, and it may be pointed out that any designation of such materials as "inclusion compounds" broadens the meaning of the term since the eluant carries away some of the original host. Moreover, the forces determining the selectivity of ions in such situations are of a different type from those determining the selectivity of neutral guest molecules. In 1937, Taylor and Urey[16,17] demonstrated that by passing aqueous LiCl through a column of Permutite there was a fractionation of the lithium isotopes. (Strictly, Permutite may not be called a zeolite because it is amorphous.) Similar work was performed with potassium and ammonium salts. Whereas with Li^+ the lighter isotope, $^6Li^+$, was held more firmly on the column, with K^+ and NH_4^+ the heavier isotopes, $^{41}K^+$ and $^{15}NH_4^+$, were held more firmly. The cumulative effect of the column led to isotope abundance ratios about 25% different from the natural abundance ratios. Further work on the fractionation of Li^+ isotopes has been carried out more recently in Czechoslovakia[18] and the USSR.[19]

3. Crown ethers as hosts

Certain species can, in liquid solution, form inclusion compounds with other neutral or ionic species. Among the first of such host materials were the cyclodextrins, considered elsewhere in this work (see Chapter 8, Volume 2), but of more recent discovery are species such as cryptates, spherands and crown ethers (see Chapters 9 and 10, Volume 2). Jepson and DeWitt[20] have discovered that the crown ethers can be used to fractionate the calcium ion isotopes. They used dicyclohexyl 18-Crown-6 (2) and dibenzo 18-Crown-6 (3). When 2 or 3, dissolved in chloroform or methylene chloride, are shaken with aqueous $CaCl_2$, the Ca^{2+} ions are distributed between the organic and aqueous phases in such a way that the $^{44}Ca^{2+}/^{40}Ca^{2+}$ ratio is different in the two layers, the heavier isotope being concentrated to a small extent in the aqueous phase. The Ca^{2+} ions in the organic layer form an inclusion species with the crown ether. They report a separation factor of 1.001 per unit mass, but do not attempt an explanation. It appears that, since this system can be adapted to a closed reflux system it has the potential for isotope separation. Their results invite further study with other guest–host combinations.

A different approach was used by Schmidhalter and Schumacher[21] to enrich potassium isotopes using the *cis-syn-cis* isomer of dicyclohexyl-18-Crown-6 (L). The isotopes were distributed between an ion exchange resin and a CH_3OH/H_2O solvent according to the equilibrium:

$$^{41}K^+_{res} + {}^{39}KL^+ \rightleftharpoons {}^{39}K^+_{res} + {}^{41}KL^+$$

The lighter isotope is preferentially retained by the resin with a separation factor ranging from 1.00118 at $-10°$ C to 1.00074 at $+10°$ C. The temperature dependence yields an enthalpy of exchange of -13 ± 4 J mol^{-1}. The authors suggest that the isotopic selectivity arises from the small zero point energy differences arising from *second* neighbour O atom interactions which perturb the (K–O) vibrations slightly, and that the method has the potential for isotope separation.

References

1. W. D. Schaeffer, W. S. Dorsey, D. A. Skinner and C. G. Christian, *J. Am. Chem. Soc.*, 1957, **79**, 5870.
2. (*a*) S. E. Ofodile, R. M. Kellett and N. O. Smith, *J. Am. Chem. Soc.*, 1979, **101**, 7725; (*b*) S. E. Ofodile and N. O. Smith, *J. Phys. Chem.*, 1983, **87**, 473.
3. S. E. Ofodile and N. O. Smith, *Anal. Chem.*, 1981, **53**, 904.
4. W. Kemula and D. Sybilska, *Nature* (*London*), 1960, **185**, 237.
5. J. Lipkowski, K. Lesniak, A. Bylina and K. Duszczyk, *J. Chromatogr.*, 1974, **91**, 297.
6. J. Lipkowski, K. Suwinska, G. D. Andreetti and K. Stadnicka, *J. Mol. Struct.*, 1981, **75**, 101.
7. G. D. Andreetti, G. Bocelli and P. Sgarabotto, *Cryst. Struct. Commun.*, 1972, **1**, 51.
8. E. R. de Gil and I. S. Kerr, *J. Appl. Crystallogr.*, 1977, **10**, 315.
9. S. Ohkoshi, Y. Fujita and T. Kwan, *Bull. Chem. Soc. Jpn.*, 1958, **31**, 770.
10. I. D. Basmadjian, *Can. J. Chem.*, 1960, **38**, 141.
11. I. D. Basmadjian, *Can. J. Chem.*, 1960, **38**, 149.
12. V. S. Parbuzin, G. M. Panchenkov, B. Kh. Rakhmukov and G. V. Sandul, *Zh. Fiz. Khim.* 1967, **41**, 193.
13. A. M. Tolmachev, T. V. Zotova and L. N. Grishina, *Zh. Fiz. Khim.*, 1968, **42**, 1263.
14. M. Scaringella and F. Botter, *J. Chim. Phys. Phys.-Chim. Biol.*, 1981, **78**, 7.
15. H. Illy and F. Botter, *J. Chim. Phys. Phys.-Chim. Biol.*, 1981, **78**, 17.
16. T. I. Taylor and H. C. Urey, *J. Chem. Phys.*, 1937, **5**, 597.
17. T. I. Taylor and H. C. Urey, *J. Chem. Phys.*, 1938, **6**, 429.
18. O. Matousova, *Sb. Vys. Sk. Chem.-Technol., Praze, Mineral.*, 1968, **10**, 113.
19. V. A. Chumakov, V. I. Gorshkov, A. M. Tolmachev and V. A. Fedorov, *Vestn. Mosk. Univ., Khim.*, 1969, **24**, 22.
20. B. E. Jepson and R. DeWitt, *J. Inorg. Nucl. Chem.*, 1976, **38**, 1175.
21. B. Schmidhalter and E. Schumacher, *Helv. Chim. Acta*, 1982, **65**, 1687.

9 · ENANTIOMERIC SELECTIVITY OF HOST LATTICES

R. ARAD-YELLIN
Weizmann Institute of Science, Rehovot, Israel

B. S. GREEN
Weizmann Institute of Science, Rehovot and Israel Institute for Biological Research, Ness-Ziona, Israel

M. KNOSSOW AND G. TSOUCARIS
Centre Pharmaceutique, Chatenay-Malabry, France

1. Introduction

There has been continuing interest in asymmetric synthesis and in the study of the discrimination between enantiomeric chiral species because of both the practical applications and the analogy with biological systems. The main effort has been directed towards devising new chemical systems in reactions occuring in solution. However, it has been shown that the rigidity and order of molecular arrangements in the crystalline state can provide a useful way of achieving highly specific syntheses and efficient chiral discriminations.

INCLUSION COMPOUNDS III
ISBN 0-12-067103-4

In order to use the solid state for enantiomeric separations a method has to be found of reliably achieving a chiral crystalline environment, a goal which is challenging, since the majority of racemic chiral substances tend to give achiral crystals. Chiral inclusion compounds provide an answer to this problem for a great variety of molecules. These compounds contain chiral cavities in which the included molecules are in a defined chiral environment and enantiomers should therefore be differentiated and also react at different rates.

Four of the host molecules which give chiral inclusion compounds have been extensively studied and will be described in this review. Cyclodextrins are chiral molecules which form host-guest complexes in solution as well as in the crystalline phase. Choleic acids involve chiral molecules which form inclusion compounds only in the solid state, urea is a non-chiral molecule which forms chiral inclusion compounds and tri-o-thymotide is a labile chiral molecule which spontaneously resolves to give chiral inclusion compounds.

Chiral inclusion compounds appear to be particularly attractive media for effecting chiral discrimination for the following reasons: (a) The guests accommodated by a given host generally vary widely in functional groups and sometimes also in size and shape. (b) The unvarying structure of a given host lattice (which allows different guests to be accommodated in identical cavities in the crystal of the host) enables the rationalization and even the prediction of the degree of enantiomer specificity as well as the correlation of host and preferred guest chirality. (c) Since the guest–host interactions are relatively weak, and no specific polar or apolar functionality is needed for the formation of the inclusion compounds, the possibilities of resolving molecules which can otherwise only be resolved with great difficulty or not at all are opened up. (d) Since resolution involves only crystallization, which can be performed even at low temperatures, the isolation and investigation of labile chiral species may also be achieved by this method.

The study of chiral inclusion compounds might provide a better understanding of the weak chiral interactions responsible for chiral discrimination and asymmetric synthesis. Compared with the more complex biological systems, such as enzymes or antibodies, inclusion complexes have the advantage of a much shorter time required to collect and interpret X-ray data of high precision. The chemical aspects of specific biological features such as the hydrophobic environment may be more easily studied by trying to understand host-mediated chiral discrimination or chemical reaction studies. A model or rationale can also be more easily achieved and tested by the design, synthesis, and study of new host molecules.

2. Cyclodextrins

Cyclodextrins (CDs) are especially versatile members of the inclusion compound family: they are able to form complexes in solution as well as in the crystal state. There do not seem to be restrictions on the nature of possible guests, as seen from the variety of molecules studied so far, e.g. esters, acids, alcohols, sulphoxides, sulphinates, phosphinates, amines, ketones. (See Chapter 8, Volume 2.) The enantiomeric selectivity is exhibited in both the liquid and solid states in different ways: enantiomeric separation by fractional crystallization; asymmetric synthesis; modification of the guest chiroptical and other spectroscopic properties. Among CD molecules, α- and β-CD have been most extensively studied and the results reported here have been obtained with these hosts. Figure 1 represents a model of α-CD from X-ray studies.

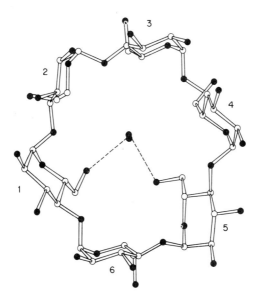

Fig. 1. α-CD conformation in the α-CD water complex perpendicular to the line connecting the two water molecules.[72] β-CD and γ-CD are composed of 7 and 8 glucoside units, respectively.

2.1. Enantiomer separation by crystallization

Selective precipitation by CD often provides a solid phase enriched with one antipode, and a supernatant liquid enriched with the other one. Some results are summarized in Table I. Most of the guests so far studied, with the exception of methyl isopropylsulfinate (23) and methyl (8) and ethyl (9) isopropylphosphinate were found to be resolved with small to medium selectivity. The average optical purity over 43 reported guests is 10.6%. However, this method of enantiomer purification is attractive because cyclo-

Table 1. Enantiomeric selectivity of different guest molecules in β-cyclodextrin

			α_D^{25}	Optical purity (%)	Absolute configuration of preferred enantiomer
			Carboxylic acids and esters		
1	atrolactic acid		−2.90	5.57	
2	dibromosuccinic acid		12.10	8.18	
3	dibromocinnamic acid		7.71	11.33	
4	phenylbromoacetic acid		−0.92	5.75	
5	menthyl chloroacetate		−1.72	2.21	
6	menthyl acetate		−1.42	1.78	
7	ethyl mandelate		−5.90	3.32	
	R′	R^2	Phosphorus derivatives R′OP(R^2)OH		
8	Pr^i	Me	−17.30	66.50	
9	Pr^i	Et	−15.18	60.00	
10	Et	Et	−5.06	23.80[a]	
11	Et	Ph	12.08	28.80[a]	
			Sulphoxides and sulphinates R′S(R^2)O		
12	Bz	Me	−8.50	8.00	R
13	Bz	Bu^t	45.00	14.50	R
14	Ph	Me	6.50	4.40	R
15	Ph	Et	16.10	9.10	R
16	Ph	Bu^n	15.30	9.20	R
17	Tol	Me	11.50	8.00	R
18	Ph	Bu^t	−1.90	1.10	S
19	Tol	Bu^t	−12.30	6.40	S
20	Me	OBu^t	−19.90	12.40	S
21	Pr^i	OMe	14.40	12.80	R
22	Me	OBu^i	10.10	8.70	R
23	Me	OPr^i	−165.91	68.20	S

[a] Inclusion in α-CD

dextrins are optically pure molecules, all the precipitating crystals have the same chirality and the guest optical purity can be determined with polycrystalline samples.

2.1.1. Carboxylic acid esters

For a series of different esters the absolute configuration of the preferentially included enantiomer in β-CD is the same, and this suggests a simple steric mechanism of discrimination.[1] The sign of the optical rotations of menthyl esters of chloroacetic (5) and acetic (6) acids are reversed when the temperature in which the complexes are formed is raised (for 5: $-1.7°$ at $0°$ C, $+3.7°$ at $50°$ C; for 6: $-1.4°$ at $0°$ C, $0.0°$ at $50°$ C, $1.5°$ at $58°$ C). This sensitivity of enantiomeric discrimination may be due to a conformational change of the guest at different temperatures or to the value of the entropy term.

2.1.2. Phosphorus and sulphur derivatives

Cyclodextrins (α and β) exhibit small to medium selectivity toward phosphinates and related molecules. Van Hooidonk *et al.*[2] showed that, in aqueous alkaline solution, the (S)-(+)-enantiomer of isopropyl methylphosphonofluoridate (Sarin) is preferentially included in α-CD prior to phosphorylation. Benschop and Van den Berg[3] investigated a series of phosphinates (Table I).

(8)   **(23)**

Cyclodextrin complexes of several sulphoxides, sulphinates and thiosulphinates have been studied by Mikolajczyk and Drabowicz[4,5] (Table I). Most of these derivatives exhibit a rather low chiral discrimination (except for 23). The stereospecificity of inclusion of sulphoxides into β-CD could be rationalized in terms of the steric bulk as well as the spatial arrangement of substituents connected to the chiral sulphur atom. Steric requirements increase in the order $ME < Bz < Bu^t$ and thus the preferred enantiomers of **12** and **13**, which have a similar spatial arrangement, are both of R configuration. Similarly, the R enantiomers of the pair **(14)**, **(17)** and the S enantiomers of **(18)**, **(19)** are expected to be preferentially complexed by β-cyclodextrin. The authors suggest that the main driving force for complex formation is the inclusion of the aromatic ring into the CD cavity. Further stabilization is achieved by hydrogen bonding between the CD OH groups and the sulphinyl oxygen atom. As expected, the enantiomer of the phosphorus analogue of **8** selected by β-CD has a configuration closely related to that of the preferred enantiomer of **23**.

2.2. Preferential binding in solution and preferential inclusion on crystallization

Rassat and his associates[6,7,8] have used the ESR technique to study the enantiomeric selectivity of both paramagnetic and diamagnetic molecules. The nitroxide biradical **(24)** has been complexed with β-CD as a pure enantiomer (3'R, 1'R) (derived from (R)-(+)-3-methylcyclohexanone) and as a racemic mixture. The values of the association constants (at $20°$ C) were measured by ESR techniques. These values are: $K(d) = 24$; $K(l) = 12$; $\rho = K(d)/K(l) = 2$. The ratio of association constants of the biradical with β-CD was compared to the same ratio ρ' for 3-methylcyclohexanone which was obtained by selective precipitation. By addition of a racemic compound to a solution of β-CD, a precipitate is formed and filtered. The filtrate was extracted and the optical rotatory power α' of the extracted material was measured. The association constant was found to be $\rho' = 1.7 \pm 0.25$. This value, which is the same order of magnitude for racemic 3-methylcyclohexanone and the biradical derived from 3-methylcyclohexanone, indicates that the chiral recognition in both systems is based on the 3-methylcyclohexane moiety.[8]

(3'R1'R)

(24)   **(25)**

The association of diamagnetic molecules to β-CD was studied by a displacement method using the paramagnetic nitroxide biradical. It has been found that the intensities of certain lines in the ESR spectrum of the biradical decreased more significantly when (+)-fenchone (**25**) was added than with (−)-fenchone, a result which indicates that (+)-fenchone is better complexed with β-CD than (−)-fenchone. A precipitation experiment was in agreement with this result: precipitation of the β-CD complex from a racemic solution of fenchone left a solution which was enriched with (−)-fenchone.

2.3. Thermodynamic data and models for chiral discrimination

There has been considerable interest in analysing the driving forces for cyclodextrin complex formation, which has been attributed to various factors: generation of activated water, change of conformational energy of the host on complexation and change in the degrees of motional freedom of the guest. Calculations taking into account what has been termed "hydrophobic interactions" have given a reasonable estimate of the energy of complexation in solution. No calculation has thus far been performed to evaluate the energy involved in chiral discrimination or crystalline inclusion compound formation.

Experimental information about the binding mode has been obtained by Cooper and MacNicol[9] by combining spectroscopic and calorimetric data for several chiral benzene derivatives complexed with α-CD. The spectroscopic method is a displacement technique based on competition for CD of the examined molecules with a powerfully bound guest, p-nitrophenolate. The subsequent variation of the CD–p-nitrophenolate concentration after the addition of benzene derivatives is evaluated by difference absorption spectra. Double reciprocal plots of the spectral titration with α-CD of p-nitrophenolate in the presence and absence of the chiral inhibitors allow the determination of the dissociation constants. Microcalorimetry has been used to determine the heat of complex formation and to provide a direct thermal titration of the complexes. A general model suggests that the insertion of the aromatic ring, and subsequent interaction with the CD, are the major driving forces for complex formation. The chiral substituents, at the edge or outside the cavity, may still interact with the rim of the CD torus. The resulting chiral discrimination in complexation and crystallization depends on the exact geometry and strength of this chiral interaction. It is not therefore surprising that enantiomeric selectivity is generally small or medium (Table 2).

Although the view that complex formation is mainly influenced by "hydrophobic interactions" has been supported by theoretical calculations, by crystal structure determination for various guests and by spectroscopic evidence such as the Intermolecular Nuclear Overhauser Effect, this view

Table 2. Thermodynamics of binding of chiral guest molecules to α-cyclodextrin

Guest	Enantiomeric form	$10^3\,K_{diss}(dm^{-3}\,mol^{-1})$		$-\Delta H(KJ\,mol^{-1})$
Phenylalanine	D	48.5[a]	55.3[b]	16.3
	L	63.0	64.6	15.5
α-Methylbenzylamine	+	38.5		15.0
	−	29.8		15.5
Mandelate	D	126.0	129.0	12.9
	L	121.0	129.0	14.2
Phenyltrifluoroethanol	D	17.6	24.9	14.6
	L	17.5	19.6	12.5
Amphetamine	D	41.8		12.5
	L	39.6		12.5
p-Nitrophenolate		0.53		

[a] Obtained from spectral inhibition titrations.
[b] Obtained from calorimetric titrations.

does not allow a sufficiently precise picture, which explains the chiral discrimination, to be drawn. There are various reasons for this uncertainty: burying hydrophobic parts of a guest does not allow the unequivocal determination of guest conformation, which is important in a study of complexation, and although the predominant host–guest conformation may be correctly determined by the above mentioned methods, this conformation may not be the one which causes discrimination. It cannot be excluded that another minor conformation is highly important and responsible for the observed effects. Additional experimental evidence such as crystal structure determination of complexes with both enantiomers where an efficient discrimination is observed (e.g. with guests 8, 9 or 23) would help to provide a better insight into chiral discrimination by CD, both in the solid state and in solution.

2.4. Induced chiroptical phenomena

The chiroptical properties (optical rotation dispersion (ORD), and circular dichroism (CID)) of the guest may be modified upon inclusion. These complexes provide a very simple and sensitive method of detecting CD guest association and present a direct way to evaluate chiral discrimination and a tool to estimate the association constant K. The physical nature of the observed modifications may be of several kinds: asymmetric perturbation of the electronic distribution in the guest molecule (electronically induced CID), stabilization of one or several chiral conformers, and chiral modifications of the guest's geometry (geometrically induced CID). It often happens

for molecules with very low barriers of internal rotation, that chiral conformers are not stable enough in solution, even at the lowest accessible
temperature, to exhibit a detectable optical rotation or circular dichroism
spectrum. But, association with an auxiliary stable chiral molecule stabilizes
these conformations and allows the determination of their chiroptical
properties. Complexation with CD has the advantage of forming compounds,
with presumably similar structural characteristics, with several members of
the same chemical family, and of being transparent for $\lambda > 220$ nm.
Moreover, the possibility of determining the chiral characteristics in both
solution and the solid state may lead to configurational correlations.

The ORD curves as well as the absorption spectra of several achiral dyes,
such as congo red, are modified upon inclusion in α-,β- and γ-CD.[10] Induced
CID arising from the $n - \pi^*$ transition appears for several derivatives of
benzophenone, benzobenzoic acid,[11,12] cyclohexanone,[13] barbiturates,[14] ben-
zil and biacetyl.[15] Modification of the CID spectra of chiral molecules such
as nucleic acid monomers[16] has also been observed.

The same enantioselectivity appears on crystallization and in the induced
CID. When the crystalline CD complex of 2-methylcyclohexanone was
analysed, it was found to contain an excess of the R-(+) enantiomer, the
sign of the induced CID of the racemic solution of this compound in the
presence of CD was found to be positive.[13]

All the above molecules can be included in the CD cavity. This is not
possible for bilirubine and biliverdine whose dimensions are much larger
than that of the CD cavity. They exhibit, however, the most intense induced
spectrum known, with $\theta_{max} \sim 20\,000$ deg dmol^{-1}. The complex can be
described as a "sitting on top" model, by the formation of a CD dimer
allowing a larger cavity or by a solvent-type effect. In any case CID probably
arises from a geometrical origin, i.e. one chiral conformation is more
stabilized than the others. It is interesting to note that the association
constant is rather weak, an observation which can be correlated with the
failure, up to now, to obtain a crystalline complex.

2.5. Chemical reactivity

Chemical reactions of guest molecules inside the cavities of a host may be
more sensitive to a small geometric or electronic chiral influence of the
cavity than a physical process such as enantiomeric separation. This is
primarily because of the kinetic aspect of the process which may be very
weakly, or not at all, related to the thermodynamic equilibrium of the
host-guest complex. The equilibrium dissociation constant K_d with β-CD is
identical for the enantiomers of (+) and (−) (**26**),[17] but the rate of hydrolysis

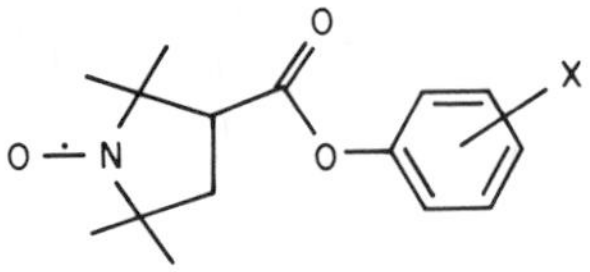

a. X = m−NO$_2$

b. X = p −NO$_2$

(26)

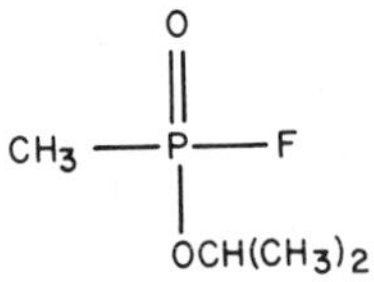

(27)

is 6.9 fold greater for the (−) enantiomer. For the methylphosphonate (27), the corresponding effect is 36.6 fold.[18] Interestingly, the equilibrium constant K_d is higher for the enantiomer for which the reaction is slower.

Enantiomeric preference is also shown in a solid state reaction within the β-CD complex cavities.[19] When the β-CD phenylethylmalonic acid (28) complex was heated at 100° C in the crystalline state, CO_2 was evolved. Among the extracted products 2-phenylbutyric acid (29) was formed with a 7.2% excess of the S-(+) enantiomer. This asymmetric synthesis is due to a crystal influence upon specific host-guest interactions because: (i) complexation of racemic (29) with β-CD and extraction of the guest from the resulting crystals led to no enantiomeric enrichment; (ii) heating of an aqueous solution of (28) and β-CD led to decarboxylation but the product (29) was racemic.

Relatively high enantiomer discrimination has been observed in chemical reactions of β-CD in solution.[20]

The recent reports of the binding of two guest molecules, either identical or different, within the cavity of γ-CD in solution[68-70] and the successful design of specifically modified cyclodextrin molecules[71] suggest further broad possibilities for chiral discrimination and reactions in these complexes.

3. Deoxycholic acid

The naturally occurring steroid, deoxycholic acid (DCA), was used in the earliest reported work[21] involving enantiomer separation by clathrate crystallization. (The inclusion compounds of DCA with guest substances are called choleic acids. See Chapter 7, Volume 2). The compounds which have been partially resolved by this method include methylethylacetic acid, 2-phenyl-1-butanol, and camphor.[21] Further work on the scope of the guests which may be resolved by this method and structural studies on diastereomeric complexes would be most interesting. Recent work on choleic acid has been mainly concerned with chemical reactivity. In a series of elegant experiments combining chemical studies and low-temperature X-ray analysis, stereospecific photochemical reactions between the host lattice molecules and included ketones have been described.[22–24] These reactions involve hydrogen abstraction from the host DCA molecules by the photoexcited ketones which then add to form a new carbon-carbon bond (Fig. 2). In each such reaction a new chiral centre is created on the steroid and, if the ketone is prochiral, a new chiral centre is also formed at the carbonyl carbon atom. The ratio of diastereomers varies with the guest, being highly selective for acetophenone. The selectivity can generally be understood in

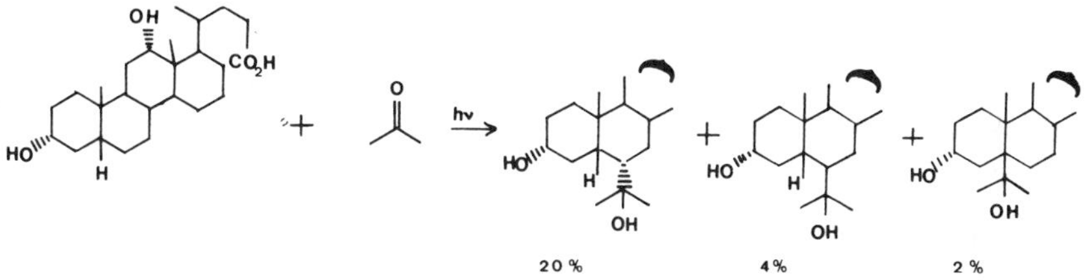

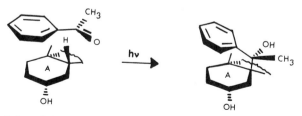

Fig. 2. The photochemical reaction between the host and the guest in the DCA–acetone inclusion compound.

terms of the orientations and distances of the carbonyl chromophore and the potentially reactive H–C groups on the steroid. Perhaps the most intriguing aspect of these studies was the discovery of a topochemical, strictly lattice-controlled, reaction in the DCA–acetophenone complex wherein the stereochemical outcome is the opposite of that anticipated on the basis of bond formation along the path of least motion. Thus, the 10:4 DCA–acetophenone complex gave, on irradiation under argon, only one photoproduct of addition to position 5 (Fig. 3). The absolute configuration

Fig. 3. Model describing abstraction of a H atom from the steroid and formation of a new chiral centre in the DCA–acetophenone inclusion compound.

of the new chiral carbon was determined by the crystal structure analysis at $-170°$ C. Two crystallographically independent guest molecules form a chain in the channel but they both occupy almost identical sites, and their C = O groups are in proximity to the steroidal H5 atom to be abstracted. The unexpected fact is that the ketone adds from that face of the aceto-phenone which is the most distant from the steroid in the starting structure. This implies the need for unusual motion of the guest acetyl group on reaction, as suggested by a detailed X-ray analysis on a crystal which had undergone a partial (40%) reaction. Both independent molecules had reacted, thus establishing the topochemical character of the reaction.

Under the same circumstances, the methylethylketone–DCA inclusion compound leads to both diastereomers. This is due to the fact that, in this structure, the guest molecules form centrosymmetric pairs within the channel so that the reactive centre of the steroid (here C5 and C6) is exposed equally well to the opposite faces of the ketone molecule. It is to be noted that precise determination of the crystal structure of the DCA-diethyl ketone inclusion compound has clearly shown the necessity for a correct distance and orientation prior to a host-guest reaction. This is exemplified by Fig. 4 which shows interatomic distances in the DCA-diethylketone inclusion

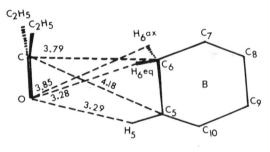

Fig. 4. Orientation and distances between the atoms of the carbonyl group of diethylketone and sites C5-H, C6-H_{eq} and C6-H_{ax} in the DCA inclusion compound.

compound. The ketone oxygen O is 3.3 Å from H6eq and H5 but 3.9 Å from H6ax. The distance from the ketone C to C6 is 3.8 Å while the distance from the ketone C to C5 is 4.2 Å. Indeed, the experiment (irradiation under argon) confirms that the reaction gives only the addition product at 6eq.

Irradiation in the presence of oxygen can yield oxidized guest and/or host. The reactions are dependent on the guest and are therefore assumed to be a function of guest functionality and orientation. Thus, DCA inclusion compounds with methylethylketone, diethylketone, cyclohexanone, methyl-propylketone, or propiophenone afford 5-β-hydroxy-DCA, while the compounds with acetone, acetophenone, p-fluoro-, p- or m-chloroacetophenone do not. Oxidation of a prochiral guest was obtained during the irradiation of the indanone–DCA compound in the presence of oxygen. The product, 3-hydroxy-indanone, was found to be optically active. The optical yield in this asymmetric synthesis has not yet been reported.

4. Urea

Urea is an achiral, planar molecule which is the building block of the well known chiral inclusion compounds of space group P6$_1$22 (or P6$_5$22). Guest molecules are found in channels which run through the crystal parallel to the 6$_1$-axis. (See Chapter 2, Volume 2). The chiral environment about the guest is created by the helical packing of the urea host (Fig. 5). As the diameter of the channel is 5.5 Å, only "slim" molecules are incorporated. The length of the guest molecules plays an important role. The minimum length for simply branched molecules with which inclusion compounds are formed is about 13 Å; for more branched derivatives, the minimum length is higher. The urea molecules in every single inclusion compound crystal, are arranged in only one enantiomeric helix. When a polycrystalline sample

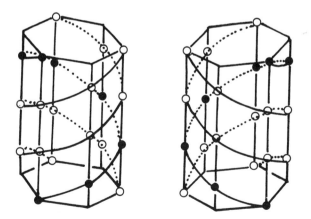

Fig. 5. Idealized representation of urea arrangement in the hexagonal lattice.

is precipitated, a racemic mixture of both crystal forms will generally be expected. Thus, unlike the inclusion compounds whose hosts are chiral, in order to use these complexes for resolution one must somehow create an initial asymmetry in the urea inclusion compound, so that only one crystal configuration prevails. A crystal having a given configuration can accommodate oppositely configurated guests, thus generating the diastereomeric pairs P-crystal.(R)-guest, and P-crystal.(S)-guest, for example. These pairs can be separated by fractional crystallization. A difficulty with these inclusion compounds is the problem of determining the absolute configuration of the urea lattice: as the chirality can only be observed in the solid state one must either use crystallography or a spectroscopic method such as solid-state circular dichroism. Despite these problems, Schlenk carried out extensive studies on several aspects of chiral discrimination in urea inclusion compounds.[25–29]

4.1. Guest enantiomeric excess

The following methods have been used to induce preferential formation of homochiral urea inclusion compounds. Even if no special method is used to induce resolution, the crystals which are obtained from a racemic guest solution might exhibit a small optical rotation by accidental inoculation with unknown chiral seeds.

4.1.1. Crystallization from optically enriched guest
A slight excess of one antipode over the other in the solution used for crystallization may result in preferential creation of one crystal configuration. This crystal grows and preferentially incorporates one enantiomer, say the D-enantiomer. The solution which is enriched with the opposite, L-enantiomer induces a new crystallization where crystals of opposite configuration, with excess of L-enantiomer are formed and so on. This swing process can be repeated by feeding in racemic guest. The process applied to n-decyl 2-chloro-4-methylpentanoate yielded a guest which had $\alpha_D = -3.18°$ for the first, 5.15° for the second, and −5.40° for the third precipitated crystals.[26]

4.1.2. Crystallization by inoculation
By using a small quantity of seed crystals, crystallization of a desired crystal configuration may be achieved. Using this method 95.6% of (+)-2-chlorooctane was obtained after several crystallizations. This technique was used to provide a definition for the relative configuration of the crystal lattice. Host

lattice configuration A was defined as that which accommodates (−)-2-butyl decanoate, host lattice B that which incorporates the (+) antipode.[26]

4.1.3. Crystallization on asymmetric surfaces
Contact with asymmetric surfaces may induce the formation of only one configuration of the urea inclusion compounds. Experiments using 2-butyl caproate showed that the A configuration was exclusively achieved with silk acetate, silk fibroin and human hair. The B configuration was obtained with pepsin.[26]

4.1.4. Crystallization with optically active cosolutes
The use of sugars as cosolutes may favour one lattice configuration. It was shown that the A configuration is achieved when D(+)-glucose or D(+)-galactose are used while configuration B is induced by D(−)-arabinose, D(+)-xylose, L-(−)-fucose and (+)-tartaric acid. In each case the other antipode induced the opposite configuration of the host lattices.[26]

4.1.5. Exchange
An interesting possibility of replacing a given guest with another, without collapse of the chiral host lattice, has been realized. Typically, a short chain paraffin molecule (heptane or octane) is replaced by a longer molecule. The rate of guest exchange is described in Table 3 and is seen to depend on both the chain length of the departing guest as well as that of the incoming guest.[26]

The possible multiple use of a urea inclusion compound crystal is intriguing and merits closer investigation, first, to ensure that collapse of the host lattice followed by recrystallization is not taking place and, second, to attempt to successively resolve different guests with the same crystal.

Table 3. Exchange (%) of guest $CH_3CHClCH_2COO(CH_2)_mCH_3$ with n-alkanes (C_nH_{2n+2}) after 4 h (20° C)

m	n = 7	8	10	12
5	46	47	26	6.6
6	45	37	24	—
7	59	43	26	—
8	70	54	—	—
9	70	64	39	15
10	—	75	42	—
11	—	—	44	—

4.2. Correlation of host lattice and guest chirality

It was observed that within a homologous series there is a correlation between the chirality of the guest and the host lattice configuration. For example, for 28 different alkyl esters of the general formula $CH_3(CH_2)_nCH(CH_3)COO(CH_2)_mCH_3$ (**30**), with $n = 1, \ldots 9$ and $m = 5, \ldots 13$, the (+) antipode induced a type-A helix structure. A correlation was also observed when CH_3 was replaced by Cl or SH, but discrepancies appeared with CH_3 or Cl in the β position.[27] As a consequence of these observations it is important to note that, with the correct precautions taken into consideration, urea inclusion compounds may be used for assigning absolute configurations to guests of unknown configuration.[29]

The ratio between host and guest is an integer when the guest's length is an integer multiple of the distance between two urea molecules along the channel, i.e. 1.834 Å. For esters **30**, this corresponds to chains with 12, 15, 25 and 28 atoms. The molecular mechanism for chiral recognition based on a lock and key picture requires strictly integer host:guest ratio. Clearly, for a non-integer ratio, the same guest atoms never "see" the same urea environment except for very rare, accidental fitting. However, the homologous series rule, verified in most series, denotes identical chiral discrimination for all molecules in the channel, including the non-integer cases. A tentative explanation is based upon the helicity of the channel walls, which would print a definite screw sense along the channel. A model which is based on the known structure of the host lattice and an examination of the space available to a chiral guest allows a comparison between the relative fit of enantiomeric guest molecules within the available void space in a given chiral channel to be made. The determination of chiral preference by this model is restricted by the lack of structural information regarding the position of the guest within the urea channel and by the disorder of the guests which thus experience a variety of different environments in any single crystal.

5. Tri-*ortho*-thymotide

Tri-o-thymotide (TOT, Fig. 6a) is a host molecule which forms inclusion compounds with a great variety of guests.[30-37] Two types of inclusion compounds are obtained: (a) cage-type (clathrates), space group P3₁21, are formed with molecules smaller than 9 Å in their extended form (Fig. 7); (b) channel-type, space groups P6₁, P6₂, P3₁, are formed with long chain molecules. Solvent-free TOT crystallizes in crystals which belong to the

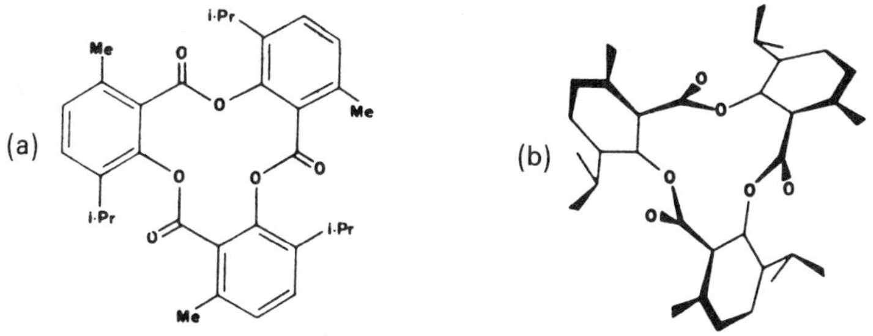

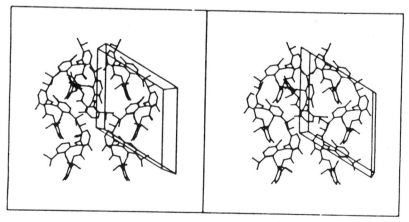

Fig. 6. The structure of TOT (a) and the propeller conformation found in the solid state (b).

racemic space group $Pna2_1$[30,35] but in the presence of potential guests, even achiral molecules, spontaneous resolution of the labile chiral molecule occurs,[31-34] and chiral inclusion compounds are obtained. This observation has been confirmed by detailed X-ray analysis.[35,36] In solution, TOT exists in helical or propeller conformations, the latter being predominant.[38] A rapid interconversion between the M- or P-propeller occurs, the racemization barrier being 88 KJ mol^{-1}. In the crystal, TOT acquires the propeller conformation (Fig. 6b). (A related molecule, N,N',N''-trimethyltrianthranilide, crystallizes in the helical conformation. See also Chapter 6, Volume 2.)[39]

Fig. 7. Stereoview of the TOT molecules which form the walls of the cage in TOT-2-bromobutane. The molecules are viewed down the c axis and the top molecule has been removed for clarity. A unit cell is plotted to indicate the direction of the crystallographic axes.

Powell had pointed out in 1952[33] the possibility of using TOT inclusion compounds as a resolving agent. We have repeated his preliminary experiments and extended the scope of this research.[40–42] Such experiments have also been carried out by Gerdil and Allemand.[43–45]

5.1. Determination of enantiomeric excess and host-guest correlation of configuration

A variety of racemic guest substances were crystallized with TOT and the guest enantiomeric excess in each single crystal was established along with the correlation of configuration between host and guest. The large rotation of TOT, $[\alpha]_D \sim 70°$, allows the chirality of TOT to be determined unambiguously on even a small sample. The chirality and enantiomeric excess of the guest was determined by several different methods: direct V.P.C. on chiral phases, NMR using chiral shift reagents, or polarimetrically. The

Table 4. Enantiomeric excess of guest and correlation of guest and host chirality in P-(+)-TOT clathrate crystals

Guest	Guest e.e (%)	Guest configuration	Ref.
Cage			
2-chlorobutane	32	S-(+)	40
	45		43
2-bromobutane	34	S-(+)	40
	35		43
2-iodobutane	<1	—	42
trans-2,3-dimethyloxirane	47	S,S-(−)	40
trans-2,3-dimethylthiirane	30	S,S-(−)	40
propylene oxide	5	R-(+)	40
2-methyltetrahydrofuran	2	S-(+)	40
methyl methanesulphinate	14	R-(+)	40
ethylmethylsulphoxide	83[a]	S-(+)	44
2-butanol	<5	S-(+)	43
2-aminobutane	<2	—	43
Channel			
2-chlorooctane	4	S-(+)	40
2-bromooctane	4	S-(+)	40
3-bromooctane	4	S-(+)	42
2-bromononane	5	S-(+)	40
2-bromododecane	5	S-(+)	40

[a] From X-ray measurements.

results are presented in Table 4, where the enantiomeric selectivity is seen to be dependent on several factors:

(a) Cavity symmetry, guest symmetry and guest enantiomeric excess: *trans*-2,3-dimethyloxirane and *trans*-2,3-dimethylthiirane have molecular symmetry 2 which is also the symmetry of the cavity where they are located in the crystal;[35] these guests were found to exhibit high enantiomeric selectivity, suggesting that the coincidence of guest and cavity symmetry favours discrimination.

(b) The absolute configuration of the preferred guest enantiomer is such that homologous atoms or groups in related molecules occupy similar positions in space. Therefore, for a host crystal of given chirality and within a family of similar molecules, the absolute configuration of the preferred guest enantiomer can be predicted.

(c) The guest's size is an important factor; the smallest molecule studied, propylene oxide, and the largest, 2-methyltetrahydrofuran, exhibit the smallest discrimination.

It is important to emphasize that the data presented in Table 4 represent the enantiomeric excess in a crystal grown from a racemic solution of guest. Enantiomeric enrichment can be considerably increased when optically active solutions of guests are used. Enhancements of enantiomeric excess on crystallization of TOT from 2-bromooctane of varying initial optical activity are given in Table 5.

Table 5. *Optical purity enhancement on crystallization from optically active 2-bromooctane*

	Optical purity of guest in solution			
	0%	12%	35%	50%
Optical purity of guest in the inclusion compound	5%	30%	66%	80%
Enantiomer ratio in solution	50:50	56:44	65:35	75:25
Expected enantiomer ratio in the inclusion compound		58:42	69:31	77:23
Actual enantiomer ratio in the inclusion compound	53:48	65:35	83:17	90:10

It is noteworthy that channel inclusion compounds can display a significant enhancement of optical purity. There are apparently other factors besides the inherent chiral discrimination of the growing TOT host lattice and the ratio of the guest enantiomers in the initial solution, which influence the final enantiomeric selectivity. Perhaps the high concentration of an enantiomer in solution organizes the enantiomers at the growing crystal in such a way as to exclude the minor enantiomer.

Finally, when a large quantity of an optically active guest is needed, one can seed a solution of TOT in the guest to be resolved with a powdered single crystal. When this was done, the optical purity of the guest in the resulting crystals was 85% of that of a single crystal.

5.2. The cage structure

Several crystal structures of cage clathrates have been determined by X-ray diffraction and some characteristic features of the host and guest conformations deduced from them. It was found that the conformation of TOT is isomorphous over the different crystal structures. The cage which is formed by eight host molecules (Fig. 7) appears as a deformed ellipsoid in which differences in axial lengths (Fig. 8) prevent random orientation of guests.

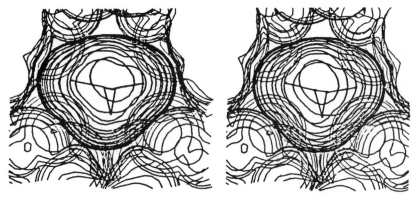

Fig. 8. Stereoview of the contours of the van der Waals envelope of TOT and of the van der Waals envelope of the volume accessible to guest molecules, viewed along the c axis.

The cage is flat on one side along the 2-fold axis and has a bulge on the other side. The hetero atoms of 2,3-dimethyloxirane, 2,3-dimethylthiirane and 2-bromobutane are situated in this bulge.

The different guests were found to be disordered in the cages. The methyl groups of the two disordered conformations of the major enantiomer in TOT·2-bromobutane and TOT·2,3-dimethylthiirane (Fig. 9) are much closer to each other than the corresponding non-terminal atoms. There is some

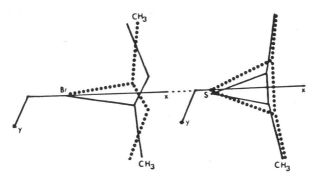

Fig. 9. View of guest molecules along *c.*

flexibility in the location of the heteroatoms, which allows either enantiomer to be accommodated in the same cage and reduces the observed enantiomeric purity. One would anticipate that if the asymmetric centre was more restricted in its position, a more efficient discrimination could be achieved. Despite the disorder present in all the crystals studied, the determination of the absolute configuration of TOT has been accomplished. It was found that (+)-TOT has the configuration of a *P*-propeller.

5.3. Discussion of chiral discrimination

Several requirements must be met for an efficient chiral discrimination, the major one being that the cage must be chiral and sufficiently rigid. Indeed, a small "chiral" departure from a non-chiral void may be sufficient for discrimination of an appropritate guest provided that the walls are rigid. A familiar image could be a soft glove, even with pronounced geometrical chirality, as opposed to an almost symmetrical but rigid boot. Next, the fit of the guest must be as tight as is compatible with the cavity. Indeed propylene oxide is much less discriminated than other molecules having five non-hydrogen atoms. However, above a certain size limit, the discrimination is again poor (e.g. 2-iodobutane, and 2-methyltetrahydrofuran); this is a puzzling result and structure determinations are called for to help elucidate this.

The calculation of packing energies, using van der Waals potentials have been performed to provide an indication of the chiral bias.[43] Such calculations have shown that the association (M)-TOT.(R)-2-bromobutane is more favourable by 14.6 KJ mol^{-1} than the diastereoisomer (M)-TOT.(S)-2-bromobutane. This value is considerably higher than those obtained by calorimetry on cyclodextrins (Table 2), and higher than the values expected on the basis of the observed discrimination. For 2-butanol an additional parameter is involved: internal rotation about the C–O bond. The ease of inclusion of different conformers is very dependent on the internal orientation of the OH group. Here the calculations suggest that both R and S enantiomers can be included with energies that differ by less than 1.0 KJ mol^{-1}, thus pointing to a greatly reduced stereoselectivity, as is observed experimentally.

Finally, it should be noted that temperature also influences clathrate chirality. This is illustrated by the following racemization of methyl methanesulphinate within the TOT cage.[42] Single crystals of the inclusion compound were heated and the guest enantiomeric excess (e.e.) was determined. The unheated crystals contained an e.e. of $14 \pm 1\%$. Crystals heated to 115° C for up to 12 h, showed no change in e.e.; however, on heating at 125° C for 12 h racemization proceeded and the e.e. fell to zero. In solution, racemization already takes place at 115° C. Thus the host affords stabilization of the guest to racemization. Powder photography before and after the heating showed that the integrity of the cage clathrate structure has not been destroyed, and it is therefore concluded that enantiomerization takes place within the TOT cage cavities. The flexibility of the TOT molecule is perhaps an important factor. When sufficient thermal energy is introduced, the motion of the TOT molecules allows enantiomerization of the guest; the walls of this thermally excited cavity no longer impose any chiral discrimination and the guest e.e. falls to zero. On cooling, the racemization barrier of the guest is passed before the "thermal barrier" of the TOT cavity, so that when the crystal is cooled to room temperature no enantiomeric excess is observed.

6. Miscellaneous chiral inclusion compounds

We report here new chiral inclusion compounds whose applications have not yet been extensively developed.

6.1. Dianin's compound and 'Hexahost' inclusion compounds

The optically pure Dianin's molecule,[46a] as well as chiral analogues of cycloveratril,[47] lack the ability to form inclusion compounds (See Chapters 1 and 4, Volume 2). However, synthetic quasi-racemates related to Dianin's compound, do form clathrates[46b] of space group R3. However, inclusion of racemic ethyl bromopropionate results in negligible enrichment (0.5%) in one enantiomer. The chiral hexahost 31 also forms clathrates (space group

P2$_1$).[48] It can accommodate two pairs of dimeric acetic acid molecules. (See Chapter 5, Volume 2.)

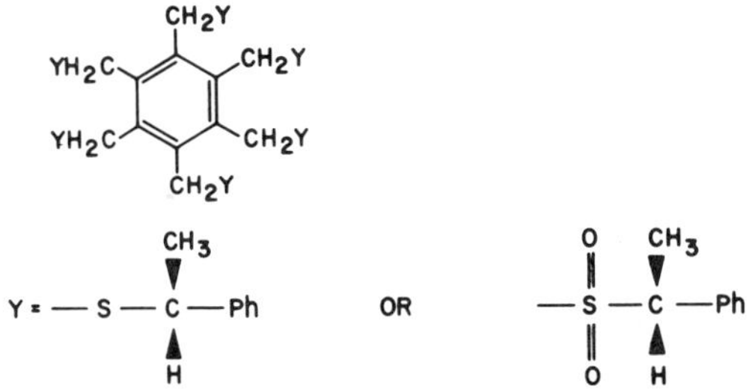

(31)

6.2. Anthroates

1-Phenylpropyl- and 1-phenylbutyl-9-anthroate (**32**) form channel inclusion compounds with *n*-hexane[49] (space group P6$_5$). The diameter of the channel along the 6$_5$ axis is approximately 5 Å. The host molecules are interconnected by hydrophobic anthracene and phenyl-anthracene contacts.

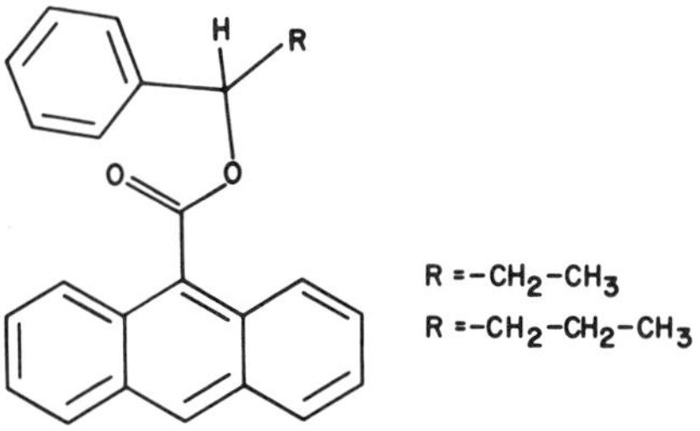

R = -CH$_2$-CH$_3$
R = -CH$_2$-CH$_2$-CH$_3$

(32)

6.3. Bicyclic diol derivative

The host **33** forms channel compounds with a trigonal network of hydrogen bonds, running along the channel axis (space group P3$_1$21 or P3$_2$21).[50] The channel walls are lined only with hydrocarbon H atoms and include a variety of guests (chloroform, toluene, dioxane, acetone, ethyl acetate).

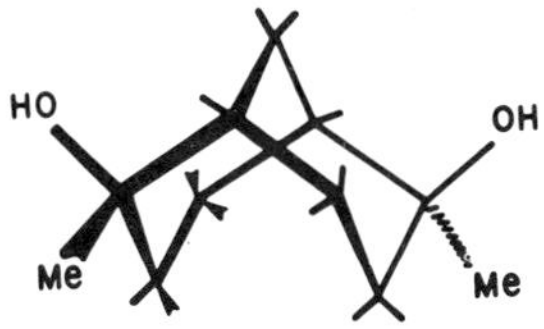

(33)

6.4. Tetraarylethane derivative

Derivative **34** forms inclusion compounds with a variety of guests (benzene, cyclohexylamine, 2-butanone, cyclododecene and ethyl acetate).[51] The main interest in this compound lies in the successful, although partial, separation of the *host* enantiomers by inclusion of an optically active *guest*, α-pinene.

(34)

5.5. Purification of antipodes via Ni(amine)$_4$(NCS)$_2$ clathrates

Optical isomers of most α-arylalkylamines can be efficiently purified through inclusion compound formation by their coordination complex [Ni(NCS)$_2$(amine)$_4$].[52] Complexes of this type form inclusion compounds with a wide variety of aromatic compounds. When the coordination complex is prepared from a solution in which one of the antipodes predominates, the amine co-ordinated into the inclusion compound is preferentially racemic, and the antipode in excess is left optically pure in the solution. (See Chapter 4, Volume 1).

7. Chiral discrimination in diastereomeric pairs

Inclusion compounds are not the only host lattices exhibiting enantiomer selectivity towards guest molecules. The selectivity is in fact a phenomenon pertinent to the formation of diastereomeric pairs, widely used, since Pasteur, in resolution. When the resolving chiral molecule is considerably larger than the molecules of the racemic mixture to be separated, the resemblance with inclusion compounds is more pronounced. This is the case with certain polypeptides, antibiotics, crown ethers, cryptates etc. Although the main practical interest lies in phenomena pertinent to the liquid state, the determination of crystal structures by X-ray diffraction is one of the most powerful methods of establishing the structures of complexes. As with cyclodextrins, there are good reasons to believe that the main features of the crystalline structure are maintained in solution.

7.1. Host lattices in general diastereomeric pairs

The subject has been extensively reviewed by Jacques *et al.*[53] Diastereomeric pairs differ in their physical properties: melting points (differences can reach 70° C), solubilities (factors up to 5), densities (a few %). The differences reflect distinct structural features in the two diastereomers. In the classical cases of acid-base pairs, in contrast to inclusion compounds, the main forces are directional; specific and strong electrostatic forces or hydrogen bonds. However, there exists an important analogy with inclusion compounds; it often happens that for a given base (resp. acid), the relative configuration of a series of associated acids (resp. bases) is similar.[53] As an example, Fig. 10 shows the spatial relations between the structures of a diastereomeric

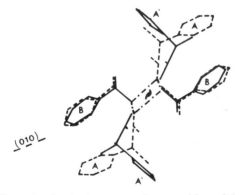

Fig. 10. α-Methylbenzylamine hydratropate. Superposition of the (001) projections of the p and n salt structures in the vicinity of the helical binary axis parallel to the common *c* axis: (—) p salt; (– – –) n salt.

pair of α-methylbenzylamine hydratropate,[54] illustrating the structural origin of enantiomeric selectivity. The crystal is made of infinite columns of hydrogen bonds between acid and base molecules along a 2_1 axis. The amine molecules are roughly superimposed, but the important difference in the positioning of the aromatic ring of the acid prevents the cocrystallization of the two salts and thus makes their separation possible by crystallization.

The salt of lasalocid (**35**) with α-methyl-*p*-bromobenzylamine (**36**) is monomeric.[55] The enantioselectivity can reach 100% after three crystallizations, and is appreciable for several other amines. The amine molecule is

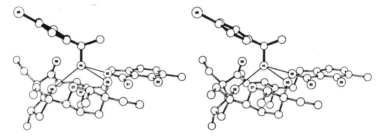

(35)

bound to **35** by 3 strong hydrogen bonds (Fig. 11); its phenyl ring is oriented so that it is situated over a depression of the lasalocid molecule forming a lipophilic pocket. The position of the smallest constituent (H) attached to C* also appears to be dictated by steric requirements.

Fig. 11. Crystal structure of the α-methyl-*p*-bromobenzylamine (solid bonds) salt of lasalocid (open bonds). View is perpendicular to the amine C–N bond.

7.2. Chiral crown ethers and cryptates

A crown ether or a cryptate containing one or two chiral groups (dilocular system) becomes a chiral host. Cram and his associates have extensively studied such host-guest systems both in solution and in the crystalline state.[56–58] Thus, (RR)-**37** or (SS)-**37** in CDCl₃ at 0° C preferentially extracts

(37)

(R)- and (S)- enantiomers of the amine ester salt $RCH(COOCH_3)NH_3^+.PF_6^-$ respectively in D_2O. A catalytic resolving machine was designed based on stereoselective transport of the two enantiomers in D_2O, in contact with separate $CDCl_3$ layers containing (RR)- and (SS)-**37** respectively. **37** attached to a macroreticular polystyrene resin was used to resolve chromatographically a variety of aminoacids and esters. The values of the ratio of the association constants for different enantiomers with **37** run from 1.5 to 14. The corresponding enantiomeric difference in free energies $-\Delta(\Delta G°)$ varies from 1.2 to 6.3 KJ mol^{-1}.

Enantiomeric selectivity is also exhibited in the crystalline state, leading to a complete separation of diastereomers. This remarkable property has been utilized to separate the host enantiomers in view of their introduction into the enantiomer resolving machine. Selectivity in both solution and crystalline states involves the same configurational bias: when racemic host **37** was treated with $D-C_6H_5CH(COOH)NH_3^+ClO_4^-$ (**38**) pure (SS). (D)-**38** crystallized.

7.3. Enantiomeric selectivity in chromatographic columns and related host lattices

Gas-liquid chromatography has been developed to achieve enantiomeric separation. The molecular mechanism of the chiral recognition in the important family of amides was established with models derived from X-ray analysis of related crystals.[59-61] It has been shown that a particularly stable structure is obtained for amides by formation of stacks in which the molecules are hydrogen bonded along a 5 Å translation axis (Fig. 12a). This model explains the different interactions of these amides with pairs of enantiomers, by assuming that a guest molecule intercalates within the H-bonded array of the host matrix. Figures 12b and 12c represent the intercalation model for host N-lauroyl-(R)-α-1-naphthylethylamine and the two enantiomers of N-trifluoroacetyl-α-phenylethylamine. In the R-R' diastereomeric packing, the methyl group of the guest points away from the stack and does not interfere with it. In the R-S' pair, the methyl positioning leads to a severe overcrowding as it points to the naphthyl ring of the host, resulting in a lower stability of the R-S' system. The validity of the model in solution is supported by experimental evidence briefly quoted here: (a) the short range order along the 5 Å stack is maintained in the melt of the amide crystals; (b) the formation of solid solutions indicates that a guest molecule may pack within the solid matrix of the host without causing major changes in the packing arrangement; the maximum solubility is 11% in the R-R', 5% in the R-S' system, indicating a better fit in the former.

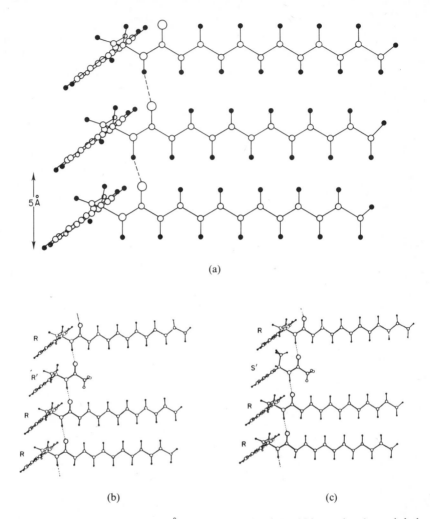

Fig. 12. (a) H bonding along a 5 Å translation axis of (R)-N-lauroyl-α-1-napththyl-ethylamine. (b) A chiral guest molecule N-trifluoroacetyl-α-phenylethylamine inserted into the H-bonded stack of (R)-N-lauroyl-α-1-naphthylethylamine: guest of configuration R. (c) As in (b) but with guest of configuration S.

7.4. Asymmetric polymerization

Perhydrotriphenylene (PHTP) forms inclusion compounds with several substances having a linear structure and in particular with macromolecular molecules.[62] Optically active PHTP forms inclusion compounds which are very similar to those of racemic PHTP (space group P6₃ for the optically pure and P6₃/m for the racemic inclusion compounds). Polymerization of 1,3-pentadiene in the chiral channel of the optically pure PHTP resulted in the formation of optically active polymer. (See Chapter 10, Volume 3).

Polymerization in liquid crystals has also been reported.[63]

7.5. Chiral recognition in two dimensional lattices

Artificial monolayers are often used as models for biomembranes. The importance of the chirality of the packing of monolayer molecules has been discussed recently by studying surface tension and force-area curves of racemic and enantiomeric monolayers of *N*-α-methylbenzylstearamide.[64–66] The significance of these observations is relevant in discussing the interaction between biological chiral molecules and biological membranes.

We would also like to mention an observation on asymmetric absorption of alanine by quartz.[67]

8. Conclusion and perspectives

Clathrates and other host lattices are part of the physicochemist's arsenal for enantiomeric selectivity. Certainly, the overall results indicate a rather low percentage of efficiently discriminated guests. But the accumulated experience for several hosts allowed a first step towards the understanding of the mechanism and rationalization. The success of a particular guest interaction depends on several parameters, summarized below. The behaviour towards enantiomer separation and chemical reaction are quite different and will be discussed separately.

8.1. Enantiomer separation

Cages versus channels. Discrimination in TOT channels is generally lower than in cages. Crystal data are not available for chiral discrimination in cyclodextrins, which also crystallize in both cage and channel types. The guests' size is of course a fundamental parameter. Intuitively, the size must not be too different from that of the cavity, too small or too large a guest size resulting in poor discrimination. However, one should not forget that hosts exhibit a certain degree of flexibility. This can be a small, but significant molecular flexibility, as in TOT and CD, a different degree and hydration pattern as in CD, or a flexibility in the van der Waals contacts as in all hosts. In the urea host lattice, where cohesion is assured by hydrogen bonds, the flexibility is probably the least. Thus one should experiment, even in an *a priori* unfavourable case of guest dimensions. It should be noted that the length of long chain guests in channel compounds seems to be an unimportant or negligible factor, despite the incommensurability between the lattice repeat and guest repeat patterns, as discussed in the urea and TOT Sections.

Guest symmetry is a significant factor, as far as it coincides with the symmetry of the corresponding crystallographic site. A question often raised is the guest's disorder within the cavities or channels. This is a very common trend in clathrates, and occurs normally when the guest's symmetry does not comply with that of the crystallographic site; it may also occur when these coincide (TOT.trans-2,3-dimethylthiirane, for instance). The unexpected fact is that quite high discrimination appears even in cases of disorder (as shown in TOT and, most probably, in CD and urea).

The guest's chemical functions certainly play an important role. It is essential in crown ethers and cryptates, and it still may play a role in TOT, although constituted by "neutral" molecules held together by van der Waals forces. Finally, temperature affects the host/guest relationship, as exemplified by the racemization experiment in TOT. This is clearly connected with the "soft chiral glove" as opposed to the "rigid, almost achiral boot" pictures. It might seem appropriate to suggest operating at the lowest temperature in order to enhance selectivity, but on the other hand, the disorder of the guest is certainly reduced at low temperatures.

The overall enantioselectivity is reflected in the differential free energy $\Delta(\Delta G°)$ between diastereomeric (R)-Host·(S)-Guest and (R)-Host·(R)-Guest. To our knowledge no such measurements have yet been performed on inclusion compounds. In solution these values are 0.4–0.8 KJ mol^{-1} in cyclodextrins and 1.2–6.3 KJ mol^{-1} in crown ethers. In several cases (CD and crown ethers) it has been shown that the configuration of the preferentially precipitated enantiomer in the inclusion compound is identical to that included in solution.

8.2. Asymmetric synthesis

The requirements for an efficient asymmetric synthesis in solution are quite different from those in the crystalline phase. The crystalline environment does not intervene through its overall asymmetry; only parts of the host molecules play an important role directly. Excluding the fascinating reactions of external (gas or liquid) reagents,[73] there are two types of reaction in inclusion compounds: a host-guest reaction (exemplified in choleic acids), and a reaction which involves only the guest (e.g. in TOT and CD). A thorough study of the reactions in choleic acids has highlighted the importance of the guest's position and orientation towards the host.

The reactions within the cavities or channels of clathrates are interesting not only in providing novel examples of asymmetric synthesis, but also in contributing to the understanding of general reaction mechanisms. Indeed, inclusion compounds provide an exceptional advantage over reactions in solution or in the gas phase, as they allow exact knowledge of the molecular environment before and after the reaction.

References

1. F. Cramer and W. Dietsche, *Chem. Ber.*, 1959, **92**, 378.
2. C. van Hooidonk and J. C. A. E. Breebaart-Hansen, *Recl. Trav. Chim. Pays-Bas* 1970, **89**, 289.
3. H. P. Benschop and G. R. Van den Berg, *J. Chem. Soc., Chem. Commun.*, 1970 1431.
4. M. Mikolajczyk, J. Drabowicz and F. Cramer, *J. Chem. Soc. Chem. Commun.* 1971, 317.
5. M. Mikolajczyk and J. Drabowicz, *J. Am. Chem. Soc.*, 1978, **100**, 2510.
6. J. Martinie, J. Michon and A. Rassat, *J. Am. Chem. Soc.*, 1975, **97**, 1818.
7. J. Michon and A. Rassat, *J. Am. Chem. Soc.*, 1979, **101**, 4337.
8. J. Michon and A. Rassat, *J. Am. Chem. Soc.*, 1979, **101**, 995.
9. A. Cooper and D. D. MacNicol, *J. Chem. Soc., Perkin Trans, 2*, 1978, 760.
10. K. Sensse and F. Cramer, *Chem. Ber.*, 1969, **102**, 509.
11. S. Takenaka, N. Matsuura and N. Tokura, *Tetrahedron Lett.*, 1974, **26**, 2325.
12. N. Matsuura, S. Takenaka and N. Tokura, *J. Chem. Soc., Perkin Trans, 2*, 1977, 1419.
13. M. Otagiri, K. Ikeda, K. Uekama, O. Ito and M. Hatano, *Chem. Lett.*, 1974, 679.
14. A. L. Thakkar, P. B. Kiehn, J. H. Perrin and W. L. Wilham, *J. Pharm. Sci.*, 1972, **61**, 1841.
15. G. le Bas, C. de Rango and G. Tsoucaris, *in Proceedings of the First International Symposium on Cyclodextrins, Budapest, Hungary, 1981* (ed. J. Szetli), D. Reidel Publ. Co., Holland, 1982, p. 245.
16. C. Formoso, *Biochem. Biophys. Res. Commun.*, 1973, **50**, 999.
17. K. Flohr, R. M. Paton and E. T. Kaiser, *J. Am. Chem. Soc.*, 1975, **97**, 1209.
18. C. van Hooidonk, *Recl. Trav. Chim. Pays-Bas*, 1972, **91**, 1103.
19. B. S. Green, R. Arad-Yellin and M. Knossow, unpublished work.
20. G. L. Trainor and R. Breslow, *J. Am. Chem. Soc.*, 1981, **103**, 154.
21. H. Sobotka and A. Goldberg, *Biochem. J.*, 1932, **26**, 905.
22. C. P. Tang, Ph.D. Thesis, The Feinberg Graduate School, Weizmann Institute, Rehovot, Israel 1979.
23. R. Popovitz-Biro, H. C. Chang, C. P. Tang, N. R. Shochet, M. Lahav and L. Leizerowitz, *Pure Appl. Chem.*, 1980, **52**, 2693.

24. R. Popovitz-Biro, C. P. Tang, H. C. Chang, M. Lahav and L. Leizerowitz, *Nouv. J. Chim.*, 1982, **6**, 75.
25. W. Schlenk, *Analyst*, 1952, **77**, 870.
26. W. Schlenk Jr., *Justus Liebigs Ann. Chem.*, 1973, 1145.
27. W. Schlenk Jr., *Justus Liebigs Ann. Chem.*, 1973, 1156.
28. W. Schlenk Jr., *Justus Liebigs Ann. Chem.*, 1973, 1179.
29. W. Schlenk Jr., *Justus Liebigs Ann. Chem.*, 1973, 1195.
30. S. Brunie and G. Tsoucaris, *Cryst. Struct. Commun.*, 1974, **3**, 481.
31. W. Baker, B. Gilbert and W. D. Ollis, *J. Chem. Soc.*, 1952, 1443.
32. A. C. D. Newman and H. M. Powell, *J. Chem. Soc.*, 1952, 3747.
33. H. M. Powell, *Nature (London)*, 1952, **170**, 155.
34. D. Lawton and H. M. Powell, *J. Chem. Soc.*, 1958, 2339.
35. S. Brunie, A. Navaza, G. Tsoucaris, J. P. Declercq and G. Germain, *Acta Crystallogr.*, 1977, **B33**, 2645.
36. D. J. Williams and D. Lawton, *Tetrahedron Lett.*, 1957, 111.
37. R. Arad-Yellin, S. Brunie, B. S. Green, M. Knossow and G. Tsoucaris, *J. Am. Chem. Soc.*, 1979, **101**, 7529.
38. A. P. Downing, W. D. Ollis and I. O. Sutherland, *J. Chem. Soc.*, B, 1970, 24.
39. W. D. Ollis, J. A. Price, J. Stephanidou-Stephanatou and J. F. Stoddart, *Angew. Chem. Int. Ed. Engl.*, 1975, **14**, 169.
40. R. Arad-Yellin, B. S. Green and M. Knossow, *J. Am. Chem. Soc.*, 1980, **102**, 1157.
41. R. Arad-Yellin, B. S. Green, M. Knossow and G. Tsoucaris, *Tetrahedron Lett.*, 1980, **21**, 387.
42. R. Arad-Yellin, B. S. Green, M. Knossow and G. Tsoucaris, *J. Am. Chem. Soc.*, 1983, **105**, 4561.
43. R. Gerdil and J. Allemand, *Helv. Chim. Acta.*, 1980, **63**, 1750.
44. (a) R. Gerdil and J. Allemand, *Tetrahedron Lett.*, 1979, 3499; (b) J. Allemand and R. Gerdil, personal communication.
45. (a) J. Allemand and R. Gerdil, *Cryst. Struct. Commun.*, 1981, **10**, 33; (b) J. Allemand and R. Gerdil, *Acta Crystallogr.*, 1982, **B38**, 1478.
46. (a) M. J. Brienne and J. Jacques, *Tetrahedron Lett.*, 1975, 2349; (b) A. Collet and J. Jacques, *Isr. J. Chem.*, 1976–1977, **15**, 82.
47. A. Collet and J. Jacques, *Tetrahedron Lett.*, 1978, **15**, 1265.
48. A. Freer, C. J. Gilmore and D. D. MacNicol, *Tetrahedron Lett.*, 1980, **21**, 205.
49. M. Lahav, L. Roitman, L. Leiserowitz and C. P. Tang, *J. Chem. Soc., Chem. Commun.*, 1977, 928.
50. R. Bishop and I. Dance, *J. Chem. Soc., Chem. Commun.*, 1979, 922.
51. K. S. Hayes, W. D. Hounshell, P. Finocchiaro, K. Mislow, *J. Am. Chem. Soc.*, 1977, 4153.
52. M. Hanotier-Bridoux, J. Hanotier and P. de Radzitzky, *Nature (London)*, 1967, **215**, 502.
53. J. Jacques, A. Collet and S. H. Wilen, *Enantiomers, Racemates and Resolutions*, John Wiley and Sons, New York (1981).
54. M. C. Brianso, *Acta Crystallogr.*, 1976, **B32**, 3040.
55. J. W. Westley, R. H. Evans and J. F. Blount, *J. Am. Chem. Soc.*, 1977, **99**, 6057.
56. M. Newcomb, J. L. Toner, R. C. Helgeson and D. J. Cram, *J. Am. Chem. Soc.*, 1979, **101**, 4041.
57. S. C. Peacock, D. M. Walba, F. C. A. Gaeta, R. C. Helgeson and D. J. Cram, *J. Am. Chem. Soc.*, 1980, **102**, 2043.
58. I. Goldberg, *J. Am. Chem. Soc.*, 1980, **102**, 4106.

59. S. Weinstein, B. Feisbush and E. Gil-Av, *J. Chromatogr.*, 1976, **116**, 97.
60. S. Weinstein and L. Leiserowitz, *Acta Crystallogr.*, 1980, **B36**, 1406.
61. S. Weinstein, L. Leiserowitz and E. Gil-Av, *J. Am. Chem. Soc.*, 1980, **102**, 2768.
62. M. Farina, G. Audisio and G. Natta, *J. Am. Chem. Soc.*, 1967, **89**, 9071.
63. L. Liebert, L. Strzelecki and D. Vacogne, *Bull. Soc. Chim. Fr.*, 1975, 2073.
64. N. Theobald, J. N. Shoolery, C. Djerassi, T. R. Erdman and P. J. Scheuer, *J. Am. Chem. Soc.*, 1978, **100**, 5575.
65. E. M. Arnett and O. Thompson, *J. Am. Chem. Soc.*, 1981, **103**, 968.
66. E. M. Arnett, J. Chao, B. J. Kinzig, M. V. Stewart, O. Thompson and R. J. Verbiar, *J. Am. Chem. Soc.*, 1982, **104**, 389.
67. S. Furuyama, H. Kimura, M. Sawada and T. Morimoto, *Chem. Lett.*, 1978, 381.
68. A. Veno, K. Takahashi, T. Osa, *J. Chem. Soc., Chem. Commun.*, 1980, 921.
69. K. Kano, I. Takenoshita, T. Ogawa, *Chem. Lett.*, 1980, 1035.
70. A. Veno, K. Takahashi, Y. Hino and T. Osa, *J. Chem. Soc., Chem. Commun.*, 1981, 194.
71. I. Tabushi, *Acc. Chem. Res.*, 1982, **15**, 66.
72. P. C. Manor and W. Saenger, *J. Am. Chem. Soc.*, 1974, **96**, 3630.
73. Y. Tanaka, H. Sakuraba and H. Nakanishi, *J. Chem. Soc., Chem. Commun.*, 1983, 947.

10 · INCLUSION POLYMERIZATION

M. FARINA

Università di Milano, Milan, Italy

1. Introduction

Of all the innumerable organic reactions, addition polymerization is endowed with a series of characteristics which make it particularly interesting when carried out inside inclusion compounds. It produces linear macromolecules whose form fits in well with the channels available inside the crystal structure, and the reaction takes place without the formation of by-products, the elimination of which would lead to serious diffusion problems in the solid state. Furthermore, the reaction proceeds according to a chain mechanism whereby a very small number of elementary acts (initiation reactions), commonly caused by radiation, gives rise to an extremely large number—many hundreds or thousands—of reactions between guest molecules (propagation reaction). The ratio between polymer yield and initiation reaction may thus reach very high values. Finally, the macromolecular nature of the product facilitates separation between host and guest by selective dissolution or by sublimation, even when the chemical constitution of the two components is very similar.

INCLUSION COMPOUNDS III
ISBN 0-12-067103-4

Though it can hardly be considered to be a method suitable for large-scale preparation of polymers, inclusion polymerization has been quite thoroughly studied during the past quarter of a century because of the particular structural properties of the polymers obtained (for some it represents the only way of synthesis) and as a versatile example of a solid-state reaction.

Numerous reviews have been dedicated to this subject, to which the reader may refer for detailed information.[1-4] In this chapter I intend to deal with the fundamental aspects—structural and thermodynamic—of the process, mainly using polymerization in the channels of perhydrotriphenylene as an example.

2. Historical survey

Interest in inclusion polymerization began to develop in the second half of the 50's, as an alternative method to Ziegler–Natta coordination polymerization to obtain highly stereoregular polymers. The first example was reported by Clasen[5] in 1956 with polymerization of 2,3-dimethylbutadiene included in thiourea. No initiating agent was used and polymerization was extremely slow (weeks and months). The polymer was characterized solely by its melting point.

The potential of inclusion polymerization was clearly illustrated by the later research of Brown and White[6,7] on inclusion compounds with urea and thiourea. Most of the results obtained on inclusion polymerization are to be found in their papers which, for their wide scope and completeness of investigation, constitute an essential point of reference. Brown and White sampled over 175 compounds as guests for the formation of inclusion compounds and more than 30 as monomers for inclusion polymerization, and observed that high-energy radiation (such as β, γ, or X-rays) are excellent initiators of polymerization and that the polymers obtained present considerable chemical and steric regularity. For example the polybutadiene (1)[7] obtained in urea (NH_2–CO–NH_2) or the poly-2,3-dimethylbutadiene

(1)

(2)[6] obtained in thiourea (NH_2–CS–NH_2) have a 1,4-*trans* structure and are crystalline. The latter polymer has a melting point about 50° higher than that obtained by Clasen. On a quite different plane but of equal importance

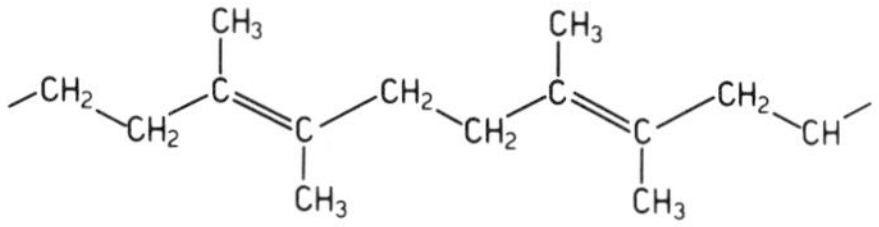

(2)

is White's observation,[7] to which we will return later, that the polybutadiene obtained in urea has a melting point higher than the conventional crystalline polymer, and the connection of this phenomenon with an extended-chain macrostructure of the polymer.

Polymerization of vinylchloride in urea has been studied by White[7] and by Sakurada and Nanbu.[8] A syndiotactic structure (3) has been attributed to the polymer, much more regular than the structure observed in polymers obtained by other methods. This is the first example of an inclusion polymerization of a vinyl monomer in which control is exerted over the steric configuration of the tetrahedral carbon atoms.

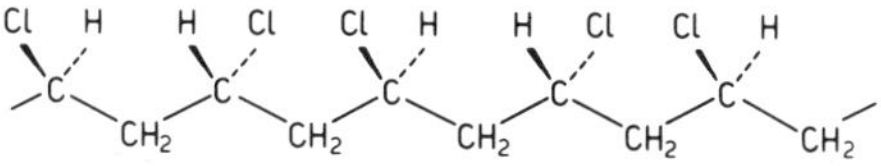

(3)

In 1963 at Milan Polytechnic a new host for inclusion polymerization was synthesized: the *trans-anti-trans-anti-trans* stereoisomer of perhydrotriphenylene (PHTP), a saturated tetracyclic hydrocarbon of formula $C_{18}H_{30}$ (4).[9] This compound presents a marked tendency to polymorphism and to

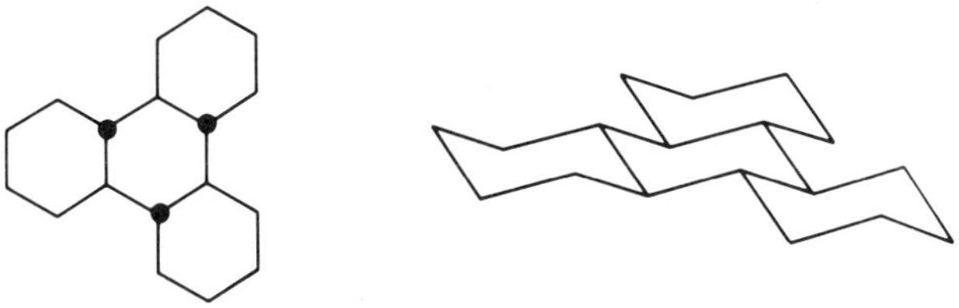

(4)

the formation of crystalline adducts with numerous substances of different form, dimensions and polarity[10] (see Chapter 3, Volume 2). Diffractometric investigation showed such adducts to be channel-type inclusion compounds, in many ways analogous to the previously reported urea and thiourea inclusion compounds (see Chapter 2, Volume 2). However, some important features distinguish PHTP from these well known hosts: the greater thermal stability of the inclusion compounds, many of which have a congruent melting point higher than that of either constituent in the pure state; the complete miscibility in the liquid phase with the numerous guests with which experiments have been carried out; and a greater structural flexibility, which makes it possible to include linear, branched and cyclic molecules.

These facts lead to an enormous increase in the ease with which inclusion compounds are formed, both from a thermodynamic and from a kinetic point of view, and, in particular, to the uncommon ability to form very stable inclusion compounds with macromolecular substances (polyethylene, polybutadiene, polyisoprene, polyoxyethylene, etc.), many of which present congruent melting at temperatures between 125° C, the melting point of pure PHTP, and 180° C.

To compare the ease of formation of these compounds, we may recall that the preparation of inclusion compounds between urea and volatile guest compounds takes place at low temperature (between −78° C and 0° C) in the presence of small quantities of promoters (methanol) and requires a time which may run from many hours to days and weeks. In contrast, preparation of PHTP inclusion compounds can be carried out over a wide range of temperatures and in very short times (from seconds to minutes).[2,4,10]

Two methods of preparation of PHTP inclusion compounds with polymerizable guests are generally used, which are based on solid–liquid and solid–vapour reactions: wetting the pure PHTP crystals with liquid guest or even leaving the PHTP in a guest–vapour atmosphere, within suitable temperature and pressure ranges, leads to inclusion compound formation in a very short time.[11]

Since I was a member of Natta's group, the idea of testing PHTP as a host for inclusion polymerization came almost automatically, the first positive results were presented in 1965 at the Prague Symposium on Macromolecules.[12]

Inclusion polymerization in PHTP is from many points of view similar to that in urea and thiourea; it is promoted by γ or X-rays and gives rise to polymers which have a very high degree of regularity and are often crystalline. Of particular interest is the polymerization of 1,3-*trans*-pentadiene which made it possible to obtain the first example of a highly isotactic polymer, specifically, the 1,4-*trans*-isotactic polymer (**5**). With respect to examples already known, inclusion polymerization in PHTP has

(5)

a wider range of feasibility both with regard to the choice of monomers (in fact both butadiene and 2,3-dimethylbutadiene polymerize in PHTP, whilst the former monomer polymerizes only in urea and the latter only in thiourea) and to the experimental conditions (polymerization takes place at room temperature or even above, whilst in urea low-temperature conditions must be used).

The progress of the reaction can be determined by examining the melting behaviour. The PHTP–monomer adduct examined in the open air (hot-plate microscope or unsealed DSC sample holder) melts at 125° C (pure PHTP) because of the evaporation of the guest during heating. After polymerization, partial melting is generally observed at 125° C (pure PHTP) followed by a second endothermic phenomenon between 170 and 180° C due to the melting of the PHTP–polymer inclusion compound.

The most important result obtained by inclusion polymerization in PHTP was the achievement of the first asymmetric polymerization in the included state (which was also the very first solid-state asymmetric synthesis).[13] PHTP is a chiral compound (point group D_3); it has been obtained in optically active forms.[14–16] The single enantiomers also form inclusion compounds; however, unlike the adducts of the racemate, these have a crystal lattice which is chiral (space group $P6_3$ instead of $P6_3/m$). Polymerization of 1,3-*trans*-pentadiene in optically active PHTP yielded an isotactic polymer endowed with weak optical activity, from which optically active methylsuccinic acid was obtained by oxidative cleavage.

$(R)(-)$PHTP + 1,3-*trans*-pentadiene →

$-CH_2-CH=CH-CH(CH_3)-CH_2-CH=CH-CH(CH_3)-CH_2-CH=CH-CH(CH_3)-$

$(S)(+)$ polypentadiene $\quad |\alpha|_D = +2.7$

→ $HOOC-CH(CH_3)-CH_2-COOH$

$(S)(-)$ methylsuccinic acid $\quad |\alpha|_D = -1.2$

A second important finding concerns the mode of polymerization. Instead of irradiating the host–monomer inclusion compound directly, it is possible to irradiate the host alone in the crystal state.[17] The successive addition of monomer in the liquid or gaseous phase, over a suitable pressure and temperature range,[11] makes it possible to obtain the polymer with a yield lower than that obtained by the direct method, but with a higher molecular weight. The feasibility of this process is linked with the stability of the active species in the crystal state and with the ease with which inclusion compounds are formed with the monomer.

By this technique it was possible to ascertain that inclusion polymerization in PHTP is a living polymerization, in which termination and chain-transfer reactions are suppressed or considerably reduced.

More recently we have succeeded in preparing block copolymers, successively inserting two different monomers into the same previously irradiated
PHTP sample.[18] The radical nature of radiation-promoted inclusion polymerization had, in the meantime, been ascertained by ESR spectroscopy.[19]
It may thus be stated that inclusion polymerization is one of the extremely
rare examples of living radical polymerizations.

After the first discoveries, research activity was directed towards different
aims. Studies of the crystal structures of the host–monomer and host–polymer
inclusion compounds were carried out by Chatani and his group on urea
and thiourea[20–24] and by Colombo and Allegra on PHTP.[25] A study of the
phase diagrams of the inclusion compounds and the determination of the
relationship between their stability and their reactivity, was carried out in
the author's laboratory. The search for new hosts has been carried out by
various groups, of which that of Miyata and Takemoto may be mentioned
as the most active. The determination of the structure of the polymers in
terms of micro- and macro-structure, and the exploration of the possibilities
of copolymerization have been considered.

In this first part of the chapter I would like to devote some attention to
research carried out on deoxycholic acid (DCA) (6) and, more recently, on
apocholic acid (ACA) (7) and tris(*o*-phenylenedioxy) cyclotriphosphazene
(TPP) (8).

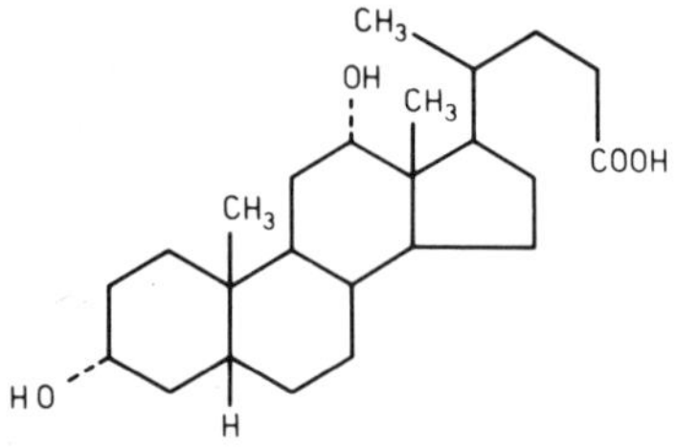

(6)

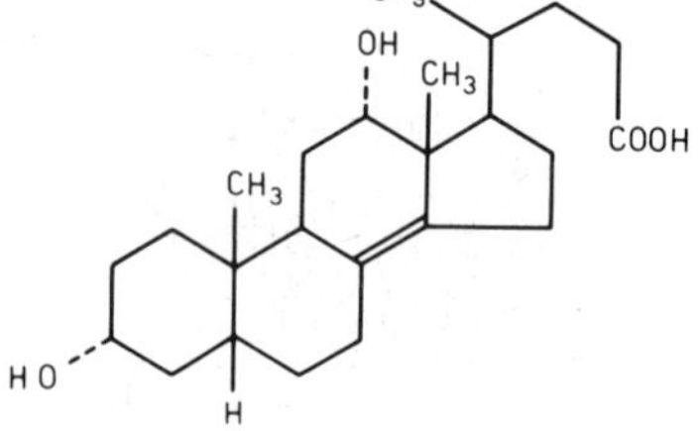

(7)

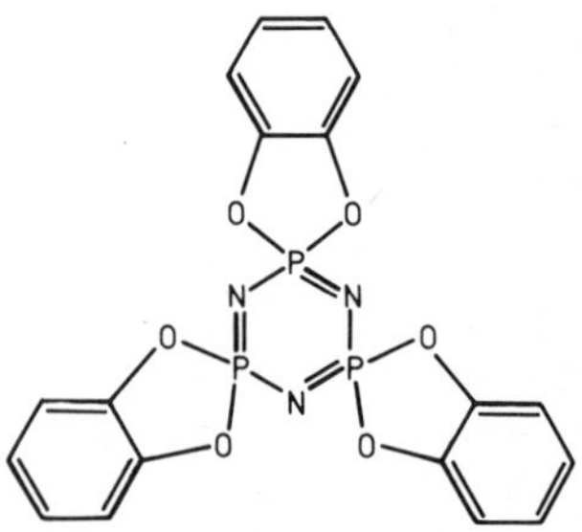

(8)

The possibility of carrying out inclusion polymerization in inclusion compounds of deoxycholic acid was revealed by Miyata and Takemoto in 1975.[26] Since then, the Japanese group has carried out a considerable amount of research in this field.[3,26-34] Many of the phenomena observed are common to the other types of inclusion polymerization; for example, the living nature of the polymerization, the influence on yield of extraneous agents co-crystallized inside the channels, and the increase in melting point after polymerization. On the other hand, inclusion polymerization in DCA is very different from those mentioned previously with regard to the selectivity with respect to the dimensions of the monomers and the regularity induced in the polymers. We may say that thiourea forms rather wide rigid channels which allow, for example, the inclusion of substituted diene monomers; urea forms narrower rigid channels suitable, for example, for butadiene; PHTP forms flexible channels suitable for linear and bulkier monomers; in all three cases there is a high degree of control over the type of polymer (in the case of the dienes there is always 1,4 instead of 1,2 polymerization). In DCA too, monomers of different bulkiness may be polymerized, but their structure is widely different: polybutadiene shows a considerable quantity (over 25%) of 1,2 units next to 1,4 units,[32] whilst poly-2,3-dimethylbutadiene is described as a highly regular 1,4-*trans* polymer.

A detailed structural analysis of DCA inclusion compounds has shown that in certain cases the guest molecules are not arranged homogeneously inside the channels.[35] Various arrangements or conformations of the included molecules might be responsible for the different degree of regularity. It is worth noting that the constitutional regularity of polybutadiene appears to be strongly influenced by the presence of inert additives in the channels. The addition of ethyl acetate, acetone, hexane or acetonitrile makes it possible to obtain pure or nearly pure polymers having 1,4 concatenation. The influence of these compounds might well be linked to the already known ability of DCA to include two different guests in its structure.[3,36,37] (See also Chapter 7, Volume 2).

As regards the ease with which inclusion compounds with the monomer are formed, DCA stands in a position intermediate between those of PHTP and urea. The greatest difficulty appears to be more of a kinetic than of a thermodynamic origin, and one of the best methods seems to be that of replacing a guest with another in a pre-formed inclusion compound. The starting point is the easily obtained DCA—acetone inclusion compound.[31]

The polymerization of *cis*- and *trans*-1,3-pentadiene in DCA has been studied by Audisio and Silvani[38] with particular reference to its asymmetric induction and to the constitutional and steric purity of the polymer. A comparative examination of the results obtained by various host–guest systems from this point of view is given later in this chapter.

The Japanese research group mentioned earlier has recently attained inclusion polymerization in apocholic acid (ACA), closely linked to deoxycholic acid both in origin and in properties.[39] The monomers used were *cis*- and *trans*-pentadiene, and *cis*- and *trans*-2-methylpentadiene.

A further example of inclusion polymerization has been obtained in recent years, using a cyclophosphazene as host. The ability of these compounds to form inclusion compounds was discovered and explored by Allcock,[40] who has provided a thorough report on his work in another chapter of this work (Chapter 8, Volume 1). Finter and Wegner used tris-(*o*-phenylenedioxy)cyclophosphazene (TPP) to polymerize butadiene and vinylchloride.[41] Although its channel size is similar to that of the urea, TPP forms inclusion compounds with the monomer much faster and more easily (for example by gas–solid reaction between the monomer and the host), and so presents considerable advantages with respect to urea, above all in the polymerization of vinylchloride.

Polymerization in TPP and in its homologous tris-(2,3-naphthalenedioxy)cyclotriphosphazene (TNP) was extended by Allcock, Ferrar and Levin to styrene monomers.[42] Particularly interesting is the synthesis of linear non-crosslinked poly-4-vinylstyrene and of stereoregular poly-4-bromostyrene. A comparison of the behaviour of 4-bromostyrene and styrene at room temperature and at $-78°$ C in TPP and TNP highlighted the existence of a series of conditions concerning the structure and mobility of the inclusion compounds, which control polymerization in these systems. (See Chapter 8, Volume 1). Lastly, I should like to mention the polymerization of vinylidene chloride and other monomers in β-cyclodextrin reported by Maciejewski.[43] In contrast with the cases mentioned previously, the polymer chains are inserted within the molecular structure of the host (in an intra-annular position). The product of the reaction thus belongs to the class of topological compounds known as rotaxanes.[44]

3. General features of inclusion polymerization

By definition, inclusion polymerization takes place within the channels existing in the host lattice of inclusion compounds. The demonstration that such a process really exists, and that it is not simply a polymerization taking place on the surface of the crystals or an energy- or electron-transfer phenomenon promoted by the presence of a cyrstalline component, was obtained for urea, thiourea and PHTP by means of two different techniques. By submitting the inclusion compounds PHTP-butadiene,[25] thioureadimethylbutadiene[21] and urea-butadiene[23] to brief exposures to X-rays, it was possible to obtain their diffraction patterns and hence to determine the structure of the compound containing the monomer. Longer exposure times brought to light the progressive transformation of the host–monomer compound into the host–polymer compound: this process takes place with a slight modification of the crystal structure of the inclusion compound which, however, keeps its form and orientation. When the repetition distance of the monomer in the channel does not coincide with that of the corresponding monomer unit, a considerable variation in the stoichiometry of the inclusion compound is found and the formation of excess host crystals is observed.

A perfectly consistent picture of all this is provided by an examination of the limiting conditions of polymerization and by a knowledge of the phase diagrams.[4,45,46] As we have already seen, polymerization in PHTP can be carried out by separating the stage in which the radicals form in the solid matrix from that in which the inclusion compound is formed. This latter may in turn be conducted in the presence of liquid or gaseous monomer and at different pressures.

Figures 1 and 2 report the yield of a series of polymerizations carried out under conditions identical with respect to all except one parameters, the only variation being in one case the temperature and in the other the pressure during the inclusion and polymerization stages.[11] To understand the sharp change in behaviour, reference must be made to the P–T phase diagram already discussed in the chapter dedicated to PHTP inclusion compounds (Chapter 3, Volume 2) and here reported in Fig. 3.[47] The phenomenon observed takes place when pressure and temperature values are reached

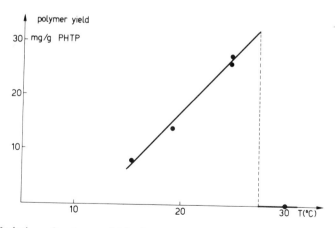

Fig. 1. Variation of polymer yield with temperature at constant monomer pressure for *trans*-pentadiene included in PHTP.

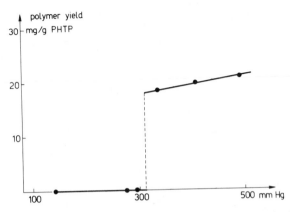

Fig. 2. Variation of polymer yield with monomer pressure at constant temperature for *trans*-pentadiene included in PHTP.

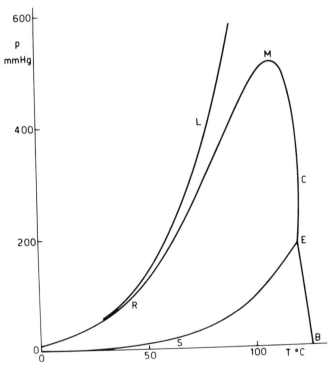

Fig. 3. Pressure–temperature projection of the solid–liquid–vapour phase diagram of the binary system PHTP–*n*-heptane.

corresponding to those of the decomposition of the inclusion compound (S–E curve). Below this curve polymerization does not take place: this is also the case if one operates with such a large excess of liquid monomer that all the solid phase dissolves (i.e., above the R–M–C–E curve of Fig. 3). It follows that the range of feasibility for polymerization coincides with the range through which the crystalline inclusion compound exists.

As regards the final stage of the process it has been observed that the maximum yield of polymer coincides with that expected for total channel filling. Furthermore, the formation of inclusion compounds containing polymers is often revealed by a rise in melting temperature.[10,12] A study of the PHTP–polyethylene phase diagram made it possible to obtain a further understanding of these aspects.[45,48] From Fig. 4 it can be observed that when the composition of the solid phase, expressed as a volume fraction (ϕ_1), is greater than that corresponding to total filling of the channels ($\phi_{1,0}$), the only phases present at equilibrium are the inclusion compound and the excess PHTP. Furthermore, the DSC curves offer no evidence of exothermic phenomena, either simultaneous with or following the melting of the PHTP, as is observed when nonequilibrium mixtures of polymer and PHTP are examined.[48]

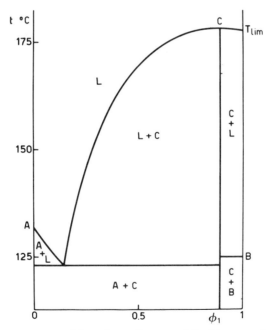

Fig. 4. Temperature-composition phase diagram of the binary system PHTP–polyethylene.

It may thus be stated that monomer inclusion always precedes polymerization and that, in general, the final product of inclusion polymerization is a crystalline host–polymer inclusion compound. The only noteworthy exceptions to this last statement relate to polyvinylchloride and polyvinylbromide in urea. As polymerization takes place in these cases decomposition of the inclusion compound is observed, with the formation of tetragonal urea crystals and free polymer.[24] This would seem to be due to the particular instability of the inclusion compound because of a mismatch between the cross-section of the chain and that of the channel. Nonetheless, even in this case we may speak of inclusion polymerization: this is shown by the structure of the polymer after removal of the urea.

With regard to the mechanism of inclusion polymerization, considerable interest has been stimulated by the nature of the initiation reaction. The presence of trapped radicals in X- or γ-irradiated inclusion compounds was ascertained by Griffith using ESR spectroscopy both in urea[49] and in PHTP.[50] Radicals derive principally from the guest component. However, evidence also exists for the presence of stable radicals formed by irradiation of pure PHTP.[17,51] The ESR spectrum of the urea–butadiene inclusion compound when irradiated and examined at very low temperature reveals the presence of crotyl radicals $CH_3-CH=CH-CH_2^{\cdot}$, obtained by the addition of a hydrogen atom into the conjugate double bonds of the monomer.[19] By heating to room temperature, the spectrum is changed considerably because the propagation reaction begins to take place. This is interpreted on the basis of the following structure of the propagating species:

$$H-(CH_2-CH=CH-CH_2-)_n-CH_2-CH=CH-CH_2^{\cdot}$$

However, experiments carried out with ^{14}C-labelled PHTP show that the prevalent, and in some cases the only, reaction is the addition of a radical derived from PHTP to the monomer:[17]

$$RH \xrightarrow{\gamma\,\text{rays}} R^{\cdot} \xrightarrow{C_4H_6} R-CH_2-CH=CH-CH_2^{\cdot}$$

In any case, the radical nature of the polymerization appears certain.

Various authors have underlined the presence of considerable post-irradiation polymerization phenomena.[2,3] Especially in the case of PHTP, the situation varies according to the monomer examined. With pentadiene the yield grows considerably with polymerization time once irradiation has ended,[17] whilst with butadiene post-irradiation polymerization appears to be negligible.[52] This discrepancy might be explained by the greater polymerization rate of butadiene compared with other monomers, whereby the factor limiting the yield is the formation of radicals (which depends upon the radiation dose) and not the polymerization time. The stability of PHTP radicals in the crystalline state (after a rapid initial decay their concentration remains constant for many days) has made it possible to devise a method of inclusion polymerization, already mentioned several times, in which the radical formation stage is separated from the polymerization stage.[17] The differences between polymerization obtained by direct irradiation of the inclusion compound and polymerization in pre-irradiated PHTP were brought to light by using ^{14}C-labelled PHTP as host.[17] In both cases the polymer has some radioactivity, but a comparison between the osmometric and the radiochemical molecular weights (both are number-average molecular weights) shows that in the latter process there exists a correspondence close to 1:1 between labelled groups and macromolecular chains, whilst in the former process only a fraction (from about 30–50%) contains a labelled chain end. This fact can be explained by taking into consideration various processes, for example the transfer of a hydrogen atom from the host, the formation of active species involving the monomer alone, or even the chain-transfer reaction with the monomer.

It can be seen by post-irradiation polymerization that termination and chain-transfer processes, if they exist, are extremely limited. In fact, in a series of polymerizations conducted under strictly controlled conditions, the molecular weight of the polymer was found to be directly proportional to the yield,[17] a behaviour typical of living polymerizations. Furthermore, a block copolymer was obtained by the successive introduction of various monomers into the channel, with minimal quantities of homopolymer.[18,53]

With regard to the propagation reaction, considerable differences are observed between the various systems, connected with the higher or lower degree of adaptability of the shape and size of the channel to that of the guest. These factors influence the constitutional and steric regularity of the polymer; regularity is at a maximum in urea, thiourea, PHTP and TTP, and relatively low in DCA.

A study carried out on the various dienes and trienes included in PHTP has made it possible to establish some conditions for polymerization. Whilst the prime condition is the capacity of both host and monomer to form an inclusion compound, it has been observed that certain included monomers such as *trans,trans*-2,4-hexadiene or 2,5-dimethyl-2,4-hexadiene, unlike *cis-trans* and *cis-cis*-hexadiene, do not polymerize. This is not due to an intrinsic lack of reactivity (all such monomers polymerize in the presence of Ziegler–Natta catalysts) but to an unfavourable arrangement of the monomers in the channel. The knowledge of the stoichiometric host/guest ratio and of the channel shape makes it possible to determine the distance between the unsaturated carbon atoms of two successive monomers: this distance is below 5.5 Å for all monomers that polymerize, but for the two above mentioned compounds it is over 6 Å.[54] This long distance is linked with the presence of methyl groups in the *trans* position, i.e., arranged along the channel axis. It is interesting to note that the absence of head-to-head and tail-to-tail concatenations in certain polymers such as polypentadiene can be attributed to the same feature: a succession of monomers generating a head-to-head sequence would produce an excessive separation between the two unsaturated carbon atoms and halt the propagation or make it very slow. The polymerization rate also seems to depend, at least to a first approximation, upon the value of the $C\cdots C$ distance.

Similar considerations may be put forward with regard to substituted hexatrienes: the polymerization inertia of alloocimene (2,6-dimethyl-2,4,6-octatriene) included in PHTP is attributed to the presence of three methyl groups bonded to the unsaturated atoms 2 and 7.[55]

The interpretation given here must be considered as a first approximation. A more precise discussion should be in terms of radical–monomer interaction and should require a precise knowledge of the geometry of the channel at the interface between the part occupied by the polymer and that occupied by the monomer.

Examples of inclusion polymerization are presented in Table 1: they mainly include diene monomers, but also contain some significant examples of vinyl and triene monomers.

4. Control phenomena at the molecular and supermolecular level

The structures of the first polymers obtained by inclusion polymerization were determined by comparing their IR and X-ray spectra and their more common physical properties (solubility, melting point and so on) with those of polymers of known structure obtained by conventional methods. With progress in microstructural methods of investigation such as NMR- and in particular ^{13}C-NMR-spectroscopy, new information has been obtained both as to the type and the extent of the regularity of these polymers. The following discussion will, as far as possible, be in such terms.

Inclusion polymerization in urea, thiourea, PHTP and TPP is a highly-selective process at a constitutional and stereochemical level. In the case most widely studied, that of substituted butadienes, only polymers having the 1,4-*trans* structure have been obtained. The fact that some of the most important monomers used in inclusion polymerization are symmetrical — for example butadiene or 2,3-dimethylbutadiene — makes it impossible to observe the irregularity of the insertion of the monomer into the chain. For non-symmetrical monomers the situation differs according to the position of the substituents. For example in the polymerization of *trans*-pentadiene [which has a methyl group bonded to C(4)] in PHTP, only head-to-tail sequences are found, as previously described. ESR spectra taken during the polymerization of *trans*-pentadiene in pre-irradiated PHTP have shown that the growing radicals have the structure:[66]

$$R-CH_2-CH{=}CH-CH^{\cdot}-CH_3$$

As a result the insertion of the monomer unit into the chain takes place in the 1,4 and not in the 4,1 way. Other monomers having one or two methyls in position 4 (*trans*-2-methylpentadiene, 4-methylpentadiene) behave in the same way.

Among the more common monomers, only isoprene allows the detection of chain disorder. The presence of a methyl bonded in position 2 makes the two modes of insertion non-equivalent without causing the separation of the reactive atoms from each other. The polyisoprene obtained in PHTP and extracted with boiling methanol is amorphous and easily soluble; its IR spectrum coincides with that of the 1,4-*trans* polymer (which, when highly regular, is crystalline). The ^{13}C-NMR spectrum shows, however, the presence of considerable quantities ($\simeq 15\%$) of chain inversions as evidenced by two signals at 28.27 and at 38.51 ppm from TMS, together with more intense signals at 26.74 and 39.74 ppm corresponding to the head-to-tail structure.[53]

Head-to-head and tail-to-tail sequences have also been observed in the polymerization in PHTP of 2-methylhexatriene: it should be borne in mind that hexatrienes produce insoluble polymers of 1,6-*trans-trans* structure.[55] Ozonolysis of the polymer followed by reduction of the ozonides with LiAlH$_4$, also produces, in addition to ethanediol and 1,4-pentanediol, 1,4-butanediol and 2,5-hexanediol, these latter two diols being derived from the sequences =CH–CH$_2$–CH$_2$–CH= (tail-to-tail) and =C(CH$_3$)–CH$_2$–CH$_2$–C(CH$_3$)= (head-to-head).

Polymerization in PHTP of *trans*- and *cis*-pentadiene gives rise to isotactic polymers having high regularity, as revealed by X-ray, IR and NMR techniques.[12,17,53,59,60] A study of the ^{13}C-NMR spectrum, carried out under resolution enhancement conditions, has shown the presence of *mm* triads alone, both in the polymer as such and after reduction to a saturated polymer,[60] unlike the situation for the analogous polymer obtained in DCA (Table 2). An isotactic structure was attributed also to the poly-2-methylpentadiene obtained in PHTP on the basis of its X-ray fibre pattern.[54] An independent confirmation of the structure has recently been obtained and will be discussed below.[64] In the vinyl series, the most noteworthy examples of stereoregularity are those of polyvinylchloride obtained in urea[7,8] and TPP[41] and of poly-4-bromostyrene obtained in TPP.[42] The content in syndiotactic diads in polyvinylchloride has been evaluated at 86% by IR and X-ray techniques.[24,56] It must, however, be pointed out that some doubts

Table 1. Examples of inclusion polymerization

Monomer	Host	Polymer structure	Methods of characterization	References
Vinylchloride	urea	partly syndiotactic	IR, X-ray	7, 8, 24, 56
	TPP	partly syndiotactic	IR	41
Acrylonitrile	urea	partly isotactic	IR, NMR	7, 57, 58
4-Bromostyrene	TPP	stereoregular	IR, NMR	42
Butadiene	urea	1,4-trans, crystalline	DSC, IR, X-ray	7, 23
	PHTP	1,4-trans, crystalline	DSC, IR, NMR, X-ray	12, 25, 52, 59
	DCA	1,4-trans + 1,2	IR	32
	TPP	1,4-trans(+1,2) crystalline	IR, NMR	41, 42
Trans-pentadiene	PHTP	1,4-trans, highly isotactic, crystalline	DSC, IR, NMR, X-ray, ozonolysis, $|\alpha|$, hydrogenation	11, 12, 13, 17, 53, 59, 60
	DCA	1,4-trans + 1,4-cis + 1,2	NMR, ozonolysis, $|\alpha|$	38, 60, 61
	ACA	1,4-trans + 1,4-cis + 1,2	NMR, ozonolysis, $|\alpha|$	39, 62
Cis-pentadiene	PHTP	1,4-trans, isotactic crystalline	DSC, IR, NMR, X-ray	17, 59
	DCA	1,4-trans, atactic amorphous	NMR, ozonolysis, $|\alpha|$ hydrogenation	38, 60, 61
	ACA	1,4-trans + 1,4-cis	NMR, $|\alpha|$	39, 62

Compound	Host	Structure	Methods	References
Cis + *trans*-pentadiene	urea	1,4-*trans*	IR	63
	PHTP	1,4-*trans*, isotactic, crystalline	DSC, IR, NMR	59
Isoprene	DCA	1,4-*trans*, +1,2, atactic	NMR, $|\alpha|$	38
	PHTP	1,4-*trans*, amorphous	IR, NMR	12, 53, 54, 59
2,3-Dimethylbutadiene	thiourea	1,4-*trans*, crystalline	DSC, IR, X-ray	5, 6, 21
	PHTP	1,4-*trans*, crystalline	DSC, IR, NMR, X-ray	12, 54, 59
	DCA	1,4-*trans*, crystalline	IR	26, 27, 29, 31, 33
Trans-2-methylpentadiene	PHTP	1,4-*trans*, highly isotactic, crystalline	DSC, IR, NMR, X-ray, hydrogenation	54, 59, 64, 85
	DCA	1,4-*trans*	IR, NMR, ozonolysis, $|\alpha|$.	30, 34, 62, 65
Cis-2-methylpentadiene	ACA	1,4-*trans* + 1,4-*cis*	NMR, $|\alpha|$	39, 62
	DCA	1,4-*trans*	NMR, $|\alpha|$, ozonolysis	39, 62, 65
	ACA	1,4-*trans*	NMR, $|\alpha|$	39, 62
Other substituted butadienes	thiourea	1,4-*trans*	IR, X-ray	6, 20
Hexatriene and various substituted hexatrienes	PHTP	1,4-*trans*	DSC, IR, NMR, X-ray	54, 59
	DCA	1,4-*trans*	IR	26, 27, 28
	PHTP	1,6-*trans-trans*	IR, ozonolysis	55

$|\alpha|$—Optical rotatory study.

Table 2. Stereochemistry of pentadiene polymerization (pentadiene = PD)

Host	Monomer	Polymer structure			Triad composition				$|\alpha|_D$	References
		1,4-trans	1,4-cis	1,2	mm	mr+rm	rr			
(±) PHTP	trans-PD	100	0	0	100	0	0	—	12, 17, 59, 60
(−) PHTP	trans-PD	100	0	0	100	0	0	+2.7	13, 53
(±) PHTP	cis-PD	100	0	0	≃100	0	0	—	12, 17, 59
(+) DCA	trans-PD	91	6.5	2.5	34	47	19	−3	38, 60, 61
(+) DCA	cis-PD	97	3	0	52	34	14	−21	38, 60, 61
(+) ACA	trans-PD	90	10	0	nd	nd	nd	−6.5	39, 62
(+) ACA	cis-PD	90	10	0	nd	nd	nd	−3.7	39, 62

have recently been advanced as to the chemical purity and stereoregularity of polyvinylchloride produced in urea.

The X-ray pattern of polyacrylonitrile obtained in urea shows a residual orientation of the polymer crystals even after removal of urea.[7] The use of deuterated monomers has made it possible to determine the stereoregularity index by NMR:[57,58] the polymer was found to be predominantly isotactic with a content in *m* diads varying from 75 to 90%.

Although, with the exception of polypentadiene, we have no detailed analyses of the sequences in polymers obtained in DCA, both the constitutional regularity and the stereoregularity appear lower than those for the cases discussed earlier. Polybutadiene, for example, presents considerable constitutional disorder with a content in 1,2 units varying between 1% and 28% according to the conditions,[3,32] and poly-2-methylpentadiene, a polymer of great interest because of its chirality, is said to be a *trans*-rich polymer.[34] More recently, a sample endowed with a lower optical activity was found to have a nearly pure *trans* structure.[66]

A particularly instructive comparison is that between the behaviour of *trans*- and *cis*-pentadiene included in DCA.[60,61] Both yield a polymer having a predominantly 1,4-*trans* structure, but with different contents of constitutional defects and with different microstructures, as can be seen in Table 2. These facts have been related to the different bulk of the two monomers (greater for the *cis*) and hence to the different conformational freedom within the channel.

No conclusive answer has yet been formulated as regards the causes of stereoregularity in inclusion polymerization, if we exclude the aspects of asymmetric induction which will be discussed later. An ordered arrangement of monomers in the channel seems to be of fundamental importance. The attainment of a regular 1,4-*trans* structure in polydienes is related to the limits imposed by the channel walls which restrict the conformational freedom and cause the monomer to assume an *anti* conformation about the central single bond at the moment in which polymerization takes place (**9**).

$$R^\bullet \;+\; \underset{CH_2}{}^{CH}\diagdown\diagup^{CH_2}\;\underset{CH}{}\;\longrightarrow\; R\diagdown\underset{CH_2}{}\diagup^{CH}\diagdown\underset{CH}{}\diagup^{CH_2^\bullet}$$

9

According to the principle of least-motion, such a structure is more easily transformed into a 1,4-*trans* than into a 1,4-*cis* monomer unit. It is more difficult to understand the very high isotacticity observed in racemic PHTP. With regard to this, the accidental coincidence of the repeat distance of the PHTP molecule with that of the monomer unit of the 1,4-*trans* polydienes (both close to 4.8 Å) might also play a part in determining the isotactic character: each stage of addition within the channel would take place in a geometrical surrounding essentially identical to the preceding one.[2]

Cne of the most important developments of inclusion polymerization relates to the synthesis of optically active polymers starting from non-chiral molecules. Asymmetric polymerization differs from the asymmetric syntheses known in organic chemistry because of its multistep nature: hundreds, or even thousands, of asymmetric carbon atoms are contained in the same macromolecule. It thus becomes of the greatest importance to determine whether there exists any correlation between the production of such centres of stereoisomerism and, also, in what way the asymmetric induction is transmitted in the single stages of chain growth. The conditions required for chirality in macromolecular compounds and the examination of the various mechanisms are set out in numerous texts on the subject.[67-70] In the field of asymmetric inclusion polymerization a single enantiomer of a chiral host must be used and the monomers must be suitable for the formation of chiral polymeric structures. Fortunately, both conditions can be satisfied: PHTP, DCA and ACA are in fact chiral and can be obtained in optically active forms: furthermore, the 4 or 1,4 substituted butadienes constitute one of the few classes of monomer suitable for the purpose. Both the *cis*- and the *trans*-isomers of pentadiene and of 2-methylpentadiene are representatives of this class. Asymmetric polymerization of *trans*-pentadiene in PHTP dates back to 1967.[13] The greatest difficulty in this process consists of obtaining optically active PHTP. Given its saturated hydrocarbon nature, its resolution requires transformation into a carboxylic acid followed by decarboxylation, an extremely expensive process requiring a very long time (many months for a few grams of product) and a high level of manual skill. The optical purity of the polypentadiene obtained in this way is rather low, about 7%. Only in one case has a polymer of high optical activity been obtained,[2] but attempts to repeat this result have failed. The hypothesis advanced to explain this fact, which seems most attractive, is the chance formation of a metastable phase having a higher power of asymmetric induction. This hypothesis is not illogical in view of the polymorphism of PHTP.[71]

DCA and ACA are natural products consisting of a single enantiomer and thus appear as ideal hosts for asymmetric inclusion polymerization. Furthermore, their power of asymmetric induction appears to be higher than that of PHTP. The results obtained with *cis*- and *trans*-pentadiene[38,61] are illustrated in Table 2, and those with *cis*- and *trans*-2-methylpentadiene[30,34,39] in Table 3. In this latter case we find one of the highest values obtained by asymmetric polymerization. The rotary power of poly-2-methylpentadiene is highly dependent on temperature and disappears on going from room temperature to 70° C. The phenomenon is said to be slowly reversible and has been interpreted as a conformational transformation.[3,34,39]

The same polymer has been converted by ozonolysis into a low-molecular-weight derivative to determine its absolute configuration and optical purity.

Table 3. Stereochemistry of 2-methylpentadiene polymerization (2-methylpentadiene = MPD)

Host	Monomer	Polymer structure		$|\alpha|_D$	References
		1,4-*trans*	1,4-*cis*		
(±) PHTP	*trans*-MPD	100	0	—	54, 59, 64, 85
(+) DCA	*trans*-MPD	nd	nd	+250	30
(+) DCA	*trans*-MPD	≃100	0	+90	39, 62, 65
(+) DCA	*cis*-MPD	≃100	0	+320	39, 62, 65
(+) ACA	*trans*-MPD	95	5	−106	39, 62
(+) ACA	*cis*-MPD	100	0	−66	39, 62

Findings so far available are incomplete and appear rather contradictory.[62] Their clarification and rationalization will require a more detailed knowledge of the microstructure of the polymer.

The question of the origin of asymmetric induction in inclusion polymerization has been brilliantly resolved both for PHTP and for DCA by two different types of experiment. According to the general schemes of asymmetric polymerization, two different mechanisms for the transmission of chirality can be envisaged.[72,73] In the first, asymmetric induction is limited to the initiation stage: a radical, or another optically active residue induces a preferential configuration into the first monomeric unit. In the following stages there should be no further influence of chirality and the configuration of the second monomeric unit should be influenced by the first only by factors of regularity, with no regard for sign. Once there is a configurational error, the new configuration would have the same probability of propagating as the earlier one. We then speak of a process with asymmetric initiation and symmetric growth ($I_D \neq I_L$; $P_{DD} = P_{LL}$, $P_{DL} = P_{LD}$ where P represents the conditional probability of each sequence).

In the second scheme the probability of insertion of the monomer into a given configuration is different from that of its enantiomer in each stage of propagation (asymmetric growth). This may take place following an asymmetric Bernoulli distribution ($P_{LD} = P_{DD} \neq P_{LL} = P_{DL}$) in which each act is independent of the preceding one, or with a more complex distribution, for example with a first-order asymmetric Markow process, with a more or less marked influence of the last monomer unit in the chain. In this last case we have $P_{DD} \neq P_{LL}$ but also $P_{DD} \neq P_{LD}$ and $P_{LL} \neq P_{DL}$. In a process of asymmetric initiation and symmetric growth the optical purity of the polymer decreases rapidly with molecular weight (M), whilst in asymmetric growth processes the rotatory power is independent of M. Nonetheless the possibility of distinguishing between the various processes practically disappears when the isotacticity tends towards 100%.

With regard to inclusion polymerization, the extreme hypotheses are as follows: asymmetric induction may derive from the chiral radical which initiates the chain or from the chirality of the channel which influences the mode of insertion of the single monomers into the chain. In order to choose between the two possibilities, Sozzani and Di Silvestro prepared an isoprene-pentadiene block copolymer in optically active PHTP, in which the first monomer (isoprene) gives rise to a non-chiral chain many hundreds of Å long.[53] Having in this way eliminated the influence of the chiral initiator, the optical activity of the successive polypentadiene block *must* be attributed solely to the influence of the chiral crystal lattice. In effect, the rotatory power value, corrected for the non-chiral polymer fraction, is perfectly consistent with this explanation (Table 4).

Table 4. *Asymmetric block copolymerization in (−) PHTP*

Optical activity of pentadiene homopolymers	$\lvert\alpha\rvert_D = +2.7$
Optical activity of a 30:70 isoprene–pentadiene block copolymer	$\lvert\alpha\rvert_D = +1.9$
Corrected optical activity of the polypentadiene block	$\lvert\alpha\rvert_D = +2.7$

Structure of the block copolymer:

$$\text{PHTP}-(-CH_2-CH=C(CH_3)-CH_2-)_m-(-CH_2-CH=CH-CH(CH_3)-)_n-X$$

chiral   non-chiral   chiral

The hydrocarbon nature of both host and guest prevents the formation of bonds of any kind between the two components, except for weak and non-directional van der Waals forces: we may thus speak of "through-space asymmetric induction" instead of "through-bond induction" as in the more generally known cases of asymmetric synthesis.

Substantially analogous conclusions have been obtained by Audisio *et al.* from the examination of ^{13}C-NMR spectra of polypentadiene obtained in DCA.[61] The quantitative relationships between steric triads suggest the presence of a process following an asymmetric Bernoulli distribution in which each monomer unit has no interaction with the preceding ones. In this case too, induction must be attributed to the chiral molecules of the host.

As a final point in the discussion on the control phenomena existing in inclusion polymerization, I wish to illustrate briefly the morphology of native polymers obtained by dissolving the host in a solvent which neither dissolves nor swells the polymer itself.

Diffractometric studies have shown that the conformation of the polymer chain is the same in the included polymer, in the native polymer, or in that crystallized from a solution or from the melt. However, the native polymer seems to possess a different macroconformation or morphology which influences the melting behaviour and the small-angle X-ray patterns.

The first observation was made by White[7] who reported a melting point of 147–154° C for polybutadiene obtained in urea, 10–15° higher than that of the same polymer crystallized from benzene. This result has not been confirmed by further research, but some X-ray diffraction and electron microscopy findings have indicated the possible absence of the lamellar structures which are generally present in conventional crystalline polymers.[23] Analogous behaviour has also been observed in native polypentadiene obtained in PHTP.[4,74] A detailed analysis of the melting process, carried out by means of DSC, has shown that the first-melting temperature is considerably higher than that observed during the second melt (104° C against 82–84° C). The melting temperature is also highly sensitive to heating rate: a superheating of about 10° has been found on varying the temperature programme from 0.1 to 30° min^{-1}. Heat treatments consisting of partial meltings followed by crystallization at relatively high temperatures have shown that high-melting crystals cannot grow from the melt at atmospheric pressure.

This behaviour corresponds to that observed by Wunderlich for polyethylene crystallized under high pressure, for which an extended-chain morphology is accepted.[75–77] A confirmation of this hypothesis has been obtained from small- and wide-angle X-ray diffraction patterns: the former, in this case also, suggests no sign of lamellar structures, whilst the latter suggests a considerably defective but substantially monophasic paracrystalline structure.[74]

As a conclusion to the observations made by the author and by other workers, the formation of extended chain native polymers may be summed up schematically as follows (Fig. 5):

i) Insertion of the monomers into the channels of an inclusion compound;

ii) Polymerization and formation of extended chains of the polymer parallel to each other but not directly in contact, because of the presence of the host component;

iii) Removal of the host and collapse of the crystal structure. In this phase the chains still keep their orientation and extended macroconformation, but, at least in the case of polypentadiene, do not achieve ordered packing on the plane perpendicular to the chain axis.

It must be observed that the polymerization is an integral part of the process of crystallization. In effect, attempts to obtain an extended-chain morphology starting from preformed polymers by formation of an inclusion compound followed by its decomposition, have failed, at least up to the present time, both in the urea–polyethylene system[78] and in the PHTP–polypentadiene system.

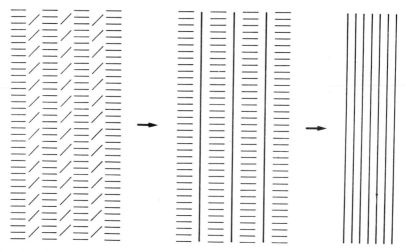

Fig. 5. Schematic representation of the process of formation of extended-chain polymers.

5. Inclusion copolymerization

The first instances of inclusion copolymerization were obtained by Brown and White[6] who used various pairs of monomers chosen from dimethyl-butadiene, dichlorobutadiene, cyclohexadiene, vinylidene chloride, cyclo-pentadiene and cyclooctadiene included in thiourea. The formation of copolymers was deduced from the examination of the IR spectra (presence of localized bands corresponding to each of the two monomer units, and band-shifts with respect to the two homopolymers in the fingerprint region), or from the presence of a marked alteration of the melting point, compared with that of the homopolymers. An attempt to copolymerize vinylchloride and acrylonitrile in urea, however, did not lead to significant results.[7]

Later, the copolymerization of acrylonitrile and ethyl acrylate in urea was described.[79] The reaction product was separated with acetone into two fractions, the composition of which was found to be constant over a large range of compositions of the monomer mixture and strongly biased towards the two homopolymers: the insoluble fraction contains over 90% of monomer units derived from acrylonitrile, and the soluble fraction about 95% of ethyl acrylate. The DSC curve of the inclusion compound before polymerization presents, in its turn, two endothermic transitions around

$-40°$ C and $-10°$ C, which may be attributed to the presence of two different types of inclusion compound, one containing mainly (or solely) acrylonitrile, the other ethyl acrylate. These findings indicate considerable difficulty in obtaining random copolymers if the right conditions for insertion of two monomers into the same channel are not found at the outset.

Inclusion copolymerization in PHTP was observed by accident during a study of the polymerization of trans-2-methylpentadiene.[54] A sample of this monomer, left for several months at room temperature, isomerized in part into 4-methylpentadiene by a 1,5-hydrogen shift: the resulting polymer was much more soluble than a true homopolymer of trans-2-methylpenta-diene, and the ^{1}H-NMR spectrum clearly showed the presence of a singlet due to the gem-dimethyl group (from 4-methylpentadiene) together with the doublet from the 2-methylpentadiene. The van der Waals shape of the two monomers is very similar, and we may thus presume there is a disordered succession of the two guests in the channel without any considerable influence on the stability. Analogous conclusions as to the isomorphism of the monomers were also reached by Chatani and Nakatani,[21] examining from a structural point of view Brown and White's results obtained in thiourea.

Some more recent research has been along the same lines. The pairs isobutene–vinylidene chloride and isobutene–1,1-dichloro-2,2-difluoro-ethylene were successfully copolymerized in urea.[80] The f_1/F_1 diagrams (which relate the monomer composition with that of the copolymer) practically coincide with that of the free (non-inclusion) copolymeriz-ation. However, other characteristics, in particular the dependence of conversion on temperature, which is analogous to that shown in Fig. 1, support the idea of a true inclusion copolymerization. In neither of these two cases is there an ideal copolymerization; that is, one following a Bernoullian distribution: in particular, the second gives an alternate copolymer for all monomer compositions containing more than 30–40% of dichlorodifluoroethylene. In other cases, however, it has been found that there is an ideal azeotropic copolymerization ($r_1 = r_2 = 1$), in which the composition of the copolymer is identical to that of the starting mixture.[81]

Copolymerization in DCA has been studied by Takemoto and Miyata[3] who assert the existence of considerable differences between inclusion copolymerization and free copolymerization: the former seems to depend on the ease with which the two monomers are included in the host, and not on the reactivity of the growing chain end.

It is not possible to fit all these findings into a single scheme; the two elements, control by the channel and reactivity of the chain ends, both seem to be operating, yet their relative importance seems to vary from case to case.

In recent years both block and random copolymerization in PHTP have been studied thoroughly. The former has made it possible to confirm the living nature of inclusion polymerization[18] and to identify that process of transmission of chirality which we called "through-space asymmetric induction" earlier in this chapter.[53] Random copolymerizaton was first conducted on pairs of monomers of similar bulkiness, such as 2-methylpentadiene and 4-methylpentadiene, and an unequivocal demonstration of its occurrence was obtained by the study of the sequences of monomer units.[46,59,82,83] The use of ^{13}C-NMR spectroscopy is particularly suitable for this purpose because of its ability to distinguish between block copolymers (which contain only the signals of the two homopolymers), alternate copolymers (whose spectra can be interpreted on the basis of $(AB)_n$ sequences) and random copolymers, which have more complex spectra. Depending on the constitution of the monomer units and the resolution of the spectrometer, the complexity of the spectrum makes it possible to determine extremely long sequences of monomer units, up to heptads and octads.[84] In the case of diene monomers, whose monomer units contain four carbon atoms in the chain, it is possible to interpret the spectrum in terms of triads: the chemical shift of the carbon atoms of each monomer unit is in general affected by the nature of the preceding and following monomer unit.

To give an example, Fig. 6 shows the ^{13}C-NMR spectrum of a copolymer obtained by irradiating an inclusion compound prepared from a 75:25 mixture of *trans*-2-methylpentadiene and 4-methylpentadiene (ten times in excess of the quantity required for complete channel filling). A total of 36 signals is observed, 9 attributed to the methyl groups, 11 to the other saturated carbon atoms (CH_2, CH, and quaternary C) and 16 to the unsaturated carbon atoms, as against the 44 signals $(12+16+16)$ to be expected for all possible triads of a head-to-tail polymer, where the stereochemical influences are taken to be negligible (even if the phenomena of steric irregularity were to be found, their effect on the chemical shift would be in any case less than 0.1 ppm, as observed in similar cases).[60] As a comparison, it may be recalled that mixtures of the corresponding homopolymers or block copolymers would have a total of 11 signals, and the alternate copolymers would have the same number, or 12 at most. Even more complex is the spectrum of the isoprene-pentadiene copolymers as a result of the presence of head-to-head and tail-to-tail sequences, forbidden in the pentadiene homopolymer, but here permitted by the presence of isoprene units which remove the unfavourable steric interactions. It is worth noting that, despite its structural disorder, the isoprene–pentadiene random copolymer obtained in (−) PHTP presents a measurable optical activity,[82] once again indicating that the control of the chirality is solely due to the effect of the chiral crystal lattice. A quantitative analysis of spectra and the application of statistical

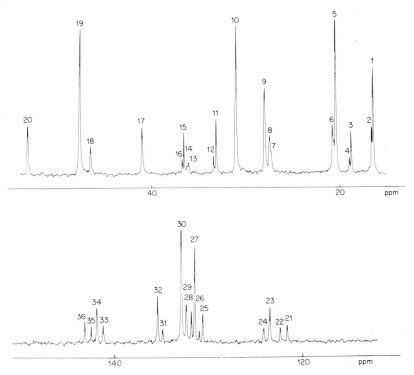

Fig. 6. ^{13}C-NMR spectrum of a *trans*-2-methylpentadiene–4-methylpentadiene copolymer obtained in PHTP.

procedures has made it possible to ascertain that there is, at least to a first approximation, a Bernoullian distribution of monomer units,[46,83] in which the probability of the existence of a unit is independent of the chemical nature of the preceding unit. In terms of copolymerization theory, the pair *trans*-2-methylpentadiene–4-methylpentadiene behaves in a manner very close to an ideal azeotropic copolymerization.

So far about a dozen diene monomer pairs have been examined (chosen among butadiene, *trans*-pentadiene, isoprene, *trans*-2-methylpentadiene, 4-methylpentadiene and 2,3-dimethylbutadiene). As research on copolymerization in PHTP continued, some of the initial hypotheses failed to find confirmation: in the first place, that regarding the isomorphism of monomers. It was, in fact, observed that a narrow monomer such as butadiene forms copolymers with bulky monomers such as 2,3-dimethylbutadiene, 2-methyl- and 4-methylpentadiene. The absence of the butadiene homopolymer, if we take into account its higher reactivity with respect to the other monomers, is proof that the two monomers are included in the same channel before polymerization.

Such observations are not as yet conclusive, and will have to be substantiated by other experiments regarding the structure and thermodynamic behaviour of three-component inclusion compounds (PHTP–guest 1–guest 2), as well as by investigations of the kinetics of monomer displacement inside the channels and between solid and fluid phases.

6. Significance and prospects of the study of inclusion polymerization

After this review of the details of inclusion polymerization, we may wonder what is the meaning of research in this field, and what developments may be possible. We may at this point distinguish between two different aspects, one the study of inclusion compounds and their solid-state chemistry, and the other their macromolecular chemistry.

As regards the former aspect, inclusion polymerization represents what is perhaps the most studied example of an organic reaction taking place inside a crystal phase. The multiplicity of examples studied so far has made it possible to identify the fundamental points of the process and to interpret them in a satisfactory manner.

A salient point appears to be the connection between the structure, the stability and the reactivity of inclusion compounds. The behaviour of the various host–guest pairs, the effect of temperature and pressure, the microstructure of the polymer and many other phenomena may be interpreted in a unified way, using at one and the same time the results of thermodynamic, diffractometric and spectroscopic investigations.

A second point of great importance is the unequivocal demonstration of the fundamental role played by the crystal lattice in inducing chirality in the polymer. This finding fits in with the current research into the feasibility of asymmetric synthesis in oriented systems, and is pertinent to the debate regarding the mimimum requisites for asymmetric synthesis. It may be noted that, given the more rigid nature of the chiral matrix, the optical purity obtained by inclusion polymerization is generally higher than that observed in reactions carried out in liquid crystals.

One chapter just being opened up, and still to be explored, regards inclusion copolymerization as a method of study of inclusion compounds containing two different guests. If we assume that the rate of polymerization is much higher than that of monomer diffusion from the outside to the inside of the channels, the succession of monomer units would represent a 'photo-

graphic image' of the situation existing in the channels before polymerization, and would supply structural information not otherwise accessible.

On the macromolecular plane, interest is at present focused on the detailed
understanding of polymer structure and on the parameters regulating it. I
feel that it will be possible to attain the synthesis of homopolymers or
copolymers having an exactly defined structure, to be used for special
purposes; for example, as reference compounds in the study of physicochemical properties and of property-structure relationships. As an example
of this kind of application we might mention the synthesis of a new type
of stereoisomer of polypropylene, hemiisotactic polypropylene, obtained by
means of inclusion polymerization. The starting point was poly-2-methyl-
pentadiene obtained in PHTP, to which a 1,4-*trans* isotactic structure had
been attributed.[54,64] Reduction with tosylhydrazide converted it into a saturated polymer constitutionally identical to head-to-tail polypropylene (Fig.
7). Structural examination reveals the existence of two series of tertiary

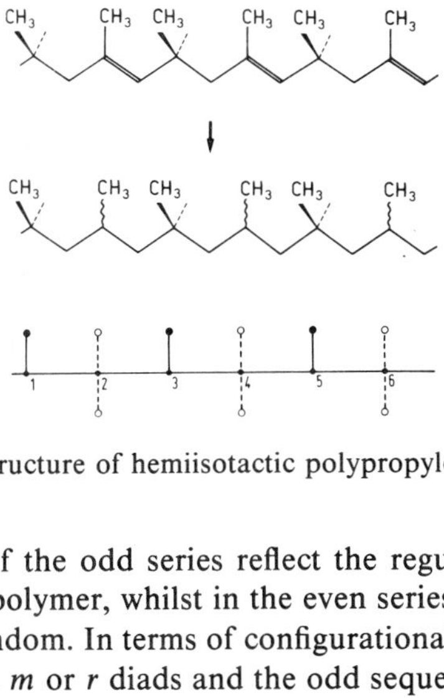

Fig. 7. Synthesis and structure of hemiisotactic polypropylene.

carbons: the atoms of the odd series reflect the regularity existing in the
starting unsaturated polymer, whilst in the even series the configuration of
tertiary carbons is random. In terms of configurational sequences it follows
that both the isolated *m* or *r* diads and the odd sequences of *m* or *r* diads,
as for example *mrmr*, *mrrrm*, *mrmmmr*, etc., are prohibited. As a result of
these selection rules, only seven of the ten pentads are permitted, with a
predetermined intensity ratio.

The ^{13}C-NMR spectrum (Fig. 8) of the polypropylene sample obtained as
described above differs considerably from that of conventionally-produced
atactic polypropylene, and is in perfect agreement with that expected for a
hemiisotactic polymer.[64] A more detailed analysis of the spectrum makes it
possible to recognize the existence of peaks attributable to nonads and even

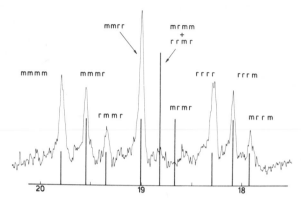

Fig. 8. ^{13}C-NMR spectrum of a hemiisotactic polypropylene recorded at 50.3 MHz and 137° C.

to undecads.[85] This experiment also provided a full demonstration of the very high isotacticity of the original poly-2-methylpentadiene (which contains over 96% of isotactic diads).[85]

The use of inclusion polymerization for these purposes is limited by the availability of suitable hosts and by the low yield which can be obtained. Nonetheless, with the use of low-cost hosts, it may be possible to carry out inclusion polymerization on a semi-preparative scale. Interest would obviously grow if, in this way, one could obtain polymers of high technological value which could not be obtained in other ways.

Speculatively, one may think of many problems as yet unresolved (to mention just one: the coexistence of low asymmetric induction and very high stereospecificity in the polymerization of pentadiene in PHTP), and many others studied only from a purely qualitative point of view.

New developments are to be expected from the use of new hosts and, possibly, of other monomers. Various research groups are active in this field and a rapid growth is to be expected. The author however feels that the study of inclusion polymerization should not be a field reserved for polymer chemists, and that other skills can and should be applied. Progress in this field can only result from research of an interdisciplinary nature.

Acknowledgement

I would like to express my deep appreciation of the work done by Dr. G. Audisio and Dr. G. Di Silvestro in the development of inclusion polymerization in PHTP. I thank Dr. P. Sozzani and Dr. M. Grassi for their active part in the recent research and present developments. The research work carried out at the University of Milan was partly financed by grants from the Italian National Research Council (CNR), Rome, and by the Italian Ministry of Education, Rome.

References

1. Y. Chatani, *Prog. Poly. Sci. Jpn.*, 1974, **7**, 149.
2. M. Farina, in *Proceedings of the International Symposium on Macromolecules, Rio de Janeiro, 1974*, (ed. E. B. Mano), Elsevier, Amsterdam, 1975, p. 21.
3. K. Takemoto and M. Miyata, *J. Macromol. Sci., Rev. Macromol. Chem.*, 1980, **C18**, 83.
4. M. Farina, *Makromol. Chem. Suppl.*, 1981, **4**, 21.
5. H. Clasen, *Z. Elektrochem.*, 1956, **60**, 982.
6. J. F. Brown and D. M. White, *J. Am. Chem. Soc.*, 1960, **82**, 5671.
7. D. M. White, *J. Am. Chem. Soc.*, 1960, **82**, 5678.
8. I. Sakurada and K. Nanbu, *Kogyo Kagaku Zasshi*, 1959, **80**, 307.
9. M. Farina, *Tetrahedron Lett.*, 1963, 2097.
10. M. Farina, G. Allegra and G. Natta, *J. Am. Chem. Soc.*, 1964, **86**, 516.
11. M. Farina, G. Audisio and M. T. Gramegna, *Macromol.*, 1971, **4**, 265.
12. M. Farina, G. Natta, G. Allegra and M. Löffelholz, *J. Polym. Sci.*, 1967, **C16**, 2517.
13. M. Farina, G. Audisio and G. Natta, *J. Am. Chem. Soc.*, 1967, **89**, 5071.
14. M. Farina and G. Audisio, *Tetrahedron Lett.*, 1967, 1285.
15. M. Farina and G. Audisio, *Tetrahedron*, 1970, **26**, 1827.
16. M. Farina and G. Audisio, *Tetrahedron*, 1970, **26**, 1839.
17. M. Farina, U. Pedretti, M. T. Gramegna and G. Audisio, *Macromolecules*, 1970, **3**, 475.
18. M. Farina and G. Di Silvestro, *J. Chem. Soc., Chem. Commun.*, 1976, 816.
19. T. Ohmori, T. Ichikawa and M. Iwasaki, *Bull. Chem. Soc. Jpn.*, 1973, **46**, 1383.
20. Y. Chatani, S. Nakatani and H. Tadokoro, *Macromol.*, 1970, **3**, 481.
21. Y. Chatani and S. Nakatani, *Macromol.*, 1972, **5**, 597.
22. Y. Chatani and H. Tadokoro, *J. Macromol. Sci., Phys.*, 1973, **B8**, 203.
23. Y. Chatani and S. Kuwata, *Macromol.*, 1975, **8**, 12.
24. Y. Chatani, K. Yoshimori and Y. Tatsuta, *Am. Chem. Soc., Polym. Preprints*, 1978, **19**(2), 132.
25. A. Colombo and G. Allegra, *Macromol.*, 1971, **4**, 579.
26. M. Miyata and K. Takemoto, *J. Polym. Sci., Polym. Lett. Ed.*, 1975, **13**, 221.
27. M. Miyata and K. Takemoto, *J. Polym. Sci., Polym. Symp.*, 1976, **55**, 279.
28. M. Miyata and K. Takemoto, *Angew. Makromol. Chem.*, 1976, **55**, 191.
29. M. Miyata, K. Morioka and K. Takemoto, *J. Polym. Sci., Polym. Chem. Ed.*, 1977, **15**, 2987.
30. M. Miyata and K. Takemoto, *Polym. J.*, 1977, **9**, 111.
31. M. Miyata and K. Takemoto, *Makromol. Chem.*, 1978, **179**, 1167.
32. M. Miyata and K. Takemoto, *J. Macromol. Sci., Chem.*, 1978, **A12**, 637.
33. M. Miyata, K. Shinmen and K. Takemoto, *Angew. Makromol. Chem.*, 1978, **72**, 151.

34. M. Miyata, K. Shinmen and K. Takemoto, *Am. Chem. Soc., Polym. Preprints,* 1979, **20**(1), 716.
35. M. Lahav, personal communication.
36. S. Candeloro De Sanctis, V. M. Coiro, E. Giglio, S. Pagliuca, N. V. Pavel and C. Quagliata, *Acta Crystallogr.,* 1978, **B34**, 1928.
37. S. Candeloro De Sanctis, E. Giglio, F. Petri and C. Quagliata, *Acta Crystallogr.,* 1979, **B35**, 226.
38. G. Audisio and A. Silvani, *J. Chem. Soc., Chem. Commun.,* 1976, 481.
39. M. Miyata, Y. Kitahara and K. Takemoto, *Polym. Bull.,* 1980, **2**, 671.
40. H. R. Allcock, *Acc. Chem. Res.,* 1978, **11**, 81.
41. J. Finter and G. Wegner, *Makromol. Chem.,* 1979, **180**, 1093.
42. H. R. Allcock, W. T. Ferrar and M. L. Levin, *Macromol.,* 1982, **15**, 697.
43. M. Maciejewski, *J. Macromol. Sci., Chem.,* 1979, **A13**, 77.
44. M. Maciejewski and Z. Durski, *J. Macromol. Sci., Chem.,* 1981, **A16**, 441.
45. M. Farina and G. Di Silvestro, *Gazz. Chim. Ital.,* 1982, **112**, 91.
46. M. Farina, G. Di Silvestro and P. Sozzani, *Mol. Cryst. Liq. Cryst.,* 1983, **93**, 169.
47. M. Farina and G. Di Silvestro, *J. Chem. Soc., Perkin Trans. 2,* 1980, 1406.
48. M. Farina, G. Di Silvestro and M. Grassi, *Makromol. Chem.,* 1979, **180**, 1041.
49. O. H. Griffith and H. H. McConnell, *Proc. Natl. Acad. Sci. USA,* 1962, **48**, 1877.
50. O. H. Griffith, *Proc. Natl. Acad. Sci. USA,* 1965, **54**, 1296.
51. G. R. Luckhurst, personal communication.
52. M. Löffelholz, M. Farina and U. Rossi, *Makromol. Chem.,* 1968, **113**, 230.
53. P. Sozzani, G. Di Silvestro and M. Farina, *Makromol. Chem., Rapid Commun.,* 1981, **2**, 51.
54. M. Farina, G. Audisio and M. T. Gramegna, *Macromol.,* 1972, **5**, 617.
55. M. Farina, G. Di Silvestro, U. Pedretti and G. Frumusa, *Chim. Ind. (Milan),* 1973, **55**, 159.
56. H. U. Pohl and D. O. Hummel, *Makromol. Chem.,* 1968, **113**, 203.
57. T. Yoshino, H. Kenjo and K. Kuno, *J. Polym. Sci., Polym. Lett.,* 1967, **5**, 703.
58. K. Matsuzaki, T. Uryu, M. Okada and H. Shiroki, *J. Polym. Sci., A1,* 1968, **6**, 1475.
59. P. Sozzani, G. Di Silvestro, M. Grassi and M. Farina, to be published.
60. L. Zetta, G. Gatti and G. Audisio, *Macromol.,* 1978, **11**, 763.
61. G. Audisio, A. Silvani and L. Zetta, *Macromol.,* in press.
62. M. Miyata, A. Wada and K. Takemoto, *Preprints of the International Symposium on Macromolecules, Strasbourg,* 1981, p. 528.
63. V. S. Ivanov, T. A. Sukhikh, Yu. V. Medvedev, A. Kh. Breger, V. P. Osipov and V. A. Goldin, *Visok. Soyed.,* 1964, **6**, 782; Engl. Trans., 1964, **6**, 856.
64. M. Farina, G. Di Silvestro and P. Sozzani, *Macromol.,* 1982, **15**, 1451.
65. M. Miyata, Y. Kitahara and K. Takemoto, *Polym. Journ.,* 1981, **13**, 111.
66. P. Sozzani, G. Di Silvestro and F. Morazzoni, to be published.
67. M. Farina and G. Bressan, in *The Stereochemistry of Macromolecules,* Vol. 3, (ed. A. D. Ketley), M. Dekker Publ. New York, 1968, p. 181.
68. E. Selegny ed., *Optically Active Polymers,* D. Reidel Publ., Dordrecht, 1979.
69. M. Farina, in *Giulio Natta: Present Significance of His Scientific Contribution,* (eds. S. Carrà, F. Parisi, I. Pasquon and P. Pino), Editrice di Chimica, Milan, 1982, p. 173.
70. M. Farina, in *Macromolecole Scienze e Technologie* Vol. 1, (ed. F. Ciardelli), Pacini Publ., Pisa, 1983, p. 22.
71. G. Allegra, M. Farina, A. Immirzi,, A. Colombo, U. Rossi, R. Broggi and G. Natta, *J. Chem. Soc. (B),* 1967, 1020.

72. H. L. Frisch, C. Schuerch and M. Szwarc, *J. Polym. Sci.*, 1953, **11**, 559.
73. T. Fueno and J. Furukawa, *J. Polym. Sci.*, *A1*, 1964, **2**, 3681.
74. M. Farina and G. Di Silvestro, *Makromol. Chem.*, 1982, **183**, 241.
75. B. Wunderlich and T. Arakawa, *J. Polym. Sci.*, *A2*, 1964, **2**, 3697.
76. E. Hellmut and B. Wunderlich, *J. Appl. Phys.*, 1965, **36**, 3039.
77. B. Wunderlich and C. M. Cormier, *J. Phys. Chem.*, 1966, **70**, 1844.
78. K. Monobe and F. Yokoyama, *J. Macromol. Sci., Phys.*, 1973, **8**, 295.
79. T. Maekawa, M. Kawasaki, K. Hayashi and S. Okamura, *J. Macromol. Chem.*, 1966, **1**, 507.
80. Ch. Schneider and H. P. Bohlmann, *Preprints of the International Symposium on Macromolecules, Florence, 1980*, vol. 2, p. 134.
81. L. Kiss and S. Polgár, *Proceedings of the Third Tihany Symposium on Radiation Chemistry*, 1972, p. 565.
82. G. Di Silvestro and P. Sozzani, *Preprints of the 5th AIM Meeting, Milan, 1981*, p. 55.
83. P. Sozzani, G. Di Silvestro and M. Farina, *Preprints of the 2nd International Symposium on Clathrate Compounds and Molecular Inclusion Phenomena, Parma, 1982*, p. 97.
84. C. Corno, A. Priola and S. Cesca, *Macromol.*, 1980, **13**, 1314.
85. G. Di Silvestro, P. Sozzani, B. Savaré and M. Farina, to be published.

11 · INDUSTRIAL APPLICATIONS OF CYCLODEXTRINS

J. SZEJTLI

Chinoin Pharmaceutical and Chemical Works, Budapest, Hungary

1. The cyclodextrins

1.1. Molecular encapsulation with cyclodextrins

Cyclodextrins have for long been known to be capable of forming inclusion complexes[1,2,3] (See Chapter 8, Volume 2) but their industrial application was hindered by three factors: (a) cyclodextrins were available only in small amounts as fine chemicals at rather high prices, (b) as a consequence of incomplete toxicological studies, adverse toxicity was reported and (c) the accumulated knowledge on cyclodextrins was not sufficient for industrial technologies.

As a result of intensive research during the past decade—competent and reassuring toxicity studies and the realization of industrial-scale production of β-cyclodextrin–there is no longer any barrier in the way of widespread utilization of cyclodextrins in the pharmaceutical, food, chemical and other industries.

INCLUSION COMPOUNDS III
ISBN 0-12-067103-4

The structure of β-cyclodextrin and the molecular dimensions of all three practically important cyclodextrin-molecular capsules are shown in Fig. 1.

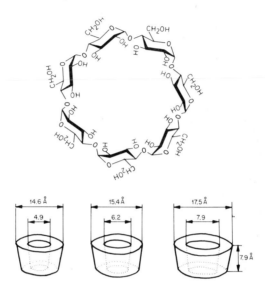

Fig. 1. Structure of β-cyclodextrin, and molecular dimensions of α-, β- and γ-cyclodextrins.

Molecules, or functional groups of molecules being less hydrophilic than water, can be included in the cyclodextrin cavity in the presence of water, if their molecular dimensions correspond to those of the cyclodextrin cavity. In aqueous solution the slightly apolar cyclodextrin cavity is occupied by water molecules, an energetically unfavoured process (polar–apolar interaction). These water molecules are therefore readily substituted by appropriate "guest molecules" which are less polar than water. Cyclodextrin (CD) is the "host" molecule, and the "driving force" for complex formation is the substitution of the high-enthalpy water molecules by an appropriate "guest" molecule.

Cyclodextrin complexes are relatively stable, their water solubilities are strongly reduced compared to pure cyclodextrins so they rapidly separate from the solution in crystalline form. One, two or three CD molecules contain one or more entrapped "guest" molecules; this is the essence of "molecular encapsulation". This process often advantageously modifies various physical and chemical properties of the encapsulated molecule—as will be illustrated for drugs, flavour substances, pesticides, vitamins, etc.

The syntheses of numerous cyclodextrin derivatives have already been reported.[3] Almost all derivatives retain their complex-forming capacity, however complex stability is modified with respect to the non-substituted CD.

Almost all industrial applications of cyclodextrins involve complexation. In many cases complexes are separated in more or less pure form and utilized as crystalline substances (drug and flavour complexes), while in other cases complexation is only a transient state, and becomes apparent through the final result (CD-catalysis, separation of mixtures).

At present only a very small proportion of the potential fields of cyclodextrin utilization is known and applied; a larger proportion has been discovered but remains to be exploited but the greater part of the potential fields of cyclodextrin application are as yet unknown. It seems that we see only the tip of the iceberg, and have only a faint idea of what lies below.

Besides molecular encapsulation (= cyclodextrin complexation) another similar modern method is microencapsulation. These two methods are not interchangeable. Their applications only yield the same effect in certain cases. Table 1 illustrates the essential differences between the two techniques.[4]

1.2. Preparation of cyclodextrins

The production of cyclodextrins consists of two phases. First the cyclodextrin–transglycosylase enzyme (CTG–enzyme) is produced by *Bacillus macerans* (or *Klebsiella pneumoniae* or an *Alcalophyl Bacillus* strain). The cell-free filtrate of the culture media contains the crude enzyme which has to be concentrated and purified.

Treatment of a partially pre-hydrolysed starch with this enzyme gives a mixture of α-, β- and γ-cyclodextrin together with a series of linear dextrins. Cyclic and linear products of enzymatic degradation of starch then have to be separated, for this purpose different procedures are used. Separation of the mixture can be achieved by the addition of an appropriate organic solvent (e.g. tetrachloroethane) a process resulting in the precipitation of the mixed crystalline complexes of the solvent, while linear dextrins remain in the mother liquor. After filtration the crystalline complexes are decomposed by steam distillation which removes the included organic solvent. The mixture of cyclodextrins is then fractionated by repeated dissolution and reprecipitation with different complex forming organic substances, which selectively form crystalline complexes either with α-, β- or γ-cyclodextrin.[5]

Table 1. Comparison of microencapsulation and molecular encapsulation

	Microencapsulation	Molecular encapsulation
Industrial application	Since 1960	Expected after 1982
Essence of technology	Application of a solid cover over the outer surface of the granules or drops of the substance to be encapsulated (physical process)	Formation of crystalline inclusion complexes based on apolar-apolar inclusion interactions (chemical process)
Material of the capsule	Polymers	Cyclodextrins
Ratio of the substance encapsulated	50–95%	5–25%
Number of molecules in the capsule	10^{10}–10^{13}	0.5–2, generally 1
Sensitivity to pressure	Microcapsules containing liquids are fragile	Not sensitive
Interaction between the capsule wall and the substance encapsulated	Negligible	Cyclodextrin in crystalline state inhibits decomposition of the substance encapsulated
Effect on the dissolution of the substance encapsulated	Retarded dissolution	Dissolution rate and solubility limit are increased
Release of the substance encapsulated	By diffusion, dissolution, enzymatic degradation or breaking of the wall	In aqueous media the dissociation equilibrium is established immediately
Pharmacokinetic action	Undesired side effects are reduced, absorption is retarded and can be programmed	Bioavailability is usually enhanced on account of the molecular disperse state

Alternative processes for the isolation of cyclodextrins from the conversion mixture utilize glucoamylase. This enzyme converts linear dextrins to glucose, and β-cyclodextrin is then obtained due to its lower solubility by direct crystallization.[6,7] While such a process avoids the use of organic solvents, it encounters the drawback that only a certain amount of the β-cyclodextrin formed is separated as a crystalline substance, the rest remains in the mother liquor together with the much more soluble α- and γ-cyclodextrin.

Under technical conditions higher yield—and lower costs of production— can be achieved by controlled conversion. The production of the desired cyclodextrin can be promoted; e.g. if the enzymatic conversion is performed in the presence of toluene at an appropriate pH, temperature and reaction time, the main product is β-cyclodextrin, because it forms an insoluble complex with toluene and, being precipitated, the continuous removal of β-cyclodextrin from the system shifts the equilibrium in favour of this product.[8] Similarly the addition of decane leads to an increased production of α-cyclodextrin.[9]

1.3. Metabolism and toxicity

The precondition for the application of a new compound either in pharmaceuticals or in foods is to have detailed and reassuring metabolism and toxicity studies.

The first published information[5] in 1957 was rather discouraging. Admixing β-cyclodextrin to the food of rats resulted in the death of the animals. However, the cyclodextrin samples used probably contained toxic substances (chlorinated or aromatic solvents?) since recent toxicological studies have definitely disproved those results.[10]

Six-month oral chronic toxicity of β-cyclodextrin was studied in rats up to 1.6 g/body weight kg/day, and up to 0.6 g/body weight kg/day in dogs.[10,11] No sign of toxic effects was observed in respect of weight gain, food consumption or in clinico-biochemical values. Pathologic and histopathologic investigations following six-month treatment have not revealed any signs indicative of toxicity in the digestive organs, central nervous system, cardiovascular system or in any other organs tested. β-Cyclodextrin also showed no embryotoxic effect. Chromosomal tests performed in rats treated for 6 months did not reveal an increase in the incidence of spontaneous aberration and no gene mutation inducing effect was found.[12] Orally administered β-cyclodextrin can thus be considered as a nontoxic substance.

It is thought that only an insignificant amount of β-cyclodextrin is absorbed from the intestinal tract in an unchanged form. In *in vitro* experiments (reverted rat intestinal sack) about 5% of ^{14}C-labelled glucose passed through the intestinal wall within 30 min, while less than 0.1 % of radioactivity appeared in the incubation medium in the case of ^{14}C-labelled β-cyclodextrin. While glucose is rapidly metabolized by intestine homogenate of the rat (demonstrated by oxygen uptake in a Warburg apparatus), the metabolism of maltose and starch are slower, and β-cyclodextrin seems to be completely resistant under such conditions.[13]

Cyclodextrin is metabolized, but probably the intestinal flora is responsible for the degradation. In spite of all efforts to detect intact β-cyclodextrin in blood following oral administration of the ^{14}C-labelled substance, no significant radioactivity could be found at R_t values which correspond to β-cyclodextrin on HPLC records.[13]

Following oral administration of ^{14}C-labelled glucose, starch, or β-cyclodextrin to rats, the blood radioactivity and $^{14}CO_2$ radioactivity of exhaled air were recorded.[13,14,15]

After administering glucose about 2% of the input radioactivity was detected (calculating for 10 ml blood) in the blood within 10 min after administration. When labelled β-cyclodextrin was administered only 0.5% of the input radioactivity could be detected in the blood and only between the 4th and 10th hour after administration.

The amount of exhaled radioactivity was practically identical in rats treated orally with ^{14}C-glucose, ^{14}C-starch or ^{14}C-β-cyclodextrin in a 24 h period (58–64% of administered radioactivity). The maximum radioactivity was detected in the first two hours with labelled glucose and starch but only between the 4th and 8th hours with labelled β-cyclodextrin[13] (Fig. 2). No significant difference was found in the tissue distribution of radioactivity, except that higher radioactivity was found in the caecum of the animals treated with labelled β-cyclodextrin.[13,14]

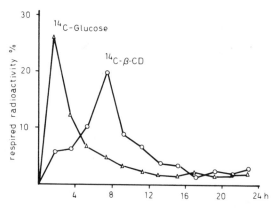

Fig. 2. Radioactivity exhaled by rats after oral administration of ^{14}C–β-cyclodextrin (36 mg kg^{-1}) and ^{14}C-glucose (13 mg kg^{-1}).[13]

The source of the radioactivity in the organism after oral treatment with ^{14}C-labelled β-cyclodextrin is probably due to the glucose and its metabolites formed from cyclodextrin by the effect of amylases of the colon bacterial flora.

2. Cyclodextrins in foods, cosmetics and toiletry

The major advantages of the application of CD-complexation in foods, cosmetics and toiletry are as follows:
Protection of the active ingredient(s) against:
  oxidation,
  light induced reactions,
  decomposition and thermal decomposition,
  loss by evaporation and sublimation.
Elimination (or reduction) of:
  undesired tastes/odours,
  microbiological contaminations,
  fibres/other undesired components,
  hygroscopicity,
Technological advantages:
  stable, standardizable compositions,
  simple dosing and handling of dry powders,
  reduced packing and storage costs,
  more èconomical technological processes, manpower saving,

2.1. Stabilization of food flavour

Stabilization of food flavours and fragrances seems to be the largest market for cyclodextrins in the 1980s.

Aroma-containing raw materials of natural origin (overwhelmingly of plant origin) are widely used in human nutrition. A number of disadvantages are brought about by their direct utilization, e.g.:[3]

(a) their preparation and processing is usually laborious (e.g. cleaning and chopping of onion);

(b) raw materials are not of constant composition, their aroma content depends on harvesting, processing, size, sort and growing site of the product;

(c) natural products can often contain microbiological contaminations and sometimes parasitic infestation;

(d) their aroma content decreases on storage, and can also involve an undesirable alteration in the ratio of certain components;

(e) storability is limited; the storage of natural sources of certain aromas constitute serious storage problems (room, optimal temperature and humidity, etc.).

Liquid aroma extracts have long been known, and marketed. The stabilization of most of them has not been solved so far, moreover the concentrates are often less stable than the original aromas. Their scope is restricted and they are only applied in certain branches of the food industry (e.g. in tinned foods). Utilization of such extracts in households is rather limited.

The most obvious carrier-substance in powder-flavour products is dextrin; it is cheap and non-toxic. Despite numerous favourable properties however its hygroscopicity is a great disadvantage. Cyclodextrins and their complexes are not hygroscopic; and this is important, especially at high relative humidity, e.g. under tropical conditions.

The loss of flavour substances on solids (by volatility, oxidation, chemical decomposition, etc.) is enhanced by their dispersion on a large surface.

Losses due to volatility and oxidation are not negligible even in the case of microencapsulated products. Release of the aroma occurs only after dissolution of the wall of the microcapsule. Thus, proper release requires dissolution in hot water. That is why in foods consumed without cooking the flavours are not always properly released on chewing. This limits the application of microencapsulated aroma products.

A very promising method for the stabilization of food flavours and fragrances is their complexation with cyclodextrins, which has important advantages in a number of areas.

In households it simplifies cooking by providing a wider choice of tastes and aromas. The dispensing units (e.g. tablets) containing various aromas can be stored in a small place for a long time without risking any loss of the active ingredient.

In catering, by the elimination of manual processing of raw materials containing aroma substances, work and storage place can be saved, transport is simple, loss on storage can be avoided and a wider variety of tastes and aromas provided. By using proper recipes identical taste and aroma can be provided independently of the location.

In dietetics and hospitals, the consumption of fibrous and seed shape aroma carrying raw materials which irritate the gastro-intestinal tract can be avoided. Appetizing aromas can also be used in hospitals. Many aromas which would otherwise be forbidden can be consumed by people on diets, the majority of digestion problems being caused not by the aroma itself, but by its carrier which is usually of plant origin.

In tinned food and the meat industry stable aroma products which resist environmental effects can be used in constant composition and in easily processable forms for dehydrated soups or sausages without the risk of microbiological contamination.

The preparation of β-cyclodextrin complexes of food flavours and fragrances is a relatively simple process[16] which has already been realized on an industrial scale.

The aroma content of these complexes is between 6–15% w/w, usually 8–10% (Table 2.). They are very stable in the dry state, their oxygen uptake measured by the Warburg method is less than 10% of that of the free substances (Fig. 3.). At 150° C under vacuum only 25–30% of the active ingredient was lost after 24 h, whereas none of the components of spice aroma substances could be detected under the same circumstances.

Heat treatment studies proved that aroma inclusion complexes, when stored in a sealed container at room temperature, lose no more than 5% of their active ingredient content even after two years. The heat stability of these complexes can be illustrated very convincingly by TAS*-chromatography[17] or by TFG* plates (Fig. 4.). Extremely heat-resistant

Table 2. Active ingredient content of flavour complexes [16]

Flavourent (source)	% Guest molecule content of the β-CD complex determined by GLC
Vanillin (synthetic)	6.20
Dill oil (*Anethum graveolens*)	6.92
Coriander oil (*Coriandrum sativum L.*)	7.72
Marjoram oil (*Majorana hortensis*)	8.00
Sage (*Salvia officinalis L.*)	8.20
Raspberry oil (*Rubus idaeus L.*)	8.66
Benzaldehyde	8.70
Lemon oil (*Citrus medica*)	8.75
Cinnamon oil (*Cinnamonum cassia Blume*)	8.76
Carrot oil (*Daucus carota*)	8.82
Anise oil (*Illicum Verum Hook. fil.*)	9.00
Orange oil (*Citrus aurantium*)	9.20
Thyme oil (*Thymus sp.*)	9.60
Peppermint oil (*Mentha piperita*)	9.70
Sweet cummin oil (*Foeniculum dulce*)	10.00
Celery oil (*Apium cepa L.*)	10.00
Garlic oil (*Allium sativum*)	10.20
Onion oil (*Allium cepa L.*)	10.20
Tarragon oil (*Artemisia dracunculus*)	10.23
Caraway oil (*Carum carvi L.*)	10.50
Basil oil (*Ocinum basilicum*)	10.72
Bay leaf oil (*Laurus nobilis*)	10.80
Mustard oil (*Sinapis alba*)	10.92
Fume aroma (synthetic)	12.20

* TAS = Thermomikro Abtrennverfahren von Substanzen = thermomicro separation method of substances.
* TFG = thermofractionating chromatography.

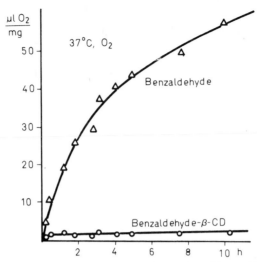

Fig. 3. The oxygen consumption of free and complexed benzaldehyde by the War-burg method.[16]

flavourants can be prepared (for baking and roasted foods) by coating the complexes with hydrogenated animal or vegetable oils or fats.[18]

A vanillin–glucose mixture and the vanillin–β-CD complex stored in open Petri-dishes at room temperature showed that after 240 days practically

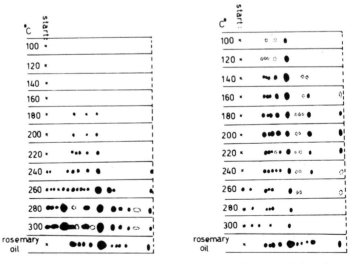

Fig. 4. Thermofractionation of rosemary oil (right) and its β-cyclodextrin complex (left).[17]

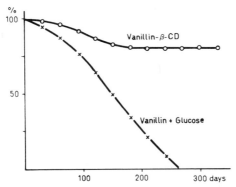

Fig. 5. Loss of vanillin from a vanillin-glucose mixture and from a vanillin–β-cyclodextrin complex (vanillin content 6.2% in both cases).[16]

no detectable aroma substance was left in the glucose mixtures, whereas the cyclodextrin complexes lost no more than about 20% of the aroma content even after a much longer time (Fig. 5.). It is likely that this 20% loss of vanillin was due to incomplete complex formation (possibly some molecules were only adsorbed on the surface of the complex crystals) because continuation of the observation for two additional years led to no further loss of vanillin.[16]

Aroma substances usually consist of several components and therefore it is important that all these components should be incorporated into the complex with their composition unchanged. As demonstrated in Fig. 6 this requirement can be satisfied to a fair degree and therefore, the aromatizing effect of aroma complexes is no different from that of the original material.

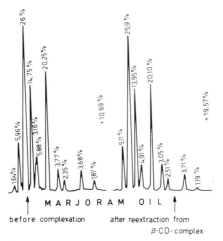

Fig. 6. Gas chromatograms of free and complexed marjoram oils.[21]

To alter the original composition however would definitely be occassionally advantageous. By appropriately selecting the host:guest ratio, one or more components can be relatively enriched in the complexed (or in the non-complexed) fraction. For example on complexing a fatty acid mixture, β-CD prefers the more unsaturated molecules.[19]

The water solubility of β-cyclodextrin complexes is usually only 0.1–0.035 g 100 ml^{-1} at room temperature, but this is always sufficient to make the taste of the aroma component which gets released on the dissociation of the dissolved complex detectable.

The taste of β-cyclodextrin cannot be detected in the taste of the complexes and it affects neither the taste nor the plasticity of foods flavoured with complexes. Meals prepared with aroma complexes were received favourably by a panel of consumers. The equivalent weights of some flavour-complexes and the corresponding natural spices are illustrated in Table 3.

Table 3. Equivalent amounts of flavour-β-cyclodextrin complexes and of the corresponding natural spices, in various foods, determined by sensory trials

1 g flavour-β-cyclodextrin complex is equivalent to:	
Onion	130–500 g
Dill	150–300 g
Garlic	33–100 g
Cummin	3–100 g
Marjoram	3–5 g

Chewing gum is easier to blend with aroma complexes—e.g. peppermint-cyclodextrin complex powder—than with liquid aromas. In this form the aroma remains in the chewing gum longer in storage and provides a longer lasting intensive taste.

Tea of poor quality can also be flavoured by adding to the tea leaves β-cyclodextrin complexes containing various amounts of bergamot, jessamine, lemon or peppermint. Either the granulated aroma complexes are added to the content of the tea bags or master mixtures are prepared and mixed with the tea leaves before packing. The aromatized tea samples have been preferred with high significance to the original non-aromatized tea in both organoleptic scoring and triangular tests.[20]

Preliminary examinations of dehydrated soups prepared with β-cyclodextrin–aroma complexes are also very promising.

Many examples of the application of flavour complexes in food processing and results of consumer trials have been reported by Lindner *et al.*[21]

In addition to applications in the food industry, β-cyclodextrin–flavour complexes could be used to advantage in clinical nutrition as has been found at the Department of Metabolic Diseases, Institute of Nutrition, Bratislava.[22] None of the 307 patients on different clinical diets who consumed food flavoured with various β-cyclodextrin–flavour complexes complained of any symptoms or unpleasant effects. Subjective repugnance has not been observed, even in cases when in clinical diet, unusual flavours had been used (mustard, garlic, thyme). Thus β-cyclodextrin–flavour complexes proved to be acceptable ingredients in clinical dietetic food preparations.[21]

Table 4 illustrates examples of the preparation and application of various flavour–cyclodextrin complexes.

Table 4. Examples of the stabilization of flavours by cyclodextrin complexation

Substances	Reference
Flavour extracts (e.g. fruit juices, essential oils of spices, meat extracts, etc.)	16, 23, 24, 25
Farnesane	26
Limonene	27
Citral and citronellol	28
Butter flavour	29, 30
Menthol (e.g. in hard candies)	31
Aromatizing of tea (with e.g. bergamot oil or jasmin oil complex)	20
Aroma powder for instant beverages (e.g. instant martini powder)	33
Horse raddish and mustard powder	34
Retention of volatile components of soy sauce on freeze-drying	35
Ethanol-complex	32

2.2. Elimination of unpleasant tastes and odours

The ratio of free to complexed guest molecules in an aqueous cyclodextrin solution depends on several factors; the most important ones are the stability (or dissociation) constant of the complex, temperature, and concentrations of both components. In cold, concentrated solutions the equilibrium is shifted towards complexation (and crystallization), while in warm, dilute solutions included guest molecules are released. Therefore when flavour complexes contact saliva at body temperature taste perception is instantaneous. In warm, dilute aqueous solutions there is little hope of eliminating tastes and odours. There were efforts e.g. to eliminate the phenol-like unpleasant bitter components from coffee and tea (which are formed on keeping the beverages at 90° C for hours) by adding 0.1% β-cyclodextrin to the hot beverages.[36] The published favourable results of sensory analysis could not be reproduced in the author's laboratory.

However, at lower temperatures, and at higher cyclodextrin concentrations unpleasant tastes and odours can be masked by cyclodextrin complexation.

Milk casein hydrolysate is a readily digestible protein source, but its bitter taste is a great problem. Adding 10% β-cyclodextrin to the protein hydrolysate the bitter taste can be eliminated[37,38] This method provides a way for the utilization of proteins which are otherwise useless for alimentary purposes. The bitter taste of grapefruit juice decreased substantially when 0.3% β-cyclodextrin was added prior to a heat treatment (95° C, 10 min).[35]

Similarly the bitter taste of ginseng extract[39] and propylene glycol[40] was also eliminated by cyclodextrin.

The characteristic odour of mutton and fish,[41] the disagreeable odours of bone powder (used as Ca-supplement in animal fodders),[42] and sodium caseinate solution[43] etc. can be removed by cyclodextrin complexation. Ethanolic-aqueous solutions of β-cyclodextrin[44] or the iodine–β-cyclodextrin complex[45] can be utilized as a deodorant. Cyclodextrin can be utilized as a mouth deodorant to mask the smell of fish, garlic etc.

Soya bean products free from a grassy smell and stringent taste are obtained by mixing with cyclodextrin,[47] and soya bean lecithin kneaded with β-cyclodextrin forms an odourless powder, which can be used in nutrition.[48] Yeast extract treated with β-CD improves in colour, decreases in hygroscopity and loses its odour.[49] Microbial cells (e.g. butyric acid, lactic acid bacteria or yeasts) yield odourless cell-powder on mixing and drying with β-CD.[50]

2.3. Improvement of food quality and technology

The addition of cyclodextrins or their appropriate complexes to food products can, in many cases, favourably modify important physical properties e.g. water retention, emulsion stability, texture, etc.[51] Table 5 summarizes the literature data.

β-cyclodextrin itself cannot be considered as a tasteless or only slightly sweet substance. Its taste threshold value is lower than that of sucrose. (Detection: 0.039% and 0.27%, recognition: 0.11% 0.52%, respectively.) An aqueous solution of 0.5% β-cyclodextrin is as sweet as sucrose, and an 2.5% solution as sweet as a 1.71% solution of sucrose.[105] Therefore when β-cyclodextrin is used in food processing, its sweetness cannot be ignored. Sucrose sweetness and β-cyclodextrin sweetness are additive.

Table 5. Examples of the improvement of food quality and technology by utilizing cyclodextrins

Application/Prevention	Reference
Formation of precipitate in soft drinks	52, 55
in canned bamboo shoots	53
in canned citrus products	54, 105
Deliquescing of sweets	56
Staling of noodles	57, 58
Preparation	
Emulsified oils	59, 60, 61, 70
Sausage casing	65
Baking powder	66, 67
Solidified sesame oil	68
Phenylalanine-free diet	69
Readily soluble sweetener tablets	71
Improvement of quality	
For stabilizing flavours	3, 21, 29, 75, 68
For stabilizing ascorbic acid	72
In ice creams	30
For increased water retention	73, 74, 75
In soft drinks (taste richness, mouth-feel)	52, 105
In whipping egg-white	64
Food preservation	
Iodine complex	76, 77, 78, 79
Ethanol complex	32
Cyclodextrins in preserving compositions	80

No significant changes in the tastes of saccharin, quinine, caffeine, xylose, or organic acids (e.g. ascorbic acid) was observed in the presence of β-cyclodextrin. The sweetness of neohesperidin dihydrochalcone was reduced to about 58% and 25% in the presence of 0.1% and 0.5% of β-cyclodextrin. Similarly the bitterness of naringin was reduced to about 50% by adding 0.5% β-cyclodextrin to the solution.[105] The taste of foods containing taste substances with low threshold values may be modified by a relatively low level of cyclodextrin.

The addition of 0.005–1% β-cyclodextrin to canned citrus products prevents precipitate formation, which is mainly due to the precipitation of bitter-tasting hesperidin and naringin.[54] Similarly the formation of white precipitate in bamboo shoots canned in water is prevented by adding 0.01–2% β-cyclodextrin.[53] Sweetening agents chalcone and dihydrochalcone easily precipitate from soft-drinks, especially on cooling. On adding β-cyclodextrin to the drink no precipitation could be observed.[52,55]

Fondants and sugars are prevented from deliquescing by mixing with a β-CD–fatty acid complex.[56] Adding 20 g β-cyclodextrin to the mixture of

1600 g flour, 16 g NaCl and 800 ml water, the quality of the kneaded noodles is improved and their freshness preserved.[57,58]

Cyclodextrin can be utilized for the preparation of stable water-in-oil emulsions, e.g. for salad dressings.[59,60,61] The consistency and flavour of soy sauce,[62] and the stability of natural food colouring agents[63] can be improved by cyclodextrins.

In processed meat products cyclodextrin improves water retention and texture,[64] and cyclodextrin-containing adhesive can be used to bind the edges of laminated cellophane sheets to make a sausage casing.[65] The cyclodextrin complex of an edible surfactant,[66] or the CO_2-α-cyclodextrin complex[67] can be utilized as an emulsifying or foaming baking-powder for sponge-cakes. β-cyclodextrin complexed sesame oil is a solid product that can be used in instant food products.[68]

Biscuits made with β-cyclodextrin stabilized butter flavour and stored for 2 months at low temperature had a butter flavour, whereas biscuits containing non-complexed aroma had no butter flavour after 2 weeks under similar conditions.[29]

For phenylketonuric dietetic food phenylalanine and tyrosine can be removed with good selectivity from casein hydrolysates with β-cyclodextrin–polymer.[69] In this case hydrolysis is performed with hydrochloric acid, therefore the taste of such hydrolysates is buillon-like while the conventional sulphuric acid hydrolysed and ion-exchange separated hydrolysates are characterized by disagreable tastes. On adding cyclodextrins to emulsified foods[73] or cheese,[74] water-retention and storage life increases.

Whipping 1 g eggwhite gives 7.36 ml eggwhite foam, while on adding 0.75% β-cyclodextrin, the volume of foam increases to 10.72 ml. After 1 month's storage at $-15°$ C the volumes were found to be 6.03 and 11.2 ml, respectively.[64]

Various cylcodextrin complexes can be utilized in foods as antiseptic or conserving agents. 0.1% iodine–β-cyclodextrin[76] inhibits putrefaction for 2 months at 20° C[78] in frozen marine products[77] or in fish paste.[78] No growth of Aspergillus, Penicillium or Trichoderma strains was observed on the surface of the iodine complex containing vinylacetate polymer or polypropylene foil.[78,79] Similarly the ethanol complex in ethanol permeable films prevents mold growth in packaged foods.[32]

2.4. Cyclodextrins in cosmetics and toiletries

Rapid growth is expected in the utilization of cyclodextrins in cosmetics and toiletries. Table 6 gives some examples. This field shows practically unlimited versatility without any toxicological limitations.

Table 6. Examples of the utilization of cyclodextrins in cosmetics and toiletries

Application	Reference
In cosmetic preparation	
Dentifrices	81, 82, 83, 93
Deodorants, antiseptics	44, 45, 46, 76, 79, 87
Emulsified cosmetic bases	84, 85, 86
Stabilization of royal jelly	92
Treatment of acne	99
In laundry	
As a defoaming agent	94
Fragrance complexes in washing powder	96
In spray starch pastes	95
In toiletries	
In incenses, solid perfumes, bath preparations	88, 89, 90, 91
In fragrant candles	97

Cyclodextrin-complexed fragrances can be utilized in solid perfumes, in fragrant candles,[97] in incenses,[88-91] and also in detergents.[96,98] Cyclodextrins form complexes with detergent molecules, and can act as defoaming agents.[3] On adding cyclodextrin to the rinsing water in a laundry, the last traces of the detergents become fixed, and this defoaming effect may help to reduce water usage in laundries. Different types of detergents are not however complexed to the same degree. On adding β-cyclodextrin complexed fragrances to detergents, fragrances were released most easily by anionic surface agents, while nonionic aliphatic ethylene oxide type and quaternary ammonium derivative type surfactants resulted in a much reduced release.[98] The degree of evaporation from the dry complexes depends more on the relative humidity than on the temperature (Table 7). Under usual storage conditions cyclodextrin complexation of perfumes provides good stabilization when mixed with detergent powders, and almost quantitative release upon contacting water.

2.5. Cyclodextrins in tobacco products

Tobacco aromas are usually mixed directly to tobacco. In the course of processing and storage these volatile aroma substances are gradually lost. In the form of cyclodextrin inclusion complexes the aroma substances

Table 7. Stabilization of perfume in dry washing powder by cyclodextrins under various storage conditions (KOCH[98]). (Residual perfume concentrations in percentage of original in head space over aqueous solutions after dissolving the washing powder)

Storage conditions temperature/relative humidity	Storage time (days)	Storage in open flask perfume			Storage in closed flask perfume		
		Sprayed on detergent	mixed with detergent		Sprayed on detergent	mixed with detergent	
			as β-CD complex	as α-CD complex		as β-CD complex	as α-CD complex
38° C/75%	1	13.2	92.5		92.4	100	
	3	0.5	27.3	56	61.6	56.8	
	7	0.3	3.4	52	33.1	50.4	100
	28	—		49	2.6	23.4	91
50° C/15%	0.125	19.7					
	1	0.4	100		61.2	96.1	
	3	—	100	99	22.1	63.1	
	7	—	86.4		2.1	42.6	
	28	—	72.5	96	—		100

— not detectable.

remain unchanged until they become liberated by burning the tobacco.[100,101] In menthol cigarettes menthol can be stabilized in the form of its β-cyclodextrin complex.

Nicotine forms a crystalline complex with β-cyclodextrin. This complexing property of β-cyclodextrin can also be utilized in cigarette smoke filters.[104] On impregnating the cellulose filters of cigarettes, a large proportion of nicotine and tar can be removed from the inhaled smoke.[102] Cyclodextrin polymers can also be utilized in smoke filters. The first and last third of the filter consist of cellulose acetate, the middle of a cyclodextrin-powder polymer prepared from raw cyclodextrin conversion mixture. The gas chromatogram of the smoke passing through this filter differed considerably from that of a control.[103]

3. Applications of cyclodextrins in pharmaceuticals

3.1. Principles and advantages of drug complexation

During the last decade the rapidly increasing activity in this area is reflected in the number of papers and patents dedicated to the pharmaceutical application of cyclodextrins.

The main results of these investigations can be summarized as follows:

(a) For reasons of solubility and toxicity β-CD (up until now (1982) the only industrially produced CD) can be utilized in oral, rectal and dermal preparations.

(b) The main features of β-CD complexation are: stabilization of labile compounds, conversion of liquids into microcrystalline solids, and enhancement of bioavailability.

(c) The accumulated knowledge in this field has allowed us to set limits to the groups of drugs, which can be transformed into CD-complexes and which are worth complexing.

The active ingredients of about 10% of orally administered drugs seem to be complexable; in about half of these cases one or more physical or chemical property can be modified sufficiently to make CD complexation worthwhile.[106]

Taking all factors into account, in at least 2% of all tablets produced the presence of CD as a complexing agent (or auxiliary substance) is favourable. It means that, in the not too distant future, a major part of the β-CD market might be consumed by the pharmaceutical industry for the production of oral preparations.

Inclusion complex formation of a drug results in the modification of its physical and chemical properties. These modifications are very often advantageous;

(a) In the formulation of drugs, liquid compounds can be transformed into a crystalline form which is suitable for tablet-manufacturing, any bad taste and smell can be masked and incompatible compounds can be mixed when one of the components is protected by inclusion complex formation,

(b) Improvement of physical and chemical stability of drugs. Volatile compounds can be stabilized against loss by evaporation, compounds are protected from oxidation by air, the rates of decomposition, disproportionation, polymerization, autocatalytic reactions etc, are considerably decreased and sensitivity to light, gastric acid, etc. is reduced,

(c) The bioavailability of poorly soluble drugs can be enhanced. Solubility in water as well as the rate of dissolution of poorly soluble substances can be increased. Following oral administration of poorly water soluble drugs higher blood levels can be achieved if they are complexed with cyclodextrin.

3.2. Limiting factors of drug complexation

Not all drugs are suitable for cyclodextrin complexation.[3,107] Three fundamental factors limiting the application of CDs in oral preparations are as follows:

(a) complexability (expressed as the stability constant)
(b) stoichiometry of the complex
(c) required dose of the drug.

The most important parameters that determine the complexability of a given molecule are its hydrophobicity, relative size and geometry in relation to the CD cavity.

Large hydrophilic organic molecules (e.g. water soluble peptides, proteins), small, highly water soluble, strongly hydrated molecules (e.g. glucose) and ionized molecules (e.g. quaternary ammonium salts) cannot be complexed. Inorganic compounds are not suitable for cyclodextrin inclusion complexation; they form only outer-sphere, or hydroxo-complexes. Only apolar molecules (or functional groups of molecules) can be included into the cyclodextrin cavity, provided that their diameter does not exceed the size of the CD–cavity (see Fig. 1.).

The extent of complexation in aqueous media (i.e. the dissociation equilibrium) is characterized by the stability (or dissociation) constant of the complex. Complexes of stability constants only between 200–10 000 seem to be suitable for pharmaceutical utilization since very labile complexes lead to the premature release of the drug and very stable complexes lead to retarded or incomplete release of the drug in the organism.

Within certain groups of compounds some correlation exists between the binding strength and some measure of the guest molecule's hydrophobicity or other physical-chemical characteristics, (e.g. polarizability, molar volume, molar refraction, or partition coefficient of the guest in an apolar solvent: water system). No obvious correlation has been found however between the physical or chemical properties of different families of guest molecules and the inclusion complex formation equilibrium constants.

Another limitation is based on the stoichiometry of the complexes. Cyclodextrin–drug complexes usually have 1:1, 2:1, or 3:2 stoichiometry, but 1:2, 2:2, 3:1, 4:1 and 5:2 ratios have also been reported.[108]

Since the drugs to be complexed have molecular weights around 120–350 while cyclodextrins have larger molecular weights (972, 1135 and 1297 for α-, β- and γ-cyclodextrins) this stoichiometry limits the amount of drug which can be practically supplied using this method. 100 mg of a complex contains only 5–25 mg of active ingredient.

Sometimes relatively small guest molecules show surprisingly unfavourable stoichiometry. For example vitamin K_3 (menadione, 2-methyl-naphthoquinone) needs three β-CD molecules to form a stable, crystalline complex.[109] The menadione content of this complex is only 4.2–4.5%, thus its practical utilization is not economic (γ-cyclodextrin however forms a 1:1 crystalline complex with menadione).

The active ingredient content of an inclusion complex of high molecular weight drugs is in the range of 15–25%, while in the case of low molecular weight guests this figure is under 15%, sometimes even less than 5%. If the single dose of a drug is not more than 25 mg then even a complex of 5% active substance content can carry the necessary dose in a single tablet of 500 mg weight. Thus, in the case of complex forming drugs, the relationship between the required dose and the molecular weight determines the feasibility of oral administration in the cyclodextrin complexed form.[3]

The correlation between drug content and the amount of the complex required for a 25 mg dose is illustrated in Table 8.[106]

Table 8. Correlation between molecular weight and the amount of 1:1 β-cyclodextrin complex which contains 25 mg drug

Molecular weight of the drug	Drug content of complex (%)	Complex needed to carry 25 mg drug (mg)
378	25	100
284	20	125
200	15	166
126	10	250
60	5	500

A 3000 I.U. vitamin D_3 tablet contains only 0.075 mg cholecalciferol; the active ingredient content of a nitroglycerin tablet is 0.5–4 mg. These and similar drugs are thus ideal for cyclodextrin complexation. If however the required dose is around several hundred milligrams, then CD–complexation is impracticable especially with low molecular weight drugs.[107]

In high doses (above 50 mg/body weight-kg) β-cyclodextrin administered subcutaneously, intraperitoneally or intravenously can induce renal damage in the rat,[110,111] therefore only its oral, rectal and dermal application is advisable.

Examples of the effects of cyclodextrin complexation on drugs are compiled in Table 9.

3.3. Examples of drug formulation and stabilization

Taking the high resistance of β-cyclodextrin against saliva-amylase and acid hydrolysis into consideration, the drug is released from the cyclodextrin complexes used in tablets by the dissociation of the complex; enzymatic or acidic degradation of the host molecule can practically be ruled out.

Frömming *et al*[112] have found that when complexes are tabletted, even without additives, hard tablets with a smooth surface can be prepared which disintegrate in water within a short time. The moisture absorption equilibrium of the salicyclic acid–β-cyclodextrin complex is established within 4–5 weeks at 20° C. Complex formation has no appreciable effect on moisture absorption.

When the tablets were heated, no sublimation of salicyclic acid was observed up to 160° C, whereas tablets prepared from 1 part of an equimolar mixture of β-cyclodextrin and salicylic acid and 4 parts of a starch-lactose mixture lose 71.5% of salicylic acid content by sublimation at 160° C.

The loss (sublimation, or escape of volatile decomposition products) can be monitored by thermogravimetry, or more specifically by the TEA-method (Thermal Evolution Analysis). Thermogravimetry measures weight changes, while the TEA method registers only the escaping organic substances with its flame-ionization detector.[3] Figure 7 demonstrates that menadione (vitamin-K_3) evaporates at 96–117° C either pure, or mixed with β-cyclodextrin, but if it is complexed with β-cyclodextrin it becomes volatile only with the thermal degradation of the carbohydrate matrix[109] (at 207–213° C menadione dimers are volatilized). Vanillin[16] and menthol[113] are further examples of compounds protected against sublimation by cyclodextrin complexation.

Unsaturated, oxygen sensitive compounds (e.g. unsaturated fatty acids, aldehydes, vitamins etc.) can be protected almost completely against

Table 9. Examples (References) of improvement of important pharmaceutical characteristics of some drugs by cyclodextrin complexation

Drugs	Improved formulation, conversion of liquids into powders, etc.	Better chemical stability, protection against decomposition, oxidation, volatility, etc.	Acceleration of dissolution and increase of solubility	Enhancement of bioavailability and biological effect
Non-steroid anti-inflammatory drugs (indomethacin, ibuprofen, ketoprofen, flufenamic-, mefenamic acid, etc.)			142, 145, 146, 159	142, 145, 146, 147, 159
Steroids			160, 161, 173	
Anthranilic acid derivatives	162		162, 163, 164	
Fat-soluble vitamins (vitamin A, D_3, K_3)		109, 114, 115, 117, 118, 177, 178	109, 140, 141, 168, 177	109, 118, 178
Prostaglandins and prostacyclin	120, 122	120, 122, 123, 154, 155, 156, 157, 158	152, 153	122, 157
Barbiturates			149, 165, 166, 167, 169, 170	149
Sulphonamides				
Cardiac glycosides		171, 172	108, 171, 172	108, 171, 172
Organic nitro compounds	126, 174, 175, 176, 179	126, 174, 175, 176		179

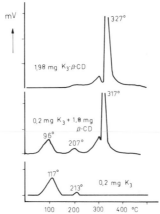

Fig. 7. Thermal evolution analysis curves of free menadione (K_3), of the mechanical mixture of β-cyclodextrin and menadione (0.2 mg K_3 + 1.8 mg β-CD) and that of the 3:1 β-cyclodextrin–menadione complex (K_3–β-CD).[109]

oxidation by atmospheric oxygen. Significant antitumour activity by benzaldehyde has been reported,[116] but without protection benzaldehyde rapidly oxidizes into the ineffective benzoic acid. Figure 3 shows the effectiveness of cyclodextrin complexation against oxidation.

Oxidation of solid, crystalline substances by atmospheric oxygen is generally not too rapid; however if they are diluted or distributed over a large surface—as e.g. in tablets—they become easily accessible for oxygen. Figure 8 demonstrates that oxygen uptake of vitamin D_3 is accelerated in a simple mechanical mixture with β-cyclodextrin, while in the complexed form it is much more stable against oxidation (Fig. 9).[117,118]

The effect of CD complexation in protecting against thermal decomposition is illustrated by vitamin D_3, the prostaglandins, and the nitrosoureas.

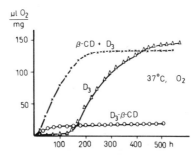

Fig. 8. Oxygen consumption of cholecalciferol (vitamin D_3) in the free form (D_3), mixed mechanically (D_3 + β-CD) or complexed (D_3–β-CD) with β-cyclodextrin.[118]

Vitamin D_3, its mixture with β-cyclodextrin, and the inclusion complex were exposed to air in a 1 mm thick layer at 80° C. No vitamin D_3 could be detected after 24 h in either the heat treated free vitamin sample, or in the mechanical mixture. The heated complex retained 49.1% of the original vitamin content even after 43 days.[117,118]

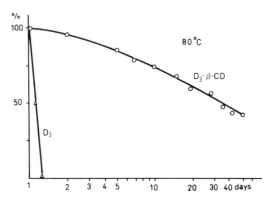

Fig. 9. Decomposition of cholecalciferol in the free and in the β-cyclodextrin complexed form at 80° C.[118]

A mixture of sodium chloride, vitamin D_3, or its cyclodextrin inclusion complex was formulated as 200 mg tablets. The vitamin D_3 content of the tablets was 1.8 mg cholecalciferol in the free state as well as in complexed form. The tablets were kept at 40, 60 and 80° C in the open air for a long period of time. The cholecalciferol content of the tablets was determined periodically. The free cholecalciferol completely decomposed at 80° C within 8 h, the decomposition was complete in 96 h at 40 and 60° C, while 80% of the complexed cholecalciferol was still present in the samples even after 30 days at all three temperatures.[118]

Prostaglandins (PGs) are very effective smooth-muscle stimulating agents even in extremely small quantities, the most active ones are however very unstable. Cyclodextrin complexation proved to be a successful method for their stabilization and enhancement of their solubility.

The prostaglandin E_2-β-cyclodextrin complex is the first marketed cyclodextrin complexed drug. The 100 mg Prostarmon-E tablet produced by ONO, Tokyo (for the induction of labour) contains 0.5 mg PGE_2 in 6 mg β-cyclodextrin. Intravenous infusions of PGE_1-α-cyclodextrin complex proved to be effective in the treatment of stenosis in extremities, thrombophlebitis, in clinical trials.[119]

16,16-dimethyl-*trans*-Δ^2-PGE_1 is more suitable for early pregnancy termination than natural prostaglandins, having reduced side-effects. Similar to the majority of natural PGs this compound is a viscous oil, which readily loses one molecule of water converting itself into the biologically inactive PGA_1-derivative. On complexing the substance with β-cyclodextrin it can be easily formed into tablets, its solubility in water is improved, and while at 60° C 50% of the noncomplexed compound is lost by decomposition in 14 days, 96% of the complexed substance remains intact under identical conditions.[120]

The structures of the α- and β-cyclodextrin complexes—and probably also the effects attained—are different. ^{13}C-NMR and relaxation time investigations showed that in the α-cyclodextrin complex ($K = 250$ M^{-1}) only the alkyl side-chain of $PGF_{2\alpha}$ is included in the cavity. In contrast, in the wider β-cyclodextrin cavity the cyclopentane ring and its immediate neighbours are accommodated, resulting in a more stable complex ($K = 1240$ M^{-1}).[121] In both cases the terminal carboxylic group is assumed to interact with the outside of the cavity through hydrogen bonding.

Within this group of compounds recent interest is centred on prostacyclin. PGI_2 is the most effective antiplatelet agent of natural origin. Its spreading application was limited by its instability, but this substantial drawback can be eliminated by CD-complexation.[122]

The stability of the PGI_2-ethylester–β-cyclodextrin complex was studied in aqueous solution at room temperature by the inhibition of rabbit platelet aggregation according to Born. Results presented in Table 10 verify that the stability of PGI_2-ethylester can be increased by cyclodextrin complexation. In aqueous solution the activity half-life of the complexed drug was increased five to eight fold compared to the free PGI_2-ethylester.[106]

Table 10. *Degradation of prostacyclin–ethylester in aqueous solution*

	Content of PGI_2 in the solid complex (%)	Concentration of PGI_2 in solution (μg ml^{-1})	Activity half-life (min)
PGI_2-Et-β-CD		10	75
complex	6.0	100	270
		1000	no decrease[a]
PGI_2-Et-β-CD		10	87
complex	4.2	100	420
		1000	no decrease[a]
PGI_2–Et		10	1
		100	45
		1000	—

[a] During the measuring period (5 h).

Further examples of the rapid degradation of PGI$_2$-methylester[123] are illustrated in Table 11. It is not yet understood why α-cyclodextrin possesses a higher stabilizing effect compared to β-cyclodextrin.

Table 11. Stability of PGI$_2$-methylester at 40°C (Hayashi et al.[123])

	Percentage remaining after (days)		
	1	7	14
PGI$_2$	—	—	—
Dissolved in tricapryline	65–70	5	—
Dissolved in N,N-dimethylacetamide	98–99	50–67	34
Dissolved in ethanol	98–99	98–99	95–97
In α-CD complex	100	95–97	90–95
In β-CD complex	80–90	67–80	50

Stable α-cyclodextrin complexes of N-nitrosourea derivatives showing antitumor activity have also been reported.[124] For example, when the cyclodextrin complex of 1,3-bis-(2-chloroethyl)-nitrosourea is stored at 60°C for 30 days in a sealed vial, 85% of the active substance content remains unchanged, whereas the free compound completely decomposes during this time. In a similar way 1-(2-chloroethyl)-3-cyclohexyl-1-nitrosourea and 1-butyl-1-nitrosourea can also be stabilized by transforming them into their α- or β-cyclodextrin complexes.

The preparation and handling of some highly explosive substances can be made risk-free by inclusion complexation. For example the nitroglycerin–β-cyclodextrin inclusion complex can be prepared in homogeneous aqueous solution and the crystalline complex will not explode.[125] Sublingual or oral pills prepared from this complex are more stable than current pills, containing nitroglycerin bound by simple adsorption.[126]

Cyclodextrin complexation can also give protection against light induced decomposition. Vitamin D$_3$ mechanically mixed or complexed with β-cyclodextrin and irradiated with a 2900 lux light of wavelength 400–600 nm for 320 h retained 45% of its vitamin content in the mixture and 95% in the complexed form.[118]

A photolysis reducing effect of cyclodextrins in aqueous solutions has also been reported. Chloropromazine, one of the most familiar tranquilizing drugs, is extremely light sensitive, yielding sulphoxide, N-oxide, and various oxidized and condensed products, depending on the experimental conditions. Cyclodextrin (corresponding to the stability constants $K_{diss,\alpha CD} = 200$; $K_{diss,\beta Cd} = 12\,000$ and $K_{diss,\gamma CD} = 1000$ mole^{-1}) decreases the production of radicals from chloropromazine, reducing the rate of photolysis. The β-cyclodextrin complex seems to be particularly resistant to photolysis. While at pH $= 4$ and $30°$ C, 21% of chloropromazine (5×10^{-4} mol dm^{-3}) decomposes, in the presence of 5×10^{-3} mol dm^{-3} dextrins the following percentages of chloropromazine are decomposed: αCD $= 18\%$, βCD $= 3\%$, γCD $= 15\%$.[121]

If the photosensitive group of the guest molecule is located outside the cavity, no protecting effect of cyclodextrin is observed, nevertheless the pathway of the photoconversion and the composition of the photoreaction products may be strongly modified by the presence of cyclodextrin. For example changes in the UV-spectrum of aqueous menadione (vitamin-K$_3$) solutions are different if the solutions are irradiated in the presence or absence of β-cyclodextrin.[127]

Alkaline hydrolysis of benzocaine becomes considerably slower in the presence of β-cyclodextrin.[128] In 1% β-cyclodextrin solution the half life of benzocaine increases fivefold. The free benzocaine undergoes hydrolysis, but the complexed form does not. Thus, the hydrolysis rate is determined by the complex equilibrium constant.

In other cases however hydrolysis rate *increases* were reported.[2,129] It has been concluded that when the hydrolysis sensitive part of the guest molecule is located inside the cavity of cyclodextrin, the rate of alkaline hydrolysis decreases. If however, incorporation is incomplete, i.e. the sensitive group is outside the cavity, the rate increases because ethereal oxygens and secondary hydroxyls also get involved in the catalysis. Thus, the rate of the alkaline hydrolysis of *p*-, *m*- and *o*-aminobenzoic acid esters is decreased by β-cyclodextrin, since the inclusion is complete and the ester group is protected against alkoxide ions derived from hydroxyls in the solution and the hydroxyls of cyclodextrin itself.[130] Although the complete atropine molecule does not fit into the cavity of β-cyclodextrin, its phenyl ester moiety is included in the cavity, thus its hydrolysis is also hindered.

Acetylsalicylic acid fits neither the α-cyclodextrin nor the β-cyclodextrin cavity completely, its carboxyl group being positioned on the edge of the cyclodextrin ring, therefore its hydrolysis is accelerated by both cyclodextrins.

Among aminobenzoates only the *p*-isomer fits the cavity of α-cyclodextrin, therefore its hydrolysis rate is reduced, while the *o*- and *m*-isomers are much too bulky for α-cyclodextrin, thus their hydrolysis is accelerated. This explains why the hydrolysis of the same ester is accelerated by one of the cyclodextrins (α-cyclodextrin in this case) and hindered by another (β-cyclodextrin in this case). All the three isomers fit the cavity of β-cyclodextrin; the actual hydrolysis rate is the sum of the hydrolysis rates of free and complexed molecules.

Certain substances of natural origin (e.g. camomile oil) cannot be stored for long even in sealed vials because the decomposition and polymerization of essential components (e.g. terpenes) is inevitable. In the cyclodextrin inclusion complex form they can be stored and applied over a long period. When kept at 150° C for 24 h in vacuum about 25–30% of the included substance is lost, while under identical conditions the uncomplexed essential oil disappears within minutes.[131]

The disagreeable smell or taste of some drugs can be reduced or almost perfectly concealed by inclusion complex formation. These products are readily acceptable e.g. as pills. Thus for example the antibacterial, antifungal allicin which is the unpleasant odour ingredient of garlic, (a well known traditional spice) can be converted into a stable inclusion complex and this is the way to eventually develop this compound into a new drug.[132]

During chronic toxicity studies experimental animals sometimes refuse to eat the drug containing feed on account of its intensive odour or throat-irritating effect. In such cases cyclodextrin complexation might be a viable method.

The volatility of highly volatile substances is strongly reduced not only in dry powderlike complexes (e.g. DDVP[133,134]) but also when cyclodextrin is dissolved in the aqueous solution of the volatile guest molecule (e.g. anethole[135]). Poultices are often applied in the treatment of certain complaints e.g. rheumatism, arthritis, bronchitis, tonsillitis, mastitis, tooth aches, parotitis, colds, etc. The odour of the drugs applied in the poultice (especially in warm poultices, cataplasms) penetrates through the cloth, and gives rise to side effects such as headache, nausea, etc. Poultices can be made partly or completely odourless by converting the active ingredients into cyclodextrin complexes[136] without any reduction of the pharmacological effects. During the application no odour develops and the water retention of the product is better than that of the conventional poultices.

The reduction of hydrophobicity by cyclodextrin complexation results in enhanced bioavailability not only in oral, but also in rectal preparations. Cyclodextrin complexed drugs in suppositories result in improved dissolution *in vitro*, and in higher blood level *in vivo*.

Suppositories containing essential oils as active ingredients gradually become soft. If the cyclodextrin complex of the essential oil is used mixed with the lipid base, no melting point decrease is observed. For example the melting point of Adeps solidus (lipoid base) is 28.6° C. If a 2 g suppository contains 0.05 g camphorae, 0.10 g Aetherolei eucalypti and 0.05 g colloidal silicic acid, the melting point decreases to 26.7° C. If camphor and eucalypt oil were applied in the form of β-cyclodextrin complexes the melting point is elevated to 30° C. Equally important is the fact that the break-strength decreased from 10.3 kg to 2.5 kg due to the active ingredients, while with their cyclodextrin complexes it actually increased to 11.0 kg.[137] Dissolution tests have shown that the release of complexed drugs in suppositories is superior to that of the conventional free-drug containing suppositories.[108,137]

3.4. Pharmacokinetic effects of drug complexation by cyclodextrins

The absorption of numerous drugs is strongly influenced by dispersity, i.e. by the crystal size.

As a consequence of cyclodextrin complex formation (molecular encapsulation) drug molecules are isolated from one another and are dispersed on the molecular level in an oligosaccharide matrix which disintegrates easily under physiological conditions. That is why drugs of low water-solubility reach significantly higher concentrations in the organism when applied in the cyclodextrin complexed form. The solvent (gastric acid or intestinal juices) need not disintegrate the crystal lattice of the pharmacon but the cyclodextrin inclusion complex crystal, which disintegrates more readily.

It is particularly important to mention that the dissolution rate of cyclodextrin-complexed drugs is much higher than that of the non-complexed ones. The wettability of the cyclodextrin complexes (= hydrophilic wrapping) is good and they reach their solubility limit usually within 5 minutes on intensive stirring (Fig. 10.). Sparingly soluble substances usually need

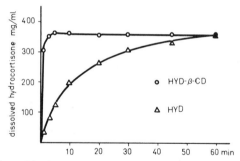

Fig. 10. Dissolution of hydrocortisone (HYD) in water at 25° C as a pure substance or as a γ-cyclodextrin complex.

more than one hour of intensive stirring to reach their solubility limit. The pseudo-first-order rate constants of dissolution calculated from the time required to reach 50% solubility (solubility half time) were increased 7–60 fold compared to uncomplexed substances.

The dissolution–absorption process of an orally administered drug can be expressed as a first approximation by the following kinetic model:

$$F \xrightarrow{k_d} GI \xrightarrow{k_a} B$$

where F = drug in the orally administered dose
  GI = concentration of the dissolved drug in the
      gastro–intestinal content
  B = concentration of the drug in the blood
  k_d = rate constant of dissolution
  k_a = rate constant of absorption

If $k_d > k_a$, the absorption is the rate limiting step. In this case cyclodextrin complexation does not promote the absorption, eventually it can decrease it. With crystalline or liquid substances which are readily dispersible and soluble in water, no enhancement of solubility and absorption is expected, rather a slight retardation. In cases when absorption is too rapid, an abrupt and excessive blood-level peak is contraindicated, complexation with cyclodextrins may be taken into consideration. For example by "cutting-off" an excessive blood level peak following oral administration, the undesired side-effects, toxicity, can be diminished.[133]

If $k_d < k_a$ the dissolution is the rate determining step. In these cases k_d has to be increased, e.g. by cyclodextrin complexation. When k_a remains constant, the blood-level (B) also has to increase. Experimental observations support this expectation.[107] Following oral application, only an insignificant amount of cyclodextrin will be absorbed simultaneously with the drug. Cyclodextrin is only a carrier agent: it transports the more or less hydrophobic guest molecule through an aqueous medium to the lipophilic membrane, which then absorbs it, since it has a higher affinity for the guest molecule than cyclodextrin itself.

Only the dissolved drug molecules are able to enter the blood circulation, get to the organs and exert their biological effect. Undissolved drug and complexed drug are considered as reservoirs, the former is generally slowly mobilizable, the latter is available more rapidly.

Rapid dissolution of a complex frequently results in a metastable state: within some minutes an unexpectedly high solubility can be observed, which later however decreases to a lower level (Fig. 11.). This lower level is determined by the concentration of the free and complexed drug which remains in solution having attained the equilibrium. A part of the dissolved complex dissociates, the released drug starts falling out—on account of its low solubility—and the picture becomes rather complicated.[138]

The partition of the drug into its different forms, such as dissolved free drug molecules, dissolved complex, solid free drug and solid complex, depends on (a) the solubility of the free drug and the complexed drug in the given medium (b) the stability constant of the cyclodextrin-drug complex, (c) the molar ratio of the components and (d) the volume of the liquid phase.

By modifying these parameters—at least hypothetically—very different blood level versus time profiles can be produced. The absorption of the

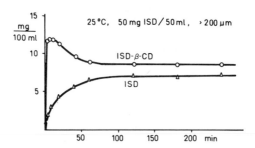

Fig. 11. Dissolution of isosorbidedinitrate (ISD) and its β-cyclodextrin complex (ISD–β-CD). The grain size was smaller than 200 μm.

same drug can be accelerated or retarded, depending on the above parameters.[138]

If the complex is very stable, or by using a high cyclodextrin concentration the equilibrium is greatly shifted towards complexation, the absorption is considerably hindered. Uekama et al.[139] reported that the absorption rate of acetohexamide in rabbit intestine (pH 7.4, 38° C) is reduced to 1/4–1/6 of its original value in the presence of a high β-cyclodextrin concentration.

If the orally administered drug does not contain an extremely large excess of cyclodextrin, no sign of reduction of k_a can be observed, consequently an accelerated absorption and improved biological effect are expected,[107] which is observed with non-steroidal antiphlogistics, barbiturates and fat soluble vitamins.

The very first absorption experiment was published in 1973 by Frömming and Weyermann.[140,141] The absorption and elimination of salycilic acid was studied on human volunteers. The subjects received 200 mg of salicyclic acid, or 2140 mg of the complex in a powder capsule with 100 ml water. Urine was collected each hour, analysed and after each sampling test subjects received 50 ml water to drink. In the first hour the absorption from complexed salicyclic acid was 220% compared to the free acid; after that it dropped to 125% up to the 6th hour. The maximum level of salicylic acid excretion is attained in the third hour in the case of the free acid, whereas its cyclodextrin complex reaches the same level as early as the first hour. After 8 h 99.1% of the dose administered in the complexed form is excreted. No excretion of the free acid can be detected after 14 h, though the total excretion is only 89.3%. When β-cyclodextrin and salicylic acid were mixed mechanically the total salicylic acid excretion exceeded the free salicylic acid level only by 4%. A significant difference was only observed after 8 h.

The percentage of absorbed salicylic acid at a given time was obtained by kinetic calculations. From free salicylic acid 24.8% appeared in the blood in the first hour whereas 43.0% appeared from the complex. In two hours the corresponding values were 56.8% and 82.5%, respectively (average of 9 persons).

Up until now (1982) this is the only published human experiment (which is explained by the fact that well-documented adequate toxicological studies became accessible only in 1981). Numerous experiments have been published however on rats, rabbits and dogs. 100 mg flufenamic acid, or 100 mg ketoprofen or 200 mg ibuprofen, or the equivalent amount of their 1:1 β-cyclodextrin complexes were administered into the stomach of male albino rabbits by Nambu *et al.*[142] The concentrations of the drugs were determined in the blood and in the urine as a function of time. Administration of the cyclodextrin complexed drugs resulted in considerably higher blood-levels, and increased cumulative urinary excretion.

Similarly studies on the absorption of crystalline mefenamic acid and its β-cyclodextrin complex in dogs showed that when equivalent doses were given orally, a significantly higher blood level could be achieved by using the complex.[143]

An indomethacin–β-cyclodextrin complex of 1:1 molar ratio could not be prepared.[144] There is also no unambiguous evidence for the 1:2 molar inclusion of indomethacin in the crystalline state. In solution however circular dichroism as well as diffusion tests provide direct evidence for inclusion complexation.[145] Oral administration of the 1:2 indomethacin–CD complex resulted in about 25% higher blood level as compared with an equivalent amount of indomethacin.

On administering ^{14}C-labelled indomethacin orally, 28% of the activity is excreted with the urine whereas on administering its β-cyclodextrin complexed form the corresponding value is 32%. The remaining activity in the faeces, or in the intestinal contents is 57% in the case of indomethacin, and only 46% in the case of the complex. In an *in vivo* experiment 56% of indomethacin introduced into a ligated section of the ileum appeared in the blood within 10 min, whereas the corresponding value in the case of the complex was 68%.[145] An unexpectedly high (by a factor of two), blood level was observed in rats after rectal application of suppositories containing indomethacin–CD complex compared to the equivalent amount of free indomethacin, in Adeps solidus lipoid base[106] (Fig. 12.). On treating rats (130–150 g males) rectally with 2 mg indomethacin/day, all the animals showed serious peritonitis, stomach and duodenal ulcers on the 7th day, while a similar 14 day treatment with 14 mg indomethacin–β-CD 1:2 complex (equivalent to 2 mg drug) caused significantly less ulceration and no peritonitis. The 2 mg dose was intolerable after the 7th day, but the complexed drug was tolerable even for 14 days.

An increased blood level must be manifested in an improved anti-inflammatory effect too. The experiment was carried out with carragheenane

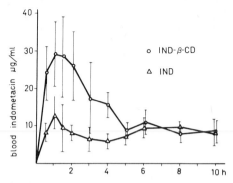

Fig. 12. Indomethacin blood levels in rats following rectal application of suppositories containing indomethacin (IND) or the equivalent amount of a 2:1 β-cyclodextrin:indomethacin coprecipitate (IND–β-CD) in Adeps solidus lipoid base.

induced oedema in rats. The leg weight increase of the control group (injected with carragheenane, but not treated with indomethacin) was taken as 100% inflammation. Weight differences in the treated groups were compared to the control, and expressed in percentages. Figure 13 demonstrates a dose-response correlation of free and complexed indomethacin providing direct evidence for the improved bioavailability.[146]

The acute toxicity of free and complexed indomethacin has been studied on mice and rats. In the 96th hour after treatment the LD_{50} (a dosage which kills 50% of the sample) was $35\,\mathrm{mg\,kg^{-1}}$ for free, and $60.0\,\mathrm{mg\,kg^{-1}}$ for complexed indomethacin in mice, 14 and $25\,\mathrm{mg\,kg^{-1}}$ respectively in rats. The fact that, notwithstanding better bioavailability (higher blood level), reduced toxicity could be observed suggests that mainly blood level

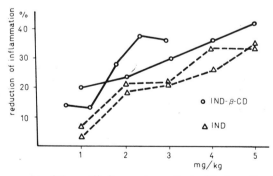

Fig. 13. Enhancement of the antiinflammatory effect of indomethacin by cyclodextrin complexation: dose-response relation of free (IND) or complexed indomethacin (IND–β-CD) (two separate experiments).

independent local effects and not systemic effects contribute to the noxious side effects.[146]

Further experiments showed that the ulcer inducing effect of indomethacin can be reduced significantly by cyclodextrin inclusion complex formation. Male and female rats were treated with indomethacin and indomethacin–cyclodextrin complex. 70% of the males and 60% of the females treated with a 5 mg kg^{-1} dose of indomethacin developed a serious ulcer, whereas in the animals treated with the complex containing the same indomethacin dose, these figures were reduced to 20% and 10%, respectively (Table 12.).

Table 12. Stomach irritating effect of 5 mg kg^{-1} day^{-1} p.o. doses on rats. The majority of rats treated with free indomethacin perished before completing the 28 day treatment

	Ulcus		Erosion	
	Male	Female	Male	Female
Control	0/10	0/10	0/10	0/10
40.3 mg kg^{-1} IND-CD				
(=5 mg IND)	2/10	1/10	4/10	4/10
5 mg kg^{-1} IND	7/10	6/10	9/10	7/10

A similar observation was published by Nambu *et al.*[147] with phenylbutazone. Rats were treated with a single 100 mg kg^{-1} dose (or with the equivalent amount of the β-cyclodextrin phenylbutazone complex) orally. 18 hours later the animals were killed, and the damage to the mucosa of the excised stomach was evaluated under a dissection microscope. The degree of injury was evaluated as follows:

0: normal
0.5: latent damage
1.0: slight damage (2 or 3 small, localised areas)
2.0: severe damage (continuous lined areas or 5–6 localised areas
3.0: very severe damage (several continuous lined areas)

The degree of injury was found to be 2.3 ± 0.5 for free phenylbutazone, while only 0.6 ± 0.2 for the complexed drug (significant by *t*-test at 5% level).

The probable explanation for these observations is that the crystals of the non-steroid antiinflammatory agents, being poorly soluble in gastric acid, remain in contact with the stomach wall for a long period of time resulting in a dangerously high concentration *in loco*. This results in the local irritation of the stomach wall and on prolonged treatment provokes an ulcer. In the complexed form the pharmacon dissolves faster and therefore shows an accelerated absorption. Moreover it does not come into direct contact with the stomach wall in the crystalline form, since it remains "encapsulated" within the cylcodextrin matrix until dissolution.

There may be a similar explanation for the fact that 10% of male rats treated with 50 mg kg^{-1} day^{-1} salicylic acid orally for 4 weeks developed an ulcer and 90% showed a slight reddening and vascular inflammation of the mucous membrane, whereas the administration of 381.5 mg^{-1} kg^{-1} day of the β-cyclodextrin–salicylic acid complex (equivalent to 50 mg drug) containing the same dose of salicylic acid caused no such gastric symptoms.[148]

Another interesting and thoroughly studied group is that of the barbiturates. Koizumi[149] compared the hypnotic potency of different barbiturates to those of their β-cyclodextrin complexes on oral administration to mice. The 50% effective dose (ED$_{50}$ determined by using 50 animals per experiment) was smaller for the complexed barbiturates than for the non-complexed ones. ED$_{50}$ of phenobarbital which shows the greatest enhancement in solubility (and consequently forms the most stable complex) was reduced most markedly by complexation. This observation was further substantiated by studying the sleeping lag and sleeping time. Barbital showed no appreciable solubility improvement, consequently no enhancement of the hypnotic effect could be observed while all the other barbiturates and their complexes exhibited a difference in sleeping lag significant at the 0.1% level. On phenobarbital treatment (solubility enhancement 8.7 fold) the sleeping lag was shortened from 75.7 ± 7.0 min to 27.0 ± 3.6 min ($p < 0.001$) by complexation (sleeping lag: time from oral administration to the loss of righting reflex). Simultaneously, sleeping time (time from loss to recovery of righting reflex) was prolonged from $122.9 \pm 20\cdot4$ min to 241 ± 29.8 min with $p < 0.001$.

On oral administration of the β-cyclodextrin complex of tritium labelled vitamin D$_3$ a considerably higher blood radioactivity level was measured, than with the equivalent dose of free ^{3}H-labelled vitamin D$_3$.[150,151] The absorption enhancing effect of complexation is highest in the first 90 min, in the case of the complex it is 2.3–2.8 times higher, expressed as the percentage of administered activity calculated for 10 ml blood. The blood level ratio decreases later concomitant with the substantial increase of the scattering of the values measured, especially in the case of the complex (Fig. 14.).

The enhanced absorption is well illustrated by the biological effects of vitamin D$_3$, which can be easily studied on vitamin deficient (rachitic) animals. Typical symptoms of rachitis (great retardation of weight gain, coarse fur, adinamy, decrease of plasma phosphate level, etc.) were induced in rats by feeding them with rachitogenic (vitamin D free) diet. Rachitic

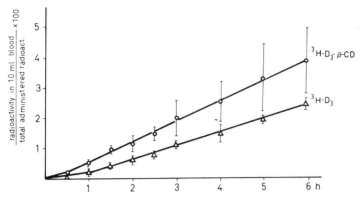

Fig. 14. Blood radioactivity of rats following oral administration of tritium labelled cholecalciferol (^{3}H–D$_3$) or its β-cyclodextrin complex.[151]

rats were treated with vitamin D$_3$ and with β-cyclodextrin complexed vitamin D$_3$.[117] Treatment began after a 48 day rachitogenic diet. The animals were given 5 I.U. vitamin D$_3$ in the first two days, thereafter 10 I.U. (free or complexed) daily. The changes in plasma phosphate concentration are demonstrated in Fig. 15. As can be seen, treatment with the complexed vitamin nearly normalizes the plasma phosphate levels by the 4th day of treatment. The effectiveness of the uncomplexed vitamin is lower.

A similar trend was observed in the plasma calcium level, and a morphological study of the ossification of bones suggested an enhanced biological effect of the complexed vitamin D$_3$ as compared to the non-complexed one.[117]

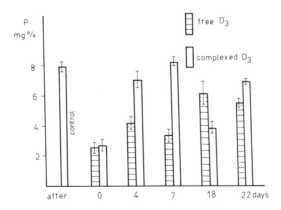

Fig. 15. Plasma phosphate level in rachitic rats treated with a 10 IU/day dose of free or complexed cholecalciferol (Vitamin D$_3$). Control: phosphate level of healthy animals.[118]

In solution the dissociation/association processes are instantaneous, the equilibrium of free and complexed drugs is established very rapidly. It is thus conceivable that tablets containing both cyclodextrin complexed and the free drug will be developed. The complex dissolves readily, dissociates and the drug very rapidly gets into the circulation while the free drug of lower solubility can act as a reservoir of the drug.

4. Applications in the chemical industry

4.1. Separation by selective CD- complexation

The composition of a mixture of various aromatic hydrocarbons can be modified—under favourable conditions even separated—by CD-complexation. This process would be important in cases where boiling points are similar, and separation cannot be performed by distillation.

The basic feature of this process is the countercurrent contacting of hydrocarbons with an aqueous solution of α-cyclodextrin in a column in which they are homogenized. At the bottom of the column the crystalline hydrocarbon–CD complex is separated. On warming, the complex dissociates, the hydrocarbons are recovered (with more or less modified composition) and the CD-solution is recycled to the column.

Separations of xylene isomers and ethylbenzene,[180,181] trimethylbenzenes,[182] isomeric alkylphenols[183], alkyltoluene isomers[184], straight and branched chain hydrocarbons[185] have been reported.

A one step separation cannot be achieved, only the enrichment of one guest molecule can be observed in the complexed fraction. For effective separation a multistage process is needed. For example by partial complexation, separation and decomposition of the crystalline complex the original composition o-, m-, p-cymene 35:27:38 was shifted to 1:2:97. A 50:50 mixture of m- and p-cresol yielded a 44:56 mixture.[183] Similar enrichment of unsaturated fatty acids was reported[186] on partial complexation of various fatty acid mixtures of natural origin.

Reiners and Birkhaug[187] elaborated a procedure for the removal of free fatty acids from vegetable oils which is based on cyclodextrin complex formation, since the usual method (saponification) has many disadvantages. Cyclodextrin was dissolved in water and emulsified in the oil, then the aqueous phase was separated. On heating the latter to boiling, the complex decomposes and after cooling, the fatty acids solidify on the top, while in the lower phase, cyclodextrin precipitates and can be reused.

Technologies involving the use of large amounts of volatile materials and their vapours (e.g. distillation, regeneration of solvents, ventilation of tanks) always bring about pollution of the atmosphere with vapours. So far the waste of volatile solvents can only be reduced either by means of intensively cooled condensers or with adsorbers filled with active carbon or silica-gel. An alternative method is the fixation of solvent vapours by CD-solution followed by continuous regeneration.[215] Large volumes of the gas or vapour phase are bubbled through a cyclodextrin solution at a temperature of complex crystallization. This temperature is specific to the given system, being controlled by the quality of the volatile compound, the cyclodextrin concentration and the incidental presence of other substances, e.g. salts. The crystalline complex is formed instantenously, the molar ratio of guest molecule: cyclodextrin is generally 0.4–1 : 1.

The crystals are continuously removed from the cyclodextrin solution (e.g. by removing the precipitated crystals by pumping, or by centrifugal separation) and the solvent concentrated in the small volume of the crystal slurry is continuously passed through a heat exchanger to transform it into a small volume of vapour which can be condensed with a small cooling capacity. Warm cyclodextrin solution is led through a counter-current heat exchanger and continuously recirculated into the absorber.[215]

4.2. Resolution of enantiomers

Cyclodextrin has a chiral structure. With a tight fitting guest molecule a pronounced stereoselectivity—depending on the circumstances of complexation—should therefore be possible.

No complete resolution has yet been reported by CD-complexation, only the enrichment of one enantiomer; with multistage processes however (e.g. with chromatography on CD-polymers, Section 4.3) complete resolution may be achieved.

Cramer and Dietsche[188] studied the effect of the host:guest molar ratio on resolution. In principle, at a 1:1 molar ratio no resolution can be expected, but since complex formation is an equilibrium reaction, the ratio of enantiomers in the complexed and free state is not identical. By using 2:1, 1:8 and 1:20 molar ratios of β-cyclodextrin–ethyl chloromandelate the optical rotations achieved were $[\alpha]_D^{25°} = -4.58°$, $-12.65°$ and $-12.68°$, respectively.

A remarkable phenomenon was observed during the resolution of racemic menthol esters. When the complex was formed at 0° C, 50° C and 58° C, the $[\alpha]_D^{25°}$ values of the incorporated component were $-1.42°$, $0.00°$ and $+1.48°$, respectively. In one of the two possible chair conformations of the menthyl group the acetyl group is in the equatorial, in the other one in the axial position. There is an 8.25 kJ mol^{-1} energy difference between the two conformers. Above a certain temperature the particular enantiomer which incorporates preferentially at lower temperatures, takes up a conformation which is no longer preferred.

Partial resolution of isopropyl methyl phosphinate was reported[189] with β-cyclodextrin. In the liquid phase the (+)-enantiomer is enriched in 17% optical purity, from the separated crystals the (−)-enantiomer can be recovered with methylene chloride–water in 66.5% optical purity. Repeating the inclusion complex formation twice 49.6% optical purity of the (+)-enantiomer can be achieved.

On preparing the β-cyclodextrin complex of 2-methylcyclohexanone the R-(+)-isomer was preferentially incorporated.[190]

In attempting[191] the optical resolution of halothane (2-bromo-2-chloro-1,1,1-trifluorethane, an anaesthetic) with α-cyclodextrin the laevorotatory molecule was complexed with a slight preference, the optical purity, however, was less than 1%.

Optical purities of the partially resolved alkyl–aryl and alkyl–benzyl sulphoxides and sulphinyl compounds do not exceed 15%.[192,193] The highest stereospecificity of inclusion was observed for $CH_3–S(O)OCH(CH_3)_2$ which was isolated after one inclusion process with 68% optical purity.[194]

Michon and Rassat[195] observed that in the ESR-spectrum of a β-cyclodextrin complexed nitroxide biradical the intensity of the peaks decreased when fenchone was added to the system. The decrease was more intensive with (+)-fenchone than with (−)-fenchone. The complexation of the (+)-enantiomer seems to be preferred.

A fuller account of the enantiomer selectivity of cyclodextrins can be found in Chapter 9, Volume 3.

4.3. Chromatography on cyclodextrin polymers

By using appropriate bifunctional reagents cyclodextrins can be polymerized into block-, foam- or bead-form polymers. These polymers show remarkable chromatographic separation properties.[196–200,202]

The separation is based either on selective inclusion (separation of different compounds or resolvation of enantiomers) or on specific affinity (separation of enzymes).

In well swelling insoluble cyclodextrin polymers cyclodextrin cavities cause the special cyclodextrin effects, whereas cavities formed by bridges and chains produce Sephadex-like (molecular sieve) effects.[201] Substances which cannot be separated by Sephadex (e.g. o- and m-dichlorobenzene) can be easily separated by β-cyclodextrin polymer. On passing a dilute aqueous solution of cinnamic alcohol and eugenol through cyclodextrin polymer only eugenol is eluted.[103] Cinnamic alcohol can be subsequently removed by washing the column with ethanol.

It is known that certain components of nucleic acids, particularly adenine bases, form inclusion complexes with β-cyclodextrin.[203] In a column filled with β-cyclodextrin–epichlorohydrin polymer nucleotides and nucleosides can be easily separated.[204]

During the chromatography of aromatic acids the undissociated molecules are bound to the cyclodextrin polymers and the extent of inclusion follows the Freundlich- or Langmuir isotherm.[198] Aromatic amino acids can be separated more efficiently (with greater retention volume differences) both from each other and from aliphatic amino acids than on dextran gels 205, 206 (Fig. 16.). Phenylalanine can be removed from casein-hydrolysate of 20% (w/w) dry substance content with a fairly good selectivity.[207]

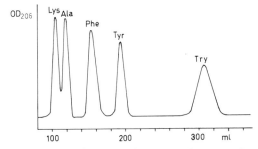

Fig. 16. Separation of lysine (Lys, 4 mg), alanine (Ala, 2 mg) phenylalanine (Phe, 0.1 mg) tyrosine (Tyr, 0.1 mg) tryptophan (Try, 0.2 mg) on a β-cyclodextrin–polymer column (1.6×88 cm; $V_t = 176$ ml, 23° C, pH 5.5, flow rate 20 ml h^{-1}).

Water-soluble organic substances (e.g. 2-naphthalenecarboxylate) can be removed from aqueous solutions (e.g. from pharmaceutical wastewater) by polystyrene–cyclodextrin derivatives.[208]

The cross-linked β-cyclodextrin polymer preferentially binds the S(+)-isomer of mandelic acid and in the first fraction the R(−)-enantiomer was obtained in 100% purity. In contrast to this, α-cyclodextrin selectively binds the R(−)-enantiomer, but the purity of the S(+)-enantiomer was not as good as that of the R(−)-enantiomer with β-cyclodextrin polymer.[209]

Zsadon *et al.*[210] reported the resolution of (+) and (−)-vincadifformine on a β-cyclodextrin bead-polymer column. The relative retention volumes were 1.9–2.1 for the (+)-, and 2.2–2.4 for the (−)-enantiomer (Fig. 17).

A special chromatographic application is based on the fact that α-cyclodextrin competitively inhibits β-amylase.[211] On coupling cyclodextrin

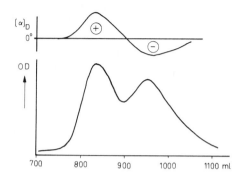

Fig. 17. Separation of (+)- and (−)-vincadifformine on a 225(=3×75)×1.6 cm β-cyclodextrin polymer column at room temperature, pH 4, flow rate 50 ml h⁻¹, detection: optical rotation and UV absorption at 340 nm.[210]

to Sepharose-6B,[212] the binding of the enzyme is biospecific, only β-amylase is retained by the polymer, α-amylase is not; the bound enzyme can be eluted with cyclodextrin.

Cyclodextrin-transglycosylase enzyme can be purified and concentrated with α-cyclodextrin covalently bound to agarose gel.[213] Other enzymes also subjected to chromatography on similar affinity-gels were obtained in high purity; (e.g. from raw pig prancrease, a preparation containing not more than 3.5% inactive protein was obtained).[214]

A fuller account of the use of cyclodextrins in chromatography can be found in Chapter 6, Volume 3.

4.4. Miscellaneous industrial applications

The number of papers and patents which are dedicated to various possible applications of the CDs continues to increase. The majority of these possibilities have not yet been exploited, this will only happen after a further substantial decrease in CD-prices.[219] Such possibilities are e.g. the utilization of the CDs in diazo-copying sheets,[216–218] as anti-foam agents in the flotation of limonite,[94] as retarders in textile dyeing processes,[220] in non-smearing aqueous ink,[221] in inks for transfer printing onto polar polymers etc.[222]

The cyclohexylamine–β-cyclodextrin complex can be utilized as a vapourizable anticorrosive agent,[223] and the hydroxypropyl–β-cyclodextrin complex in temperature indicating lyotropic cholesteric liquid crystals.[224]

Highly toxic substances can be removed from industrial effluent. For example in the mother liquor of trichlorfon (an insecticide) the non-crystallizable substance can be converted into the β-cyclodextrin complex of

DDVP (dichlorovos, also an insecticide). The complex is slightly soluble in water, therefore it easily separates from the liquid–phase. By a single treatment the toxic substance content can be reduced to one tenth of the original.[225,226]

Important properties (relative sensitivity, rapid processability and fog) of silver halide containing photographic materials can be improved by adding CDs to the photographic gelatin layers or to the processing solutions.[227] β-cyclodextrin can be applied also in organotellurium containing photothermographic layers.[228]

A process has also been reported for the production of petroleum from oil sand.[229] The petroleum content of oil sand is converted to its cyclodextrin complex, separated from the sand, and on decomposing the complex, both components, are recovered. Another process has been proposed for preparing solid fuel.[230]

The effectiveness of powder-filled fire-extinguishers can be improved by 50–200% by adding 6–15% of the phosphoric acid–β-cyclodextrin complex[231] to the conventional extinguisher powder (e.g. finely dispersed $KHCO_3$). The inorganic acid in the flame acts partly by producing CO_2 *in loco*, partly as a radical scavenger.[232]

Numerous examples have been published for the utilization of the catalytic and stereoselective effects of cyclodextrins in chemical syntheses. These aspects are discussed in Chapters 13 and 14 of Volume 3.

5. Cyclodextrins in analytical chemistry

The spots of separated substances become visible on a thin-layer chromatographic plate after spraying with an α-cyclodextrin solution and then exposing it to iodine vapour.[233,234]

Aqueous α-cyclodextrin was demonstrated to be a very effective mobile phase in thin-layer chromatography.[235,236] The chromatographic behaviour of 26 substituted phenolic and naphtholic compounds were studied using polyamide sheets. R_F values were found to be dependent on both the structural features of the separated compounds and the concentration of α-CD in the mobile phase.

The separation of various prostaglandins (PGE, PGA, PGB, PGF) by liquid chromatography is rather difficult. Being rather hydrophobic, they have long retention times, and show undesirable tail-formation. Separation is however a rapid process, with distinct sharp peaks if the mobile phase is an aqueous cyclodextrin solution.[237]

Applying β-cyclodextrin in the mobile phase, racemic phenylalanine could be resolved by reverse-phase HPLC.[238] β-cyclodextrin immobilized on a polyacrylamide or agarose gel is a good stationary phase for the separation of disubstituted benzene isomers.[239]

Aromatic compounds can be determined as phenol derivatives after oxidative hydroxylation with Hamilton's reagent (hydrogen peroxide and catechol in the presence of ferric ions), however the phenols formed can easily be further hydroxylated. In the presence of cyclodextrins the reaction is more selective; inclusion complex formation stabilizes the primary oxidation product.[240]

The pK values of different organic acids and phenols in aqueous CD-solutions change differently. This enables the selective titration of e.g. salicylic acid and p-hydroxybenzoic acid, or p- and m-nitrophenol, etc. with NaOH.[241]

In amylase activity determination cyclodextrins can be utilized either as a substrate (γ-cyclodextrin[242]), or the release of a pigment from a cyclodextrin–pigment complex (γ-cyclodextrin and orange IV[243]), can be monitored. Similarly the dihydro chlorofluorescein–β-CD complex can be used in the ultramicrodetermination of serum uric acid and glucose (by determination of the H_2O_2 formed in the presence of oxidizing enzymes, appropriate substrates and peroxidase).[244,245]

The fluorescence intensity of dansylamino acids is enhanced very effectively (up to 10 fold) by cyclodextrins both on chromatographic plates and in aqueous solutions.[246–250] In this way as little as 0.2 nmol of dansyl-amino acids can be determined.

The interference of serum components in thyroxine determination (by competitive protein-binding or immunoassay) can be inhibited by adding α-cyclodextrin (or its salicylate complex) to the sample.[251]

In the determination of serum calcium levels, free fatty acids can cause erroneous results; preincubation of the sample with α-cyclodextrin eliminates this interference.[281]

For the rapid, and specific determination of cephalosporins and penicillins the antibiotics were first hydrolyzed with β-lactamase, then the cyclodextrin–iodine complex was added and the absorbance was measured at 360 nm. A linear correlation was found between cephalotin concentration and the absorbance.[252]

Schlenk et al.[253] used β-cyclodextrin acetate, propionate, butyrate and valerate as the stationary phase in gas–liquid chromatography. It was possible to operate a column at 220 °C for 16 months without any purification. Cyclodextrin-based column packings are particularly suitable for the separation of olefins, alcohols, aldehydes and esters. Retention times of homologues follow the rule of logarithmic linearity. Kuge and Takeo[254] carried out similar experiments with cyclodextrins.

Cyclodextrin–polyurethane resins (cyclodextrin cross-linked with diisocyanates) exhibited strong interactions with guest molecules containing π-electrons or heteroatoms; therefore pyridine, tetrahydrofuran, 1,4-dioxan, etc.[255] have very long retention times in gas–solid chromatography.

Casu,[256] and Smolkova-Keulemansova and Krysl[200] have summarized the application of inclusion complexes (urea, tri-o-thymotide, cyclodextrin and Werner complexes) in separation chromatography. See also Chapter 6, Volume 3.

6. Agricultural applications

6.1. Formulation of pesticides

The effects of pesticide complexation are similar to those utilized in drug-formulations:

(a) Stabilization of labile substances against too rapid a decomposition (by light, oxygen, heat, etc.).

(b) Conversion of (volatile) liquids into microcrystalline (persistent) substances (territorially restricted, long lasting effect is ensured).

(c) Increasing water solubility, absorption and biological effects can be enhanced.

(d) Safer handling, transport and application of highly toxic or explosive substances.

(e) Release of the complexed substance occurs only on contacting with water (e.g. a complexed soil insecticide gets released only after rainfall or watering, etc.).

Pesticides (insecticides, fungicides, herbicides, plantgrowth regulators, defoliants, molluscicides, etc.) are utilized in solid formulations (wettable or dusting powder, incenses, granulated products) or liquid form (solutions, emulsions, suspensions). Cyclodextrin complexes—being slightly soluble microcrystalline substances—can be utilized in both solid formulations and concentrated aqueous suspensions. It should be taken into consideration, however, that the guest content of the complexes is only 5–25%.

The DDVP–β-cyclodextrin complex is a good example of how a short-lived liquid can be converted into a persistent solid substance by CD-complexation. DDVP (O,O-dimethyl-2,2-dichlorovinyl-phosphate) is a volatile, inflammable, explosive, liquid insecticide which acts mainly as a contact poison. Its β-cyclodextrin complex is crystalline, containing 16.2% of the active ingredient in a 1:1 molar ratio.[133,257] The loss of DDVP from the crystalline complex is so slow, that from 5 g DDVP complex only 0.14 g

DDVP was lost under permanent air circulation in 24 h at 25° C, while after 6 days 99.2% of the original DDVP content remained in the complex.

The complex has no gas-effect, nevertheless, it is active as a contact insecticide and acaricide. With the standard Petri dish method 0.1% DDVP on the etched glass-plate killed 80% of dried-bean beetles within 20 min, while the complex killed only 38% (see Table 13). After two days however, DDVP did not kill a single beetle even after 300 min contact time, whereas the complex killed 70%. Other data in Table 13 also demonstrate that the instantaneous effect of the complex is much less effective than free DDVP, but the former is much better when the duration of the effect is considered. At 0.01% DDVP concentration the differences are even more conspicuous.[133]

Table 13. *Comparison of the effect of DDVP and of its complex with β-cyclodextrin on dried-bean beetles. Death rate (%)*

Contact time (min)	0.1% DDVP				0.1% DDVP β-CD[a]			
	Time after preparation (days)				Time after preparation (days)			
	1	2	3	4	1	2	3	4
20	80	—	—	—	38	12	—	—
60	90	—	—	—	75	15	15	12
300	92	—	—	—	90	70	55	35
1440	99	30	28	20	95	93	91	84

Contact time (min)	0.01% DDVP				0.01% DDVP β-CD[a]			
	Time after preparation (days)				Time after preparation (days)			
	1	2	3	4	1	2	3	4
20	12	—	—	—	6	—	—	—
60	35	—	—	—	25	—	—	4
300	55	—	—	—	45	65	23	30
1440	70	—	—	—	94	90	87	82

[a] Based on DDVP.

Table 14 demonstrates that a 50% killing effect in the soil can no longer be attained with free DDVP after 12 days, while with the complex in a 48 h contact time a 50% killing effect can be achieved even after 48 days. Another observation was that in dry soils the complex was rather ineffective, but the increase in moisture content of the soil, for instance by irrigation, resulted in higher effectiveness. The probable explanation of this phenomenon is that under the effect of water the DDVP–cyclodextrin complex dissociates, but no DDVP is released from the dry complex.[133]

Table 14. Comparison of the effect of DDVP and of its β-cyclodextrin complex in the soil on the larvae type L_1 and L_2 of forest maybeetle and on the larvae of wire worm and house fly. [133]

Period after application (days)	Time required to achieve ED_{50} (min)	
	Complex	DDVP
1	43	31 min
12	59	<50%
20	840	<50%
25	2280	—
48	2820	—
117	<50%	—

The most thoroughly studied pesticide–cyclodextrin complexes are those of the β-cyclodextrin complexes of various pyrethrins and pyrethroides.[258–263]

Generally two β-cyclodextrin molecules are required to include one molecule of a pyrethroid. Natural pyrethrins and synthetic pyrethroids rapidly undergo photoconversion ·nd degradation under mild conditions. By β-cyclodextrin complexation the half-life of allethrin in sunlight was extended from 3 to 3.5 days, and of resmethrin from 1 to 40 days in powder formulations.

The contact activity of complexed pyrethroids is reduced as compared to the free pyrethroids. However, as stomach poisons they are active against herbivorous pest insects without affecting their natural enemies. The decreased contact activity and volatility naturally leads to a strongly reduced knock-down effect, but in some cases (e.g. Blattela germanica) improved longevity of deposits of complexed pyrethroids more than compensates for the decrease of inherent activity.[263]

Kariya[261] investigated the insecticidal activity of pyrethrin and resmethrin–β-cyclodextrin complexes on new-born larvae of various tea parasites. The activity of the complexes was identical with that of the free insecticides, but the loss of activity was slower in the case of the complexes. Against the leaf-roller parasite the complexes proved to be more effective in field experiments than the free substance.

A Teijin patent application[264] describes the preparation of pynamin–β-cyclodextrin containing a fumigant insecticide, that does not require any heating device. On contacting with atmospheric oxygen the thermogenic composition generates heat, and the released insecticide kills the mosquitoes.

The preparation of an incense stick containing pyrethroid cyclodextrin complexes combatting Blattariae has also been reported.[265]

Yamamoto et al.[262,266] described the β-cyclodextrin inclusion complexes of phosphate and thiophosphate (e.g. methylparathion) insecticides. Compared with the original compounds the complexes exhibited better heat and light stability and a prolonged insecticidal effect (Table 15).

Table 15. Residual methylparathion on cotton leaves[266] (sunlight exposure, determination by GLC)

Formulation	Percentage of original methylparathion after exposure for	
	1 h	10 days
Substance: β-CD (1:1)	71.3	47.5
Substance: β-CD (1:2)	100.0	98.3
Emulsified substance	54.4	0.3
Microencapsulated substance	73.3	1.6

Chloropicrin (CCl_3NO_2) forms in aqueous medium a crystalline complex with β-cyclodextrin which can be used as a bactericide and insecticide of prolonged activity.[267] This product can be used as a dusting material as well.

Shimada et al.[268] prepared an antifouling agent by complexing ethyl-3,7,11-trimethyl-2,4-dodecadienoate with β-cyclodextrin. Adding 2 mg of the complexed substance to 20 dm^3 water containing barnacle larvae prevented the larvae from adhering to a rotating wood panel for 1 month, whereas the panel in the control experiment was covered with barnacles. The degradation of agricultural mulching paper can be retarded by the iodine–β-cyclodextrin complex.[269] The β-cyclodextrin complex of pyrrolnitrin is a fungicide, that is stable against sunlight.[270]

Sometimes the evaporating herbicide can cause serious damage in neighbouring environments. The herbicide effect can be restricted to the selected areas by CD-complexation. For example thiolcarbamate esters (benthiocarb, molinate, etc.) are effective herbicides in rice paddies, but their migration to neighbouring environments by evaporation or by wind is not desired. Cyclodextrin complexation reduces their decomposition as well as evaporation rate. In 2 days about 60% of the free herbicide was lost, while less than 30% of the complexed herbicide was lost.[271]

The metabolism of a number of insecticides is far too rapid. This process can be retarded by appropriate enzyme inhibitors. For example the inactivation of pyrethroides by mixed-function oxydases is strongly retarded in the presence of piperonylbutoxide. The solubility of piperonylbutoxide in water is rather low: 0.066 mg ml^{-1}, while in the CD-complexed form it is about 2.5 fold higher. This is probably the explanation of the enhanced biological effect: the ED_{50} of tetramethrin synergized with piperonylbutoxide is reduced to $1/4$ by complexing the synergist with β-cyclodextrin (Table 16.).[272]

Table 16. Enhancement of synergetic effect of piperonylbutoxide by complexing with β-cyclodextrin. Percentage of paralysed and killed Drosophyla [133]

Synergent	Tetramethrin[a] dose (mg disc^{-1})						
	5	2.5	1.25	0.625	0.313	0.156	0.078
Piperonylbutoxide	88	40	14	10	0	0	0
β-CD complex (1:1)	100	90	88	24	16	4	0

[a] Tetramethrin + piperonylbutoxide 1:4 mixture on filter paper disc, percentage.

Sometimes the addition of bird-repellents to various pesticides are required to save useful birds from intoxication. Cyclodextrin complexes of unsaturated alcohols or aldehydes (e.g. 3-hexen-1-ol) mixed with seeded soya-beans were demonstrated in field experiments to have bird repellent effects.[273]

6.2. Effects of cyclodextrins and their complexes on plants

Ethylene is an extremely effective hormone-like agent in the plant kingdom. The ripening process of fruits, and the abscission of leaves, are somehow connected with intracellular production of ethylene. External ethylene is active in 0.1–1 ppm concentration.

A patent of Teijin,[274] describes the preparation of the ethylene–α-cyclodextrin complex in an aqueous medium under 1–50 atm. pressure. The crystalline ethylene–α-cyclodextrin complex (~0.7 mole ethylene per mole α-cyclodextrin) can be used for the acceleration of the ripening process of fruits. For example tomatoes sprayed with $0.5 \text{ g } 100 \text{ ml}^{-1}$ solution of the complex ripened 4 days before the controls.

The α-, β- and γ-cyclodextrin complexes of 2-chloroethane phosphonic acid are very soluble in water. The content of this ethylene releasing compound in the α-cyclodextrin complex is about 25% (~2 mole per mole α-cyclodextrin).[275] On treating plants with this complex a prolonged ethylene effect could be achieved, which proved to be advantageous e.g. in controlling the ripening of paprika. Table 17 illustrates that the defoliating effect of 2-chloroethanephosphonic acid in the α-cyclodextrin complexed form is stronger.

Table 17. Abscission effect of the free and complexed 2-chloroethanephosponic acid (CEPA) in the bean test

Treatment	Concentration	Percentage defoliation after		
		18 h	24 h	48 h
β-cyclodextrin	17.54 μg ml^{-1} [a]	0	0	0
CEPA	5×10^{-5} mol dm^{-3}	0	33	50
CEPA–β–CD complex	5×10^{-5} mol dm^{-3}	18	67	83

[a] 17.54 μg ml^{-1} corresponds to the β-cyclodextrin concentration of the complex.

Cyclodextrins are competitive inhibitors of amylolytic enzymes. This fact could probably explain that germination of starchy seeds (cereals) can be temporarily retarded by treating the seeds with cyclodextrins (Table 18.). After retarded germination however plants grow more vigorously (Table 19.) and the crop yield increases significantly.[276,277] Moreover these CD-treated plants are less sensitive to herbicides. This "antidotal effect" may be useful in pre-emergent weed-control.

Very promising effects have been observed on vegetables after treating the seeds or the developing plants with cyclodextrin. Table 20 illustrates improved yields of vegetables. Green-mass production increased (lettuce 18–30%, celery 50–60%, paprika 7–16%) as a result of CD-treatment.[278]

Table 18. Initial germination decelerating effect of cyclodextrin treatment on barley[a]

Treatment	After 3 days		After 6 days	
	Roots	Shoots	Roots	Shoots
	mm (Δ%)	mm (Δ%)	mm (Δ%)	mm (Δ%)
Control	5123	662	7714	2090
α-cyclodextrin	1056 (-79.4)	96 (-85.5)	2593 (-66.4)	754 (-63.9)
β-cyclodextrin	1808 (-64.7)	191 (-71.2)	2277 (-70.5)	556 (-73.4)

[a] The data represent the amount of corresponding values obtained with 30 plants, grains were soaked in 10^{-2} mol dm^{-3} CD-solutions for 48 h at 28° C, then washed with distilled water, and kept in darkness, in Petri dishes over humid filter paper.[277]

Table 19. Growth enhancing effects of cyclodextrin treatment, as observed 3 weeks after treatment[a]

Species	Treatment	Fresh weight of roots g (Δ%)	Fresh weight of shoots g (Δ%)	Fresh weight of whole plant g (Δ%)
Wheat	Control	3.93	1.91	5.84
	β-cyclodextrin	4.56 (+16)	2.52 (+32)	7.08 (+21)
Triticale	Control	3.67	2.63	6.30
	β-cyclodextrin	5.20 (+42)	2.91 (+11)	8.11 (+29)

[a] The data represent the results from 10 plants.[277]

Table 20. Effect of cyclodextrin treatments (pilling, spraying) on vegetable cultivation

	Lettuce		Celery		Sweet pepper		Red pepper	
	C	β-CD	C		β-CD	C	β-CD	C	β-CD
One plant yield fresh mass (g)	195.1	229.3	shoot: 18.4 root: 8.4	33.2 11.1	75.6	87.8	20.8	22.3
dry mass (g)	12.1	16.3	shoot: 3.4 root: 1.9	5.9 2.1	—	—	—	—
Germination (%)	87	97	—	—	66	73	77	89
Pigment content (%)	—	—	—	—	2.44	3.05	3.16	3.91

[a] C, control; β-CD, cyclodextrin treated.[278]

6.3. Cyclodextrin complexes in animal feeds

Fat-soluble vitamins lose activity on storage by isomerization, anhydro-vitamin formation, oxidation and photochemical reactions. This problem is also present in drugs, but it is more important in vitamin preparations applied in animal husbandry. These products are in fact mixed to a great mass of fodder, i.e. they are distributed on a large surface. Under these circumstances the decomposition of vitamins is much faster than in the neat form.

The usual stabilization method of vitamins is microencapsulation with gelatine or similar materials.

Problems may arise when microencapsulated vitamins have to be admixed to fodders. Let us assume, e.g. that 100 IU vitamin D_3 has to be mixed in 1 kg of chicken feed. For 1000 chickens for 49 days 3000 kg of nourishment is required. Since 1 IU is equivalent to 0.025 μg vitamin D_3, only 75 mg of vitamin D_3 should be mixed with 3000 kg fodder. If it was dispersed in a non-stabilized form, the vitamin would decompose in a short time. The average size of microcapsules is 0.1 mm, and the required amount of vitamin is equivalent to about 75 000 microcapsules. The even dispersion of this amount, which would ensure that each chicken should receive the required dose evenly in time, is practically impossible—it would require homogenization for an extremely long time. Since cyclodextrin inclusion complexes are easily dispersable powders consisting of particles of a few microns their homogeneous dispersion is a much easier task.

Schlenk et al.[67] were the first to apply the idea of stabilizing fat-soluble vitamins with cyclodextrins. The preparation and biological studies of β-cyclodextrin complexes of vitamin-D_3,[14-118] vitamin-K_3[140,117,109,127] and vitamin-C fatty acid esters[72,279] were reported.

A current problem of animal husbandry is that feeds made with premixes containing all the necessary vitamins, minerals and nutrients, are not sufficiently attractive for the animals. This is especially often the case with the weaning of calves. Using the anethole–β-cyclodextrin inclusion complex as an aromatizer, the fodder consumption of calves can be considerably increased. Free-flowing, light, powderlike cyclodextrin complexes can be easily dispersed in the premix or fodder. Anethole—when homogenized—oxidizes very rapidly or is wasted in other ways.

The palatability of bone-powder[42] or microbial cell-mass[50] (butyric or lactic acid bacteria) can be improved by admixing with cyclodextrin. In fish-feed containing high levels of unsaturated fatty acids, β-cyclodextrin complexation prevents the dispersion of unsaturated fatty acids into the water.[280] Gastric acid can be supplied to piglets by adding cyclodextrin-complexed inorganic acids (HCl or H_3PO_4) to their fodder.

The author hopes that this Chapter has shown that industrial uses of cyclodextrins are now being realized and that many future applications are confidently expected.

References

1. F. Cramer, *Einschlussverbindungen*, Springer Verlag, Berlin, 1954.
2. M. L. Bender and M. Komiyama, *Cyclodextrin Chemistry*, Springer Verlag, Berlin, Heidelberg, New York, 1978.
3. J. Szejtli, *Cyclodextrins and their Inclusion Complexes*, Akadémiai Kiadó, Budapest, 1982.
4. B. É. Dósa and J. Szejtli, *Magyar Kémikusok Lapja, Budapest*, 1981, **36**, 314.

5. D. French, *Adv. Carbohydr. Chem.*, 1957, **12**, 189.
6. S. Kobayashi, K. Kainuma and S. Suzuki, *Denpun Kagaku*, 1975, **22**, 6.
7. K. Horikoshi, T. Ando, K. Yoshida and N. Nakamura, Deut. Offenlegungschrift, 2,453,860, 1975.
8. H. Vakaliu, G. Seres, M. Miskolczy-Török, J. Szejtli and M. Járay, Hung. Patent, 173,825, 1977.
9. E. Flaschel, J. P. Landert and A. Renken, in *Proc. Ist. Int. Symposium on Cyclodextrins, Budapest 1981* (ed. J. Szejtli) Reidel Publ. Co., Dordrecht and Akadémiai Kiadó, Budapest, 1982, p. 41.
10. J. Szejtli and Gy. Sebestyén, *Starch*, 1979, **31**, 385.
11. T. Makita, N. Ojima, Y. Hashimoto, H. Ide, M. Tsuji and Y. Fujisaki, *Oyo Yakuri*, 1975, 449.
12. V. Gergely, Gy. Sebestyén and S. Virág, in, *Proc. Ist. Int. Symposium on Cyclodextrins, Budapest, 1981* (ed. J. Szejtli) Reidel Publ. Co., Dordrecht and Akadémiai Kiadó, Budapest, 1982, p. 109.
13. A. Gerlóczy, A. Fónagy and J. Szejtli, in *Proc. Ist. Int. Symposium on Cyclodextrins, Budapest, 1981* (ed. J. Szejtli) Reidel Publ. Co., Dordrecht and Akadémiai Kiadó, Budapest, 1982, p. 101.
14. G. H. Andersen, F. M. Robbins, F. J. Domingues, R. G. Moores and C. L. Long, *Toxicol. Appl. Pharm.*, 1963, **5**, 257.
15. J. Szejtli, A. Gerlóczy and A. Fónagy, *Arzneim. Forsch.*, 1980, **30**, 808.
16. J. Szejtli, L. Szente and E. Bánky-Elöd, *Acta Chim. Acad. Sci. Hung.*, 1979, **101**, 27.
17. (a) L. Kernóczy, P. Tétényi, E. Mincsovics and J. Szejtli, *Quart. J. Crude Drug. Res.*, 1978, **16**, 153. (b) L. Kernóczy, P. Tétényi and J. Szejtli, *Herba Hungarica*, 1980, **19**, 63.
18. Ogawa and Co. Ltd., Japan Kokai, 81,92,754 and 81,92,755, 1981.
19. J. Szejtli, E. Bánky-Elöd, Á. Stadler, P. Tétényi, I. Héthelyi and L. Kernóczy, *Acta Chim. Acad. Sci. Hung.*, 1979, **99**, 447.
20. J. Szejtli, M. Szejtli and L. Szente, Hung. Patent 180,557, 1980.
21. K. Lindner, L. Szente and J. Szejtli, *Acta Alimentaria*, 1981, **10**, 175.
22. A. Bučko (Bratislava), Personal Communication, 1979.
23. W. I. Rogers and W. M. Whaley, US Patent, 3,061,444, 1962.
24. J. Szejtli, L. Szente, R. Kolta, K. Lindner, T. Zilahy and B. Köszegi, Hung. Patent, 174,699, 1977.
25. T. Yoneda and J. Matsukura, Japan Kokai, 77,18,782, 1977.
26. K. Shimada, I. Tanaka and S. Nagahama, Japan Kokai, 75,29,724, 1975.
27. Y. Suzuki and H. Ikura, Japan Kokai, 75,83,454, 1975.
28. T. Osato, S. Takeuchi, S. Esumi and C. Higashikaze, Japan Kokai, 80,38,338, 1980.
29. T. Konishi, J. Komiya and T. Yoneda, Japan Kokai, 78,18,755, 1978.
30. Lotte, KK. Japan Kokai, 55,034,042, 1980.
31. T. Miyake, Japan Kokai, 78,124,657, 1978.
32. Asahi Denka Kogyo KK., Japan Kokai, 80,87,701, 1980.
33. Suntory Ltd., Japan Maize Products Co., Ltd. Japan Kokai, 80,114,283, 1980.
34. House Food Ind. KK., Japan Kokai, 51,012,970, 1976.
35. A. Konno, M. Miyawaki, M. Misaki and K. Yasumatsu, *Agric. Biol. Chem.*, 1981, **45**, 2341.
36. R. M. Hamilton and R. E. Heady, US Patent, 3,528,819, 1970.
37. Y. Suzuki, Japan Kokai, 75,69,100, 1975.

38. N. B. Helbig, L. Ho, G. E. Christy and S. Nakai, *J. Food Sci.*, 1980, **45**, 331.
39. Y. Akiyama, Japan Kokai, 79,80,463, 1979.
40. Asama Kasei KK., Japan Kokai, 80,71,456, 1980.
41. P. Nagano, Japan Kokai, 80,77,875, 1980.
42. T. Fukinbara, Japan Kokai, 80,44,305, 1980.
43. San-Ei Chem. Ind., Japan Kokai, 76,006,219, 1976.
44. T. Watanabe, H. Ishizone and N. Kawai, Japan Kokai, 78,41,440, 1978.
45. Toyo Ink MFG. KK., Japan Kokai, 52,015,809, 1977.
46. M. Yajima, Deut. Offenlegungschrift, 3,008,663, 1980.
47. House Food Ind. KK., Japan Kokai, 51,148,052, 1976.
48. Sugiyama Ind. Chem. Inst., Japan Kokai, 80,159,760, 1980.
49. Toyo Jozo Co. Ltd., Japan Kokai, 80,156,564, 1980.
50. Y. Miyairi, S. Mori, T. Yoshimasu and K. Horikoshi, Japan Kokai, 79,05,092, 1979.
51. N. Nakamura, *Shokuhin to Kagaku*, 1979, **21**, 110.
52. T. Nakashima, Japan Kokai, 79,145,258, 1979.
53. Takeda Chem. Ind. Ltd., Japan Kokai, 81,78,574, 1981.
54. Takeda Chem. Ind. Ltd., Japan Kokai, 81,48,849, 1981.
55. S. Hashimoto and A. Katayama, Japan Kokai, 75,35,349, 1975.
56. Y. Nawata, K. Yamamoto and M. Sano, Japan Kokai, 77,12,684, 1977.
57. T. Nakashima, Japan Kokai, 79,145,241, 1979.
58. Nippon Kayaku Co. Ltd., Japan Kokai, 81,109,580, 1981.
59. Teijin KK., Japan Kokai, 52,010,448, 1977.
60. Y. Nawata, T. Ofuji and M. Sano, Japan Kokai, 76,126,204, 1976.
61. M. Yanagise, Japan Kokai, 77,296,774, 1977.
62. S. Higeta, Shoyo Co. Ltd., Japan Kokai, 80,71,472, 1980.
63. F. Kawashima, Japan Kokai, 79,117,536, 1979.
64. N. Kuroda and S. Mogi, Deut. Offenlegungschrift, 2,406,376, 1974.
65. T. Fujishiro, N. Nakamura and M. Matsuzawa, Japan Kokai, 79,143,548, 1979; 79,142,282, 1979.
66. S. Nakamura, N. Chujo and T. Takahashi, Japan Kokai, 79,93,686, 1979.
67. W. Schlenk, US Patent, 2,830,040, 1958.
68. Nishii KK., Japan Kokai, 81,26,147, 1981.
69. M. Specht, M. Rothe, L. Szente and J. Szejtli, German (DDR) Patent, 147,615, 1981.
70. Kanegafuchi Chem. KK., Japan Kokai, 53,072,839, 1978.
71. Maruzen Chem. Co. Ltd., Japan Kokai, 81,4250, 1981.
72. M. Sudo, Japan Kokai, 54,113,455, 1979.
73. Kanegafuchi Chem. Ind. KK., Japan Kokai, 52,012,955, 1977.
74. T. Ota and F. Takeda, Japan Kokai, 81,75,060, 1981.
75. M. Hamano, M. Okayasu and H. Sugimoto, *Toketsu oyobi Kanso Kenkyukai Kaishi*, 1979, **25**, 31.
76. T. Hirose, Japan Kokai, 76,88,625, 1976.
77. Toyo Ink MFG. KK., Japan Kokai, 51,118,859, 1976; 51,101,123, 1976.
78. T. Hirose and K. Miwa, Japan Kokai, 76,101,123, 1976, and 76,101,124, 1976.
79. T. Yoshitomi, Y. Tanaka, R. Yoshida, J. Kokumai and T. Hirose, Japan Kokai, 76,140,964, 1977, and T. Yoshitomi, E. Yamazaki, Y. Tanaka and T. Hirose, Japan Kokai, 77,15,809, 1977.
80. Asahi Electrochem Ind. KK., Japan Kokai, 53,113,017, 1978.

81. S. Hashimoto, K. Inoue and T. Watanabe, Japan Kokai, 77,34,941, 1977, and S. Hashimoto, Japan Kokai, 80,05,484, 1980.
82. Sun Star Dentifrice, Japan Kokai, 52,057,338, 1977.
83. Sun Star Dentifrice, Japan Kokai, 51,121,531, 1976.
84. Kanebo Co. Ltd., Japan Kokai, 80,64,511, 1980.
85. T. Yokota and K. Kawahara, Japan Kokai, 79,70,434, 1979.
86. H. Yokota and K. Ito, Japan Kokai, 80,11,527, 1980.
87. I. Horiuchi, Japan Kokai, 53,041,440, 1978.
88. (a) S. Hirano, J. Tsumura, I. Imazaki, M. Ohuchi and H. Kito, Japan Kokai, 75,64,320; 75,64,418, 1975. (b) S. Hirano, I. Tsumura, I. Izeki, H. Kawamura and M. Ohhara, Japan Kokai, 75,63,126, 1975. (c) S. Hirano, J. Tsumura, I. Imazaki, M. Ohuchi and T. Yoneda, Japan Kokai, 77,30,822, 1977.
89. Kokando Co. Ltd., Japan Kokai, 54,080,433, 1979; 80,78,965,1980.
90. S. Masuda, T. Ito, M. Matsuda and H. Kitano, Japan Kokai, 79,80,433, 1979.
91. T. Fukunaga, Japan Kokai, 78,113,040, 1978.
92. Y. Akiyama and K. Miyao, Japan Kokai, 77,114,060, 1977.
93. MSC. YG., Japan Kokai, 53,015,467, 1978.
94. I. Honma, Japan Kokai, 76,133,960, 1976.
95. Y. Nakagawa and I. Honma, Japan Kokai, 76,99,195, 1976.
96. J. Koch, German Patent, 3,020,269, 1981.
97. Kyoshin Co. Ltd., Japan Maize Products Co. Ltd., Japan Kokai, 81,11,955, 1981.
98. J. Koch, in, *Proc. Ist. Int. Symposium on Cyclodextrins, Budapest, 1981* (ed. J. Szejtli) Reidel Publ. Co., Dordrecht and Akadémiai Kiadó, Budapest, 1982, p. 487.
99. J. Koch, *Deut. Offenlegungschrift*, 2,947,742, 1979.
100. A. Bavley and E. W. Robb, US Patent, 3,047,431, 1962.
101. E. W. Robb, J. J. Wesbrook and A. Bavley, *Tobacco Sci.*, 1964, **8**, 3.
102. K. Miwa and Y. Tanaka, Japan Kokai, 75,125,100, 1975; 76,32,799, 1976.
103. S. A. Buckler, R. F. Martel and R. J. Moshy, US Patent, 3,472,835, 1964.
104. K. Arakawa, *Deut. Offenlegungschrift*, 2,527,234, 1974.
105. J. Toda, M. Misaki, A. Konno, T. Wada and K. Yasumatsu, in, *The Quality of Foods and Beverages, Proc. Symp. Int. Flavor Cont.* (eds. G. Charalambous and G. Ingle•.) Academic Press, New York, 1981, p. 19.
106. Á. Stadler-Szöke and J. Szejtli, in, *Proc. Ist. Int. Symposium on Cyclodextrins, Budapest, 1981* (ed. J. Szejtli) Reidel Publ. Co., Dordrecht and Akadémiai Kiadó, Budapest, 1982, p. 377.
107. J. Szejtli, *Starch*, 1981, **33**, 387.
108. K. Uekama, *Yakugaku Zasshi*, 1981, **101**, 857.
109. J. Szejtli, É. Bolla-Pusztai, M. Tardy-Lengyel, P. Szabó and T. Ferenczy, *Pharmazie*, 1983, **38**, 189.
110. D. W. Frank, J. E. Gray and R. N. Weaver, *Amer. J. Path.*, 1976, **83**, 367.
111. J. Serfözö, P. Szabó, T. Ferenczy and A. Tóth-Jakab, in, *Proc. Ist. Int. Symposium on Cyclodextrins, Budapest, 1981* (ed. J. Szejtli) Reidel Publ. Co., Dordrecht and Akadémiai Kiadó, Budapest, 1982, p. 123.
112. K. H. Frömming, R. Sandman and I. Weyermann, *Deut. Apoth. Ztg.*, 1972, **112**, 707.
113. Y. Akiyama, Japan Kokai, 77,114,014, 1977.
114. H. Ikura, K. Takeuchi and S. Nakabachi, Japan Kokai, 76,128,517, 1976.
115. A. Shima and K. Ikura, Japan Kokai, 77,130,904, 1977.
116. S. Takeuchi, N. Kochi, K. Sakaguchi, K. Nakagawa and T. Mizutani, *Agric. Biol. Chem.*, 1978, **42**, 1449.

117. J. Szejtli and É. Bolla, *Starch*, 1980, **32**, 386.
118. J. Szejtli, É. Bolla, P. Szabó and T. Ferenczy, *Pharmazie*, 1980, **35**, 779.
119. Jap. Pharm. Rev., 1980, 1, March.
120. K. Uekama, F. Hirayama, Y. Yamada, K. Inaba and K. Ikeda, *J. Pharm. Sci.*, 1979, **68**, 1059.
121. K. Uekama, *Jap. J. Antibiotics*, 1979, **32**, Suppl. S-103.
122. J. Szejtli, M. Szejtli, Gy. Cseh and I. Stadler, Hung. Patent, 179,141, 1977.
123. M. Hayashi, K. Shuto, O. Takatsuki, Y. Iijima and S. Hikone, Deut. Offenlegungschrift, 2,819,447, 1978.
124. T. Nagai and Y. Murata, Japan Kokai, 48,075,526, 1973.
125. Á. Stadler-Szöke and J. Szejtli, *Acta Pharm. Hung.*, 1979, **49**, 30.
126. E. Akito, Y. Nakajima and M. Horioka, Japan Kokai, 75,129,520, 1975.
127. J. Szejtli, É. Bolla-Pusztai and M. Kajtár, *Pharmazie*, 1982, **37**, 725.
128. J. L. Lach and T. F. Chin, *J. Pharm. Sci.*, 1964, **53**, 942.
129. M. L. Bender, *Trans. N.Y. Acad. Sci.*, 1967, **29**, 301.
130. T. F. Chin, P. H. Chung and J. L. Lach, *J. Pharm. Sci.*, 1968, **57**, 44.
131. L. Szente, M. Gál-Füzy and J. Szejtli, in, *Proc. Ist. Int. Symposium on Cyclodextrins, Budapest, 1981* (ed. J. Szejtli) Reidel Publ. Co., Dordrecht and Akadémiai Kiadó, Budapest, 1982, p. 431.
132. J. Szejtli, L. Szente, G. Kulcsár and Gy. Körmöczy, Hung. Patent, 177,081, 1978.
133. L. Szente and J. Szejtli, *Acta Chim. Acad. Sci. Hung.*, 1981, **107**, 195.
134. S. Miyamoto, A. Mifune, Y. Okada and T. Yoneda, Deut. Offenlegungschrift, 2,422,316, 1974.
135. J. Szejtli, *Starch*, 1977, **29**, 26.
136. K. Noda, K. Furuya, S. Miyata, S. Tosu and T. Yoneda, Deut. Offenlegungschrift, 2,356,098, 1974.
137. J. Szejtli, L. Szente, I. Apostol and A. Gerlóczy, Hung. Patent, 180,183, 1980.
138. I. Habon and J. Szejtli, in, *Proc. Ist. Int. Symposium on Cyclodextrins, Budapest, 1981* (ed. J. Szejtli) Reidel Publ. Co., Dordrecht and Akadémiai Kiadó, Budapest, 1982, p. 413.
139. K. Uekama, N. Matsuo, F. Hirayama, T. Yamaguchi, Y. Imamura and H. Ichibagase, *Chem. Pharm. Bull.*, 1979, **27**, 398.
140. K. H. Frömming and I. Weyermann, *Arzneim.-Forsch.*, 1973, **23**, 424.
141. K. H. Frömming, in, *Proc. Ist. Int. Symposium on Cyclodextrins, Budapest, 1981* (ed. J. Szejtli) Reidel Publ. Co., Dordrecht and Akadémiai Kiadó, Budapest, 1982, p. 367.
142. N. Nambu, M. Shimoda, Y. Takahashi, H. Ueda and T. Nagai, *Chem. Pharm. Bull.*, 1978, **26**, 2952.
143. Sankyo KK., Japan Patent, 1,048,420, 1976.
144. M. Kurozumi, N. Nambu and T. Nagai, *Chem. Pharm. Bull.*, 1975, **23**, 3062.
145. J. Szejtli and L. Szente, *Pharmazie*, 1981, **36**, 694.
146. J. Szejtli, L. Szente, Á. Dávid, S. Virág, Gy. Sebestyén and A. Mándi, Hung. Patent, 176,215, 1978; Deut. Offenlegungschrift, 2,746,087, 1979.
147. N. Nambu, K. Kikuchi, T. Kikuchi, Y. Takahashi, H. Ueda and T. Nagai, *Chem. Pharm. Bull.*, 1978, **26**, 3609.
148. J. Szejtli, A. Gerlóczy, Gy. Sebestyén and A. Fónagy, *Pharmazie*, 1981, **36**, 283.
149. K. Koizumi, M. Miki and Y. Kubota, *Chem. Pharm. Bull.*, 1980, **28**, 319.
150. A. Fónagy, A. Gerlóczy, P. Keresztes and J. Szejtli, in, *Proc. Ist. Int. Symposium*

on *Cyclodextrins, Budapest, 1981* (ed. J. Szejtli) Reidel Publ. Co., Dordrecht and Akadémiai Kiadó, Budapest, 1982, p. 409.
151. J. Szejtli, A. Gerlóczy and A. Fónagy, *Pharmazie*, 1983, **38**, 100.
152. K. Uekama, F. Hirayama, K. Ikeda and K. Inaba, *J. Pharm. Sci.*, 1977, **66**, 706.
153. K. Uekama and F. Hirayama, *Chem. Pharm. Bull.*, 1978, **26**, 1195.
154. M. Hayashi and T. Ishihara, Deut. Offenlegungschrift, 2,128,674, 1971.
155. M. Hayashi and T. Nishibori, Japan Kokai, 72,39,057, 1972.
156. N. Hayasaki, T. Tsutomu, T. Matsumoto and K. Inaba, Deut. Offenlegungschrift, 2,353,797, 1974.
157. Gy. Blaskó, E. Nemesánszky, G. Szabó, I. Stadler and L. Pálos, *Thrombosis Res.*, 1980, **17**, 673.
158. K. Uekama, F. Hirayama, T. Wakuda and M. Otagiri, *Chem. Pharm. Bull.*, 1981, **29**, 213.
159. Y. Hamada, N. Nambu and T. Nagai, *Chem. Pharm. Bull.*, 1975, **23**, 1205.
160. K. Uekama, F. Tujinaga, M. Otagiri, F. Hirayama and M. Yamasaki, *Int. J. Pharm.*, 1982, **10**, 1.
161. I. Stadler, J. Szejtli, I. Habon, Gy. Hortobágyi and I. Kolbe, Swiss Patent Appl., 6222/82.
162. S. Kawamura, M. Murakami, H. Kawada and J. Terao, Japan Kokai, 94,108, 1975.
163. K. Ikeda, K. Uekama and M. Otagiri, *Chem. Pharm. Bull.*, 1975, **23**, 201.
164. Yamanouchi Pharm. KK., Japan Kokai, 75,89,516, 1975.
165. K. Koizumi and K. Fujimura, *Yakugakuzasshi, J. Pharm. Soc. Jpn*, 1972, **92**, 32.
166. K. Koizumi, K. Mitsui and K. Higuchi, *Yakugakuzasshi, J. Pharm. Soc. Jpn*, 1974, **94**, 1515.
167. T. Miyaji, Y. Kurono, K. Uekama and K. Ikeda, *Chem. Pharm. Bull.*, 1976, **24**, 1155.
168. K. H. Frömming, *Pharm. Unserer Zeit.*, 1973, **2**, 109.
169. K. Uekama, F. Hirayama, M. Otagiri, Y. Otagiri and K. Ikeda, *Chem. Pharm. Bull.*, 1978, **26**, 1162.
170. K. Uekama, F. Hirayama, N. Matsuo and H. Kainuma, *Chem. Lett.*, 1978, 703.
171. K. Uekama, T. Fujinaga, M. Otagiri, H. Seo and M. Tsuruoka, *J. Pharm. Dyn.*, 1981, **4**, 726.
172. K. Uekama, T. Fujinaga, F. Hirayama, M. Otagiri, H. Seo and M. Tsuruoka, in, *Proc. Ist. Int. Symposium Cyclodextrins, Budapest, 1981* (ed. J. Szejtli) Reidel Publ. Co., Dordrecht and Akadémiai Kiadó, Budapest, 1982, p. 399.
173. I. Udvardy-Nagy, G. Hantos, Zs. Vida, I. Stadler, Á. Balázs, I. Bartho, M. Trinn, J. Szejtli and I. Habon, French Demande, 8216234, 1983.
174. J. Szejtli, I. Stadler, Á. Balázs, G. Nagy, I. Remport and Zs. Budai, Hung. Patent, 176,074, 1978.
175. A. Stadler-Szöke and J. Szejtli, *Acta Pharm. Hung.*, 1979, **49**, 30.
176. Zeria Shinyaku Kogyo KK., Japan Kokai, 81,92,221, 1981.
177. J. Pitha, *Life Sci.*, 1981, **29**, 307.
178. J. Pitha, S. Zawadzki, F. Chytil, D. Lotan and R. Lotan, *J. Natl. Cancer Inst.*, 1980, **65**, 1011.
179. Nitto Electric Industrial Co. Ltd., Japan Kokai, 81,123,912, 1981.
180. C. G. Gerhold and D. B. Broughton, US Patent, 3,456,028, 1969.
181. W. K. T. Gleim, R. C. Wacker and F. C. Ramquist, US Patent, 3,465,055, 1969.
182. Y. Suzuki and T. Maki, Japan Kokai, 75,151,827, 1975.
183. Y. Suzuki, T. Maki and K. Mineta, Japan Kokai, 75,151,833, 1975.

184. Y. Suzuki, T. Maki and K. Mineta, Japan Kokai, 75,96,530, 1975.
185. Y. Suzuki and T. Maki, Japan Kokai, 75,151,804, 1975.
186. J. Szejtli, É. Bánky-Elöd, I. Stadler, P. Tétényi, I. Héthelyi and L. Kernóczy, Hung. Patent, 174,279, 1977.
187. R. A. Reiners and F. J. Birkhaug, US Patent, 3,491,132, 1970.
188. F. Cramer and W. Dietsche, Chem. Ber., 1959, 92, 1739.
189. H. P. Benschop and G. R. Van der Berg, J. Chem. Soc. Chem. Commun., 1970, 1431.
190. M. Otagiri, K. Ikeda, K. Uekama, O. Ito and M. Hatano, Chem. Lett., 1974, 679.
191. J. Knabe and N. S. Agarwal, Deutsch. Apoth. Ztg., 1973, 113, 1449.
192. M. Mikolajczyk, J. Drabowicz and F. Cramer, J. Chem. Soc. D., 1971, 317.
193. M. Mikolajczyk and J. Drabowicz, Tetrahedron Lett., 1972, 2379.
194. M. Mikolajczyk and J. Drabowicz, J. Am. Chem. Soc., 1978, 100, 2510.
195. J. Michon and A. Rassat, J. Am. Chem. Soc., 1979, 101, 4337.
196. J. Solms and R. H. Egli, Helv. Chim. Acta, 1965, 48, 1225.
197. N. Wiedenhof, Starch, 1969, 21, 163.
198. N. Wiedenhof and R. G. Trieling, Starch, 1971, 23, 129.
199. B. Zsadon, M. Szilasi, K. H. Otta, F. Tüdös, É. Fenyvesi and J. Szejtli, Acta Chim. Acad. Sci. Hung., 1979, 100, 265.
200. E. Smolková-Keulemansová and S. Krýsl, J. Chromatogr., 1980, 184, 347.
201. N. Wiedenhof, J. N. J. J. Lammers, C. L. Van Panthaleon and B. Van Eck, Starch, 1969, 21, 119.
202. É. Fenyvesi, B. Zsadon, J. Szejtli and F. Tüdös, Ann. Univ. Sci. Budapestiensis R. Eötvös Nom., 1979, 15, 13.
203. J. L. Hoffmann and R. M. Bock, Biochemistry, 1970, 9, 3542.
204. J. L. Hoffmann, J. Macromol. Sci. Chem., 1973, 7, 1147.
205. B. Zsadon, M. Szilasi, J. Szejtli, É. Fenyvesi and F. Tüdös, Starch, 1979, 31, 11.
206. Y. Mizobouchi, M. Tanaka and T. Shono, J. Chromatogr., 1981, 208, 35.
207. L. Szente, J. Szejtli, M. Specht and M. Rothe, British Patent, 2,066,265, 1979.
208. I. Tabuse, N. Shimizu and K. Yamamura, Japan Kokai, 79,60,761, 1979.
209. A. M. Harada, M. Furue and S. Nozakura, J. Polym. Sci., Polym. Chem. Edn, 1978, 16, 189.
210. B. Zsadon, M. Szilasi, F. Tüdös and J. Szejtli, J. Chromatogr., 1981, 208, 109.
211. J. A. Thoma, D. E. Koshland, J. Ruscica and R. Baldwin, Biochem. Biophys. Res. Commun., 1963, 12, 184.
212. P. Vretblad, FEBS Lett., 1974, 97, 86.
213. E. László, B. Bánky, G. Seres and J. Szejtli, Starch, 1981, 33, 281.
214. Á. Hoschke, E. László and J. Holló, Starch, 1976, 28, 426.
215. Zs. Budai and J. Szejtli, Magyar Kémikusok Lapja, 1981, 36, 248.
216. T. Goto, Japan Kokai, 74,120,629, 1974.
217. A. T. Noguchi, T. Kazami, M. Sasaki, Y. Tsujimoto, T. Yammamuro and T. Saito, Japan Kokai, 76,03,220, 1970.
218. S. Iwata, Japan Kokai, 80,09,834, 1980.
219. W. Saenger, Angew. Chem. Int. Ed. Engl., 1980, 19, 344.
220. T. Shibusawa, T. Hamoyose and M. Sasaki, Nippon Kagaku Kaishi, 1975, 12, 2171.
221. Pilot Ink Co. Ltd., Japan Kokai, 81,14,569, 1981.
222. Nihon Shashin Insatsu KK., Japan Kokai, 81,36,556, 1981.
223. T. Hirose and K. Miwa, Japan Kokai, 76,108,641, 1976.
224. J. Maeno, Deut. Offenlegungschrift, 2,704,776, 1977.

225. J. Szejtli, L. Szente, G. Kis, K. Jakus, G. Horváth and B. Radványi, Hung. Patent, 174,699, 1977.
226. L. Szente and J. Szejtli, *Acta Chim. Acad. Sci. Hung.*, 1981, **107**, 195.
227. M. Fujiwara, S. Matsuo, M. Kawasaki, Y. Kaneko, T. Masukawa and K. Oishi, Japan Kokai, 78,48,735, 1978.
228. Ricoh Co. Ltd., Japan Kokai, 80,113,036, 1980.
229. I. Shibanai, K. Horikoshi and N. Nakamura, Eur. Pat. Appl., 6,855, 1980.
230. I. Shibanai, K. Horikoshi and N. Nakamura, Eur. Pat. Appl., 14,228, 1980.
231. J. Szejtli, Zs. Budai, Cs. Pap and A. Kerekes, Hung. Pat. 182,217, 1980.
232. A. Kerekes, L. Teke, J. Szejtli and Zs. Budai, Hung. Pat. Appl., 623/81, 1981.
233. H. P. Kaufmann and H. Wessels, *Fette-Seifen-Anstrichm.*, 1964, **66**, 81.
234. H. Wessels and N. S. Rajagopal, *Fette-Seifen-Anstrichm.*, 1969, **71**, 543.
235. W. L. Hinze and D. W. Armstrong, *Anal. Lett.*, 1980, 13, 1093.
236. W. G. Buckert, C. N. Owensby and W. L. Hinze, *J. Liq. Chromatogr.*, 1981, **4**, 1065.
237. K. Uekama, F. Hirayama and H. Kainuma, *Chem. Lett.*, 1977, 1393.
238. J. Debowski, D. Sybilska and J. Jurczak, *J. Chromatogr.*, 1982, **237**, 303.
239. M. Tanaka, Y. Mizobuchi, T. Sonoda and T. Shono, *Anal. Lett.*, 1981, **14**, 281.
240. K. A. Connors and K. S. Albert, *Anal. Chem.*, 1972, **44**, 879.
241. K. A. Connors and J. M. Lipari, *J. Pharm. Sci.*, 1976, **65**, 379.
242. Y. Yokobayashi, Japan Kokai, 79,118,297, 1979.
243. Iwashiro Seiyaku Co. Ltd., Japan Kokai, 80,108,299, 1980.
244. T. Kato, Y. Hiraga, Y. Takahashi and T. Kinoshita, *Chem. Pharm. Bull.*, 1979, **27**, 3073.
245. T. Kinoshita, T. Kato and M. Kato, Japan Kokai, 79,43,791, 1979.
246. T. Kinoshita, F. Iinuma and A. Tsuji, *Biochem. Biophys. Res. Commun.*, 1973, **51**, 666.
247. T. Kinoshita and F. Iinuma, Japan Kokai, 73,43,393, 1973.
248. T. Kinoshita, F. Iinuma and A. Tsuji, *Chem. Pharm. Bull.*, 1974, **22**, 2413.
249. T. Kinoshita, F. Iinuma and A. Tsuji, *Chem. Pharm. Bull.*, 1974, **22**, 2421.
250. T. Kinoshita, F. Iinuma, K. Atsumi, Y. Kanada and A. Tsuji, *Chem. Pharm. Bull.*, 1975, **23**, 1156.
251. E. F. Ullmann and J. E. Lavine, U.S. Patent, 4,121,975, 1978.
252. Y. Takasaki, Y. Suzuki and H. Ohmori, Japan Kokai, 77,90,991, 1977.
253. W. Schlenk, J. L. Gellerman and D. M. Sand, *Anal. Chem.*, 1962, **34**, 1529.
254. T. Kuge and K. Takeo, *Agric. Biol. Chem.*, 1968, **32**, 753.
255. Y. Mizobuchi, M. Tanaka and T. Shono, *J. Chromatogr.*, 1980, **194**, 131.
256. B. Casu, *Cronache di Chimica*, 1967, **15**, 13.
257. S. Miyamoto, A. Mifune, Y. Odaka and T. Yoneda, British Patent, 1,453,801, 1974.
258. A. Mifune, Y. Katsuda and T. Yoneda, Deut. Offenlegungschrift, 2,357,826, 1974.
259. K. Shimada, J. Tanaka and T. Fukuda, Japan Kokai, 75,46,826, 1975.
260. Y. Katsuda and S. Yamamoto, Japan Kokai, 76,81,888, 1976.
261. A. Kariya, *Chagyo Gijutsu Kenkyu*, 1977, **52**, 8.
262. I. Yamamoto, K. Ohsawa and F. W. Plapp, *Nippon Noyaku, Gakkaishi*, 1977, **2**, 41.
263. I. Yamamoto and Y. Katsuda, *Pestic. Sci.*, 1980, **11**, 134.
264. Teijin Ltd., Japan Kokai, 80,149,202, 1980, and 80,149,203, 1980.
265. Y. Katsuda, Japan Kokai, 52,054,021, 1977.

266. I. Yamamoto, A. Shima and N. Saito, Japan Kokai, 76,95,135, 1976.
267. Y. Suzuki, H. Iwasaki and F. Kamimoto, Japan Kokai, 75,89,306, 1975.
268. K. Shimada, I. Tanaka, T. Fukuda and S. Nagahama, Japan Kokai, 75,40,726, 1975.
269. Toyo Ink Mfg. KK., Japan Kokai, 51,118,643, 1976.
270. Mikata Shokai KK., Japan Kokai, 80,149,204, 1980.
271. Mikasa Chem. Ind. Co. Ltd., Japan Kokai, 80,81,806, 1980.
272. J. Szejtli, Zs. Budai, E. Radvány-Hegedüs, L. Papp, Gy. Körmöczy and Cs. Pap, Hung. Patent Appl., 3597/82, 1982.
273. R. Kaneki, I. Yamamoto and T. Udagawa, Japan Kokai, 78,101,531, 1978.
274. Teijin KK., Japan Kokai, 50,058,226, 1975.
275. Zs. Budai and J. Szejtli, Acta Chim. Acad. Sci. Hung., 1981, 107, 231.
276. J. Szejtli, M. Tétényi and P. Tétényi, Deut. Offenlegungschrift, 2,920,568, 1979.
277. J. Szejtli and M. Tétényi, Die Nahrung, 1981, 25, 765.
278. P. Tétényi, in, Proc. Ist. Int. Symposium on Cyclodextrins, Budapest, 1981 (ed. J. Szejtli) Reidel Publ. Co., Dordrecht and Akadémiai Kiadó, Budapest, 1982, p. 501.
279. Toyobo Co. Ltd., Japan Kokai, 80,92,702, 1980.
280. Kanegafuchi Chem. Ind. Co. Ltd., Japan Kokai, 80,120,755, 1980.
281. D. L. Witte, B. J. Pennell, J. K. Pfohl and R. D. Feld, Am. J. Clin. Pathol., 1981, 76, 86.

12 · CYCLOAMYLOSE-SUBSTRATE BINDING

R. J. BERGERON

University of Florida, Gainesville, USA

1. Cycloamyloses

In 1891 Villiers isolated a group of unusual non-reducing oligosaccharides from *Bacillus macerans* grown on a medium rich in amylose.[1] However, it was not until sometime later that Schardiger[2] accomplished the definitive structural elucidation of these compounds, showing them to be cyclic oligosaccharides containing from six to twelve α-1,4 linked glucose units (Fig. 1).

1.1. Structural features

The important structural features to notice in these compounds are the toroidal shape, hydrophobic cavity and outer surface, and hydrophilic faces (Fig. 2). Because of the apparent lack of free rotation about the glycosidic bond which connects the glucose units, the cycloamyloses are not perfectly

INCLUSION COMPOUNDS III
ISBN 0-12-067103-4

R. J. Bergeron

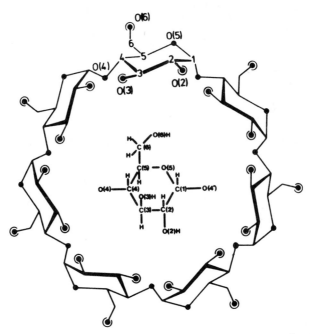

Fig. 1. Cycloheptaamylose. Reproduced with permission from the Journal of Chemical Education.

cylindrical molecules but are somewhat cone shaped. The 6-hydroxyl face is the narrow side while the 2,3-hydroxyl face is somewhat wider. (See Fig. 1 for the numbering scheme). X-ray studies[3] have provided some information about the diameter of several of these oligosaccharides (Table 1). In addition, the X-ray analysis revealed the glucose rings to be in the C1(D) chair conformation. Furthermore, both NMR and ORD studies have shown that the glucose residues maintain this C1(D) conformation in solution[4] although there must certainly be more conformational mobility in solution

Table 1. Molecular dimensions of cycloamyloses

Cycloamylose	Number of glucose residues	Cavity dimensions (Å)	
		Diameter	Depth
Cyclohexaamylose (α-CD)	6	~4.9	~7.9
Cycloheptaamylose (β-CD)	7	~6.2	~7.9
Cyclooctaamylose (γ-CD)	8	~7.9	~7.9

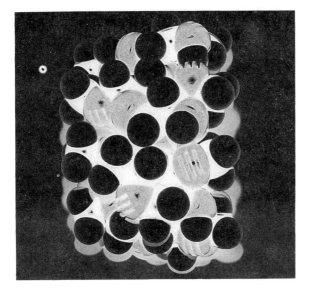

Fig. 2. Front and side view of cyclohexaamylose. Reproduced with permission from the Journal of Chemical Education.

than in a solid matrix. Although a slight increase in the *gauche–gauche* conformer population was observed for cyclohexaamylose (α-CD) relative to methyl α-glucopyranoside, proton chemical shifts and coupling constants observed for α-CD and the methyl pyranoside in aqueous solution at 100 and 220 MHz demonstrated that the C-1 conformation is essentially maintained by both the monomer and hexamer. It has been suggested that the coupling constant data are in agreement with major contributions from *gg* and *gt* to the C(5)–C(6) conformer blend of the hexamer and monomer, although the macrocyclic ring formation does enhance the contribution from *gg*. The picture that emerges is one which allows the individual units of α-CD to be regarded as rigid building blocks.

The X-ray analysis further revealed that the 3-hydroxyl hydrogen of each glucose residue is hydrogen bonded to the 2-hydroxyl oxygen of an adjacent ring. This hydrogen bonding also seems to persist in solution. Even in dimethyl sulphoxide, a strong competitor for intramolecular hydrogen bonding, the 3-hydroxyl proton NMR signal at 5.2δ is unchanged.[5] It is this intramolecular hydrogen bonding which has been suggested to account, in part, for the toroidal shape of the cycloamyloses.

1.2. Inclusion complexes

The most interesting characteristic of these cyclic oligosaccharides is their ability to complex a variety of guest molecules in their cavities. (See Chapter 8, Volume 2). These guest molecules can range in size from noble gases to fatty acid coenzyme A derivatives,[6] the stability of the resulting complexes varying with the size of both guest and host (Table 2). If a substrate is too large, it will simply not fit into the cavity and therefore will not bind to the cycloamylose. Conversely, if the substrate is too small it will pass in and out of the pore with little apparent binding.[7] In fact, this "best fit" phenomenon is used in separating the polysaccharides from each other.[7] Cyclooctaamylose, for example, can be precipitated from aqueous solution

Table 2. *Dissociation constants of cyclohexaamylose* (α-CD) *complexes*[7]

Acid anion guest	Dissociation constant
Acetate	>1.0
Propionate	$5.7 \pm 2.0 \times 10^{-1}$
Isobutyrate	$2.2 \pm 0.3 \times 10^{-1}$
Pivalate	$2.0 \pm 0.6 \times 10^{-1}$
Benzoate	$8.1 \pm 1.0 \times 10^{-2}$
p-Toluenesulphonate	$6.0 \pm 2.0 \times 10^{-2}$
p-Benzoylbenzoate	$4.0 \pm 0.5 \times 10^{-2}$
Cyclohexanecarboxylate	$1.9 \pm 0.3 \times 10^{-2}$
p-Phenylbenzoate	$1.7 \pm 0.2 \times 10^{-2}$
m-Chlorobenzoate	$1.2 \pm 0.2 \times 10^{-2}$
Adamantanecarboxylate	$7.0 \pm 2.0 \times 10^{-3}$
m-Chlorocinnamate	$7.0 \pm 1.0 \times 10^{-3}$
p-Chlorobenzoate	$6.0 \pm 2.0 \times 10^{-3}$
p-Chlorocinnamate	$5.1 \pm 1.0 \times 10^{-3}$

with anthracene, while cycloheptaamylose and cyclohexaamylose remain in solution simply because the anthracene is too large to effectively penetrate their cavities to form the necessary insoluble complex.

The fact that the guest molecule is actually contained within the cavity was first shown by X-ray studies.[8] However, further proof was required before it could be established that this was the case in solution. While a more detailed discussion of molecular disposition within these complexes, as determined from NMR spectroscopic data, will be taken up at a later point, it suffices to say at this point that, when the guest molecule is aromatic, the host–guest disposition in solution is easy to clarify.[9] Since both the 3 and 5 methine protons of the glucose units are pointing inside the cavity, upon complexation of an aromatic substrate, these protons as observed in their NMR spectra are strongly shielded, indicating that they are in the magnetic field of the aromatic pi electron cloud (Table 3). For such shielding to occur, of course, the aromatic substrate must be contained within the cavity in virtually every case. The formation of inclusion complexes therefore appears to occur in solution as well as in a solid matrix. Furthermore, in many cases, after the inclusion complex has formed the cycloamylose catalyses some specific reaction of the enclosed substrate e.g., hydrolysis of *p*-nitrophenylacetate.

As a result of their ability to form such inclusion complexes and to catalyse the reactions of many guest molecules, a great deal of effort has been invested in developing the cycloamyloses as enzyme active site models.[7] (See Chapters 13 and 14, Volume 3). The central theme in these studies has been to expand the number and nature of reactions they may catalyse, as well as to improve their catalytic ability by chemical modifications.[10–13] While such endeavours have met with great success, relatively little has been done as far as the investigation of the nature of the driving forces responsible for substrate binding is concerned. The cycloamyloses offer a particularly good system for such an investigation since the cycloamylose

Table 3. *Substrate induced shifts ($\Delta\delta$) for cycloheptaamylose (β-CD) protons*[9]

Substrate	H-1	H-2	H-3	H-4	H-5	H-6
Benzoic acid	+0.04	+0.04	+0.16	+0.03	+0.19	+0.05
m-Hydroxybenzoic acid	+0.04	+0.04	+0.11	+0.04	+0.19	+0.09
p-Hydroxybenzoic acid	+0.04	+0.04	+0.14	+0.04	+0.21	+0.06
Methyl *p*-hydroxybenzoate	+0.04	+0.03	+0.14	+0.03	+0.21	+0.05
Phenol	+0.06	+0.08	+0.09	—	+0.26	+0.17
m-t-Butylphenol	+0.05	+0.04	+0.20	+0.03	+0.20	+0.13
p-t-Butylphenol	+0.04	+0.04	+0.21	+0.02	+0.03	+0.11

substrate binding constants, as well as the thermodynamic parameters for inclusion, can easily be measured.[6] In addition, it is now possible to modify these oligosaccharides and determine the effect of these structural changes on binding.[9] Because of the symmetry of the cycloamyloses, it is further possible spectroscopically to determine easily changes that occur in the cavity when the molecule binds a substrate.[12] Thus, these features render the cycloamyloses excellent candidates for clathrate binding studies focused on the elucidation of the driving forces responsible for complexation.

2. Early explanations of binding forces

Although there has been a great deal of speculation, the driving forces for cycloamylose substrate complexation have not been well understood until recently.[7,14,15] A major problem is that any explanation[39] of these forces must account for the large differences in binding constants observed for various cycloamylose substrate complexes. The differences in complex stabilities occur not only as the substrate changes but also as the host changes; these dissociation constants can differ by as much as a factor of 6000 (Table 4). There were two popular early explanations for the driving forces responsible for complexation: one involving release of enthalpy-rich water from the cycloamylose cavity;[7] the other, relief of cycloamylose strain energy.[16,17]

Table 4. *Comparison of dissociation constants of cyclohexa(α)- and cyclo-hepta(β)-amylose complexes*

Cycloamylose	Substrates	Dissociation Constant (M)
Cyclohexa-	Sodium propionate	$5.7 \pm 2.0 \times 10^{-1}$
	Methyl orange	1.1×10^{-4}
	p-Nitrophenyl acetate	1.2×10^{-2}
Cyclohepta-	p-Nitrophenyl acetate	6.1×10^{-3}

2.1. Enthalpy-rich water

Bender first suggested that the water molecules associated with the cavity are enthalpy-rich because they cannot have a full complement of hydrogen bonds; thus, the inclusion of a substrate is favoured by release of this high energy water. This down-hill process would then be expected to be associated with a favourable enthalpy term. In fact, measurement of the thermodynamic parameters for the inclusion of various compounds has shown such favourable enthalpy terms exist.

The volumes of the cyclohexa-, cyclohepta- and cycloocta-amylose cavities, and thus the number of water molecules they can accommodate, vary considerably.[39] Assuming the cycloamyloses are nearly cylindrical and employing the molecular dimensions of Cramer,[6] we have *approximated* the volumes, and therefore the number of water molecules that could be contained by each cavity (Table 5). From these calculations it is clear that expanding the cyclic oligosaccharide by one glucose unit at a time could substantially increase the volume of high energy cavity water. However, as the cavity becomes larger and larger, the water displaced should approach bulk water in its energy content and the driving force for complexation should disappear. It is unclear precisely how large the cavity must become before this might be observed.

Table 5. Approximate cycloamylose cavity volumes and calculated number of water molecules contained in solution.[39]

Cycloamylose	Volume $(\text{Å})^3$	Number of H_2O molecules
Cyclohexaamylose (α-CD)	176	6
Cycloheptaamylose (β-CD)	346	11
Cyclooctaamylose (γ-CD)	510	17

In the case of cyclohexaamylose, the fact that small substrates like sodium-propionate,[10] which would displace little cavity water, bind weakly, and larger substrates like *p*-nitrophenylacetate[7] which would displace more cavity water, bind tightly, is consistent with these enthalpy-rich water arguments. Furthermore, the observation that cycloheptaamylose binds a large number of phenyl esters more tightly than cyclohexaamylose is consistent with the idea that, although the substrate fits into the cycloheptaamylose cavity more loosely, it could displace more high energy water. The observation that cyclooctaamylose binds aromatic substrates more weakly than cycloheptaamylose is consistent with the idea that the critical cavity volume might have been reached and passed. The water displaced from the cyclooctaamylose cavity is becoming more like bulk water with the increasing cavity size and the driving force begins to disappear *i.e.*, the water within the cyclooctamylose cavity is no longer "enthalpy rich". Thus, it would seem that water molecules confined to a space larger than the 3.5×10^{-22} cm^3 cavity of the cycloheptaamylose energetically begin to approach those of bulk water.

Clearly, two factors here are at odds. As the cavity increases in size, the enthalpy-rich water displaced would become more like bulk water, but more of this lower energy water would be released. In the case of cyclooctaamylose, even though penetration might be maximal, the water displaced would be of low energy and thus there would be less of a driving force for complexation.

2.2. Release of strain energy

Unfortunately, observed trends in binding which might be attributed to enthalpy-rich water can just as easily be explained by release of strain energy.[39] X-ray studies demonstrated that one of the glucose rings of cyclohexaamylose is orthogonal to the others.[16,17] The 6-hydroxyl group of this ring is pointing into the cavity and thus serves as a hydrogen bond acceptor for cavity water. Extrapolating from these studies in a matrix to solution phenomenon, Saenger has suggested[16] that the strain energy associated with this orthogonal glucose ring is released on complexation of a substrate. Small substrates would, of course, be expected to be less effective in releasing this strain energy since they would not need as much space within the cavity as would larger substrates. The fact that cycloheptaamylose binds the same substrates more tightly than cyclohexaamylose or cyclooctaamylose would have to mean its orthogonal form is of higher energy than that of the hexamer or octamer.

It is possible to get some idea of the extent to which the release of strain energy plays a role in the overall driving forces for complexation. If strain energy is important, changing the 6-hydroxyl position, making it either more or less difficult for a glucose unit to attain its orthogonal position, should alter the ability of the cycloamylose to bind. A comparison, then, of how well the modified cycloamylose and the unmodified cycloamylose bind to the same substrate should indicate the importance of strain. Of course, in order for the data from such an experiment to be meaningful, modification of the 6-hydroxyls must be complete. Failure to accomplish this would mean that the energy required for any remaining unmodified glucose unit to attain orthogonality would be different from that required for the modified units. Thus, the source of strain energy would be unclear.

Space-filling models clearly indicate that methylation of all of the 6-hydroxyls, because of steric hindrance, would make it more difficult for any glucose unit to attain orthogonality. This is true no matter what the orientation of the methyls. However, it is likely that all of the methyl groups would be pointing into the centre of the cavity in order to optimize the hydrophobic interaction between them.[13] Such an orientation would, of course, generate even more steric hindrance to the attainment of orthogonality. However, the methyl groups are not sufficiently large to prevent completely the attainment of orthogonality. This means that if one of the 6-O-methylated rings is orthogonal to the others at the expense of steric interaction, the methylated cycloamylose will be of higher energy, and tighter substrate binding should be observed if release of strain energy is important.

A comparison[39] of the binding of *p*-nitrophenolate to cyclohexaamylose with its binding to dodecakis-2,6-O-methylcyclohexaamylose revealed that the 2,6-O-methylated compound was only 2.9 ± 0.08 times more effective in binding. This number hardly suggested that strain energy was of any great significance. However, at this point it still was not clear if the methylation of the 2-hydroxyls, as well as the 6-hydroxyls, rendered the results inaccurate. In particular, as mentioned earlier, it has been shown by both X-ray[18] and NMR[19] studies that the 3-hydroxyl hydrogens of the glucose units are strongly hydrogen bonded to the 2-hydroxyl oxygens of adjacent units. Further, Rao's[20] calculations have shown that this hydrogen bonding plays a significant role in determining the potential energy of the cycloamyloses. For example, it lowers the energy of cyclohexaamylose by 84 kJ mol^{-1} and the energy of cycloheptaamylose by 163 kJ mol^{-1}. If this hydrogen bonding is also significant in aqueous solution, methylation of the 2-hydroxyl groups could easily affect the mobility of the glucose units by restricting rotation about the C(1)–O–C(4') glycosidic bonds. Any effect on such mobility would influence the strain energy and therefore, the binding ability. To verify that this hydrogen bonding was insignificant, and therefore the binding constant measurements meaningful, we completely removed the hydrogen bond donors, the 3-hydroxyls. This was accomplished by selective and complete 3-O-methylation.

Because of the fact that hydrogen bonding is more important in the structure of cycloheptaamylose by 79 kJ mol^{-1} than in cyclohexaamylose, it seemed that if this hydrogen bonding was of any significance in binding a substrate in aqueous solution, the heptamer would be the place to look for it. We chose to examine several cycloheptaamylose systems to see if this complete 3-O-methylation had any effect on the binding ability of these oligosaccharides. Cycloheptaamylose and its analogues were further chosen because of their synthetic accessibility. The heptakis-3-O-methylcycloheptaamylose was generated in a four step sequence.[21] The 2- and 6-hydroxyls were protected by exhaustive allylation with 3-bromopropene in dimethyl sulphoxide and dimethylformamide with barium oxide and barium hydroxide octahydrate as base. The tetradecakis-2,6-O-allylcycloheptaamylose was then methylated with methyl iodide in dimethylformamide using sodium hydride as base. The resulting tetradecakis-2,6-O-allylheptakis-3-O-methylcycloheptaamylose was then isomerized to the corresponding vinyl ether which was cleaved under neutral conditions with mercuric chloride and mercuric oxide in a water-acetone mixture to give the desired 3-*O*-methylated product.

Comparison[39] of the binding of *p*-nitrophenolate to cycloheptaamylose with its binding to tetradecakis-2,6-*O*-methylcycloheptaamylose and heptakis-3-*O*-methylcycloheptaamylose, reveals only small differences among the three, as summarized in Table 6. Tetradecakis-2,6-*O*-methylcyclohepta-

Table 6. Comparison of the dissociation constants (M) of cycloamylose complexes

Cycloamylose	*p*-Nitrophenolate dissociation constant[a]	Dissociation constant[b] ratio
Cyclohexaamylose[c] (α-CD)	$4.0 \pm 0.8 \times 10^{-4}$	
Dodecakis-2,6-*O*-methylcyclo-hexaamylose	$1.4 \pm 0.2 \times 10^{-4}$	2.9 ± 0.08
Cycloheptaamylose (β-CD)	$1.4 \pm 0.3 \times 10^{-3}$	
Tetradecakis-2,6-*O*-methylcyclo-heptaamylose	$1.4 \pm 0.5 \times 10^{-3}$	1.0 ± 0.41
Heptakis-3-*O*-methylcycloheptaamylose	$1.6 \pm 0.4 \times 10^{-3}$	1.0 ± 0.36

[a] All cycloamylose values determined spectroscopically at 25°, $I = 0.5$, pH, 11.0.
[b] Ratio of the dissociation constants of modified to unmodified cycloamyloses.
[c] Literature value 3.55×10^{-4} M.

amylose binds the substrate about as well as the unmodified cycloheptaamylose but slightly more effectively than heptakis-3-*O*-methylcycloheptaamylose. While it is true that unlike 2,6-*O*-alkylated cyclohexaamylose, the 2,6-alkylated heptamer binds the same as the unalkylated analogue, what is of concern here is how the 3-*O*-alkylated case compares with the parent oligosaccharide. The 3-*O*-methylated case binds *p*-nitrophenolate only slightly less effectively than the unmodified oligosaccharide. If strain energy is at all important, this implies that it is easier for a ring to attain orthogonality and therefore, that the overall system would have less strain energy. However, this difference is particlarly small and hardly enough to disqualify the earlier findings with the dodecakis-2,6-*O*-methylcyclohexaamylose.

Overall, it is clear from this work that relief of ring strain as the driving force for inclusion is not of great importance. It is much more likely that a variety of factors are involved in the overall energetics of inclusion: high energy water, London dispersion forces and possibly, some relief of ring strain. However, before a more quantitative picture of solution binding can be produced, a more accurate knowledge of cycloamylose-substrate disposition is clearly essential, as all of the above arguments depend critically upon how deeply the substrate penetrates the cycloamylose cavities.

3. Elucidation of cycloamylose-substrate disposition

It is impossible to understand cycloamylose-substrate binding forces if the position adopted by the substrate within the cavity is at all unclear. The extent to which sodium *p*-nitrophenolate penetrates the cycloheptaamylose and cyclohexaamylose cavities has been defined by NMR studies of the complexes in aqueous solution. Measurements of changes in the ^{1}H NMR spectra of both the sodium *p*-nitrophenolate guest and the cyclohexaamylose host, along with an intermolecular nuclear Overhauser effect, reveal that this guest only partially penetrates the cyclohexaamylose cavity and does so nitro end first. With cycloheptaamylose, sodium *p*-nitrophenolate penetrates more deeply, but the orientation may be less specific. These findings are in accord with the notion that *both* London dispersion forces *and* possibly removal of high energy water contribute to substrate binding.

3.1. Effect of substrate inclusion on cycloamylose ^{1}H NMR spectra

^{1}H NMR spectra were obtained[33] for a number of samples prepared with different ratios of *p*-nitrophenolate to cycloamylose. The spectrum of free cycloheptaamylose in D_2O was initially assigned by Demarco and Thakkar.[9] These authors found that upon addition of any of a number of aromatic substrates, the H-3 and especially the H-5 resonances of cycloheptaamylose shift upfield because of the diamagnetic anisotropy[27] of the included benzenoid guests. We also observed this upfield shift of the two interior methine protons, H-3 and H-5, of cycloheptaamylose upon addition of sodium *p*-nitrophenolate. However, the situation was considerably different for the cyclohexaamylose case.

The ^{1}H NMR spectrum of cyclohexaamylose, shown in Fig. 3a is almost identical to that of cycloheptaamylose. The assignments were confirmed by decoupling experiments at 250 MHz. However, the effect of *p*-nitrophenolate upon the cyclohexaamylose spectrum differs quite markedly from the situation with cycloheptaamylose. When sodium *p*-nitrophenolate is added to solutions of the cyclic hexamer, large changes occur in the chemical shift of H-3 only; the chemical shift of H-5 is affected very little. This result is

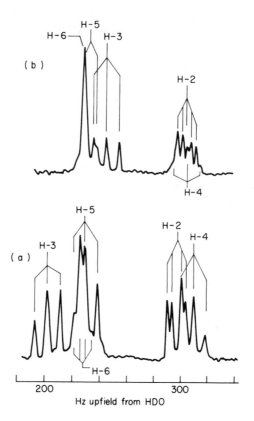

Fig. 3. ^{1}H correlation NMR spectra at 250 MHz. of (a) 0.005 mol dm^{-3} cyclohexaamy-lose and (b) 0.005 mol dm^{-3} cyclohexaamylose-0.005 mol dm^{-3} sodium p-nitro-phenolate. Both samples were prepared using pD 11 phosphate buffer in D$_2$O as described in ref. 33. The resolution in both spectra was digitally enhanced: spectrum (b) is presented with a smaller vertical scaling factor. Reproduced with permission from Bioorganic Chemistry.

illustrated in Fig. 3b which shows the 250 MHz spectrum of cyclohexaamy-lose (except for the anomeric proton resonance), which is 80% bound with p-nitrophenolate (assuming an association constant of the 1:1 complex of 4.5×10^3 M^{-1}). Comparing this spectrum to the one below it, that of free cyclohexaamylose, it can be seen that, while most of the cyclohexaamylose resonances have shifted somewhat upfield relative to the internal lock frequency (the HDO signal), only the resonance of H-3 has shifted more than 0.1 ppm. In particular, the resonance of H-5 has shifted only slightly, if at all.

3.2. Effect of cycloamyloses on substrate ^1H NMR spectra

Figure 4 shows the effect of increasing the fraction of *p*-nitrophenolate bound to cycloamylose upon the ^1H chemical shifts of *p*-nitrophenolate.[33] Large downfield shifts are induced for the resonances of both sets of *p*-nitrophenolate protons by cyclohexaamylose. The *meta* proton resonance of *p*-nitrophenolate is shifted most drastically, approximately twice as much as is the *ortho* proton resonance for a given fraction of *p*-nitrophenolate bound to cyclohexaamylose. By comparison, the binding to cycloheptaamylose has relatively little effect on the chemical shifts of *p*-nitrophenolate. As shown in Fig. 4, in this case the *ortho* proton resonance is shifted slightly downfield, while the *meta* proton resonance is shifted slightly upfield.

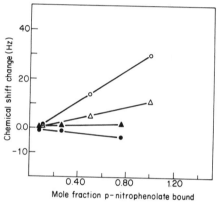

Fig. 4. Effects of cycloheptaamylose and cyclohexaamylose complexation of sodium *p*-nitrophenolate on the ^1H NMR spectrum of sodium *p*-nitrophenolate. Samples were prepared as described in ref. 33. The sodium *p*-nitrophenolate concentrations were varied between 0.005 and 0.050 mol dm^{-3} and the cycloamylose concentrations between 0.005 and 0.02 mol dm^{-3} for cyclohexaamylose and 0.005 and 0.017 mol dm^{-3} for cycloheptaamylose. The cycloamylose-induced changes in chemical shifts of the sodium *p*-nitrophenolate are plotted relative to free sodium *p*-nitrophenolate: ($\triangle$,$\bigcirc$) cyclohexaamylose system: ($\blacktriangle$,$\bullet$) cycloheptaamylose system: ($\triangle$,$\blacktriangle$) *ortho* protons of sodium *p*-nitrophenolate: ($\bigcirc$,$\bullet$) *meta* protons of sodium *p*-nitrophenolate. Reproduced with permission from Bioorganic Chemistry.

3.3. Intermolecular nuclear Overhauser enhancements

A ^1H homonuclear Overhauser experiment was done on samples of *p*-nitrophenolate plus cyclohexaamylose at both 100.1 and 250 MHz. In both cases, substantial enhancement of the *meta* proton resonance of *p*-nitro-

phenolate was observed upon saturation of the cyclohexaamylose resonances upfield of the HDO resonance. In these experiments there was no
significant effect on the intensity of the *ortho* proton resonance of *p*-
nitrophenolate. In Fig. 5 are shown 100.1 MHz spectra from such a NOE
measurement for a sample in which the *p*-nitrophenolate was 99% bound
to cyclohexaamylose; the measured enhancement was 34% for the *meta*
proton resonance and 1% for the *ortho* resonance. At 250 MHz, the greater

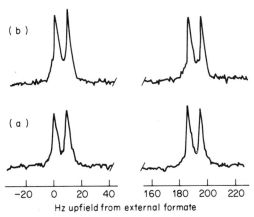

Fig. 5. A ^{1}H homonuclear intermolecular NOE experiment on sodium *p*-nitropheno-
late and cyclohexaamylose: (a) second radiofrequency far from any resonances of
sample: (b) second radiofrequency saturating cyclohexaamylose C–H resonances.
In each spectrum the left-handed doublet is due to *p*-nitrophenolate *meta* protons
and the right-hand doublet to *p*-nitrophenolate *ortho* protons. The sample was prepared using pD 11 phosphate buffer as described in ref. 33, 0.020 mol dm^{-3} in cyclohexaamylose and 0.005 mol dm^{-3} in *p*-nitrophenolate. Both spectra were obtained
at 100.1 MHz in Fourier transform mode using 65 pulses and the procedure described
in ref. 33. Reproduced with permission from Bioorganic Chemistry.

separation of the peaks made it possible to saturate particular individual
cyclohexaamylose resonances. As shown in Table 7 irradiation of the H-3
resonance of cyclohexaamylose produced the largest enhancement of the
p-nitrophenolate resonances.[33]

3.4. Evaluation of spectral data

From the observed changes[33] in the chemical shifts for both the *p*-nitro-
phenolate guest and the cyclohexaamylose host as the host : guest ratio was
varied, as well as from the known forward and reverse rate constants for

Table 7. 250 MHz intermolecular nuclear Overhauser enhancements

m[a]	Cyclohexaamylose proton(s) irradiated[b]	p-Nitrophenolate NOE's[c]	
		Meta	Ortho
0.86[d]	H-3	8	−2
0.99[e]	H-3	9	2
0.99[e]	H-5	3	0
0.99[e]	H-2, H-5	0	0

[a] Mole fraction of p-nitrophenolate bound to cyclohexaamylose.
[b] Second radiofrequency was centred at the resonance position of the indicated proton, using the minimum power level necessary to saturate that resonance.
[c] Enhancements in percents; estimated accuracy ±3%.
[d] Sample composition: cyclohexaamylose, 0.005 mol dm^{-3}; p-nitrophenolate, 0.005 mol dm^{-3}; solvent, pD 11 buffer as described in [33].
[e] Sample composition: cyclohexaamylose, 0.020 mol dm^{-3}; p-nitrophenolate, 0.005 mol dm^{-3}; solvent, pD 11 buffer as described in [33].

the association equilibrium $(5.2 \times 10^8 \, M^{-1} s^{-1}$ and $1.3 \times 10^5 \, s^{-1}$, respectively),[6] it is clear that the system is within the NMR chemical shift fast-exchange limit.[24] That is, the p-nitrophenolate resonances appear at the average of the chemical shift of free p-nitrophenolate and the chemical shift of p-nitrophenolate bound in each possible orientation to cyclohexaamylose, weighted by the fractional population of p-nitrophenolate molecules in each environment. Likewise, each of the cyclohexaamylose resonances occurs at its fast-exchange position, weighted by the fraction of empty cyclohexaamylose molecules and the fraction of cyclohexaamylose molecules which have p-nitrophenolate guests.

Considering first the observation that only the resonance of H-3 of cyclohexaamylose shifts significantly upfield when p-nitrophenolate is present in the solution, we conclude that p-nitrophenolate only partially penetrates the cyclohexaamylose cavity. If full penetration had occurred, the resonance of H-5 of cyclohexaamylose would also be shifted upfield at least to some extent, and this was not observed. However, it is clear from the shift of H-3 that it is indeed the "open" face of cyclohexaamylose which is the site of inclusion of p-nitrophenolate. The conclusion that only partial penetration of the cyclohexaamylose cavity by p-nitrophenolate occurs is consistent with model building, which indicates that, although possible, complete penetration of the cyclohexaamylose cavity by p-nitrophenolate is unlikely. The substrate will not fit entirely into the cavity without considerable distortion of the cyclic oligomer. Furthermore, only two general orientations of the p-nitrophenolate with respect to the cavity allow any penetration at all, namely, the orientations which have either the hydroxyl or the nitro end of the guest projecting into the cavity.

In order to ascertain what, if any, orientational preference exists, we turn next to consideration[33] of the changes in the p-nitrophenolate spectrum caused by binding to the cyclohexaamylose. Here, we must first account for the marked downfield shifts observed for each of the p-nitrophenolate resonances and then for the fact that the resonances of the *meta* protons are shifted about twice as much as those of the *ortho* protons. Similar downfield shifts have been reported by MacNicol[25] for the *ortho* and *meta* resonances of an aqueous solution of p-cymene and cyclohexaamylose, although no assignments of the aromatic peaks were made. Such downfield ^{1}H magnetic resonance shifts can be induced in the spectrum of one molecule when binding to another by several physical mechanisms: diamagnetic anisotropy of particular bonds or regions of the host,[26] van der Waals shifts,[23] or steric perturbations.[28] From the limited amount of experimental data available, while it is not possible to assign the observed downfield shifts in p-nitrophenolate and p-cymene caused by complexation of cyclohexaamylose to a particular mechanism, two conclusions are inescapable. First, whatever the mechanism, there must be intimate contact between the p-nitrophenolate protons whose resonances are being shifted and an area of the host molecule which is causing the downfield shift. Second, and more important, is the implication that the complexation of p-nitrophenolate to cyclohexaamylose is not random but does, in fact, have a strong orientational preference. If the complexation occurred randomly between the two possible orientations, both *ortho* and *meta* protons of p-nitrophenolate would exhibit approximately equal shifts. Since the *meta* protons, however, are shifted much more than the *ortho* protons, then presumably, it is the *nitro* end of p-nitrophenolate which preferentially enters the cyclohexaamylose cavity.

The intermolecular nuclear Overhauser experiments described previously, and listed in Table 7, identify the cause of the differential line broadening of the p-nitrophenolate resonances and confirm the nature of the preferred orientation of p-nitrophenolate in its complex with cyclohexaamylose. While the theory of intermolecular NOE's in rapidly exchanging systems has been presented by Noggle and Schirmer,[22] Balaram *et al.*[29] considered the specific case of observing the resonances of a small molecule exchanging between its environment free in solution and a position bound to a macromolecule while saturating resonances of the macromolecule. However, the most complete discussion of intermolecular NOE's in bimolecular systems in equilibrium has been presented by Bothner-By and Gassend.[30]

Briefly, the most significant points to keep in mind for our purposes, regarding intermolecular NOE's are that, as with intramolecular NOE's, the magnitude of the observed effect depends upon the relative extent to which the nucleus whose resonance is being observed, is relaxed by the nucleus whose resonance is being saturated. Furthermore, while quantitative interpretation of intermolecular NOE's in exchanging systems may in general be quite complex, the situation is simplified when, as in the present case, the chemical shift fast-exchange approximation holds. The most important conditions for the validity of the interpretation which follows are: first, that the bound lifetime of the species observed in the NOE experiment be long enough for significant intermolecular relaxation to occur; and second, that the mole fraction of the observed species in the bound environment be large enough that the NOE intensity changes will be seen in the averaged (free plus bound) spectrum.

Since the resonances of the *meta* protons of *p*-nitrophenolate showed enhancements when the cyclohexaamylose resonances were saturated, the cycloamylose protons contribute substantially more to the relaxation of the *meta* protons than to the *ortho p*-nitrophenolate protons. This also establishes that it is indeed increased intermolecular relaxation of the *meta p*-nitrophenolate protons, and not exchange broadening, which causes the observed greater increase in the linewidth of these resonances in the presence of cyclohexaamylose. It is well known that internuclear dipole–dipole relaxation rates have a $1/r^6$ dependence on the distance r between the relaxed nucleus and the nucleus causing relaxation. Thus, the preferred orientation of *p*-nitrophenolate in the bound complex is firmly established as the one having the nitro group pointing into the cavity, for only in this orientation can the differential line broadening NOE's be explained.

The NOE's at 250 MHz listed in Table 7 also offer further confirmation of the proposal that the depth of penetration of *p*-nitrophenolate into the cyclohexaamylose cavity is not enough to bring the protons of *p*-nitrophenolate close to H-5 of the cyclohexaamylose.[33] Irradiation of H-5 caused significantly smaller enhancements of the *p*-nitrophenolate resonances than did saturation of H-3 of cyclohexaamylose, implying a substantially greater distance between H-5 and the *p*-nitrophenolate protons. Furthermore, nonspecific contact between the exterior of the cyclohexaamylose molecule and the protons of *p*-nitrophenolate can be eliminated as causing relaxation of *p*-nitrophenolate since saturation of H-4 and H-2 caused no enhancement of *p*-nitrophenolate at all. Balaram *et al.*[29] showed, in the case of small molecules binding to macromolecules, that, as the quantity $2\pi\nu_0\tau_c$ increases, where τ_c is the rotational correlation time of the macromolecular complex and ν_0 is the resonance frequency, the NOE would be diminished and eventually become negative i.e., the signal of the nucleus being observed would *decrease* in intensity when resonances of the macromolecule were saturated. The molecular weight of the cyclohexaamylose *p*-nitrophenolate complex (about 1100) is such that in aqueous solution τ_c of the complex is calculated from microviscosity theory to be about 1.1×10^{-10} s, assuming the solution is sufficiently dilute that the viscosity is not greatly increased above that of water, so $2\pi\nu_0\tau_c \simeq 0.18$ for $\nu_0 = 250$ MHz, and $2\pi\nu_0\tau_c \simeq 0.07$ for $\nu_0 = 100$ MHz. This would account for the substantially lower NOE's observed at the higher frequency. This variation in magnitude of the NOE's observed at the two resonance frequencies, combined with the large down-field shift of the p-nitrophenolate resonances upon complexation, enables an additional comment about the nature of the bound complex to be made. These data suggest the tentative conclusion that p-nitrophenolate binds quite rigidly to cyclohexaamylose, in the sense that during the lifetime of an individual bimolecular complex, a substantial reduction of rotation of the p-nitrophenolate within the cavity occurs. This conclusion will be verified by measuring the extent to which the rotational correlation time of p-nitrophenolate changes upon going from its solution environment to that of the bound complex. Such ^{13}C NMR relaxation time measurements, which enabled the change in correlation time to be determined, will be discussed in the final section of this review, that which discusses the dynamics of complexation.

3.5. Cycloheptaamylose-sodium p-nitrophenolate binding[33]

For reasons identical to those discussed above with respect to the complexation of p-nitrophenolate to cyclohexaamylose, it is evident that the association equilibrium between p-nitrophenolate and cycloheptaamylose is also within the NMR chemical shift fast-exchange limit. The fact that the resonances of both interior methine protons of cycloheptaamylose, H-5 and H-3, are shifted upfield upon addition of p-nitrophenolate is strong evidence that with this cycloamylose, p-nitrophenolate fully penetrates the cavity. Furthermore, the fact that both the *ortho* and *meta* proton resonances of p-nitrophenolate are broadened also indicates that the protons in both positions on the aromatic ring of the guest molecule experience equal intermolecular dipolar relaxation in the complex. However, since no large shifts in the p-nitrophenolate spectra are observed upon complexation with cycloheptaamylose, we infer that in this case the distance between guest and host is greater than it was in the complex of p-nitrophenolate with cyclohexaamylose. Also, from the data plotted in Fig. 4 it is not evident that there is a strong orientational preference for p-nitrophenolate in the cycloheptaamylose cavity.

4. Physical factors influencing binding forces

As mentioned initially, any explanation of inclusion driving forces must, of course, be consistent with the large differences in dissociation constants observed for various cycloamylose substrate complexes. The data[33] in Table 8 suggest that, as might be expected, both the size and charge of the substrate are important in binding.

Table 8. Cycloamylose-substrate dissociation constants

Cycloamylose	Substrate	Dissociation constant (M)
Cyclohexaamylose (α-CD)	Sodium propionate	$5.7 \pm 2.0 \times 10^{-1}$
	p-Nitrophenyl acetate	1.2×10^{-2}
	Sodium p-nitrophenolate	$4.0 \pm 0.8 \times 10^{-4}$
Cycloheptaamylose (β-CD)	p-Nitrophenyl acetate	6.1×10^{-3}
	Sodium p-nitrophenolate	1.4×10^{-3}

Of particular interest is the fact that sodium p-nitrophenolate binds more tightly to cyclohexaamylose than does neutral p-nitrophenylacetate. This could be interpreted as solely an increase in the dipole–induced dipole interactions, the London dispersion forces between the guest and host caused by an increase in the substrate's dipole moment as its charge increases. However, there are discrepancies when one compares the binding of the same substrate to different cavities; for example, p-nitrophenylacetate binds more tightly to cycloheptaamylose than to cyclohexaamylose, while sodium p-nitrophenolate binds more effectively to cyclohexaamylose than to cycloheptaamylose. An explanation which disposes of those apparent anomalies calls on *both* London dispersion forces and expulsion of high-energy water from the cycloamylose cavity. However, the validity of such an explanation depends entirely on an understanding and consideration of where the substrates are located in the respective cycloamylose cavities in solution.

As discussed above, sodium p-nitrophenolate can only penetrate the cyclohexaamylose cavity effectively in one of two different orientations (Fig. 6): either oxygen or nitro end first. A third orientation with penetrating *ortho* and *meta* positions is unreasonable, simply because very little of the substrate would fit into the cavity. However, as fully discussed, our ¹H NMR studies clearly indicated that sodium p-nitrophenolate penetrated the cyclohexaamylose cavity at the wide 2,3-hydroxyl side, nitro end first and

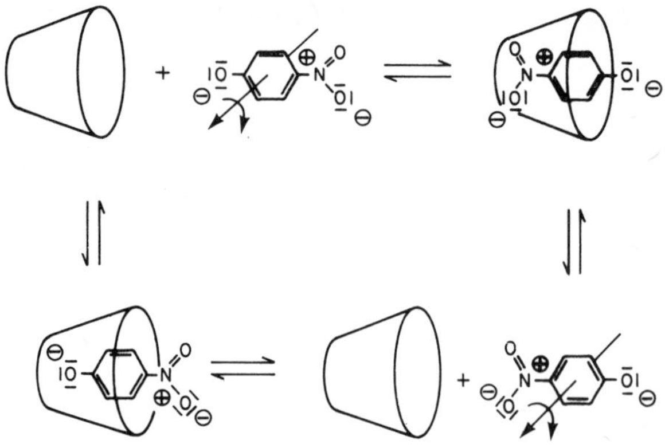

Fig. 6. The two most likely dispositions for cycloamylose–sodium p-nitrophenolate complexes. Reproduced with permission from the Journal of the American Chemical Society.

only to the extent that the *meta* protons were in close proximity to the cycloamylose H-3 protons.

The significance of this orientation to substrate binding is, of course, relevant to the nature of the binding forces. Two possibilities exist with respect to substrate orientation: either the sodium p-nitrophenolates or p-nitrophenols can only bind in the cavity nitro end first, or this orientation represents an energetically slightly more favourable disposition and, if sterically prohibited, the substrate can reorient itself and penetrate the cavity hydroxyl oxanion end first and bind. From the data[32] in Table 9, it is clear that the substrate can only bind in the cavity nitro end first. Using sodium p-nitrophenolate as the standard, introduction of methyl groups at the 2 and 6 positions of p-nitrophenol weakens the binding only slightly by a factor of about 2.4. However, introduction of a single methyl group at the 3 position of p-nitrophenol weakens the binding substantially by a factor of about 100, while introduction of methyl groups at both the 3 and 5 positions completely inhibits binding by the usual mechanism. With the neutral compounds, introduction of methyl groups at the 2 and 6 positions of p-nitrophenol actually enhances the binding slightly by about a factor of 4; however, introduction of methyl groups at the 3 or at the 3 and 5 positions of p-nitrophenol prevents any binding of the substrate in the cavity. In each case, the phenol binds in the cavity more weakly than the corresponding anion. Sodium p-nitrophenolate binds about 130 times tighter than the phenol, while sodium 3-methyl-4-nitrophenolate has a K_D of $4.2 \pm 0.4 \times 10^{-2}$ M and the corresponding phenol does not bind at all.

Table 9. *Cyclohexaamylose-substrate dissociation constants*[32]

Substrate	$K_D^{(a)}$ (M)	$K_D^{(b)}$ (M)	$K_D^{(c)}$ (M)
OH—C₆H₄—NO₂ (4-nitrophenol)	5.3×10^{-2}		
O⁻—C₆H₄—NO₂ (4-nitrophenolate)	4.0 ± 0.8 $\times 10^{-4}$	3.7 ± 0.2 $\times 10^{-4}$	6.3 ± 0.4 $\times 10^{-4}$
2,6-dimethyl-4-nitrophenol (OH, CH₃, CH₃, NO₂)	1.76 ± 0.18 $\times 10^{-3}$		
2,6-dimethyl-4-nitrophenolate (O⁻, CH₃, CH₃, NO₂)	9.42 ± 1.1 $\times 10^{-4}$	1.96 ± 0.53 $\times 10^{-4}$	1.42 ± 0.58 $\times 10^{-3}$
2-methyl-4-nitrophenol (OH, CH₃, NO₂)	N.B.		
2-methyl-4-nitrophenolate (O⁻, CH₃, NO₂)	4.2 ± 0.4 $\times 10^{-2}$	3.5 ± 0.16 $\times 10^{-2}$	$>10^{-1}$

Table 9—(continued)

Substrate	$K_D^{(a)}$ (M)	$K_D^{(b)}$ (M)	$K_D^{(c)}$ (M)
OH / CH$_3$—(ring)—CH$_3$ / NO$_2$	N.B.		
O$^-$ / CH$_3$—(ring)—CH$_3$ / NO$_2$	N.B.	N.B.	N.B.

(a) (UV), $25 \pm 0.1°$ C).　　　　　N.B.—No Binding
(b) (NMR, $23 \pm 0.5°$ C).
(c) (opt.rot., $25 \pm 0.1°$ C).

Although solvation of the hydroxyl oxanion of the sodium p-nitrophenolates would be best served by having it point out into solution, the neutral compounds should certainly be able to penetrate and bind in the cavity hydroxyl end first; however, they do not. This means either the phenolic hydroxyl cannot penetrate the cavity for solvation reasons or the substrate dipole has a strong orientational preference.

4.1. Significance of substrate molecular orientation

The extent to which cyclohexaamylose binds p-nitrophenol, 2,6-dimethyl-4-nitrophenol, 3,5-dimethyl-4-nitrophenol, 3-methyl-4-nitrophenol, and their corresponding sodium phenolates has thus been shown to be dependent on how effectively the substrate can penetrate the cavity nitro group first.[32] These results both bear out earlier intermolecular nuclear Overhauser effect studies on cycloamylose-substrate penetration and establish asymmetric penetration as a general phenomenon experienced by substituted p-nitrophenol and sodium p-nitrophenolate guests.

The three different explanations for cycloamylose-substrate binding energy: release of cycloamylose strain energy,[16] release of high-energy cavity water, and London dispersion forces interactions,[7,31] can now be evaluated. Although the first two proposals are somewhat difficult to separate because they predict similar phenomenological results, the third concept is easy to partition from the first two. Both the strain energy and high energy water theories suggest that no matter how the substrate penetrates the cavity, as long as it relieves the ring strain or displaces the high-energy cavity water, it should bind. They also suggest that the more effective a substrate is at performing these tasks, the tighter it should bind. For example, the observation that benzoic acid binds about 90 times more effectively in the cavity than phenol would be explained by assuming benzoic acid penetrates with the carboxyl group first, just as the phenol would be expected to penetrate with the hydroxyl end first, but the carboxyl group is more effective at "filling the cavity."

However, the binding constants for the *p*-nitrophenol and the *p*-nitrobenzoic[15] acid cyclohexaamylose complexes reveal some disturbing inconsistencies. The *p*-nitrophenol binds in the cavity more tightly than phenol while the *p*-nitrobenzoic acid binds more weakly than benzoic acid. Although the nitro group of the *p*-nitrophenol would "fill the cavity" somewhat more effectively, and therefore the phenol would bind more tightly, the fact that *p*-nitrobenzoic acid binds more weakly than benzoic acid defies any explanation of this type. Furthermore, neither theory suggests that *p*-nitrophenolate anions should bind more tightly than the corresponding neutral species.

The above findings then, are very difficult to rationalize in terms of both the high-energy water picture as well as in terms of the strain energy model. It is clear that *p*-nitrophenol-like substrates are willing to bind in the cavity in only one orientation, nitro end first, and anything that prevents this weakens binding substantially. Furthermore, we continue to see the *p*-nitrophenolate anions binding more tightly than the corresponding neutral compounds. After consideration of all these findings, the picture that emerges is one which minimizes the importance of both high energy cavity solvent and strain energy and amplifies the role of London dispersion forces or some dipole–dipole interaction. It is, of course, important to recognize that the London dispersion forces' contribution to the overall binding energy will vary with the polar nature of the substrate and, thus, will be less important for less polar substances.

4.2. Significance of charge

The significance of dipole–induced dipole interactions having thus been established, one fact is undeniable: the charge on the substrate as well as its direction of penetration must be of great importance in regulating the stability of the cycloamylose-substrate complexes formed.[32,33] The relationship between these factors still requires clarification e.g., sodium p-nitrophenolate binds 130 times more tightly in the cycloamylose cavity than the neutral phenol, while just the opposite is true of benzoic acid and its anion with the carboxylate anion binding 82 times more loosely.[32,14] An understanding of this apparent anomaly would help clarify the relationship between the direction of substrate penetration and charge on substrate binding. We therefore extended our investigations to the latter system.

Both benzoic acid and sodium benzoate were first shown to form 1 : 1 AB complexes with cyclohexaamylose.[48] The direction of sodium benzoate and benzoic acid penetration of the cyclohexaamylose cavity in aqueous solution was then determined by a ^{1}H NMR study of the respective cycloamylose complexes. The protons inside the cyclohexaamylose cavity, the C-3 and C-5 methine protons, experience the largest changes in chemical shift on benzoic acid complexation. When the cyclohexaamylose is 83% bound by benzoic acid, the C-3 methine protons move upfield approximately 87 Hz, while the C-5 methine protons move downfield 24 Hz. However, the chemical shift of the C-1 anomeric protons and the C-6 methylenes hardly change at all. The shielding of the cycloamylose's C-3 methine protons is possibly a result of their being within the magnetic field of the benzoic acid's aromatic pi cloud; however, the mechanism for the C-5 methine deshielding is not so obvious. This C-5 methine deshielding has also been observed for the p-iodoaniline and sodium 2,6-dimethyl-4-nitrophenolate complexes of cyclohexaamylose.[32,34] There are a number of possible mechanisms which could explain this: diamagnetic anisotropy of particular bonds or regions of the host,[26] van der Waals shifts,[23] or steric perturbation.[28] However, because of the diverse electronic nature of the guest molecules causing the deshielding, it seems likely that either the van der Waals shifts or steric perturbation, not diamagnetic anisotropic shielding, are responsible for the shifting.

The fact that the C-6 methylenes are not being shifted at all confirms two structural features about the complex: first, that the C-6 methylene protons do not lie within the deshielding magnetic field of the benzoic acid's aromatic pi cloud; and secondly, that they are too far from the carboxyl group to experience van der Waals or steric perturbation induced shifts. Lehn has shown that there is considerable freedom of movement about the cyclohexaamylose C(5)–C(6) bond.[35] Therefore, the absence of any change in the chemical shift of the C-6 methylene protons cannot be attributed to the protons being held "pointing away" from the substrate.

The most notable feature of the changes in the ^{1}H NMR spectrum of benzoic acid on cycloamylose complexation is in the relative magnitudes of the changes in the *ortho* and *meta* protons. The *ortho* protons which penetrate the cavity sustain about 2.5 times the change in chemical shift of the *meta* protons (Fig. 7). A similar phenomenon was observed for the

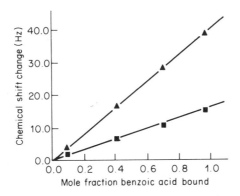

Fig. 7. A plot of the change in chemical shifts of the aromatic protons of benzoic acid vs. the percent bound benzoic acid: (▲) *ortho* protons; (■) *meta, para* protons (from ref. 48). Reproduced with permission from the Journal of the American Chemical Society.

changes in the chemical shifts of the *meta* and *ortho* protons of sodium *p*-nitrophenolate on cyclohexaamylose complexation i.e., the penetrating *meta* protons experience about 2.8 times the change in chemical shift of the *ortho* protons (Fig. 4). The downfield ^{1}H magnetic resonance shifts of the substrate could, as with the cycloamylose's C-5 methine proton downfield shifts, be explained by several physical mechanisms: diamagnetic anisotropy of particular bonds or regions of the host,[26] van der Waals shifts,[23] or steric perturbation.[28] However, due to the limited amount of experimental data available, assignment of a particular mechanism was again not possible.

Based on changes in ^{1}H chemical shifts of the host and guest molecules and on an intermolecular nuclear Overhauser effect, we were able to establish conclusively that the sodium *p*-nitrophenolate is sitting in the cyclohexaamylose cavity nitro end first at the 2,3-hydroxyl side. In view of these findings, our observation that the benzoic acid *ortho* protons sustained the largest change in chemical shift on cyclohexaamylose complexation suggested it might be penetrating the cavity carboxyl group first. This idea was borne out by an intermolecular nuclear Overhauser experiment in which, upon irradiation of the cycloamylose, the *ortho* protons' area was enhanced by 31% while the *meta–para* proton multiplet remained effectively unchanged.

The sodium benzoate-cyclohexaamylose complex is substantially less stable than the corresponding benzoic acid complex. This is reflected in the magnitude of the chemical-shift changes generated in the host and guest molecules on complexation. However, the dissociation constant determined[48] by measuring the changes in the sodium benzoate's ^1H chemical shifts ($1.02 \pm 0.3 \times 10^{-1}$ M) is in good agreement with the literature value ($8.1 \pm 1.0 \times 10^{-2}$ M).[14] Although the direction of the changes in chemical shift of the guest molecules is the same for the cycloamylose-sodium benzoate and cycloamylose-benzoic acid complexes, there are some differences in the magnitude of the changes in chemical shifts. As with benzoic acid, the sodium benzoate's *ortho* protons are shifting more than the *meta* protons. However, the ratio of *ortho* to *meta* proton shifts for the fully bound substrate is greater for the benzoic acid complex, 2.56 vs. 1.41. This can be interpreted as a more random penetration of the cyclohexaamylose cavity by the sodium benzoate. Penetration of the cavity by the carboxylate anion is likely to be somewhat less favourable than penetration by the carboxyl group because of the ion's solvation requirements.

4.3. Benzoic acid vs sodium benzoate-cyclohexaamylose binding

The NMR evidence clearly supports the idea that both benzoic acid and sodium benzoate penetrate the cavity at the 2,3-hydroxyl side, carboxyl group first, although the sodium benzoate penetration is more random.[48] In light of this finding, the difference in dissociation constants between the cyclohexaamylose-benzoic acid and the cyclohexaamylose-sodium benzoate complexes is understandable. The energy required to take a charged species from a medium of high dielectric constant and insert it into a medium of lower dielectric constant might well make up the difference in binding energy between the neutral and charged substrates. In a study on the effect of various solvent systems on the visible spectra of cycloamylose-substrates, Bender and coworkers showed that the environment experienced by the substrate bound in the cycloamylose cavity could be approximated by *p*-dioxan.[14] In keeping with this observation, we felt that the difference in the free energies of solution of sodium benzoate in pH 11.00 phosphate buffer, ΔG_1, and in dioxan, ΔG_2, as compared with the difference in the free energies of solution of benzoic acid in pH 3.00 phosphate buffer, ΔG_3, and in dioxan, ΔG_4, (Table 10), should reflect the difference in sodium benzoate and benzoic acid cyclohexaamylose binding constants. These numbers would then provide some idea of the magnitude of the insertion

Table 10. Free energies of solution for benzoic acid and sodium benzoate in various solvents at 24.5° C.[48]

ΔG_n	$\Delta G(kJ\ mol^{-1})$	Solute	Solvent	pH
ΔG_1	−3.030	$C_6H_5CO_2Na$	$H_2O, I = 0.5$	11.00
ΔG_2	+20.90	$C_6H_5CO_2Na$	$C_4H_8O_2$	
ΔG_3	+9.91	$C_6H_5CO_2H$	$H_2O, I = 0.5$	3.00
ΔG_4	−2.771	$C_6H_5CO_2H$	$C_4H_8O_2$	

energy described above. The results suggest that movement of benzoic acid out of pH 3.00 phosphate buffer into *p*-dioxan, a cyclodextrin-like environment, is favoured by ΔG_4-$\Delta G_3 = -12.68$ kJ mol^{-1}, a free energy very close to the free energy of formation for the cyclohexaamylose-benzoic acid complex, -14.63 kJ mol^{-1}. However, movement of sodium benzoate out of the phosphate buffer solution into dioxan is an unfavourable process by ΔG_2-$\Delta G_1 = +23.93$ kJ mol^{-1}. This, of course, means that the insertion energy is greater for the charged species, sodium benzoate, by 36.61 kJ mol^{-1}. This difference, although much larger than the difference in the free energies of formation for the cyclohexaamylose-benzoic acid and cyclohexaamylose-sodium benzoate complexes, 10.91 kJ mol^{-1}, can be attributed to the fact that the sodium benzoate would not be completely stripped of its water of solvation when in the cyclohexaamylose cavity, no matter how it penetrates the cavity. If the sodium benzoate is sitting in the cavity at the 2,3-hydroxyl side, carboxylate first, water molecules could still partially solvate the carboxylate anion and the sodium cation at the 6-hydroxyl side of the cavity, thus lowering the "insertion energy". Alternatively, if the sodium benzoate is oriented such that only the aromatic ring is penetrating the cycloamylose cavity, solvation of the anion and cation would not be substantially different from solvation of the ions in bulk solvent. It is important to point out that an X-ray study supports this suggestion for cycloamylose-benzoic acid disposition.[7]

4.4. Depth of substrate penetration of carboxylic acids

Further studies[49] on the binding of carboxylic acids in the cycloamylose cavity verified the importance of carboxylate hydration. The geometries and thermodynamic stabilities of the 3,5-dimethyl-4-hydroxybenzoic acid, 3,5-dimethyl-4-hydroxycinnamic acid, and 3,5-dimethyl-4-hydroxyhydrocinnamic acid cyclohexaamylose complexes were evaluated[49] at pH = 7.60 and 12.00.

These substrates can bind in the cyclohexaamylose cavity in either of two different orientations: carboxylate anion first (geometry A) or methyl and hydroxyl first (geometry B) (Fig. 8). Binding may occur exclusively in one of these two orientations or, alternatively, it can be bimodal with a certain mole fraction of the substrate bound in each orientation. In geometry A,

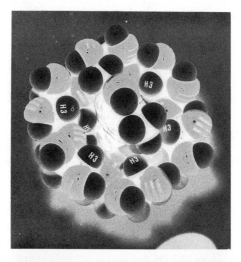

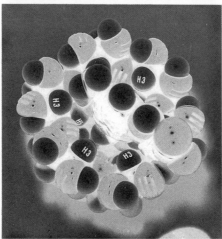

Fig. 8. CPK space-filling models of the two possible orientations of 3,5-dimethyl-4-hydroxycinnamic acid in the cyclohexaamylose cavity at the 2,3-hydroxyl side. (A) The substrate's carboxyl group is penetrating the cavity. (B) the substrate's methyl and hydroxyl groups are penetrating the cavity (from ref. 49). Reproduced with permission from Bioorganic Chemistry.

the H-5 protons are in intimate contact with the substrate's carboxylate or carboxylate side chain. In geometry B, the H-5 protons do not make contact with the substrate. A consequence of this is that for the same mole fraction of bound cycloamylose, the H-5 protons should be shifted more; the Q values, defined as the change in the chemical shift of a given proton at 100% bound, i.e., the maximum possible chemical shift of a proton, should be more positive in geometry A than in geometry B.

The stability of a particular cycloamylose substrate geometry at a given pH is determined by the stability of the substrate anion in that geometry. When the substrates bind in the cavity at $pH = 7.60$, the carboxylate anions are more completely solvated in geometry B than in geometry A. However, the carboxylate anion can be, at least partially, solvated in geometry A by solvent water in the cavity. Furthermore, as the carboxylate side chain increases in length, its interaction with the cavity improves binding, and brings the carboxylate anion closer to the back of the cavity where it can be more effectively solvated. At $pH = 12.00$, geometry B is likely to be of high energy since in this orientation the phenolate oxanion cannot be solvated. This means that in going from mono- to dianions, if the substrate is binding in geometry A, neither the free energy of formation nor the Q values for the cycloamylose's H-3 and H-5 protons should change significantly. However, if the substrate is bound in geometry B, the free energy of formation should increase drastically while the Q values should still remain the same. If a certain mole fraction is bound in each geometry i.e., binding is bimodal, the apparent free energy of formation should increase along with the Q values for the cycloamylose's H-5 protons.

The expected magnitude of such an increase in the apparent free energy of formation can be predicted from our previous studies of the benzoic acid and sodium benzoate-cyclohexaamylose complexes.[48] The free energies of formation for these complexes are $-16.55 \pm 0.29 \, kJ \, mol^{-1}$ and $-5.64 \pm 0.88 \, kJ \, mol^{-1}$, respectively.[48] It is clear that when benzoic acid's neutral carboxyl group, which has been shown definitely to lie within the cyclohexaamylose cavity, becomes charged, the dissociation constant of the resulting complex increases almost 100-fold. Furthermore, it is important to recognize that even if the sodium benzoate is also binding in the cyclohexaamylose cavity carboxylate first, the anion can be at least partially solvated by water molecules at the back of the cavity. Our free energy of solution studies clearly indicated that if the carboxylate anion were not solvated at all, the free energy of formation of the sodium benzoate complex would be some $36.49 \, kJ \, mol^{-1}$ less favourable than that for the benzoic acid complex.[48] In geometry B at $pH = 12.00$, the phenolate oxanion binding in the cyclohexaamylose cavity cannot be solvated at all. Therefore, at $pH = 12.00$, the difference in ΔG_f's between geometries A and B should be well in excess of 24 kJ mol^{-1}. The data[49] given in Table 11 are in accord with these predictions.

The 3,5-dimethyl-4-hydroxybenzoic acid and 3,5-dimethyl-4-hydroxyhydrocinnamic acid protons were all deshielded on complexation by cyclohexaamylose. As mentioned previously, these downfield ^{1}H NMR shifts can be induced in the spectrum of one molecule when binding in another by several physical mechanisms. However, once again, from the limited amount of experimental data available, it was not possible to definitely assign the observed downfield shifts to a particular mechanism.

The changes induced in the ^{1}H NMR spectrum of 3,5-dimethyl-4-hydroxycinnamic acid on cycloamylose complexation are very different from those observed for the other substrates. The cinnamic acid's α protons are shielded at both pH's. In addition, at pH (pD) = 12.00 they experience substantial line broadening. Prior to this study, we and others had always observed that substrate protons were deshielded on cycloamylose complexation and line broadening, if it occurred, was very slight.

Studies of chemical shift changes of the free substrate protons as a function of pH reveal that the α protons are very sensitive to charge variation in the molecule. In going from pH (pD) = 3.00 to 7.60, the α proton shifts only -0.010δ. However, on going from pH (pD) = 7.60 to 12.00, the α proton shifts -0.190δ.

The cycloamylose-induced shifts in the substrate's α proton at pH (pD) = 7.60 were too small to measure. However, at pH (pD) = 12.00, a Q value of $-0.057 \pm 0.002\delta$ was obtained. Again, as with the unbound substrate, the largest changes are occurring at pH (pD) = 12.00. The shielding of the substrate's α protons on cycloamylose complexation suggests that the charge on the α carbon, and therefore on the α proton, is somehow enhanced relative to that on the uncomplexed substrate. This can be related to the poor solvation of charge on the α carbon when inside the host's cavity where no solvent molecules are available.

The linewidths of the 3,5-dimethyl-4-hydroxycinnamic acid at pH (pD) = 12.00 increased significantly with the fraction of substrate bound to cycloamylose. No internal linewidth standard was available since the possibility existed that any small molecule added to the solutions to provide such a standard would compete with the substrate in binding to the cycloamylose. Thus no perfectly quantitative linewidth comparisons are possible between one sample and the next. However, for the binding of 3,5-dimethyl-4-hydroxycinnamic acid to cyclohexaamylose at pH (pD) = 12.00, it was clear that in each sample the signal area associated with the aromatic and β protons broadened significantly. For example, under the spectrometer conditions used, including an exponential weighting factor for free induction decay of -2.0 s, the total linewidths of the aromatic, β and α protons are

Table 11. Free energies of formation for the various cycloamylose substrate complexes and Q values (ppm) for the substrate's protons[a].

Substrate	Aromatic Q	Methyl Q	α Vinyl Q	β Vinyl Q	α-CH_2 Q	β-CH_2 Q	ΔG_f (kJ mol^{-1})
(structure: H_3C, HO, CO_2^-, CH_3)	+0.127±0.002	+0.014±0.005	—	—	—	—	ΔG_f^{nmr} = −7.15±1.05
(structure: H_3C, HO, CO_2^-, CH_3)	+0.130±0.005	+0.127±0.010	—	+0.096±0.002	—	—	ΔG_f^{nmr} = −14.34±0.63; ΔG_f^{uv} = −12.37±0.50
(structure: H_3C, O^-, CO_2^-, CH_3)	Too broad	Too broad	−0.057±0.002	Too broad	—	—	ΔG_f^{nmr} = −7.73±1.16; ΔG_f^{uv} = −8.11±1.17
(structure: H_3C, HO, CO_2^-, CH_3)	+0.030±0.002	+0.038±0.001	—	—	—	—	ΔG_f^{nmr} = −10.41±0.84; ΔG_f^{uv} = −8.78±1.25
(structure: H_3C, O^-, CO_2^-, CH_3)	+0.098±0.002	+0.045±0.001	—	—	−0.011±0.001	−0.057±0.002	ΔG_f^{nmr} = −6.52±1.67; ΔG_f^{uv} = −6.56±3.93

[a] As determined from the data analysis described. Downfield shifts are indicated by positive Q values.

approximately 2 Hz each for a sample of 0.006 mol dm^{-3} 3,5-dimethyl-4-hydroxycinnamic acid (pD = 12.00). Under the same spectrometer conditions, the spectrum of a sample 0.006 mol dm^{-3} in 3,5-dimethyl-4-hydroxycinnamic acid and 0.040 mol dm^{-3} in cyclohexaamylose i.e., 49% substrate bound to cyclohexaamylose, shows that the aromatic and β proton signals merge providing a single signal 15 Hz in width while the α proton signal increases by 4.5 Hz.

Two factors could be contributing to the differential line broadening observed in 3,5-dimethyl-4-hydroxycinnamic acid at pH (pD) = 12.00. First, there could be exchange broadening, a contribution to the linewidth caused by the process of chemical exchange itself. Second, the differential broadening could be due to a greater transverse relaxation rate of the aromatic and β protons of the bound substrate relative to that of the α protons. This could occur if the orientation of the substrate protons in the complex is such that the aromatic and β protons are closer to particular protons of cyclohexaamylose causing them to experience enhanced nuclear dipole relaxation relative to the relaxation rate of the α proton. Although this is currently under investigation, we do not presently have sufficient information to assign the observed broadening to either of these mechanisms.

A comparison[49] of the binding of 3,5-dimethyl-4-hydroxybenzoic acid and 3,5-dimethyl-4-hydroxycinnamic acid in the cyclohexaamylose cavity at pH (pD) = 12.00, i.e. in the form of their dianions (Table 11), clearly demonstrates the importance of the position of the carboxylate anion at pH (pD) = 12.00 relative to pH (pD) = 7.60. The depth of penetration of the benzoate carboxylate is clearly limited by the 3- and 5-methyl groups. The cinnamate carboxylate can penetrate the cycloamylose deep enough to hydrogen bond to either the 6-OH groups or water molecules at the back of the cavity while the benzoate cannot. Furthermore, at pH 12.00 the 2-OH groups are at least partially ionized which would render any geometry B complex unstable. A geometry in which the carboxylate and a hydroxyl oxanion interact would clearly be a high-energy geometry. The pH 12.00 data then can be understood if geometry B is of importance in 3,5-dimethyl-4-hydroxybenzoate binding at pH 7.60.

It is apparent that 3,5-dimethyl-4-hydroxybenzoic acid binds in the cyclohexaamylose cavity at pH = 7.60, but not at pH = 12.00, while both the geometries and stabilities of the 3,5-dimethyl-4-hydroxycinnamic acid and 3,5-dimethyl-4-hydroxyhydrocinnamic acid-cyclohexaamylose complexes are only marginally pH sensitive. These facts strongly suggest that the position of the carboxylate anion in the cyclohexaamylose cavity is of great importance in binding. This implies that carboxylate anion solvation is probably the single most important factor in regulating both the geometries and stabilities of cyclohexaamylose carboxylate anion complexes.[49]

5. Thermodynamic arguments

The thermodynamics of cycloamylose-substrate binding can be analysed[47] in terms of the Nemethy–Scheraga hydrophobic bond picture, an analysis which also points to a substantial contribution from van der Waals–London dispersion forces interaction. In addition, the origins of the rather "unusual" entropy term associated with cycloamylose-substrate binding can be assigned to rotational restrictions imposed on the cycloamylose cavity on substrate complexation. The results are also in agreement with the importance of dipole interactions.

5.1. Entropy of association

The entropies of association for the sodium *p*-nitrophenolate cyclohexaamylose, dodecakis-2,6-*O*-methylcyclohexaamylose, cycloheptaamylose, and tetradecakis-2,6-*O*-methylcycloheptaamylose complexes are listed in Table 12. The negative values for the cyclohexaamyloses suggest that the entropy of formation is largely associated with a loss in the cycloamylose's mobility and not with the normal increase in solvent entropy associated with hydrophobic bond formation. However, the positive entropies of formation for the cycloheptaamylose complexes suggests there is very little loss in the oligosaccharide's mobility on complexation.

Most of the water in the cycloamylose cavity as well as some of the water surrounding the sodium *p*-nitrophenolate guest molecule is stripped away on complex formation. Based on calculations of the number of water molecules contained in the cycloamylose cavities when in aqueous solution[35] and on the direction and depth of sodium *p*-nitrophenolate penetration as determined using NMR,[33,37] as mentioned previously, one can estimate the number of water molecules released to the bulk solvent on complexation (Table 13). With the Nemethy–Scheraga expression, $\Delta S_w^0 / \Delta Y^s = 2.80$ eu, it is possible to approximate the entropy changes, ΔS_w^0, associated with the displacement of a particular number, ΔY^s, of water molecules from between two hydrophobes when the hydrophobes come together (Table 13). It is clear from the fact that these entropies are positive that the ΔS associated with the interaction of the molecules themselves is even more negative than the experimentally determined values indicate, which is probably due largely to a loss of rotational freedom in the cycloamylose molecules.

Models show that movement about the glycosodic linkage of cyclohexaamylose is restricted when a large guest molecule such as sodium *p*-nitrophenolate "fills" the cavity and could thus explain the negative entropy of cycloamylose-substrate association. The entropy of complexation

Table 12. Thermodynamic parameters for sodium p-nitrophenolate-cycloamylose complexes[a]

Cycloamylose	K_D at 25°C (M)	ΔG^o (kJ mol⁻¹)	ΔH (kJ mol⁻¹)	ΔS (eu)
Cyclohexaamylose (α-CD)	5.0×10^{-4}	-18.68 ± 0.12	-37.87 ± 0.25	-65.6 ± 0.8
Dodecakis-2,6-O-dimethylcyclohexaamylose	1.27×10^{-4}	-21.11 ± 0.17	-42.05 ± 0.33	-70.2 ± 1.2
Cycloheptaamylose (β-CD)	1.59×10^{-3}	-15.97 ± 0.12	-15.84 ± 0.35	$+0.4 \pm 0.8$
Tetradecakis-2,6-O-dimethylcycloheptaamylose	1.24×10^{-3}	-16.64 ± 0.17	-13.79 ± 0.33	$+9.6 \pm 1.2$

[a] Values are given for 25°C and represent averages for three sets of runs.

Table 13. Nemethy–Scheraga parameters for sodium p-nitrophenolate-cycloamylose complexes[a]

Cycloamylose	$\Delta H_w^{o(b)}$	$\Delta S_w^{o(c)}$	ΔY^s	$\Delta F_w^{o(b)}$	$(\frac{1}{2})E_{rw}\Delta Y^{s(b)}$	$Z_r E_r^{(b)}$	$\Sigma\Delta F_{rot}$	$\Delta F_{h\phi}^{o(b)}$	$\Delta F_{obs}^{(b)}$
Cyclohexaamylose (α-CD)	14.04	58.9	21	−4.038	10.0	−21.94	ΔF_{rot}^2	> -15.67	-18.60 ± 2.5
Dodecakis-2,6-O-dimethylcyclohexaamylose	18.06	75.62	27	−5.191	11.3	−28.21	ΔF_{rot}^1	> -22.11	-21.40 ± 0.8
Cycloheptaamylose (β-CD)	27.42	114.53	41	−7.82	27.42	−42.84	ΔF_{rot}^4	> -23.24	-16.05 ± 0.46
Tetradecakis-2,6-O-dimethylcycloheptaamylose	34.78	145.63	52	−9.998	34.78	−54.3	ΔF_{rot}^3	> -29.55	-16.59 ± 0.46

[a] $\Delta F_{h\phi}^0 = \Delta F_w^0 - (\frac{1}{2})E_{rw}\Delta Y^s + Z_r E_r + \Sigma\Delta F_{rot}$

[b] Values are given in kilojoules per mole at 25°C.

[c] Values are given in entropy units at 25°C.

would be expected to be less negative for the cycloheptaamylose than for the cyclohexaamylose sodium *p*-nitrophenolate complexes because of the larger diameter of the cycloheptaamylose cavity. Any movement about the glycosidic linkages is less restricted by sodium *p*-nitrophenolate penetration of the cycloheptaamylose cavity than by penetration of the cyclohexaamylose cavity and would result in a smaller negative entropy term contribution to the overall entropy of complexation. Furthermore, since the substrate penetrates the cycloheptaamylose cavity further than it does the cyclohexaamylose cavity, releasing a larger number of water molecules to the bulk solvent, the penetration will be associated with a larger positive entropy component.

To further verify that these ideas about the differences in the relative contribution of the entropy term to the overall free energy of complexation for the cycloheptaamylose and cyclohexaamylose complexes were correct, we examined the entropy of complexation for the sodium *p*-nitrophenolate, 2,6-*O*-permethylated cycloamylose analogs. The magnitude and direction of the effect of cycloamylose methylation on the entropy of substrate complexation was in keeping with the above entropy arguments for each of the cycloamyloses.

The difference in the entropy of complexation between the cyclohexaamylose-sodium *p*-nitrophenolate complex and the dodecakis-2,6-*O*-methylcyclohexaamylose-sodium *p*-nitrophenolate complex was nearly within experimental error. In the dodecakis-2,6-*O*-methylcyclohexaamylose complexation of sodium *p*-nitrophenolate, the cyclohexaamylose 2-*O*-methyl groups can make two different and opposing contributions to the entropy of formation. The stripping away of solvent molecules from between the 2-*O*-methyls and the substrate will result in a positive entropy contribution, while a negative contribution will arise from rotational restrictions placed on the methyl groups by the substrate. Both of these effects are likely to be small and cancel one another to some extent.

Any rotation of the 2-*O*-methyl groups toward the centre of the cyclohexaamylose cavity would be hindered by the presence of the sodium *p*-nitrophenolate guest, resulting in a greater negative entropy of association. However, the movement of the methyl groups is already quite restricted by the neighbouring hydroxyls. Consequently, any restrictions placed on the cavity methyls by the substrate are likely to be small. Furthermore, simple calculations show that only an additional five water molecules would be released to the bulk solvent from the methylated cycloamylose relative to the unmodified cycloamylose.

The positive difference in the entropy of complexation between the cycloheptaamylose and the tetradecakis-2,6-*O*-methylcycloheptaamylose-sodium p-nitrophenolate complexes, although small, suggests the importance of the release of cycloamylose cavity water to the bulk solvent.

Because of the increased diameter of the cycloheptaamylose and therefore the increased distance between sodium p-nitrophenolate and the 2-O-methyl groups, there is no rotational restriction placed on the methyls at all by the guest molecules. However, the stripping of the water molecules from between the methyls and guest molecules, about 11 water molecules, should result in a positive entropy contribution. Using the previous expression, $\Delta S_w^0 / \Delta Y^s = 2.80$ eu, this would mean approximately an additional 30.9 eu. The observed difference, although smaller than this, is in the correct direction and of reasonable magnitude (Table 13). Finally, regarding any restriction of the substrate, it is likely that translational losses are about the same in both the cycloheptaamylose and cyclohexaamylose complexes. Although rotational losses should be greater for the cyclohexaamylose complexes because the cavity is smaller and the substrate fits more tightly, dynamic coupling experiments have shown that aromatic rings rotate freely in the cavity.[35]

5.2. Enthalpy of association

Water structure plays an integral role in the overall ΔG of formation of the "hydrophobic bond," contributing to both the favourable ΔS of association as well as to the ΔH component.[47] Because of the dependence of this water structure on temperature, the ΔH of hydrophobic bond formation is also temperature dependent, with the result that van't Hoff plots for such associations should be curved and not linear. This dependence of ΔH on temperature for hydrophobic associations has been shown both theoretically and experimentally.[38] However, it is clear from Fig. 9 that the van't Hoff plots for the various cycloamylose–sodium p-nitrophenolate complexes are not curved, suggesting that the driving forces for cycloamylose-substrate complexations are not simple entropy-controlled hydrophobic bond forces.[47]

The enthalpy contribution[47] from the stripping away of the water molecules between the cycloamylose and substrate molecules on complexation can be approximated from the relationship[16] $\Delta H_w^0 / \Delta Y^s = 0.67$ kJ, where ΔY^s has the same meaning as previously described (Table 13). The positive sign of the calculated ΔH_w^0 clearly indicates that the experimental ΔH of association between the two molecules themselves would be even more negative than determined, by the amount $-|\Delta H_w^0|$. If the water molecules in the cycloamylose cavity were very "enthalpy-rich," a negative ΔH of association would be consistent with the expulsion of this cavity

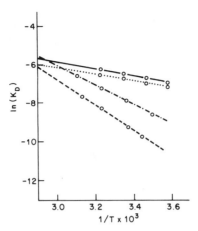

Fig. 9. van't Hoff plots for sodium *p*-nitrophenolate complexes: — · —, cyclo-hexaamylose; - - - -; dodecakis-2,6-*O*-methylcyclohexaamylose; ———, cyclohep-taamylose; · · · · · ·; tetradecakis-2,6-*O*-methylcycloheptaamylose. Ref. 47. Repro-duced with permission from Bioorganic Chemistry.

water upon substrate complexation. There are, however, a larger number of cycloamylose-substrate binding constants inconsistent with this idea. For example, if expulsion of cavity water was a major driving force in complexa-tion, although sodium isobutyrate, sodium cyclohexanecarboxylate, and sodium *p*-nitrophenolate all have substantially different geometries, they would all displace about the same amount of water from the cyclohexaamy-lose cavity and thus should have similar binding constants: however, they do not. The dissociation constants of these three complexes are $2.2 \pm 0.3 \times 10^{-1}$, $1.9 \pm 0.3 \times 10^{-2}$, and $5.0 \pm 0.8 \times 10^{-4}$ M, respectively.[7] It is, of course, true that sodium isobutyrate takes up less space than the other substrates, but the remaining spaces are too small for water molecules to occupy.

If, however, London dispersion forces contribute to the binding of the substrate, the negative enthalpy of association ΔH_a, the difference in the ΔH_a values between the cycloheptaamylose and cyclohexaamylose com-plexes, and the effect of cycloamylose methylation on the ΔH_a values are understandable. Because of the r^{-6} dependence of these forces on the distance, r, between the two interacting species, any substrate which fits into and binds to cyclohexaamylose is likely to bind more weakly in the cycloheptaamylose cavity due to the greater diameter of the heptamer's cavity and thus the greater distance between the host and guest molecules. Methylation of the cyclohexaamylose should therefore result in a more negative ΔH of complexation, while methylation of the cycloheptaamylose should have little effect on the ΔH of complexation (Table 12).[47]

5.3. Hydrophobic bond analysis

According to the Nemethy–Scheraga picture,[16] the free energy of formation of the hydrophobic bond $\Delta F_{h\phi}^0$ can be separated into two components ΔF_w^o and ΔF_h^o, a contribution from the change in water structure and a contribution from a change in the state of the hydrophobes respectively i.e., $\Delta F_{h\phi}^o = \Delta F_w^o + \Delta F_h^o$. The ΔF_w^o term can be estimated from the approximation $\Delta F_w^o / \Delta Y^s = -0.192 \text{ kJ mol}^{-1}$, while the ΔF_h^o term is broken up into three components: $(\frac{1}{2}) E_{rw} \Delta Y^s$, the energy loss of the water-solute interactions; $Z_r E_r$, the hydrocarbon interaction energy; and $\Sigma \Delta F_{rot}$. The most troublesome aspect of such an approximation is choosing the correct values for the above-defined terms. The ΔY^s and, therefore, the Z_r components $(Z_r = (\frac{1}{2}) \Delta Y^s)$ can be easily estimated from earlier NMR studies[36] and our calculations on the number of water molecules contained in the cycloamylose cavities (Table 13).[33,39]

Sodium p-nitrophenolate penetrates the cyclohexaamylose cavity from the wide 2,3-hydroxyl side, nitro-end first, to the extent that the *meta* protons of the substrate do not enter the cavity beyond a point where they are in contact or nearly in contact with the cyclohexaamylose C-3 hydrogens. However, the same substrate penetrates the cycloheptaamylose cavity completely. This means that approximately 6 and 11 molecules of water are displaced from the respective cavities. Using literature values for the number of water molecules surrounding the benzyl group,[38] the number of water molecules stripped from the phenolate surface was calculated by assuming that its insertion is approximated by inserting the benzyl group completely into the cycloheptaamylose cavity or partially into the cyclohexaamylose cavity. The results of the calculations are indicated in Table 13.

Although it is impossible to accurately evaluate the ΔF_{rot} component, certain qualitative observations can be made. Nemethy and Scheraga have pointed out that ΔF_{rot} is controlled by ΔS_{rot} and that the positive ΔH_{rot} term is very small. Based on our earlier arguments regarding the ΔS_{rot} losses for the various cycloamylose-substrate complexes, the ΔF_{rot} order for the sodium p-nitrophenolate complexes should be dodecakis-2,6-O-methylcyclohexaamylose $(\Delta F_{rot}^1) \geqslant$ cyclohexaamylose $(\Delta F_{rot}^2) >$ tetradecakis-2,6-O-methylcycloheptaamylose $(\Delta F_{rot}^3) >$ cycloheptaamylose (ΔF_{rot}^4). The data in Table 13 then establishes that the stability of the cycloamylose-sodium p-nitrophenolate complexes should be in the order tetradecakis-2,6-O-methylcycloheptaamylose $>$ cycloheptaamylose $>$ cyclohexaamylose $\geqslant$ dodecakis-2,6-O-methylcyclohexaamylose. However, the observed stability of these complexes is quite different from the predicted stability.

Although the ΔY^s and therefore the Z_r terms are certainly accurate, the $\Delta F_w^o / \Delta Y^s = -0.192$ ratio may be somewhat small. This ratio should be the same or very similar for each of the cycloamylose systems considered. However, the E_r terms are likely to differ substantially from one cycloamylose complex to the next. The E_r term, or the energy of interaction between the hydrophobes, is a van der Waals–London dispersion forces interaction and can be described by a Lennard–Jones 6–12 potential function, $U(r) = \varepsilon_o[2(r_o/r)^6 - (r_o/r)^{12}]$, showing a strong inverse dependence on internuclear distances.[38] With the cycloheptaamylose system, 6.2 Å in diameter, the distance between the cavity wall and the substrate is greater than in the cyclohexaamylose system, 4.9 Å in diameter; consequently the E_r for cycloheptaamylose should be much smaller than the E_r for cycloheptaamylose, and this would explain the observed order for the cycloamylose-substrate dissociation constants.

This analysis[47] is consistent with our earlier NMR and binding constant studies of a number of cycloamylose-substrate complexes.[32] The observed thermodynamic binding parameters show that cycloamylose-substrate binding energy is not associated with a normal hydrophobic interaction i.e., an entropy controlled phenomenon, but rather an enthalpy controlled equilibrium. Furthermore, a qualitative analysis of the binding in terms of the Nemethy–Scheraga hydrophobic bond picture suggests that van der Waals–London dispersion forces interactions are largely responsible for the differences in binding energy between the cyclohexa- and cycloheptaamylose complexes.

6. Dynamic properties of cycloamylose complexes

Although the picture we have painted of the cycloamylose-substrate driving force is now more complete, we have said nothing about the nature of the complex's dynamic properties. The thermodynamic complexation parameters shed no light upon the tightness with which the substrate fits into the cavity nor how much mobility it has i.e., how free is it to rotate or move in the cavity. In evaluating which system to consider, the nitrophenolates come immediately to mind simply because of the extensive data available on them.

Because of this, the dynamic coupling of sodium 4-nitrophenolate and sodium 2,6-dimethyl-4-nitrophenolate to cyclohexaamylose has been determined[46] using ^{13}C NMR. The coupling has been evaluated in terms of the geometry of the respective complexes as determined by intermolecular Overhauser effects.

6.1. Intermolecular homonuclear Overhauser effects

A ^{1}H homonuclear Overhauser experiment was performed[46] on the cyclo-hexaamylose–sodium 2,6-dimethyl-4-nitrophenolate complex. Irradiation of the C-3 methine protons at 100.1 MHz resulted in a substantial enhancement in the phenolate's *meta* protons of $22 \pm 1\%$. However, due to the width of the decoupler's power curve, it was not possible to separate potential contributions from the C-5 methines. Although the C-6 methylene protons also fall under the power curve, because of the apparent distance between the methylene[32] and substrate protons, as evidenced by the lack of shielding in the C-6 methylenes at high substrate concentrations, any NOE contributions from these are unlikely.

6.2. ^{13}C Spin-lattice relaxation times[46]

The spin lattice relaxation times for the ^{13}C nuclei of free cyclohexaamylose, free sodium 4-nitrophenolate, free sodium 2,6-dimethyl-4-nitrophenolate, and the respective phenolate cyclohexaamylose complexes are given in Table 14. The spin-lattice relaxation times, T_1, for the ^{13}C nuclei of free phenolate or free cyclohexaamylose are dependent on the rate of reorientation of the molecules in solution, τ_c. Anything which influences this rate of reorientation e.g., viscosity and/or molecular association, will affect the spin-lattice relaxation time.[40] Therefore, the T_1 values given in Table 14 for the sodium phenolates, cyclohexaamylose, and their corresponding complexes have been viscosity corrected. These corrected values are given as $(T_1)^v_{free}$ and $(T_1)^v_{obsd}$ in Table 15. The $(T_1)^v_{free}$ values refer to the viscosity corrected T_1's for the free substrate and cyclohexaamylose. The $(T_1)^v_{obsd}$ values refer to the viscosity corrected T_1 values for the corresponding complexes prior to 100% bound corrections.

Because the rates of reorientation between the free and bound states of the host and guest molecules differ, the T_1 values observed for each ^{13}C nucleus are dependent upon the mole fraction of substrate bound, α. Since the rate of dissociation is small compared to the rate of reorientation, the T_1 values for 100% bound can be obtained from the relationship[35,41]

$$(T_1)^{-1}_{obsd} = \alpha (T_1)^{-1}_{complex} + (1 - \alpha)(T_1)^{-1}_{free} \tag{1}$$

The viscosity-corrected $(T_1)^v_{free}$ values obtained for cyclohexaamylose, sodium 4-nitrophenolate, and sodium 2,6-dimethyl-4-nitrophenolate, as well as the viscosity-corrected values, $(T_1)^v_{obsd}$, for cyclohexaamylose and the sodium nitrophenolates in their respective complexes were used to calculate values of $(T_1)^v_{complex}$, the spin–lattice relaxation times for 100% bound.

...nitrophenolate (1), and sodium 2,6-dimethyl-4-nitrophenolate (II).[46]

^{13}C spin-lattice relaxation times (s)

Compound	Concentration	1	4	3	5	2	6	$\langle T_1 \rangle_{1-5}$
	0.1	0.123	0.133	0.137	0.142	0.119	0.067	0.131
[αCD, I]	0.1, 0.2	0.108	0.118	0.120	0.127	0.127	0.073	0.120
[αCD, II]	0.1, 0.2	0.116	0.119	0.115	0.119	0.116	0.074	0.117

		2,6	3,5					$\langle T_1 \rangle_{2-6,3-5}$
I	0.10	4.67	4.93					4.80
	0.15	3.62	3.59					3.61
	0.20	3.42	3.28					3.35
[I, αCD]	0.2, 0.1	1.38	1.38					1.38

		3,5	2,6-CH$_3$					
II	0.10	1.93	4.19					
	0.15	1.36	3.38					
	0.20	1.41	3.52					
[II, αCD]	0.2, 0.1	0.434	1.59					

Table 15. *Viscosity-corrected spin-lattice relaxation times and rotational correlation times for both free and bound cyclohexaamylose (α-CD) and the sodium nitrophenolates (I and II)[46]*

Compound	Concentrations	η (cP)	$(T_1)_{free}$	$(T_1)^v_{free}$	$(T_1)_{obsd}$	$(T_1)^v_{obsd}$	$(T_1)^v_{complex}$	τ^v_c (ps)
				α-CD Overall				
[αCD]	0.1	1.265	0.131	0.166				296
[αCD, I]	0.1, 0.2	1.357	0.131	0.166	0.120	0.163	0.163	302
[αCD, II]	0.1, 0.2	1.394	0.131	0.166	0.117	0.163	0.163	302
				Primary Hydroxyl				
[αCD]	0.1	1.265	0.067	0.085				289
[αCD, I]	0.1, 0.2	1.375	0.067	0.085	0.073	0.099	0.099	249
[αCD, II]	0.1, 0.2	1.394	0.067	0.085	0.074	0.103	0.103	240
				Substrate Overall				
[I]	0.10	0.974	4.80	4.68				10.5
[I]	0.15	0.985	3.61	3.55				13.9
[I]	0.20	1.012	3.35	3.39				14.5
[I, αCD]	0.2, 0.1	1.357	4.80	4.68	1.38	1.88	1.17	42.0
[I, αCD]	0.2, 0.1	1.357	3.61	3.55	1.38	1.88	1.27	38.7
[I, αCD]	0.2, 0.1	1.357	3.35	3.39	1.38	1.88	1.28	37.9

Substrate Overall

[II]	0.10	0.958	1.93	1.84				26.7
[II]	0.15	0.969	1.36	1.32				37.3
[II]	0.20	1.030	1.41	1.45				33.9
[II, αCD]	0.2, 0.1	1.394	1.93	1.84	0.434	0.606	0.362	136
[II, αCD]	0.2, 0.1	1.394	1.36	1.32	0.434	0.606	0.393	125
[II, αCD]	0.2, 01	1.394	1.41	1.45	0.434	0.606	0.382	129

Methyls

[II]	0.10	0.958	4.19	4.01				4.09
[II]	0.15	0.969	3.38	3.27				5.02
[II]	0.20	1.030	3.52	3.62				4.53
[II, αCD]	0.2, 0.1	1.394	4.19	4.01	1.59	2.21	1.52	10.8
[II, αCD]	0.2, 0.1	1.394	3.38	3.27	1.59	2.21	1.67	9.83
[II, αCD]	0.2, 0.1	1.394	3.52	3.62	1.59	2.21	1.59	10.3

6.3. Correlation times

Measurements of the NOE's of protonated carbons indicate that dipole–dipole interactions are of major importance in the relaxation of ^{13}C nuclei.[35,42] However, in the special case of CH$_3$ groups, spin rotation as well as dipole–dipole relaxation contributes to the T_1 values of ^{13}C nuclei.[35,42] The intermolecular dipole–dipole relaxation of a spin $\frac{1}{2}$ nucleus I_1 separated a distance r from a nucleus I_2, can be expressed as

$$(T_1)^{-1}_{D-D} = 3\gamma_1^2\gamma_2^2\hbar^2 I_2(I_2+1)r_{1,2}^{-6}\tau_c \tag{2}$$

where γ_1 and γ_2 are the gyromagnetic ratios for the respective nuclei and τ_c is the reorientation rate or correlation time.[43,22] The viscosity-corrected $(T_1)^v_{free}$ values for cyclohexaamylose, sodium 4-nitrophenolate and sodium 2,6-dimethyl-4-nitrophenolate (0.10, 0.15), as well as the viscosity-corrected $(T_1)^v_{complex}$ values generated from Equation 1 for the sodium phenolates and cyclohexaamylose in their corresponding complexes, were used to calculate the respective τ_c's from Equation 2. The resulting τ_c values are given in Table 15.

6.4. Internal rate of rotation of sodium phenolate bound in the cyclohexaamylose cavity

Component analysis of local molecular motions has been accomplished[46] by the use of correlation times obtained from NMR line shape analysis of CHD–proton resonances.[44] Similarly, the analysis of local molecular motions can be accomplished using correlation times determined from ^{13}C spin–lattice relaxation times. In general, when the rotation is occurring among a large number of equilibrium positions, the rate of internal rotation, τ_i^{-1}, around a bond which is in turn rotating at a overall rate, τ_m^{-1}, results in an effective correlation time, τ_c, given by the relationship[44]

$$\tau_c = A\tau_m + B\left(\frac{1}{\tau_m}+\frac{1}{\tau_i}\right)^{-1} + C\left(\frac{1}{\tau_m}+\frac{4}{\tau_i}\right)^{-1} \tag{3}$$

In this equation A, B, and C are the geometric parameters:

$$A = \tfrac{1}{4}(3\cos^2\theta - 1)^2$$
$$B = \tfrac{3}{4}\sin^2 2\theta$$
$$C = \tfrac{3}{4}\sin^4\theta$$

where θ is the angle between the relaxation vector of the main field gradient and the principal axis of rotation. The principal axis of rotation for the sodium nitrophenolate-cyclohexaamylose complexes passes through the C_6 symmetry axis of the cyclodextrin cavity and the C_2 symmetry axis of the substrate i.e., through the C-1 and C-4 carbons of the sodium phenolate molecules. The relaxation vector for the sodium phenolates lies along the C–H bonds and makes an angle of 60 or 120° with the rotational axis of these substrates inside the cyclohexaamylose molecule.

The rate of internal rotation for the phenyl ring of the sodium nitrophenolate was calculated[46] from the overall correlation time, τ_m, of the cyclohexaamylose and the effective correlation time, τ_c, of the sodium phenolates at 100% substrate bound. From the values of the effective local correlation time, τ_c, and the overall correlation time, τ_m, it is possible to determine the degree of coupling between these two rotations. The coupling coefficient, ξ, is the ratio τ_c/τ_m and theoretically may vary from 0.11 to 1.[44] When the rate of overall rotation is approximated by the rate of internal rotation, the value of ξ approaches 1, its maximum coupling value. However, if the rate of overall rotation is slow compared to that of internal rotation, ξ approaches its minimum coupling value. The values[46] of ξ as well as τ_i^{-1} are given in Table 16.

Table 16. *Coupling coefficients and internal rates of rotation for the sodium 4-nitrophenolate- and sodium 2,6-dimethyl-4-nitrophenolate-cyclohexaamylose complexes*

Compound	ξ	τ_i (s)	τ_i^{-1} (s^{-1})
	0.14	6.68×10^{-11}	1.50×10^{10}
	0.13	5.99×10^{-11}	1.67×10^{10}
	0.13	5.83×10^{-11}	1.72×10^{10}
	0.45	4.15×10^{-10}	2.41×10^{9}
	0.41	3.52×10^{-10}	2.84×10^{9}
	0.43	3.75×10^{-10}	2.67×10^{9}

6.5. The effects of binding on the spin-lattice relaxation and correlation times of cyclohexaamylose.[46]

Although there has been some disagreement in the past as to the assignments of the ^{13}C NMR spectrum of cyclohexaamylose, in this investigation we use those of Colson, *et al.*[45] On this basis, the carbon signals in going from the least to the most shielded are assigned as C-1 < C-4 < C-3 < C-2 < C-5 < C-6.

The ^{13}C nuclei spin-lattice relaxation times, T_1's, for C-1 through C-5 as determined in a 0.10 mol dm^{-3} solution in phosphate buffer (pH 9.64; $I = 0.5$ at $25 \pm 1°$ C) are all equal within experimental error and thus are combined as the average $\langle T_1 \rangle_{1-5}$, 0.131 s (Table 14). The T_1 for the C-6 primary hydroxyl carbon determined under identical conditions is somewhat shorter, 0.067 s. The T_1's for the free cyclohexaamylose were determined by other workers at $33 \pm 2°$ C to be 0.144 s for C-1 through C-5 and 0.105 s for C-6; however, the pH was not given nor was the viscosity taken into account. Consequently, no comparison can be made between these values and those reported here.[35] The overall viscosity-corrected correlation time, τ_c^v (Table 15), for free cyclohexaamylose (0.10 mol dm^{-3}; pH 9.64; $I = 0.5$ at $25 \pm 1°$ C) is 296 ps, while the viscosity-corrected effective local correlation time, τ_c^v, for the C-6 primary hydroxyls is 289 ps. The coupling coefficient, ξ, the ratio of the C-6 effective local correlation time to the overall correlation time, is 0.98. This clearly indicates strong coupling between the two rotations.

The overall spin-lattice relaxation time for cyclohexaamylose in both the sodium 4-nitrophenolate and the sodium 2,6-dimethyl-4-nitrophenolate complexes is 0.163 s (0.10 mol dm^{-3} cyclohexaamylose, 0.20 mol dm^{-3} sodium phenolate; pH 9.64; $I = 0.5$ at $25 \pm 1°$). These values are within experimental error of $(T_1)_{free}^v$ for cyclohexaamylose. The spin-lattice relaxation times for the primary hydroxyls of cyclohexaamylose in the sodium 4-nitrophenolate and the sodium 2,6-dimethyl-4-nitrophenolate complexes are 0.099 and 0.103 s, respectively (0.10 mol dm^{-3} cyclodextrin, 0.20 mol dm^{-3} sodium phenolate; pH 9.64; $I = 0.5$ at $25 \pm 1°$ C). Similarly, these values are within experimental error of their corresponding $(T_1)_{free}^v$ values.

6.6. The effects of binding on the spin-lattice relaxation and correlation times of the sodium phenolates[46]

The ^{13}C NMR spectrum of sodium 4-nitrophenolate was assigned in an earlier paper.[37] From lower to higher field, the ^{13}C resonances are C-3 = C-5 < C-2 = C-6.

The absence of NOE enhancements in the ^{1}H decoupled spectrum of the *ipso* carbons of sodium 2,6-dimethyl-4-nitrophenolate as well as the large chemical shift differences between the aromatic and methyl ^{13}C nuclei made the assignment of the C-3, C-5, and methyl carbons fairly simple. The ^{13}C resonances from lower to higher field are C-3 = C-5 ≪ methyl carbons.

The overall T_1's for the free sodium 4-nitrophenolate (pH 9.64; I = 0.5 at $25 \pm 1°$ C) were determined from the average of the T_1's observed for the C-2, C-3, C-5 and C-6 carbons since all were within experimental error of one another ($\langle T_1 \rangle_{2-6,3-5}$ = 4.80 s at 0.10 mol dm^{-3}, 3.61 s at 0.15 mol dm^{-3} and 3.35 s at 0.20 mol dm^{-3}). The T_1 for free sodium 2,6-dimethyl-4-nitrophenolate (0.10, 0.15, and 0.20 mol dm^{-3}; pH 9.64; I = 0.5 and $25 \pm 1°$ C) was determined from the T_1's observed for the C-3 and C-5 carbons ($\langle T_1 \rangle_{3-5}$ = 1.93 s at 0.10 mol dm^{-3}, 1.36 s at 0.15 mol dm^{-3} and 1.41 s at 0.20 mol dm^{-3}). (Table 14). It is clear that the viscosity-corrected $(T_1)_{free}^v$ values for the sodium 4-nitrophenolates show a decrease in going from 0.20 to 0.10 mol dm^{-3} solutions. However, because of experimental error a clear relationship is difficult to establish at all three concentrations.

As a result of complexation with cyclohexaamylose the values obtained for the sodium nitrophenolates $(T_1)_{complex}^v$ calculated from Equation 1 are significantly smaller than their $(T_1)_{free}^v$ values. In the sodium 4-nitrophenolate-cyclohexaamylose complex the T_1 values decrease by a factor of 2.6 while in the sodium 2,6-dimethyl-4-nitrophenolate-cyclohexaamylose complex the T_1 values decrease by a factor of 3.4.

The τ_c^v values for both free sodium 4-nitrophenolate and sodium 2,6-dimethyl-4-nitrophenolate were determined at three concentrations (0.10, 0.15, and 0.20 mol dm^{-3}) from the corresponding $(T_1)_{free}^v$ values. Using Equation 1, $(T_1)_{complex}^v$ was calculated at 100% bound for each substrate at each concentration. These τ_c^v values for the substrates (effective local correlation time of substrate at 100% bound, τ_c) and the τ_c^v values obtained for cyclohexaamylose (overall correlation time of cycloamylose, τ_m) were then used to calculate τ_i^{-1} from Equation 3 (see Table 16).

The coupling coefficient, ξ, the ratio of effective local correlation time, τ_c, for sodium 4-nitrophenolate (37.9 ps) to that of the overall correlation time, τ_m, for cyclodextrin (302 ps) in the sodium 4-nitrophenolate cyclohexaamylose complex is 0.13. This indicates weak coupling between substrate and cavity. However, the coupling coefficient, ξ, of the effective local correlation time, τ_c, for sodium 2,6-dimethyl-4-nitrophenolate (125 ps) to that of the overall correlation time for cyclodextrin, τ_m (302 ps), in the sodium 2,6-dimethyl-4-nitrophenolate-cyclohexaamylose complex is at least 0.41. Thus, in this case fairly strong coupling between the substrate and cavity is indicated. Finally, the internal rate of rotation, τ_i^{-1}, for sodium 4-nitrophenolate is at least 6.2 times faster than that obtained for sodium

2,6-dimethyl-4-nitrophenolate. Only an upper limit may be calculated for the correlation time of the methyl groups of sodium 2,6-dimethyl-4-nitrophenolate since spin-rotation relaxation contributes to the relaxation of these ^{13}C nuclei.[35,42] However, it is clear that the effective correlation times of the methyl groups of free sodium 2,6-dimethyl-4-nitrophenolate increase by a factor of 2.4 on complexation with cyclohexaamylose.

6.7. Dynamic coupling of sodium 4-nitrophenolate and sodium 2,6-dimethyl-4-nitrophenolate to cyclohexaamylose[46]

The spin-lattice relaxation times for sodium 4-nitrophenolate, sodium 2,6-dimethyl-4-nitrophenolate, and cyclohexaamylose as well as those for the nitrophenolates in their respective cycloamylose complexes were all viscosity corrected. Therefore, the differences between the τ_c^v values of the free and bound states represent actual differences in rates of rotation due to complexation and not changes in the bulk viscosity. The viscosity-corrected $(T_1)_{free}^v$ values for both substrates are shortened by ~25% on increasing the concentration of free substrate form 0.10 to 0.20 mol dm^{-3}. This is probably due to self-association of the substrates. Therefore, the reported τ_c^v values are likely to be somewhat smaller than the actual values. This means that the actual coupling between the sodium nitrophenolates and cyclohexaamylose is likely to be somewhat greater than the apparent coupling.

The correlation times for free sodium 2,6-dimethyl-4-nitrophenolate are longer than those for sodium 4-nitrophenolate. This suggests that the added bulk of the methyl groups tends to slow down the rotation of this molecule in aqueous solution. In addition, comparison of the effective local correlation times reveals that, on cyclohexaamylose complexation, the rate of rotation of sodium 4-nitrophenolate decreases by a factor of 3 while that of sodium 2,6-dimethyl-4-nitrophenolate is reduced by a factor of 4. A comparison of the internal rates of rotation, τ_i^{-1}, of sodium 4-nitrophenolate and sodium 2,6-dimethyl-4-nitrophenolate (Table 16) suggests that the methyl groups of sodium 2,6-dimethyl-4-nitrophenolate are in intimate contact with the 2- and 3-hydroxyl groups of the cyclohexaamylose. Sodium 4-nitrophenolate rotates 6.2 times faster in the cyclohexaamylose cavity than sodium 2,6-dimethyl-4-nitrophenolate (1.5×10^{10} vs. 2.41×10^9 s^{-1}, respectively).[46]

In addition, it is clear that the methyl groups of 2,6-dimethyl-4-nitrophenolate rotate more slowly when the substrate is bound. This further suggests intimate contact between the substrate's methyl groups and the 2- and 3-hydroxyl groups of the cyclohexaamylose. However, because of spin rotation contributions, the apparent τ_c^v values calculated for the methyl groups of sodium 2,6-dimethyl-4-nitrophenolate are likely to be somewhat longer than the actual values.

Based on these findings, it seemed that introduction of methyl groups at the 2,3-hydroxyl side of the cyclohexaamylose cavity should introduce further steric hindrance to substrate rotation. In this regard, both the static and dynamic aspects of the sodium 4-nitrophenolate and sodium 2,6-dimethyl-4-nitrophenolate–dodecakis-2,6-*O*-methylcyclohexaamylose (DMCD) complexes were investigated.[50] The geometries of these complexes were determined based on intermolecular homonuclear nuclear Overhauser enhancements and chemical shift analyses while the dynamic parameters of complexation i.e., rotational correlation times and coupling constants, were calculated again from $^{13}C[^1H]T_1$ measurements. Comparison of these results with analogous measurements of the unmethylated cyclohexaamylose inclusion complexes of these same substrates indicates that the substrates are prevented from equivalent penetration of the methylated derivative.

6.8. Dynamic coupling control[50]

The rather low ξ values observed for both the sodium 4-nitrophenolate–DMCD and sodium 2,6-dimethyl-4-nitrophenolate–DMCD complexes examined here indicate that the complexes are of a rather "loose" nature. Especially in the case of sodium 4-nitrophenolate ($\xi = 0.19$), the substrate does not appear to be very tightly coupled to the DMCD cavity. The presence of the substrate methyl groups in sodium 2,6-dimethyl-4-nitrophenolate improves coupling to the DMCD ($\xi = 0.38$) to some extent, but the improvement is not substantial. The increase in rotational correlation times for the substrate methyl groups upon complexation suggests that they are in close contact with the face of the DMCD cavity. However, when these coupling parameters are compared to those for the sodium 4-nitrophenolate–cyclohexaamylose complex ($\xi = 0.14$) and the sodium 2,6-dimethyl-4-nitrophenolate–cyclohexaamylose complex ($\xi = 0.41$)[46] it appears that both substrates interact to about the same extent with both the methylated and unmodified cavities. In the case of sodium 2,6-dimethyl-4-nitrophenolate, penetration of the cavity such that the substrate methyl groups were in close proximity to the DMCD 3-hydroxyl would require that the substrate was, in fact, "slipping" between the DMCD 2-*O*-methyl groups. In this case, a much tighter coupling (as reflected in a ξ value approaching one) would be expected than that observed. Furthermore, a decrease in NOE's experienced by the substrate's *meta* protons is inconsistent with this type of penetration. Both the NOE data and dynamic coupling of sodium 4-nitrophenolate and sodium 2,6-dimethyl-4-nitrophenolate to the methylated cyclohexaamylose clearly

Table 17. The dissociation constants (K_D), dynamic coupling constants (ξ) and substrate nuclear Overhauser enhancements (NOE's) for the binding of sodium 4-nitrophenolate and sodium 2,6-dimethyl-4-nitrophenolate to cyclohexaamylose and dodecakis-2,6-O-methylcyclohexaamylose[50]

Host	K_D, M	ξ	% NOE 2,6	% NOE 3,5	K_D, M	ξ	% NOE 3,5
	$3.7\pm0.02\times10^{-4}$[a]	0.14^b	1[b]	34[b]	$1.96\pm0.53\times10^{-4}$[c]	0.41[a]	22[b]
	$1.4\pm0.2\times10^{-4}$[c]	0.19	16	17	$1.48\pm0.22\times10^{-4}$	0.38	15

[a] Taken from Bergeron et al.[32]
[b] Taken from Bergeron and Channing.[46]
[c] Taken from Bergeron and Meeley.[39]

indicate that the substrates are not penetrating the cavity as deeply as in the case of the unmodified cavity.

Finally, it is interesting to speculate about the correlations (or lack of) among the thermodynamic binding constants, substrate NOE's and coupling coefficients, ξ, for cycloaamylose-substrate binding.[50] Inspection of Table 17 shows that there is a decrease in K_D through the series of complexes: sodium 4-nitrophenolate–cyclohexaamylose > sodium 2,6-dimethyl-4-nitrophenolate–cyclohexaamylose > sodium-4-nitrophenolate–DMCD ≃ sodium 2,6-dimethyl-4-nitrophenolate–DMCD. However, as the NOE's reflect substrate penetration this order is reversed i.e., the magnitude of the binding constant does not correlate directly with the degree of insertion. In contrast, the coupling coefficients show no discernible trends. Thus, the dynamic aspects of molecular association, at least in these complexes, are substantially different from the picture implied by examination of the binding constants. Particularly in the case of the sodium 2,6-dimethyl-4-nitrophenolate–DMCD complex, while the possibility that the substrate methyls could slip between the DMCD rim methyls drawing the substrate more deeply into the cavity and thereby increasing dispersion interactions between guest and host exists, this does not appear to be occurring here.[50]

7. Conclusion

The cycloamyloses represent excellent systems for the evaluation of binding forces involved in inclusion complexation. Furthermore, because of their symmetrical nature a substantial amount of information about these oligosaccharides and their complexes is accessible via ^{13}C and ^{1}H NMR studies.[47–50] These studies, in addition to the standard thermodynamic parameters, provide precise information about the cycloamylose substrate disposition and dynamic coupling, unlike ultraviolet, ORD or circular dichroism information. From these investigations several observations can be made about the binding of polar molecules in the cycloamylose cavity. These substrates demonstrate substantial orientational preference in binding. When charged, they prefer to adopt geometries which allow for solvation of the charge. The binding forces are clearly related to dipole–dipole interactions and not to release of high energy cavity water or cycloamylose strain energy. Guest molecules can couple reasonably tightly with the cycloamylose motion but there is not, at present, any clear relationship between binding and coupling. Further studies will presumably reveal this relationship.

Acknowledgements

The author wishes to acknowledge Ms. Kathy McGovern for her help with this manuscript.

References

1. A. Villiers, *C.R. Acad. Sci.*, Paris, 1891, **112**, 536.
2. F. Schardiger, *Wien. Klin. Wochenschi*, 1904, **17**, 207.
3. K. Takeo and T. Kuge, *Agric. Biol. Chem.*, 1970, **34**, 1787.
4. S. Beychok and E. A. Kabat, *Biochemistry*, 1965, **4**, 2565.
5. B. Casu and M. Reggiani, *Tetrahedron*, 1968, **24**, 803.
6. F. Cramer, W. Saenger and H. C. Spatz, *J. Am. Chem. Soc.*, 1967, **89**, 14.
7. D. W. Griffiths and M. L. Bender, in *Advances in Catalysis*, Vol. 23, p. 209 (eds D. D. Eley, H. Pines and P. B. Weisz), Academic Press, New York (1973).
8. A. Hybl, R. E. Rundle and D. E. Williams, *J. Am. Chem. Soc.*, 1965, **87**, 2779.
9. P. V. Demarco and A. L. Thakkar, *J. Chem. Soc., Chem. Commun.*, 1970, 2.
10. F. Cramer and W. Kampe, *J. Am. Chem. Soc.*, 1964, **87**, 1115.
11. R. Breslow and L. E. Overman, *J. Am. Chem. Soc.*, 1970, **92**, 1075.
12. Y. Iwakura, K. Uno, F. Toda, S. Onozuka, K. Hattori and M. L. Bender, *J. Am. Chem. Soc.*, 1975, **97**, 4432.
13. B. Siegel and R. Breslow, *J. Am. Chem. Soc.*, 1975, **97**, 6871.
14. R. L. Van Etten, J. F. Sebastian, G. A. Clowes and M. L. Bender, *J. Am. Chem. Soc.*, 1967, **89**, 3242.
15. B. Casu and L. Rava, *Ric. Sci.*, 1966, **36**, 733.
16. P. C. Manor and W. Saenger, *J. Am. Chem. Soc.*, 1974, **96**, 3630.
17. W. Sáenger, R. K. McMullan, J. Fayos and D. Mootz, *Acta Crystallogr.*, 1974, **B30**, 2019.
18. R. K. McMullan, W. Saenger, J. Fayos and D. Mootz, *Carbohydr. Res.*, 1973, **31**, 211.
19. B. Casu, M. Reggiani, G. G. Gallo and A. Vigevani, *Tetrahedron*, 1966, **22**, 3061.
20. P. R. Sundararajan and U. S. R. Rao, *Carbohydr. Res.*, 1970, **13**, 251.
21. R. J. Bergeron, M. P. Meeley and Y. Machida, *Bioorg. Chem.*, 1976, **5**, 121.
22. J. H. Noggle and R. D. Schirmer, *The Nuclear Overhauser Effect*, Academic Press, New York, 1971.
23. B. B. Howard, B. Linder and M. T. Emerson, *J. Chem. Phys.*, 1962, **36**, 485.
24. B. D. Sykes and M. D. Scott, *Annu. Rev. Biophys. Bioeng.*, 1972, **1**, 27.
25. D. D. MacNicol, *Tetrahedron Lett.*, 1975, **38**, 3325.
26. J. W. Apsimon, W. G. Craig, P. V. Demarco, D. W. Mathieson, L. Saunders and W. B. Whalley, *Tetrahedron*, 1967, **23**, 2339.
27. J. A. Pople, *J. Chem. Phys.*, 1956, **24**, 1111.
28. B. V. Cheney and D. M. Grant, *J. Am. Chem. Soc.*, 1967, **89**, 5319.
29. P. Balaram, A. A. Bothner-By and R. Breslow, *Biochemistry*, 1973, **12**, 4695.
30. A. A. Bothner-By and R. Gassend, *Ann. N.Y. Acad. Sci.*, 1973, **222**, 668.

31. E. A. Lewis and L. D. Hansen, *J. Chem. Soc., Perkin Trans.* 2, 1973, 2081.
32. R. J. Bergeron, M. A. Channing, G. J. Gibeily and D. J. Pillor, *J. Am. Chem. Soc.,* 1977, **99**, 5146.
33. R. J. Bergeron and R. Rowan, *Bioorg. Chem.,* 1976, **5**, 425.
34. D. J. Wood, F. E. Hruska and W. Saenger, *J. Am. Chem. Soc.,* 1977, **99**, 1735.
35. J. P. Behr and J. M. Lehn, *J. Am. Chem. Soc.,* 1976, **98**, 1743.
36. D. W. Griffiths and M. L. Bender, *J. Am. Chem. Soc.,* 1973, **95**, 1679.
37. R. J. Bergeron and M. A. Channing, *Bioorg. Chem.,* 1976, **5**, 437.
38. G. Nemethy and H. A. Scheraga, *J. Chem. Phys.,* 1962, **36**, 3401.
39. R. J. Bergeron and M. P. Meeley, *Bioorg. Chem.,* 1976, **5**, 197.
40. G. C. Levy and I. R. Peat, *J. Magn. Reson.,* 1975, **18**, 500.
41. J. E. Anderson and D. A. Fryer, *J. Chem. Phys.,* 1969, **50**, 3784.
42. G. C. Levy, J. D. Cargioli and F. A. L. Anet, *J. Am. Chem. Soc.,* 1973, **95**, 1527.
43. T. L. James, *Nuclear Magnetic Resonance in Biochemistry*, Academic Press, New York, 1975.
44. C. H. Brevard, J. P. Kintzinger and J. M. Lehn, *Tetrahedron*, 1972, **28**, 2447.
45. P. Colson, H. J. Jennings and I. C. P. Smith, *J. Am. Chem. Soc.,* 1974, **96**, 8081.
46. R. J. Bergeron and M. A. Channing, *J. Am. Chem. Soc.,* 1979, **101**, 2511.
47. R. J. Bergeron, D. M. Pillor, G. V. Gibeily and W. P. Roberts, *Bioorg. Chem.,* 1978, **7**, 263.
48. R. J. Bergeron, M. A. Channing and K. A. McGovern, *J. Am. Chem. Soc.,* 1978, **100**, 2878.
49. R. J. Bergeron, M. A. Channing, K. A. McGovern and W. P. Roberts, *Bioorg. Chem.,* 1979, **8**, 263.
50. R. J. Bergeron and P. S. Burton, *J. Am. Chem. Soc.,* 1982, **104**, 3664.

13 · REACTIONS OF INCLUSION COMPLEXES FORMED BY CYCLODEXTRINS AND THEIR DERIVATIVES

I. TABUSHI

Kyoto University, Yoshida, Japan

1. Introduction

One of the important basic interactions between biological macromolecules is the so-called "hydrophobic interaction", the nature of which has not yet been entirely clarified. In order to elucidate the hydrophobic interaction, it is therefore necessary to investigate an extremely simplified system having an appropriate hydrophobic binding site. It is not easy, however, for chemists to find such a convenient system, since very few compounds are known to bind other molecules specifically through the hydrophobic interaction (Fig. 1). Fortunately, cyclodextrins have been shown to bind hydrophobic molecules into their cavities in aqueous solution.[1] Although cyclodextrins are naturally occurring compounds, the author first succeeded in showing that certain derivatives of higher homologs of paracyclophanes[2-5] (Fig. 2), which are entirely synthetic, can bind some organic *guests* (specific partners

INCLUSION COMPOUNDS III
ISBN 0-12-067103-4

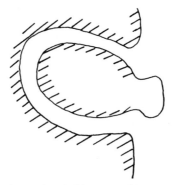

Fig. 1. Hypothetical host for hydrophobic recognition.

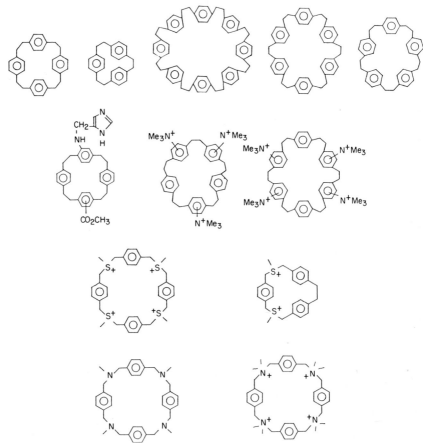

Fig. 2. Some cyclophanes which can act as hosts for organic guest molecules.

of the binding) into their cavities in aqueous solution.[6-11] (See also Chapter
11, Volume 2.) These two series of *host* compounds have a great advantage
in the study of the hydrophobic interaction in that they form 1 : 1 host–guest
inclusion complexes under normal conditions in aqueous solution. Another
advantage of using these compounds as hosts is that the geometry of the
host–guest interaction[12-14] has been extensively studied by X-ray crystallography and by spectroscopic methods. Thus, by the use of the two series
of *host* compounds as artificial hydrophobic pockets (binding sites), the
nature of the hydrophobic interaction can be elucidated in detail.

Although the hydrophobic interaction is conveniently studied by the use
of cyclodextrin and cyclophane hosts, further sophistication of "mimicking"
biological interaction by using the hosts requires the introduction of one
or more functional group(s) into the hosts. These special functional groups
"recognize" the *right* guest molecules through a specific interaction such
as ion pairing ("salt bridge" formation), hydrogen bonding, metal chelation,
etc., with a certain part of the guest molecule. However, introduction of
two or more functional groups into a single host molecule must be carried
out regiospecifically. Otherwise, a mixture of 30 regio isomers ($_{21}C_7/7$)
would be obtained by entirely nonspecific substitution: for example, in the
case where introduction of two identical functional groups into β-cyclodex-
trin is attempted then 30 isomers are obtained as shown in Equation 1:

$$\beta\text{-CD} \xrightarrow{\ 2X\ } 30 \text{ isomers of } \beta\text{-CD} \cdot X_2 \qquad (1)$$

In this chapter, the reactions of cyclodextrin inclusion complexes will be
discussed, with most attention focused on "host design" for biomimetic
reactions. The chemistry of cyclophanes will not be discussed in this chapter
since it has been discussed in a recent review.[15] Another important focus of
"guest design" in cyclodextrin chemistry will be discussed by Breslow in the
following chapter.

2. Catalysis and structure of cyclodextrin inclusion in solution

As already discussed in Chapter 8, Volume 2, the crystal structures of the
polyhydrates of cyclodextrin-guest complexes have been extensively studied
by several groups. The crystallographic results show that a cyclodextrin
usually binds a single organic guest molecule into its own cavity and the
1 : 1 complex thus formed is surrounded by many water molecules. Therefore,
the geometrical situation in the crystal state seems to be very similar to that
in homogeneous solution. This conclusion is strongly supported by many

spectroscopic measurements in aqueous solution, e.g. change in the emission maximum and intensity of fluorescence,[16] change in the absorption maximum of electronic absorption,[17] induction of circular dichroism absorption,[18] change in electron spin resonance absorption,[19] change in nuclear magnetic resonance absorption (correlation time,[20] chemical shift[21-22]), heat of mixing,[23] and many other physical properties. Theoretical calculations on cyclodextrin guest binding in aqueous solution also strongly support the above conclusion for both apolar[24] and polar guests, where consideration of minor additional polar interaction[25] may be necessary (see also Table 1):

$$\Delta G \text{ (CD binding)} = \Delta H \text{ (vdW)} + \Delta H \text{ (water state)}$$

$$+ \Delta H \text{ (polar binding)} + \Delta H \text{ (conformation)}$$

$$- T[\Delta S \text{ (water state)} + \Delta S \text{ (molecular motion)}] \quad (2)$$

where

"polar binding" includes hydrogen bonding, coordination and other electrostatic interactions

"molecular motion" includes translational, rotational and vibrational motions (or internal rotation)

This picture is also true for substituted as well as for unsubstituted cyclodextrins as shown recently by the authors for t-BuS-β-CD (Fig. 3).[26]

Table 1. Gaining and losing interactions in CD inclusion complex formation which is shown schematically at the bottom of the table

Losing	Gaining
H-bonding (●)	liquid water interaction (○)
	H–G, vdW
water assembly around G	
	H–G, restriction in motion
	H–G, polar (charge, dipole, quadrupole, etc.)
←——————→	
conformation change	

○ , ● : H_2O

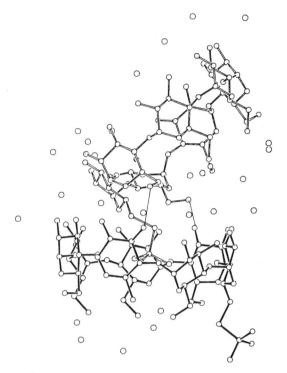

Fig. 3. X-ray crystal structure of *t*-Bus-β-CD system. ○ = water oxygen atoms.

In these circumstances, the most appropriate schematic representation of the cyclodextrin inclusion may be Fig. 4.

Formation of the inclusion complex from individual components present in aqueous solution is best interpreted by the sum of the various types of interaction listed in Table 1. Many types of interaction involved in inclusion complexes may determine the most stable H–G (Host–Guest) orientation (one or a few conformations, at most).

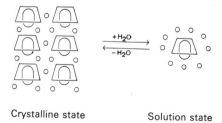

Crystalline state   Solution state

Fig. 4. Schematic representation of CD inclusion.

In the most favourable (stable) conformation, a specific part of a guest molecule takes an appropriate position for a specific hypothetical reaction which is catalyzed by a certain part of the host (Fig. 5a). Therefore if there is an appropriate functional group in the guest susceptible to the catalytic reaction, the optimum catalytic reaction must take place, since $\Delta H^{\neq}$ and $-\Delta S^{\neq}$ should be minimum in this special orientation.

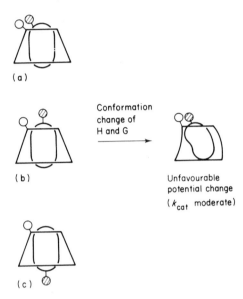

Conformation change of H and G →

(a)

(b)   Unfavourable potential change (k_{cat} moderate)

(c)

Fig. 5. Orientation of inclusion complexes: (a) best fit (k_{cat} very large); (b) poor fit (k_{cat} very small); (c) non productive binding (k_{cat} zero).

In the case of a "poor fit" between the host and the guest (Fig. 5b), any effective catalysis requires a considerable conformational change of the host as well as of the guest from their original conformations, leading to a decrease in the catalytic constant due to the unfavourable potential change involved (induced strain for example). In the case of non-productive binding, k_{cat} must be zero (Fig. 5c). Fortunately, however, almost all the rates of formation and dissociation of the inclusion complexes were found to be much faster than the rates of usual catalytic reactions (Table 2). These observations clearly indicate that every host-guest pair must have a certain probability of adopting a "fit" (best fit or poor fit) orientation, which may be equilibrated with the nonproductive orientation, if any, according to the

Table 2. Association and dissociation rate constants of typical cyclodextrin inclusion complexes

CD	Guest	k_a (s^{-1})	k_d (s^{-1} M^{-1})	Reference
β	I$^-$	3.6×10^6	6.5×10^7	51
α	$^-$O-⟨○⟩-NO$_2$	$> 10^5$	$\geqslant 4 \times 10^7$	16
β	O$_2$N-⟨○⟩-N=N-⟨○⟩-OH / CO$_2^-$	1.1×10^4	3.2×10^6	50
α	⟨○○⟩-N=N-⟨○⟩-NHMe / $^-$O$_3$S	1×10^3	1.1×10^6	16
α	⟨○○⟩-N=N-⟨○⟩-O$^-$ / $^-$O$_3$S, CH$_3$	0.28	150	16

Boltzman distribution[11] (see Equations 3 and 4 and Fig. 6).

$$\frac{[5a]}{[5c]} = \frac{g_a e^{-\varepsilon_a/kT}}{g_c e^{-\varepsilon_c/kT}} \tag{3}$$

or, in general $$\frac{[5a]}{N_T} = \frac{g_a e^{-\varepsilon_a/kT}}{\sum_i g_i e^{-\varepsilon_i/kT}} \tag{4}$$

where N_T is the total concentration of inclusion compound.

Therefore, the overall catalytic efficiency is determined not only by the catalytic constant, k_{cat}, but also by the relative population of the "active" orientation as shown in Equation 5.

$$\nu = k_{cat} \cdot K_{ass} \cdot \frac{g_a e^{-\varepsilon_a/kT}}{\sum_i g_i e^{-\varepsilon_i/kT}} \tag{5}$$

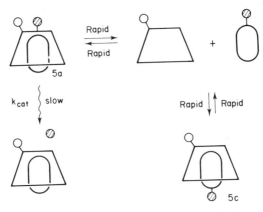

Fig. 6. Host–guest orientation determined by Boltzmann distribution.

The relative population of the "active" orientation (or any other orientation)
is determined by the "multiple recognition free energy", which will be
discussed in more detail in a later section.

Unfortunately, the catalytic constants of unsubstituted cyclodextrins
towards many reactions are much smaller than those of native enzymes,
since cyclodextrins have only poor catalytic functional groups available i.e.
primary and secondary OH groupings. In order to prepare excellent enzyme
models using cyclodextrins, it is necessary to introduce appropriate catalytic
functional group(s) onto the cyclodextrins (except for a few examples where
the parent cyclodextrins themselves are really excellent catalysts). Then, a
general strategy for the preparation of an excellent artificial enzyme seems
to be,

[excellent artificial enzyme]

= [discriminating binding site] + [efficient catalytic site]

= [multiple recognition] + [polyfunctional catalysis]   (6)

But, firstly, let us see some examples where parent cyclodextrins behave
as excellent catalysis. The most popular and important examples are hydrolysis of esters. This esterase (peptidase) model is a kind of historical monument in cyclodextrin chemistry. Fortunately, there is a good book written
by Bender and Komiyama[1] describing these model studies which allows me
to pass over this important and interesting chemistry in this chapter. The
next monument must be the discriminating para-chlorination of anisole in
the α-cyclodextrin inclusion complex.[27] This elegant chemistry is discussed
by Breslow himself in his chapter of this book. (Chapter 14, Volume 3).

Another example of a cyclodextrin catalyst acting as an efficient artificial enzyme is artificial vitamin K synthetase (Fig. 7). This catalysis is discussed here in detail, since some fundamental and important chemistry is involved.[28-30]

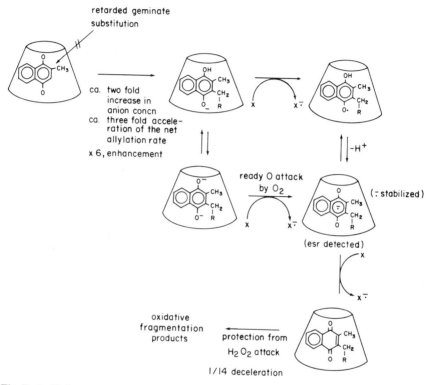

Fig. 7. Artificial vitamin-K synthetase activity of β-cyclodextrin.[30]

The first event involved is the strong binding of naphthohydroquinone to the β-cyclodextrin cavity ($K_{ass} = 490 \text{ M}^{-1}$), which was ascertained by the fluorescence enhancement (see Fig. 8). The bound naphthohydroquinone monoanion is hydrated at the oxygen atoms but not (or only poorly) hydrated at the ring carbons. This assumption is based on the X-ray crystallographic results that no water molecule is present in the cyclodextrin–aromatic inclusion complex. Thus, both ring carbons C(2) and C(3) should be activated toward electrophilic attack by an allyl bromide. However, the C(2) carbon does not have enough space to accommodate allyl bromide due to the methyl substituent. Consequently, discriminating C(3) substitution takes place. This second step is followed by oxidation, leading to the effective

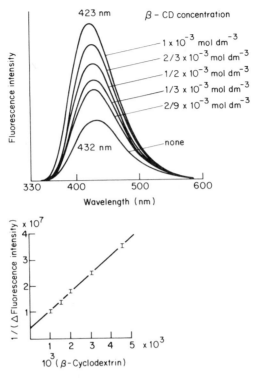

Fig. 8. Fluorescence enhancement by β-CD–naphthohydroquinone inclusion complex formation.

formation of the corresponding anion radical, which is stabilized in the cavity to a remarkable degree, as ascertained by ESR measurement. Finally reoxidation at oxygens (out of cavity) takes place selectively, giving vitamin K analogs. The final product is satisfactorily protected by formation of the inclusion complex as ascertained by independent measurements. Thus, further oxidation of the quinone with H_2O_2 is retarded by a ratio of $1/14$ in the β-cyclodextrin cavity (see Fig. 9).

3. Synthetic strategy for substituted cyclodextrins

As discussed previously, the catalytic effect of parent cyclodextrins has a serious limitation in that the functional groupings available are somewhat too weak to exhibit enzymatic activity. This limitation may be overcome by

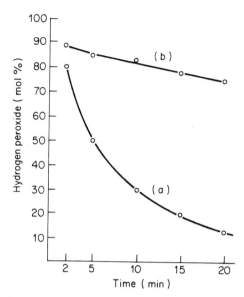

Fig. 9. Protective effect of β-CD inclusion against further guest oxidation. Quantitative determination of hydrogen peroxide produced in the oxidation of 2-methylhydronaphthoquinone ($0.005 \text{ mol dm}^{-3}$) (a) in the absence and (b) in the presence of β-CD (0.01 mol dm^{-3}).

the introduction of more potent functional groups such as imidazole or hydroxamic acid (see Table 3) based on the general strategy shown in Equation 6 (page 452).

However, many of the monofunctional artificial enzymes derived from β-cyclodextrin are still less active than the sophisticated native enzymes. Sophisticated enzymes use several functional groups for a single catalytic site, with several types of cooperation schematically shown in Fig. 10.

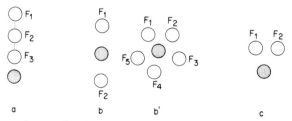

Fig. 10. Cooperation mode of functional groups in enzymatic catalysis. (a) Relay-mode (edge-on mode) e.g. chymotrypsin; (b) sandwich modes, e.g. lysozyme; (c) side-on mode, e.g. aminotransferase?

Table 3. Typical monofunctional cyclodextrins as artificial enzymes

Functional group	Position	Catalytic action	Reference
$(-NH \quad NH_2)_2 Cu$	prim	[furfuryl alcohol / difuryl ketone]	52
$-NH \quad NH \quad NH_2-Ni(Mn)$ $-NH \quad NH \quad NH_2-Ni(Mn)$	prim	[cyclohexanone dicarboxylic acid → keto carboxylic acid]	53
$-SCH_2CH_2CON(CH_3)OH$	prim	$ArOAc + H_2O$	54
$-N(CH_3)_3Cl$	prim	$ArOAc + H_2O$	55
[Fe–S cluster]	prim	$Fe^{III} + e \rightleftharpoons Fe^{II}$	56

—OPO$_3^=$	sec	PhCH$_2$SCH$_2$Ph + I$_2$ + H$_2$O → PhCH$_2$SOCH$_2$Ph	57
—OPO$_3^=$	sec	Buφ—C(=O)—CH$_2$OH + T$^+$ ⇌ Buφ—C(=O)—CHTOH	58
—OPO$_3^=$ (—NH— imidazolylethyl)	sec	[tetrahydropyranyl ether] —O—⟨C$_6$H$_4$⟩—NO$_2$ + H$_2$O	59
(imidazole)	sec	ArOAc + H$_2$O	60
—CH$_2$CON(CH$_3$)OH	sec	ArOAc + H$_2$O	61

In order to prepare such sophisticated artificial enzymes, the regiospecific and "combination specific" introduction of two (or more) functional groups (F) is necessary. Fortunately among the $3n$ ($n = 6, 7, 8$, etc.) hydroxyl groups in the cyclodextrins, the n primary hydroxyl groups are much more reactive than the $2n$ secondary hydroxyl groups, in a homogeneous organic medium.

Even so, we still have $(n - 1)$ regioisomers of unsymmetrically disubstituted cyclodextrins, as schematically depicted for β-cyclodextrin (Fig. 11). Thus, there are six unsymmetrically and three (A,B; A,C; A,D) symmetrically disubstituted β-cyclodextrins. The author's group has found that

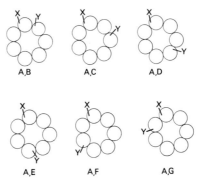

Fig. 11. Schematic representation of six regioisomers of disubstituted β-cyclodextrin.

the bifunctional modifying reagents (called "capping" reagents) listed in Table 4, having rigid skeletal structures, were very helpful for the regiospecific or regioselective bifunctionalization of β-cyclodextrins.[31] The basic idea of this bifunctionalization is that when one of the two reactive sites of a capping reagent is attached to one of the primary hydroxyl groups of the β-CD to form a covalent bond, the other reactive site is forced to come into close proximity with a certain primary hydroxyl group of β-CD due to the rigidity of the backbone of the modifying reagent (see Fig. 12). The

Fig. 12. Looper's walk mechanism of rigid capping. **OH** takes vicinal position and OH takes distal position towards the second reactive site of the rigid reagent.

Table 4. Rigid capping reagents

Loopers	AB	AC	AD	Reference
ClO_2S—⬡—CH=CH—⬡—SO_2Cl (stilbene)	0	0	100	32, 33
ClO_2S—⬡—⬡—SO_2Cl (biphenyl)	0	2	98	40[(a)]
ClO_2S——phenanthrene——SO_2Cl	0	92	8	33
ClO_2S——carbazole (N–Me)——SO_2Cl	0	90	10	
ClO_2S—⬡—CH_2—N(Me)—⬡—SO_2Cl	12	62	26	36
ClO_2S—⬡—CH_2—⬡—SO_2Cl	0	62	38	32
ClO_2S—⬡—O—⬡—SO_2Cl	0	ca. 50	ca. 50	40[(a)]

Table 4—continued

Loopers	AB	AC	AD	Reference
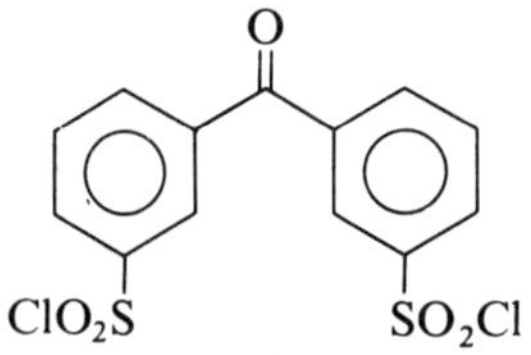 (benzophenone with ClO₂S and SO₂Cl substituents)	0	78	22	32
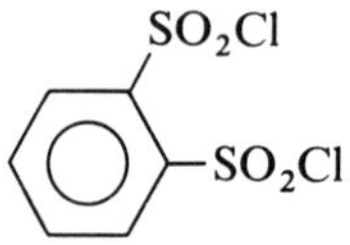 (benzene with two SO₂Cl substituents)	major	—	—	(b)
 ClOC—⟨benzene⟩—COCl	0	—	—	31

(a) This capped cyclodextrin was first synthesized by Breslow *et al.*, and the ratio of AB/AC/AD was determined by us.
(b) Unpublished result.

structure determination of these capped cyclodextrins was very difficult but was finally achieved by classical techniques, i.e. A,D-capping must be made singly[32,33] but A,C-capping may be made singly and doubly.[32,34] Single, double and triple capping are possible for A,B-capping.[35] This, combined with ^{13}C-NMR spectra where the chemical shift change of a CH$_3$ group, for example, caused by the remote substituent CH$_3$ as shown for the A,C-isomer in **1** was

$$\text{capped-}\beta\text{-CD} \xrightarrow{\text{KI}} \beta\text{-CD I}_2 \xrightarrow{\text{NaBH}_4}$$ 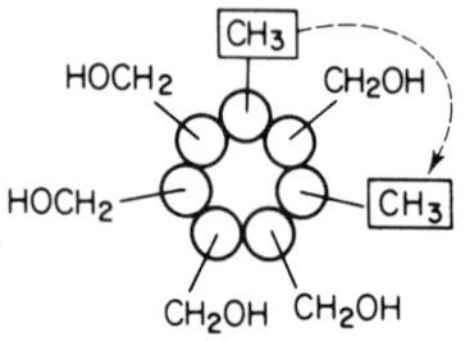 (7)

(1)

only slightly appreciable for the A,B-isomer and too small for the others, determines the structure of the regioisomers (see Fig. 13). The regioisomer ratios determined for a series of rigid capping reagents, "loopers", by the author's group are listed in Table 4.[35]

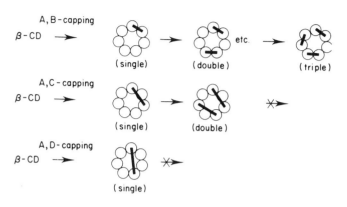

Fig. 13. Regioisomerism and multiple capping.

Another type of regioisomerism should be taken into account in the specific preparation of unsymmetrically disubstituted cyclodextrins. As discussed above, the distance between two substituents to be introduced is satisfactorily controlled and the A,B; A,C; or A,D regioisomer is prepared specifically. Each regioisomer, however, is a statistical mixture of three possible combinations, XX + XY + YY, when different substituents X and Y are to be introduced. Isolation and purification of the desired product CDXY is not at all easy due to serious contamination with other side products of similar physical properties.

The author's group has successfully solved the problem by using a combination-selective cap (or "flamingo-cap") as shown in Fig. 14.[36] The rate ratio, k_1/k_2 in Fig. 14, was found to be *ca* 13 and when the first nucleophilic substitution was carried out with N_3^- and the second with a large excess of RS^-, unsymmetrically disubstituted cyclodextrin was obtained in 59% yield together with 9.6% of symmetrically disubstituted compound, $(RS)_2CD$ (product ratio $CDX_2:CDXY:CDY_2$ was $0:86:14$).

Thus, at this stage, any cyclodextrin derivative having several different substituents at the desired positions may be prepared selectively by applying the general strategy discussed above, as illustrated in Fig. 15. Since protein–protein or protein–small molecule "total recognition" is achieved through multiple "elemental recognition" interactions (*vide infra*), any sophisticated protein model requires hydrophobic recognition sites as well as various other types of recognition site (polar, hydrogen bonding, metal coordination

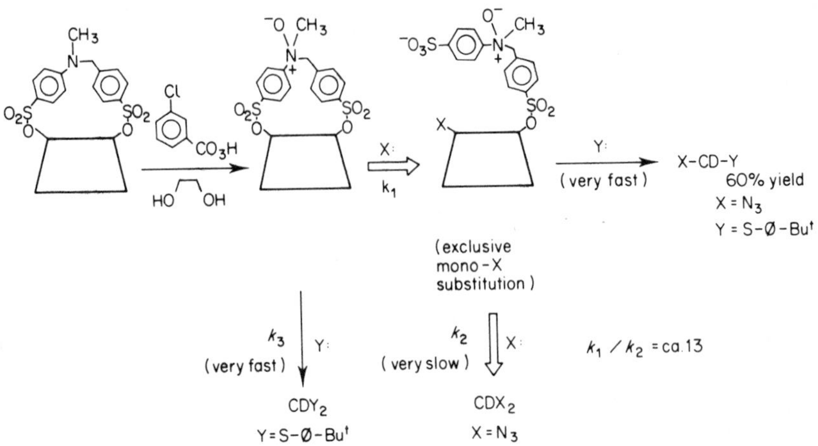

Fig. 14. Preparation of an unsymmetrically disubstituted cyclodextrin.

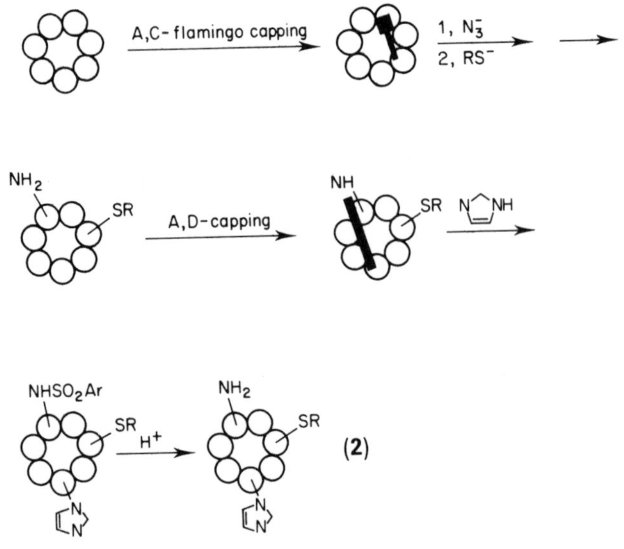

Fig. 15. Hypothetical introduction of several substituents into cyclodextrin at the desired positions.

or charge transfer) as parts of the molecule. This kind of hypothetical target molecule, (2) for example, could be prepared very readily. In this way, we can now prepare as wide a variety of protein models as we wish.

4. Biomimetic function of cyclodextrins

We now have access to any kind of desired cyclodextrin derivative, the structure of which is very similar to that of a part of a certain native globular protein. Is this structural similarity a necessary and sufficient condition to provide a certain function of a protein such as enzymatic catalytic activity, receptor activity, transport control activity or a trigger activity of stimulus-response? Unfortunately, experimental data available at present along this line of investigation are far from sufficient. Even so, we still anticipate a promising future for this field of "total functionalization" of protein activities by the use of cyclodextrin derivatives, based on the rather scarce but unique results already .obtained.

Let our discussion begin from enzyme model studies. Enzymes "recognize" the shapes and sizes of specific guests through their multiple elemental interactions and they accelerate the rates of specific reactions through appropriate alignment of the enzyme's functional groups and the substrate functional group. This situation is schematically shown, with considerable simplification, in Fig. 16, taking chymotrypsin as an example. Indole (or phenyl, or *p*-oxyphenyl) is bound tightly (based on X-ray, NMR correlation time and NMR chemical shift) to the hydrophobic pocket of the enzyme, which determines the orientation of the guest (see Fig. 16a). At the "rim" of the binding site, two serine OH groups are present at the hydrogen-bonding distance to two amide (ester) groupings of the guest. One OH

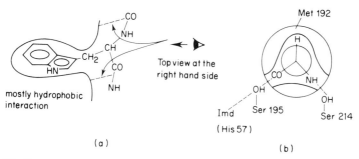

Fig. 16. Enzyme complex of chymotrypsin and a tryptophame derivative. (a) binding mode; (b) top view of the active site.

group of the two is activated by neighbouring His[57] but the other is not. This is the super-simplified aspect of the discriminating acylation step of chymotrypsin with the specific guest. And this is a kind of situation that arises in enzymatic catalysis in general. By the use of cyclodextrin derivatives, this kind of situation is, at least partly, mimicked.

Although parent cyclodextrins bind many organic substrates strongly (K_a 10^1–10^3 M^{-1}), their binding is still considerably weaker than many native enzymes. The hydrophobic recognition is remarkably enhanced by the further introduction of a hydrophobic surface to the parent cyclodextrins, as was shown for the capped β-cyclodextrins[28] (see Table 5). These observa-

Table 5. Association constants (M^{-1}) of β-cyclodextrins and guests

β-CD	$\left[O_2C-\bigcirc-CO_2\right]$ $(\beta\text{-CD})$	$\left[O_2S-\phi^P-CH_2-\phi^{P'}-SO_2\right]$ $(\beta\text{-CD})$
	58	640	1300
$\rightarrow CO_2^-$	—	—	250
$-CO_2^-$	— .	—	50 000

tions strongly support the hydrophobic recognition mechanism proposed (see Equation 2) on the one hand, and allow us to make a probable "molecular design" for artificial enzymes on the other. Fujita succeeded in showing that this strengthened hydrophobic binding affects the hydrolytic activity of capped cyclodextrin significantly where reactivity and selectivity were greatly enhanced[37,38] (see Equation 8 and Table 6).

$$ \tag{8} $$

Starting from sulphonate capped cyclodextrins, a variety of disubstituted cyclodextrins are conveniently prepared[39] as shown in Fig. 17.

Table 6. *Effective hydrolysis rates* $(k_c/K_d, M^{-1}s^{-1})$ *for substituted phenyl acetates*

X	Artificial enzyme		
	β-CD	β-CD-1°-SMe	**3**
m-NO$_2$	54.7	86.1	272
p-NO$_2$	11.2	15.8	*2260*
m-Me	2.7	4.1	17.1
p-Me	1.9	1.9	*58.4*

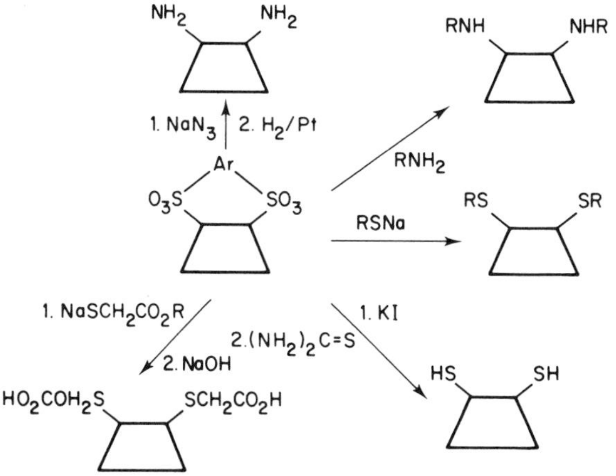

Fig. 17. Preparation of disubstituted cyclodextrins from sulphonate capped cyclodextrin.

Breslow successfully applied this type of reaction to the preparation of diimidazolyl-β-CD (Equation 9), which catalyses accelerated and discriminating hydrolysis of aromatic cyclic diphosphate. Also interesting is

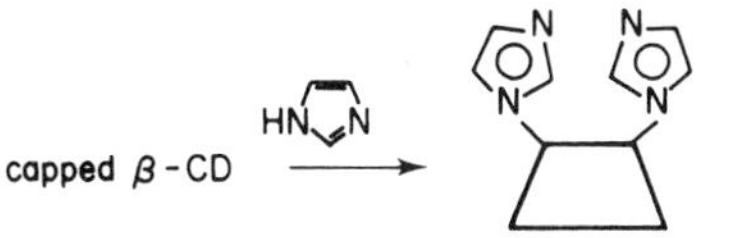

$$ (9) $$

the "bell-shaped" pH dependence of the hydrolysis rate,[40] as discussed in more detail in the following chapter.

Since the precise molecular design of hosts capable of certain (desired) molecular recognition is rather difficult in cyclodextrin chemistry, an experimental approach has been attempted by the author's group. For this purpose, the assumption is made that the total recognition free energy may be given by the sum of the free energies for the individual elemental recognitions involved. A test compound (reaction) was prepared for this purpose and also as the simplest metalloenzyme model, which has two practically independent recognition sites.[41] Measurements of the dissociation constants for a series of guests (see Table 7) indicates that the total recognition free

Table 7. Dissociation constants of holohost–substrate complexes formed from apohost **4**

$$(10)$$

Guest	Host	K_d (M)	$K_d(\beta\text{-CD})$ / K_d (holohost)
Ad(O)CO₂⁻	β-CD	1.2×10^{-3}	
	4–Zn^{2+}	3.6×10^{-6}	330
CO₂⁻	β-CD	4.3×10^{-3}	
	4–Zn^{2+}	1.9×10^{-4}	22.6
H— CO₂⁻	β-CD	7.1×10^{-3}	
	4–Zn^{2+}	5.3×10^{-4}	13.4
H— NH₂	β-CD	2.2×10^{-2}	
	4–Zn^{2+}	4.7×10^{-3}	4.7
H— OH	β-CD	2.0×10^{-3}	
	4–Zn^{2+}	2.4×10^{-3}	0.83

* $Ad(O)CO_2H$ = Adamantane-2-one-1-carboxylic acid.

energy is really given by the sum of the individual recognition energies (Equation 11). Thus, this special host for "*double recognition*" not only shows remarkable and unique substrate (shape, size or polarity) specificity ($K_{Ad(O)CO_2H*}/K_{phenol} = 665$, for example) and enzyme specificity (toward $Ad(O)CO_2H*$, $K_{CD.Zn^{++}}/K_{CD} = 330$, for example), but also affords a "building-up" principle for host design. Assume that we are looking for a specific host towards a certain bioactive compound A. As shown in Fig. 18 we should

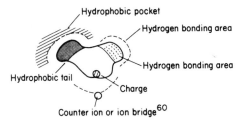

Fig. 18. Recognition counterparts for effective host-guest interaction.

choose necessary fragments as "recognition counterparts" to interact specifically with each recognition element of molecule A, where the expected total recognition energy should satisfy the relationship shown in Equation 12.

$$\Delta G \text{ (total)} = \Delta G \text{ (metal coordination)} + \Delta G \text{ (hydrophobic)} \qquad (11)$$

$$\Delta G \text{ (total)} = \sum_i \Delta G_i \text{ (elemental)} \leqslant \Delta G \text{ (assumed)} \qquad (12)$$

Successful application of this principle to the preparation of a sophisticated artificial enzyme was made by the author's group. Duplex cyclodextrin (5) has two binding sites B_1 and B_2 (not necessarily the same) connected to

$$(13)$$

each other by polar catalytic groups. This host, capable of triple recognition, binds two substrates and is useful as an artificial ligase.[42] A typical reaction catalysed by duplex cyclodextrin is amide formation from an activated carboxylic acid and an amine as shown in Equation 14.

$$\text{RCOX} + \text{R'NH}_2 \rightarrow \text{RCONHR'} \qquad (14)$$

The "building-up" principle of multi-recognition may also be helpful for chiral selection or induction. A typical example is β-CD–aminocarboxylic acid (6) overleaf, prepared recently by the author's group. Our preliminary experiments strongly suggest that a strong discriminating interaction exists between (6) and amino acids, based on circular dichroism spectra.[43]

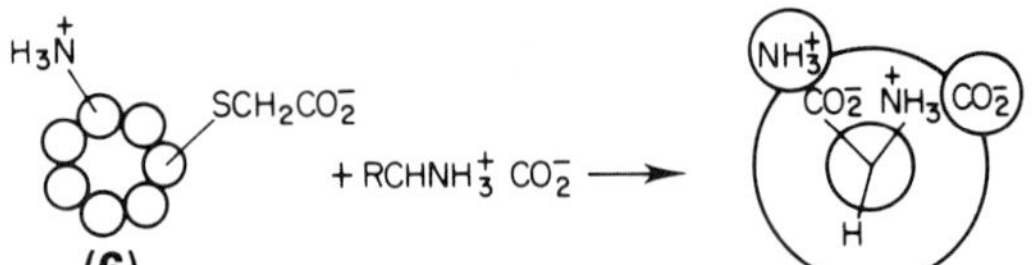

(6)

In addition to enzyme or receptor model studies, the chemistry of cyclo-dextrin inclusion will afford other protein models as well. Starting from A,C; A',C'-doubly capped β-CD, β-CD having four hydrophobic tails and two metal binding sites **(7)** was recently prepared by the author's group as

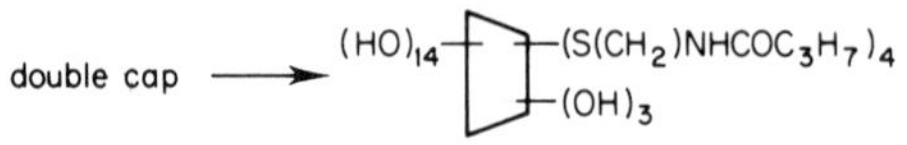

$$\text{double cap} \longrightarrow (HO)_{14} \begin{array}{c} -(S(CH_2)NHCOC_3H_7)_4 \\ -(OH)_3 \end{array} \tag{15}$$

(7)

a possible candidate for a channel forming host.[44] This compound is capable of weakly binding transition metal ions such as Cu^{2+}, Co^{2+} or Ni^{2+}. When **(7)** was incorporated into a lecithin bimolecular liposomal membrane, these metal ions were transported efficiently from the outside to the inside of the liposome where a stronger ligand was present. The rate of metal ion transport was second order with respect to **(7)**, strongly suggesting the formation of a kind of channel composed of the **(7)** dimer.

The multi-recognition principle also provides artificial systems mimicking visual pigments. A remarkable red shift was observed for retinal bound to ethylenediamino–β-CD **(8)** in aqueous acid.[45] Interestingly, the red shift observed for the "artificial protein" **(8)** is about the same in magnitude as native rhodopsins (bovine rhodopsin, 498 nm; bovine lumirhodopsin, 497 nm). Although the red shift for the native rhodopsins is mainly due to the stabilization of the delocalized positive charge in the excited state by neighbouring CO_2^-,[46,47] that for the present artificial pigment is mainly due to the destabilization of the ground state by neighbouring N^+ and also by the strain induced by the binding to some extent, as apparently shown from comparison with related compounds **(9–11)**.[48] However, it is very interesting

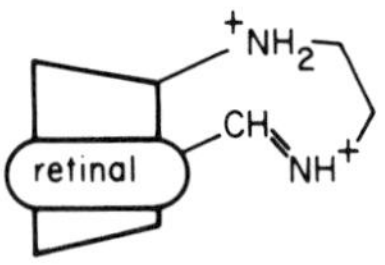

λ_{max}, 497 nm

(8)

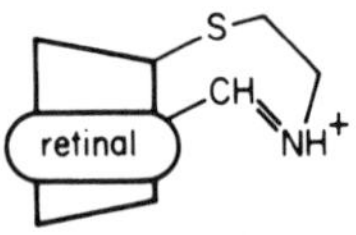

445 nm

(10)

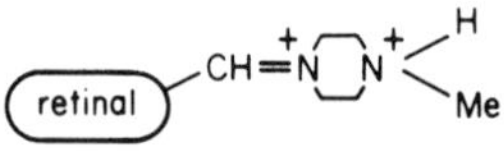

486 nm

(9)

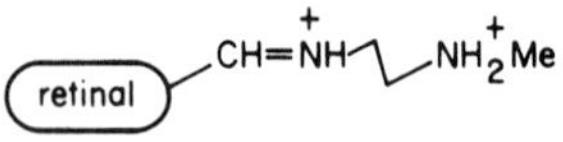

471 nm

(11)

and noteworthy that this special function of native protein is also artificially reproduced based on an extended (or generalized) mechanism. Since very efficient and specific host–guest energy transfer is already known for the benzophenone-cap,[49] the function of stimulus-response in visual systems will possibly be artificially reconstructed in an entirely different way.

5. Conclusion

At present, a general strategy and practical techniques for "host design" by the use of cyclodextrins are available. By applying these to the preparation of artificial systems mimicking various important functions of native proteins, the nature of the protein functions will be elucidated more easily and also more fundamentally. Further efforts should be concentrated in future on the application of the strategy and techniques to practical (industrial) purposes.

References

1. M. L. Bender and M. Komiyama, *Cyclodextrin Chemistry*, Springer-Verlag, 1978.
2. I. Tabushi, H. Yamada, Z. Yoshida and R. Oda, *Tetrahedron*, 1971, **27**, 4845.
3. I. Tabushi, H. Yamada, K. Matsushita, Z. Yoshida, Y. Kuroda and R. Oda, *Tetrahedron*, 1972, **28**, 3381.
4. I. Tabushi, H. Yamada and Y. Kuroda, *J. Org. Chem.*, 1975, **40**, 1946.
5. F. Imashiro, Z. Yoshida and I. Tabushi, *Tetrahedron*, 1973, **29**, 3521.
6. I. Tabushi, H. Yamada, Z. Yoshida and K. Obuchi, *Catalyst* (Jpn), 1972, **14**, 147.
7. I. Tabushi and Y. Kuroda, *Catalyst* (Jpn), 1974, **16**, 78.
8. I. Tabushi, H. Sasaki and Y. Kuroda, *J. Am. Chem. Soc.*, 1976, **98**, 5727.
9. I. Tabushi, Y. Kuroda and K. Kimura, *Tetrahedron Lett.*, 1976, 3327.
10. I. Tabushi, Y. Kimura and K. Yamamura, *J. Am. Chem. Soc.*, 1978, **100**, 1304.
11. I. Tabushi, Y. Kimura and K. Yamamura, *J. Am. Chem. Soc.*, 1981, **103**, 6486.
12. A. Hybl, R. E. Rundle and D. E. Williams, *J. Am. Chem. Soc.*, 1965, **87**, 2779.
13. P. C. Manor and W. Saenger, *Nature* (London), 1972, **237**, 392. See also Chapter 8, Volume 2 for further examples.
14. K. Harata and H. Uedaira, *Nature* (London), 1975, **253**, 190.
15. I. Tabushi and K. Yamamura, *Top. Curr. Chem.*, 1983, **113**, 145.
16. F. Cramer, W. Saenger and H. Spatz, *J. Am. Chem. Soc.*, 1967, **89**, 14.
17. R. L. van Etten, J. F. Sebastian, G. A. Clowes and M. L. Bender, *J. Am. Chem. Soc.*, 1967, **89**, 3242.
18. K. Sensse and F. Cramer, *Chem. Ber.*, 1969, **102**, 509.
19. R. M. Paton and E. T. Kaiser, *J. Am. Chem. Soc.*, 1970, **92**, 4723.
20. J. P. Behr and J. M. Lehn, *J. Phys. Colloq.* (Paris), 1973, 55.

21. P. V. Demarco and A. L. Thakkar, *J. Chem. Soc., Chem. Commun.*, 1970, 2.
22. R. J. Bergerson and R. Rowan, *Bioorg. Chem.*, 1976, **5**, 425.
23. E. A. Lewis and L. D. Hansen, *J. Chem. Soc., Perkin Trans. 2*, 1973, 2081.
24. I. Tabushi, Y. Kiyosuke and K. Yamamura, *J. Am. Chem. Soc.*, 1978, **100**, 916.
25. R. I. Gelb, L. M. Schwartz, B. Cardelino, H. S. Fuhrman, R. F. Johnson and D. A. Laufer, *J. Am. Chem. Soc.*, 1981, **103**, 1750.
26. K. Hirotsu, T. Higuchi, K. Fujita, T. Ueda, A. Shinoda, T. Imoto and I. Tabushi, *J. Org. Chem.*, 1982, **47**, 1143.
27. R. Breslow and P. Campbell, *J. Am. Chem. Soc.*, 1969, **91**, 3085.
28. I. Tabushi, K. Fujita and H. Kawakubo, *J. Am. Chem. Soc.*, 1977, **99**, 6456.
29. I. Tabushi, Y. Kuroda, K. Fujita and H. Kawakubo, *Tetrahedron Lett.*, 1978, 2083.
30. I. Tabushi, K. Yamamura, K. Fujita and H. Kawakubo, *J. Am. Chem. Soc.*, 1979, **101**, 1019.
31. I. Tabushi, K. Shimokawa, N. Shimizu, H. Shirakata and K. Fujita, *J. Am. Chem. Soc.*, 1976, **98**, 7855.
32. I. Tabushi, Y. Kuroda, K. Yokota and L. C. Yuan, *J. Am. Chem. Soc.*, 1981, **103**, 711.
33. I. Tabushi and L. C. Yuan, *J. Am. Chem. Soc.*, 1981, **103**, 3574.
34. I. Tabushi, L. C. Yuan, K. Shimokawa, T. Mizutani and Y. Kuroda, *Tetrahedron Lett.*, 1981, **22**, 2273.
35. I. Tabushi and Y. Kuroda, *Adv. Catal.*, 1983, **32**, 417.
36. I. Tabushi, T. Nabeshima, H. Kitaguchi and K. Yamamura, *J. Am. Chem. Soc.*, 1982, **104**, 2017.
37. K. Fujita, A. Shinoda and I. Imoto, *J. Am. Chem. Soc.*, 1980, **102**, 1161.
38. K. Fujita, A. Shinoda and I. Imoto, *Tetrahedron Lett.*, 1980, **21**, 1541.
39. I. Tabushi, K. Shimokawa and K. Fujita, *Tetrahedron Lett.*, 1977, 1527.
40. (a) R. Breslow, J. B. Doherty, G. Guillot and C. Lipsey, *J. Am. Chem. Soc.*, 1978, **100**, 3227. (b) R. Breslow, P. Bovy and C. L. Hersh, *J. Am. Chem. Soc.*, 1980, **102**, 2115.
41. I. Tabushi, N. Shimizu, T. Sugimoto, M. Shiozuka and K. Yamamura, *J. Am. Chem. Soc.*, 1977, **99**, 7100.
42. I. Tabushi, Y. Kuroda and K. Shimokawa, *J. Am. Chem. Soc.*, 1979, **101**, 2785.
43. I. Tabushi, Y. Kuroda and T. Mizutani, Annual Meeting of Chemical Society of Japan, Tokyo (1982).
44. I. Tabushi, Y. Kuroda and K. Yokota, Annual Meeting of Chemical Society of Japan, Tokyo (1982).
45. I. Tabushi, Y. Kuroda and K. Shimokawa, *J. Am. Chem. Soc.*, 1979, **101**, 1614.
46. K. Nakanishi, V. Galough-Nair, M. Arnaboldi and K. Tsujimoto, *J. Am. Chem. Soc.*, 1980, **102**, 7945.
47. M. Shoves, K. Nakanishi and B. Honig, *J. Am. Chem. Soc.*, 1979, **101**, 7087.
48. I. Tabushi and K. Shimokawa, *J. Am. Chem. Soc.*, 1980, **102**, 5400.
49. I. Tabushi, K. Fujita and L. C. Yuan, *Tetrahedron Lett.*, 1977, 2503.
50. N. Yoshida and M. Fujimoto, *Chem. Lett.*, 1980, 1377.
51. R. P. Rohrbach, L. J. Rodriguez, E. M. Eyring and J. F. Wojcik, *J. Phys. Chem.*, 1977, **81**, 944.
52. Y. Matsui, T. Yokoi and K. Mochida, *Chem. Lett.*, 1976, 1037.
53. I. Tabushi, N. Shimizu and K. Yamamura, *J. Am. Chem. Soc.*, 1977, **99**, 7100; 26th International Congress of PAC, Tokyo, Abstract, I, p. 44 (1977).
54. I. Tabushi, Y. Kuroda and Y. Sakata, *Heterocycles*, 1981, **15**, 815.
55. Y. Matsui and A. Okimoto, *Bull. Chem. Soc. Jpn*, 1978, **51**, 3030.

56. B. Siegel, *J. Inorg. Nucl. Chem.*, 1979, **41**, 609.
57. T. Eiki and W. Tagaki, *Chem. Lett.*, 1980, 1063.
58. B. Siegel, A. Pinter and R. Breslow, *J. Am. Chem. Soc.*, 1977, **99**, 2309.
59. Y. Iwakura, K. Uno, F. Toda and M. L. Bender, *J. Am. Chem. Soc.*, 1975, **97**, 4432.
60. Y. Kitaura and M. L. Bender, *Bioorg. Chem.*, 1975, **4**, 219.
61. I. Tabushi, K. Kiyosuke and K. Yamamura, *J. Am. Chem. Soc.*, 1981, **103**, 5255.

14 · ENZYME MODELS RELATED TO INCLUSION COMPOUNDS

R. BRESLOW

Columbia University, New York, USA

1. Introduction

There are several reasons to want to imitate enzymes. Firstly, such imitation is an important tool in understanding enzymatic reactions. By examining enzyme model systems we can obtain evidence about the possible contributions of various factors which are believed to operate in the enzymatic reaction itself, and we can also test and demonstrate our understanding by showing that the factors we invoke do indeed lead to enzyme-like behaviour. A second general reason to want to imitate enzymes is that they perform chemistry which is more attractive than the chemical reactions of the laboratory or of the chemical industry. Enzymes can achieve some very large catalytic rate accelerations under extremely mild conditions with simple acids and bases as catalytic groups. These are extremely impressive and challenge chemists to learn to do as well. Perhaps even more striking, enzymatic reactions typicaliy show selectivities which are only rarely achieved in non-biological chemistry. This includes selection of a substrate by physical recognition of the shape of that substrate, selection of a region

INCLUSION COMPOUNDS III
ISBN 0-12-067103-4

of the substrate for reaction (which might not be the most intrinsically reactive part of the substrate), and reaction with great stereochemical control which frequently makes it possible for enzymes to operate on only one enantiomer of a substrate, or to produce only one enantiomer from a pro-chiral substrate. Because this biological chemistry is so effective, it is worthwhile learning how to imitate it so as to improve the tools of modern chemistry. Even if such mimics did not contribute to our understanding of natural enzymes, they would still represent attractive approaches to the development of novel synthetic methods with the same rate and selectivity advantages which enzymes show.

At Columbia University we have been pursuing a programme aimed at the imitation of enzymatic reactions for more than 25 years. Some parts of this programme have not been focused on inclusion compounds, but instead have dealt with models for coenzyme-catalysed reactions,[1] models for metal-catalysed reactions,[2] or models for selective functionalization reactions in which rigid attached template or reagent species are utilized.[3] However, the properties of inclusion compounds were so attractive that we have used them in our enzyme model studies for most of this period.

2. Selective aromatic substitution

Our earliest published work in this area involved an examination of the use of inclusion compounds to modify an aromatic substitution reaction.[4,5] This was based on the considerations outlined above, which emphasize the attractive selectivity available to most enzyme-catalyzed reactions. It was known that aromatic compounds formed soluble complexes with the cyclo-dextrins, but no studies had been carried out on the effect of this binding on the characteristic aromatic substitution pattern of the substrate. Enzymes, in general, use hydrophobic binding as a major component of their interaction with substrates. Thus we wanted to see whether we could use the known hydrophobic binding of aromatic compounds into the cavity of cyclodextrins in order to achieve geometric control of aromatic substitution reactions. Molecular models suggested that when a substrate such as anisole binds into the cavity of a cyclodextrin its *ortho* and *meta* positions would be shielded by the cyclodextrin, while the *para*-position is potentially accessible and might be substituted in an appropriate reaction. This expectation was related to the known geometric requirements for binding of aromatic compounds into these cavities, in which *para*-disubstituted aromatic compounds were well bound compared to *ortho*-disubstituted compounds.

One immediate problem with this concept was that inclusion complexes are in reversible equilibrium with the free components. It might well be that attack on a bound anisole molecule would go preferentially to the *para*-position; but, if even this favoured part of the molecule were to be unreactive, compared with the free substrate, then all the substitution reactions might occur only with the unbound anisole in equilibrium with the complex. In this case, of course, there would be no change in the typical aromatic substitution pattern. As it turned out, for the reaction which we studied in most detail, this did not prove to be a problem. Instead of being less reactive in the complex, the anisole molecule was more reactive since cyclodextrin not only exerted geometric control but also furnished positive chemical catalysis of the aromatic substitution process.

Under ordinary conditions, chlorination of anisole by hypochlorous acid affords both *para-* and *ortho*-chloroanisole, so it represents the kind of random reaction which we hoped to make selective by the use of the geometric constraints in the complex (Fig. 1). The results were very striking.

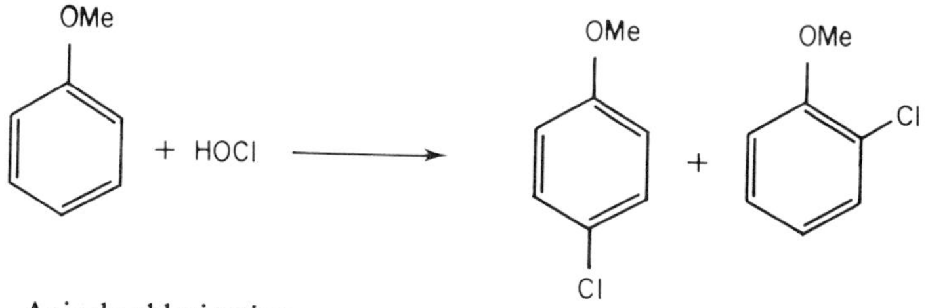

Fig. 1. Anisole chlorination.

As Table 1 shows, chlorination of anisole in the absence of a cyclodextrin under our reaction conditions gave 60% *p*-chloroanisole and 40% *o*-chloroanisole. However, when cyclohexaamylose (α-cyclodextrin) was present at various concentrations up to 10 m mol dm^{-3}, the product changed to as much as 96% *para*-substitution and 4% *ortho*-substitution. At higher concentrations of the cyclodextrin *ortho* substitution can be suppressed even further. As Table 1 also shows, a related but smaller effect is seen with cycloheptaamylose (β-cyclodextrin). At 10 m mol dm^{-3} β-cyclodextrin the product is 79% *p*-chloroanisole and 21% *o*-chloroanisole.

It is possible to determine the fraction of anisole bound under all of these concentration conditions. We did this both by direct methods, in which the change in the ultraviolet spectrum of anisole as the result of binding was used to determine the dissociation constant, and by indirect methods in which anisole binding was used to inhibit the binding and reaction of cyclodextrin with another substrate whose reaction was followed. These

Table 1. Anisole chlorination products

Concentration (m mol dm^{-3})	Fraction of anisole bound	Chloroanisole product ratio, p/o
Cyclohexaamylose (α-cyclodextrin)		
0	0	1.48
0.933	0.20	3.43
1.686	0.33	5.49
2.80	0.43	7.42
4.68	0.56	11.3
6.56	0.64	15.4
9.39	0.72	21.6
Cycloheptaamylose (β-cyclodextrin)		
0.944	0.12	1.69
1.752	0.20	1.88
2.98	0.29	2.16
5.07	0.41	2.63
7.05	0.50	3.08
10.12	0.59	3.77

two methods led to good agreement. On the basis of these determinations, the data in Table 1 include the fraction of anisole bound under our various concentration conditions with the two cyclodextrins. Strikingly, the chlorination of anisole with α-cyclodextrin gives p-chloroanisole almost exclusively even when only 72% of the anisole is actually bound in the cavity. A moment's thought makes it clear that this result is only consistent with a faster reaction within the complex than in free solution. That is, the 28% of anisole which is not bound at 10 m mol dm^{-3} α-cyclodextrin is not making its normal contribution to the product pool, but instead reaction preferentially occurs within the complex.

It was possible[5] to treat these data by a modification of the partial rate constant method. By this general technique one selects an arbitrary reaction to have a partial rate constant of 1.0, and then reaction at other aromatic positions or under other conditions can be referred to this standard.

We found that all the data in Table 1 fits such a treatment very well. The rate constant for chlorination of free anisole in the *para* position was arbitrarily assigned the standard value of 1.0, from which each *ortho* position of anisole has a partial rate constant of 0.33. Since there are two such *ortho* positions, this is simply a way of describing again the 60–40 ratio of *para* to *ortho* chlorination for free anisole under our conditions. When the cyclodextrins were present any individual product is formed by a combination of reactions of free anisole and reactions of bound anisole, so that for instance the *ortho*-chloroanisole could be formed from processes occurring both with free and bound substrate. Treating all the data of Table 1 by this model we found that within small experimental error the rate constant for *ortho*-chlorination within the complex must be zero, as is shown in Table 2. This confirmed our original prediction that the *ortho*-position within such a complex would be buried inside the cavity and inaccessible to attack by the chlorinating agent.

Table 2. *Partial rate constants for anisole chlorination*

| | k_{free} | k_{bound} | |
		Cyclohexaamylose (α)	Cycloheptaamylose (β)
para	1	5.31 ± 0.05	1.1 ± 0.05
ortho	$0.67/2 \pm 0.02$	0.01 ± 0.05	-0.02 ± 0.05

Treatment of the data of Table 1 by this model indicates that under our reaction conditions chlorination of the *para* position of anisole is 5.3 times as fast in the α-cyclodextrin complex as it is in free solution. However, this particular value is valid only for our reaction conditions, in particular for a concentration of 10 m mol dm^{-3} HOCl, for reasons which will be described below. The conclusion that α-cyclodextrin is catalysing the rate of chlorination of anisole in the *para* position was also confirmed by direct kinetic measurements. The rate increase was as expected, in full accord with the values in Table 2 which were derived only from the product distribution data. As the data in Table 2 also indicate, under these particular reaction conditions the rate of attack on the *para* position of anisole in the complex with β-cyclodextrin is essentially the same as that for attack on uncomplexed anisole.

This entire situation is more complicated than it appears at first sight, since the reactions of free anisole and complexed anisole do not have the same kinetic dependence on the concentration of HOCl. This striking result indicates a major change in the mechanism of the process, but it also indicates that the partial rate factors of Table 2 are only valid for a particular concentration of HOCl. Chlorination of anisole in free solution had a second order dependence on HOCl, while chlorination of complexed anisole had a first order dependence. Thus if the HOCl concentration were 0.1 m mol dm^{-3}, the rate acceleration for *para* chlorination in the complex with α-cyclodextrin would be 531, not 5.31, while the complex with

β-cyclodextrin would show 100-fold catalysis rather than an unchanged reaction rate. More interesting than this point, however, is the implication of this change in kinetics for the mechanisms involved.

HOCl is in reversible equilibrium with Cl_2O and water, and Cl_2O is the reagent which normally attacks anisole in aqueous solution. Thus the second order kinetic dependence on HOCl simply reflects the expected dependence of Cl_2O concentration as the result of this equilibrium. It is apparent that the chlorinating agent involved in attack on the complexed anisole must be different, and in particular it must be a species with only one chlorine not two. There were really only two reasonable possibilities.

One possibility is that for the complex, the chlorinating agent is HOCl itself, which is for some reason reactive enough with complexed anisole, but not reactive enough to attack free anisole in competition with Cl_2O. The second possibility is that the chlorinating agent is a cyclodextrin hypochlorite, formed in equilibrium. It is well known that in the presence of HOCl, alcohols are in rapid equilibrium with alkyl hypochlorites. Molecular models suggested that in a complex with cyclodextrin the hydroxyl groups rimming the cavity would be in a position to deliver a chlorine to the *para* position of bound anisole (Fig. 2). Furthermore, this mechanism would explain the catalysis, and it would also explain the different effects of α-cyclodextrin and β-cyclodextrin. The catalysis, by this model, occurs because of well known neighbouring group and entropy effects. An otherwise insufficiently reactive hypochlorite species, which in

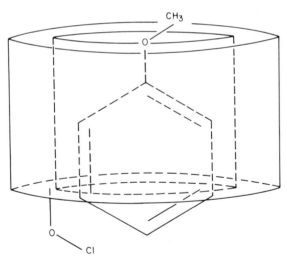

Fig. 2. The anisole–cyclodextrin complex. One of the hydroxyls which rim the cavity is shown as a hypochlorite, in a position to chlorinate the substrate.

free solution is only kinetically effective as its anhydride, Cl_2O, can be reactive enough in the complex because of the excellent proximity of a bound chlorine atom to the bound substrate. Such a picture also accounts for the poor catalysis by β-cyclodextrin, since the cavity is larger and the fit to the bound anisole is poorer. The resulting wobble in the complex should be expected to lead to a lower effective neighbouring group catalysis of the chlorination, as is observed.

This picture was checked by examining another aromatic substitution reaction in which such neighbouring group delivery of the electrophile should not occur. Diazo coupling reactions of phenols require attack by the free diazonium ion, so we examined the effect of cyclohexaamylose on the rate of coupling of phenol with *p*-diazoniumbenzenesulphonate.[5] The phenol binds to the cyclohexaamylose, but the diazonium salt does not. The observed rate of the coupling reaction was simply that expected for coupling with the free phenol in equilibrium with the complex, and the cyclohexaamylose simply inhibited this reaction by binding some of the phenol. Thus in this case in which an electrophile cannot be delivered by the neighbouring hydroxyl groups within the complex no attack at all was observed. This finding is further support for our proposition that in the chlorination reaction the chlorine atom is catalytically delivered by first binding to the hydroxyl groups (Fig. 3).

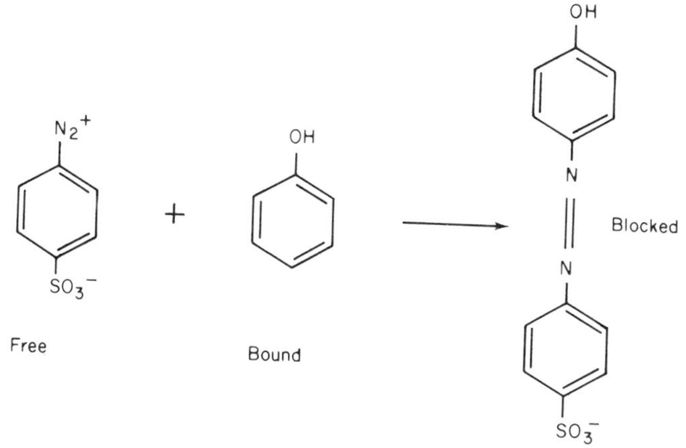

Fig. 3. Aromatic substitution which cannot be assisted by complexing with an α-cyclodextrin hydroxyl.

Studies have been done on some related substrates, with interesting results (Fig. 4). If the *para* position of anisole is blocked, as in *p*-methylanisole, then chlorination can only occur in the *ortho* position, but as with anisole we find that this chlorination is sterically blocked in the complex and the

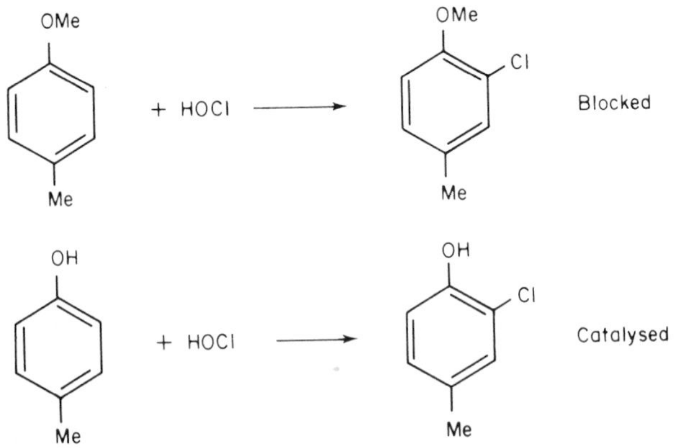

Fig. 4. Chlorination in α-cyclodextrin complexes.

entire rate of the reaction is simply that calculated for reaction of the free substrate in equilibrium. That is, cyclodextrin is again an inhibitor of this aromatic substitution process, as with the diazo coupling. However, the chlorination of p-cresol shows a different result. Again only the *ortho* position can be chlorinated, but now cyclohexaamylose turns out to be a catalyst for the reaction, as it was for anisole. Since the depth of the cyclodextrin cavity will not fully enclose the p-cresol molecule, we believe that it sits in the cavity in such a way as to bury the hydrocarbon section and let the polar phenol group project into solution. In this case the *ortho* position is raised out of the cavity within reach of the rim, and a catalysed chlorination is then reasonable, as observed. Apparently in the methyl ether of p-cresol the less polar methoxyl group does not come out into the solvent to the same extent, and thus the *ortho* positions remain buried.

Modification of the cyclodextrin catalyst, rather than the substrate, also has interesting effects. α-Cyclodextrin can be converted to the dodecamethyl derivative, in which all the primary hydroxyls and all the hydroxyls at C(2) are methylated.[6] When we examined this material with anisole under our standard conditions (10 m mol dm^{-3} HOCl) we found an even larger preference for *para* chlorination.[6] Now less than 1% of *ortho*-chloroanisole was formed, showing not only that this partially methylated cyclodextrin can perform the catalytic function but that it does so more effectively. The higher selectivity can be accommodated completely by our finding that binding is 4.4 times as strong as it is for the unmethylated α-cyclodextrin, provided that again the reaction in the complex is exclusively *para* chlorination. The increased binding is expected since the primary methyl groups can cluster in aqueous solution to produce a hydrophobic floor to the cyclodextrin cavity. We had observed such conversion of a cyclodextrin cylinder into a pocket by flexible floors previously,[7] and will discuss it later. On the more rigid and more open secondary side the methyls at C(2) are not able to cluster so as to close up the pocket, but they do of course guarantee that the only hydroxyls left for catalysis in this system are those at C(3). A molecular model of anisole inserted into this methylated cyclodextrin so as to rest on the floor established by methyl clustering put the *para* position of the anisole within contact distance of a cyclodextrin hypochlorite formed by substituting a chlorine on one of these C(3) hydroxyl groups. Thus this result is in complete agreement with our picture of the reaction. However it does not demonstrate that the C(3) hydroxyl is also the catalytic one in simple cyclodextrins in which the other hydroxyls have not been blocked.

We produced a cyclodextrin polymer by reacting α-cyclodextrin with epichlorhydrin under basic conditions.[6] Adsorption isotherm studies showed that this polymer bound anisole from aqueous solution, and it turned out that this polymer was also an extremely effective catalyst for the *para* chlorination of anisole. When a column of resin was loaded with anisole, and an aqueous solution of HOCl was then passed through, the product (washed out with tetrahydrofuran) was greater than 99% *para*-chloroanisole, with much less than 1% *ortho*-chloroanisole formed. In the column, essentially all of the anisole is bound because of the high effective concentration of cyclodextrin residues. Thus, chlorination only occurs in the cavity, and again this is a completely selective process.

Subsequent to this work, a very interesting report has appeared in which this cyclodextrin catalysed chlorination was adapted to an electrochemical process.[8] Cyclodextrin was bound to an electrode, and a solution containing anisole and Cl⁻ was anodically oxidized. Again *para*-chloroanisole was selectively formed, presumably by a mechanism related to that we have described in which the cyclodextrin hypochlorite species is generated either by anodically formed HOCl or by direct anodic chlorination of the cyclodextrin. This is a particularly attractive application of biomimetic selectivity in complexes to a practical chemical process.

3. Diels–Alder reactions in cyclodextrin complexes

In line with the philosophy outlined at the beginning of the chapter, we have also been interested in learning how to apply biomimetic principles to other important synthetic reactions. The Diels–Alder reaction is one of the most useful synthetic processes, although it does not play an important role in biochemistry. In fact the evidence that Diels–Alder reactions occur in nature at all simply comes from a consideration of the structures of some natural products which look as if they must have been derived from such reactions, and no characterized Diels–Alder-catalysing enzymes have yet been described.

In spite of this, the Diels–Alder reaction does represent a type of process which is very common in enzyme-catalysed reactions. In a reaction between two different substances a principle function of the enzyme is to gather the substances together rather than just to furnish specific catalytic groups. Since the Diels–Alder reaction is not in general subject to catalysis by simple functional groups, although it is catalysed by strong Lewis acids, it seemed an ideal candidate for mimicking this "gathering" function of an enzyme.

Our work[9] on this process started only recently, and we still have much to learn about it. However the results to date are quite interesting. As one example, the reaction of cyclopentadiene with acrylonitrile shows a 9-fold acceleration in water when 10 m mol dm^{-3} β-cyclodextrin is added, but it slows down by 20% when α-cyclodextrin is added instead. Molecular models make this result quite reasonable. When cyclopentadiene is inserted into the cavity of α-cyclodextrin, the cavity is more or less completely filled. Reaction is only possible outside of the cyclodextrin cavity, and binding of the diene slows the process. However, in the larger cavity of β-cyclodextrin the binding of a cyclopentadiene molecule occupies only a little more than half the available space, and there is room for the binding of acrylonitrile into the same cavity. Thus the Diels–Alder reaction can occur within the cavity of a single cyclodextrin molecule; the result turns out to be a catalysed process.

There are in principle several ways in which binding into the cavity could catalyse the reaction. There could be various polar catalytic effects. As a second possibility, the tendency of substances to bind into the cavities could produce a smaller effective volume over which they are distributed and thus lead to a higher rate simply because little of the substrate is present in the dilute aqueous phase. This mechanism would be related to the one which has been shown to operate in many reactions which occur faster in frozen water than in liquid water, in which small melted pools concentrate the reactants and produce higher reaction rates simply because of this higher available concentration. The third possibility is related, but more interesting. The hydrophobic cavity of cyclodextrin would, in general, be expected to bind both the diene and the dienophile to some extent, but an interesting possibility is that the half-occupied cavity carrying a cyclopentadiene molecule would be particularly good at binding the dienophile. That is, binding could be cooperative so that dienes are not randomly distributed among various cyclodextrin cavities but instead tend to concentrate together into doubly occupied complexes. We believe that this mechanism is operating in at least some of our cases. Our studies indicate that kinetic saturation of the Diels–Alder reaction rate occurs at lower concentrations of cyclodextrin than are required to saturate the binding of either diene or dienophile alone. This stronger binding of the transition state than of either individual starting material is really only consistent with cooperative binding of the two species.

Cooperative binding is expected. Thus in the work described previously, the inclusion of an additional hydrophobic surface in the methylated cyclodextrin increased the affinity of the cavity for an anisole substrate. In the present case the additional hydrophobic surface is being furnished by a bound substrate. Furthermore, we have seen other cases in which two equivalent substrates, in particular adamantanecarboxylate ions, bind cooperatively to a single cyclodextrin molecule.[10] Again, the binding of one produced an additional hydrophobic surface which assisted the binding of the second molecule. Other workers have also reported spectroscopic evidence for binding of more than one molecule to a cyclodextrin cavity.[11] While the evidence for this cooperative binding is good, and the result is reasonable, we cannot yet evaluate an even more interesting possibility. It would be extremely desirable to produce binding interactions which preferentially complex the transition state for a reaction more strongly than the two individual starting materials are bound, so as to lower the activation energy within the ternary complex. There is no evidence that this phenomenon is yet operating in our Diels–Alder catalyses.

The magnitudes of catalysis observed are deceptively small, and the real effects are much larger. This is because the reaction rate is being compared with the rate in pure water, but as we have described[9] and confirmed more recently,[12] the reaction in water shows a very special effect of its own. Dienes and dienophiles in general have significant hydrophobic surfaces, and one might expect that even in pure solvent water they would tend to associate. This is the case, and the result is significant accelerations of the Diels–Alder reaction because of an association promoted simply by the solvent water.[9] For instance, the reaction of cyclopentadiene with butenone shows more than a 700-fold acceleration in water compared with the rate in 2,2,4-trimethylpentane. The rate in methanol is intermediate, closer to that in the hydrocarbon solvent. Various controls we reported[9] make it clear that the predominant effect is hydrophobically-induced association of the two reactants.

In the cyclodextrin reaction a different hydrophobic mechanism is involved in the association, namely binding of the two components into a cyclodextrin cavity. Thus the effect of this binding on the rate constant must be very large compared with the rate for *unassociated* reactants, and the catalytic effectiveness cannot be adduced by comparison with the aqueous solution in which the reactants are associated for other reasons. As an example, the reaction of cyclopentadiene with butenone was 1500 times as fast in aqueous solution containing 10 m mol dm^{-3} β-cyclodextrin as it was in isooctane solution.[9] In several cases the actual rates achieved exceed those which would be observed for undiluted mixtures of the reactants. Thus hydrophobic binding of the reactants together into the cavity is more effective at inducing proximity than is simple mixing of the pure reagents.

Perhaps even more exciting than increases in rate would be changes in products. The Diels–Alder reaction frequently gives mixtures of *exo* and *endo* adducts, and in any case a normal Diels–Alder reaction produces racemic mixtures of optical enantiomers, when these are present. One would expect that Diels–Alder reactions occurring in the cavity of a cyclodextrin could have shape constraints which favour the formation of particular products over others, possibly including particular enantiomers since cyclodextrins are of course chiral molecules. The results to date confirm these predictions, although the effects are not yet as striking as one might hope. For instance, we looked at the reaction of the equilibrating mixture of 1-methyl and 2-methylcyclopentadiene with acrylonitrile in water at 18° C. A mixture of eight compounds is produced, disregarding optical isomers, but two of the components are the *endo* and *exo* isomers of 2-methyl-5-cyano-norbornene. In aqueous solution alone the *endo* isomer is favoured by 2 to 1, but when the reaction is performed in the presence of 10 m mol dm^{-3} β-cyclodextrin the preference reverses to a 5 to 4 ratio of *exo* to *endo* product.

The finding of a preference for *endo* addition in the simple aqueous reaction is consistent with much experience in normal Diels–Alder additions. The reversal to a preference for *exo* reaction in the presence of β-cyclodextrin shows that the cavity has imposed a somewhat different shape requirement on the reaction; in molecular models the *exo* compound seems to fit somewhat better. The effect is even more striking if the diene is *t*-butylcyclopentadiene.[12] However, a case has not yet been found in which reaction in a cyclodextrin cavity goes exclusively to a product which is formed to only a minor extent in the absence of the cavity. In several of the Diels–Alder reactions we have examined there are small asymmetric inductions, corresponding to enantiomeric excesses of a few percent. With careful design of the interactions one might hope that a reaction in such a well defined environment could lead to a major preference for one enantiomer.

4. Acylation of cyclodextrin hydroxyl groups by bound substrates

It has been known for some time that when reactive molecules are bound in a cyclodextrin cavity they can react with one of the hydroxyl groups of the cyclodextrin. Many examples are known,[13] and we will not review them here. Our work in this area has focused on a simple central issue. Many enzymes use the hydroxyl group of a serine residue to react with a bound substrate, and such reactions are frequently very rapid compared with simple bimolecular reactions in solution. Intramolecular reactions with neighbouring hydroxyl groups can show very large rate accelerations which are commonly invoked in the explanation of very fast enzymatic reactions. However, in the examples of reactions within cyclodextrin complexes, only very modest rate accelerations had been observed at the time we started our programme. The largest was a factor of 250 for the acylation of β-cyclodextrin by m-t-butylphenyl acetate, compared with the rate of hydrolysis of the same substrate at the same pH (Fig. 5).[14] Compared with enzymatic accelerations of a million-fold or more, and similar large accelerations in some intramolecular reactions, it was a curious situation that reactions within a cyclodextrin complex had such modest rate accelerations.

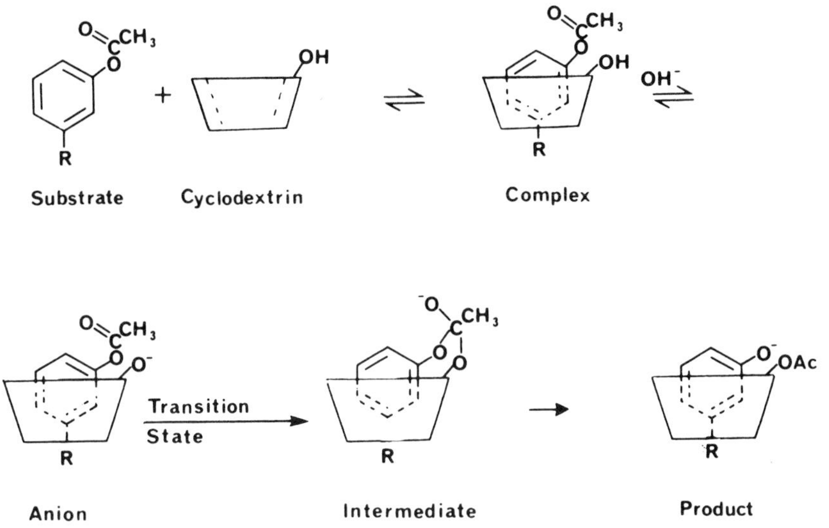

Fig. 5. Acetyl transfer.

We have taken two approaches to this question. In the first of them, we have modified the cyclodextrin unit to change the binding position of the substrates.[7,10] It was our hope to produce complexes with a better defined geometry and also with a geometry closer to the optimum needed for intra-complex reaction. In the second approach we have modified the substrate molecules in order to adapt their shapes to the requirements of the cyclodextrin system. This latter approach is in a sense easier, and corresponds to the search for an optimum substrate for a given potential catalyst rather than searching for an optimum catalyst for some predetermined reaction. Actually these two lines have at times run together, as we determined the composite effect of simultaneously optimizing a substrate and modifying the cyclodextrin cavity. Cavity modification itself was reported by us in 1975 by a process we described as "flexible capping".[7] This took advantage of the fact that β-cyclodextrin can be converted to a heptatosylate in which all of the primary hydroxyls have been substituted. Reaction of this heptatosylate with methylamine or ethylamine affords heptaamino derivatives which were then converted to the N-formyl compounds, which are water soluble. It was expected, and demonstrated, that they exist with the methyl or ethyl groups preferentially clustered in such a way as to invade the β-cyclodextrin cavity to some extent. Thus this clustering was expected to produce an intrusive flexible floor which defines the cavity, converting it from a hollow cylinder to a pocket and also making it more shallow (Fig. 6).

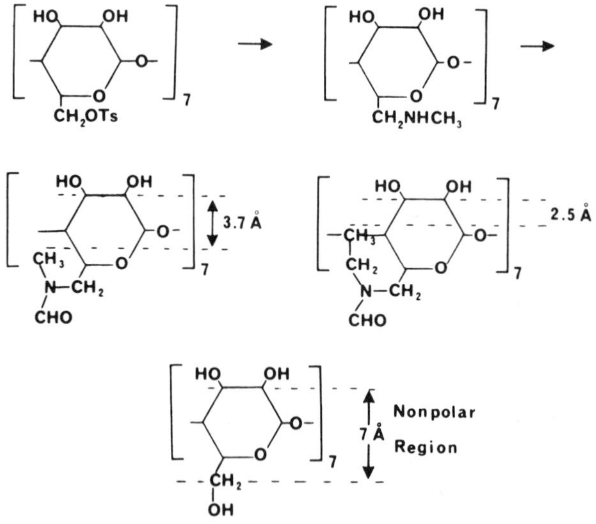

Fig. 6. The effect of flexible capping on the cyclodextrin cavity, as judged from models.

Two kinds of evidence demonstrate that this "flexible capping" does indeed produce an intrusive floor in the cyclodextrin cavity.[10] Direct evidence comes from the study of the binding of adamantanecarboxylate to cyclodextrin and to the modified molecule. We found that, as molecular models suggest, adamantanecarboxylate is too large to fit entirely into the β-cyclodextrin cavity, but instead it associates onto the cyclodextrin molecule so as to leave some residual cavity space. The result of this is that a second adamantanecarboxylate can bind to the other face of the cyclodextrin, using this residual cavity and incidentally interacting hydrophobically with the first adamantane unit (Fig. 7). (This two-to-one complexing is quite

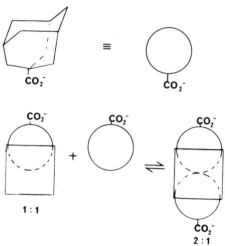

Fig. 7. Formation of a two-to-one complex.

different from that we described above for the Diels–Alder reactions, in which two small substrates can simultaneously fit side by side in the same cavity.) When the cyclodextrin derivatives which carry a flexible floor were examined, it was found that they only bind a single adamantanecarboxylate, as expected if the previously available residual section of the cavity is now occupied by the intrusive floor. Furthermore, the binding of this adamantanecarboxylate was stronger than that in the unmodified material, suggesting that it can contact the additional hydrophobic methyl or ethyl groups, which therefore must have invaded the cavity. Binding of the adamantanecarboxylate does, however, occur on the secondary face of the cyclodextrin derivative even when the floor is present, rather than in some completely independent position, since the adamantanecarboxylate on binding inhibits reactions of the cyclodextrin derivatives with bound substrates which utilize the secondary face (*vide infra*).

The other evidence comes from the effect of these intrusive floors on the rates of reaction of bound substrates. As one simple example, m-nitrophenyl acetate acylates β-cyclodextrin on the secondary face by binding in the cavity and performing an intracomplex acetyl transfer to an alkoxide group on the rim. In simple β-cyclodextrin the rate of this process, compared with the rate of simple hydrolysis of the substrate at the same pH, shows an acceleration of 64-fold because of neighbouring group attack within the complex. Models suggest clearly that a problem with this reaction is that the starting material can bind deeply into the cavity, but the tetrahedral intermediate for acetyl transfer must bind in a shallower manner, with the nitrophenyl group pulled up to an extent, partly out of the cavity. Thus if the substrate could bind in a higher initial geometry, closer to the geometry required for the transisiton state of the acetyl transfer, one would expect an improved rate. This is what was observed.[10] With the methyl capped cyclodextrin derivative, the rate within the complex was improved 10-fold so that the overall acceleration was by a factor of 660. With the ethyl capped compound we expect the floor to intrude further, and the compound to be bound in an even shallower way, with an even closer approach to the required geometry for reaction. In this case another factor of almost 2 was added, so that the overall reaction was accelerated by a factor of 1140.

The behaviour of the binding constant for complexing the substrate showed interesting changes. Binding of the substrate by the methyl capped cyclodextrin derivative was essentially equivalent to that by unmodified cyclodextrin. The additional binding expected from a contribution of the floor is more or less cancelled by a decreased binding interaction with the cavity itself because the substrate is more shallowly held. This is confirmed by the results with the ethyl capped cyclodextrin derivative which binds m-nitrophenylacetate 5 times more weakly. Here the floor is again present, but the cavity is even shallower and so even more binding interaction of the substrate with cyclodextrin itself has been lost. The results of our studies with the methyl capped compound on various other substrates, to be discussed below, are also consistent with this general picture.

All previous studies on the binding of small molecules into cyclodextrin cavities had used water as solvent, occasionally with tiny additional amounts of organic solvents accidentally present. It was widely believed that this binding was unique to aqueous solution,[13] but this was both unlikely and unfortunate. Many interesting potential substrates are not water soluble, and if we were going to pursue our studies in detail we wanted to be able to work with other solvents. Thus we undertook a study of the binding of substrates to cyclodextrins in polar solvents other than water.[15]

In dimethyl sulphoxide solution at 25°, m-t-butylphenyl acetate has a dissociation constant of 18 mM from its β-cyclodextrin complex, while in water the dissociation constant is 0.1 mM. Thus it is feasible to obtain saturation of cyclodextrin by such a substrate even in pure dimethyl sulphoxide, and easily in mixed solution with water. For instance, in 50% (v/v) water/DMSO, the dissociation constant is 2.0 mM. Although we have studied it less extensively, binding in dimethylformamide solutions seems to be similar to that in DMSO.

This finding makes it possible to examine a much wider range of substrates, which are soluble in DMSO–water solutions but not in pure water. However, the effect of the solvent is also kinetically interesting. We studied the acetyl transfer from bound *m-t*-butylphenyl acetate to β-cyclodextrin within the complex (Fig. 8).[15] In 60% DMSO/water, the rate is almost 50-fold faster

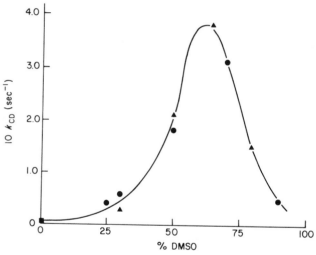

Fig. 8. The rate constant at 25.0° for deacylation of *m-t*-butylphenyl acetate at kinetic saturation with β-cyclodextrin as a function of the volume percent DMSO in water. Buffers were used, at 10 m mol dm^{-3}, which had an aqueous pH of 9.5. Circles: carbonate buffer. Triangles: borate buffer. Square: both buffers.

than it is in water, using the same buffer in both systems. Part of this effect is because of a greater basicity of the buffer in this medium, but when this effect is corrected for there are still solvent effects on the acidity of the cyclodextrin hydroxyl and on the rate constant for attack by the cyclodextrin anion on the acetyl group. Such solvent effects mirror suggestions that reaction rates may be faster in the interior of an enzyme because the environment is no longer purely aqueous. We will refer to them again in considering some of the rate comparisons to be discussed in the next section.

As we have indicated above, the best rate acceleration for reaction of a bound substrate with a hydroxyl group of the cyclodextrin molecule, compared with reaction at the same pH with simple solvent, was only 250-fold.[14] This substrate, m-t-butylphenyl acetate, can completely fill the cyclodextrin cavity, but molecular models suggest that in the tetrahedral intermediate, which must strongly resemble the transition state for the acetyl transfer, the t-butylphenyl group must be pulled substantially up out of the cavity. Judging from molecular models, we can roughly say that in the tetrahedral intermediate there would be only a 25% vertical penetration of the cavity, in contrast to the 100% vertical penetration in the complex of the starting material.[10] This is of course why modification of the cavity with intrusive floors led to rate improvement. The other approach to this problem is to synthesize and study substrates whose initial binding geometry is closer to that required for reaction. As expected, this approach also led to improvements in rates, and in some cases the results were quite spectacular. The work will be described in order of increasing effectiveness, which is not the historical order in which the work was performed.

One simple modification of the geometry of the substrate is to extend the arm connecting the carbonyl group with the t-butylphenyl group. Thus we prepared the p-nitrophenyl ester of 2-methoxy-5-t-butylcinnamate (1, Fig. 9) and examined its rate of acylation of β-cyclodextrin in 60% DMSO/water. (This medium was used for all the rest of the substrates described).[10] Note that in this case it is the acyl group which is bound, and the nitrophenyl leaving group has a lesser affinity for cyclodextrin than does the t-butylaryl group. We found that this shows a rate acceleration over the simple hydrolysis rate of 1500-fold, consistent with the prediction from molecular models

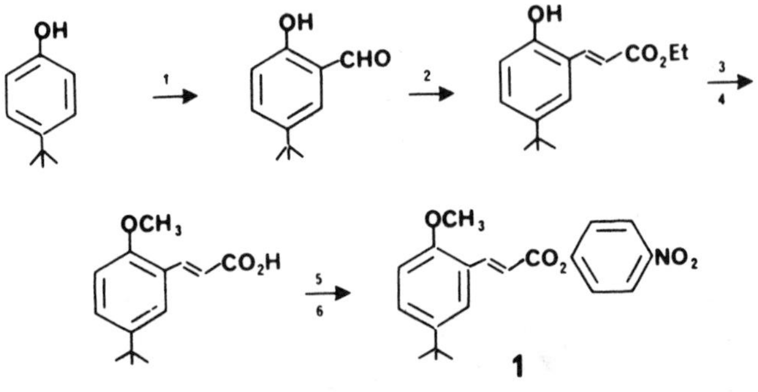

1, $NaOH/CHCl_3$; 2, $Ph_3P{=}CHCO_2Et$; 3, NaOH and Me_2SO_4; 4, KOH, then H^+; 5, $SOCl_2$; 6, p-nitrophenol in pyridine.

Fig. 9. Synthesis of a substrate.

that the substrate binding is closer to the geometry of the tetrahedral intermediate, but still not perfect.

Interestingly, when this substrate was looked at with the cyclodextrin derivative carrying a methyl intrusive floor, the rate acceleration fell to 400-fold. In our models this methyl floor decreases the depth of the cavity by approximately 50%, while the tetrahedral intermediate for substrate 1 can occupy 70% of the cavity. Thus the combination of this new substrate and an intrusive floor have overcompensated, raising the original binding geometry more than was required to fit the transition state well. To make a more gentle change we modified this substrate slightly by preventing it from binding quite so deeply in the original complex. This was achieved by attaching a projection on the substrate molecule in the double acrylate ester 2 (Fig. 10). Models show that it is not possible for the *t*-butylaryl

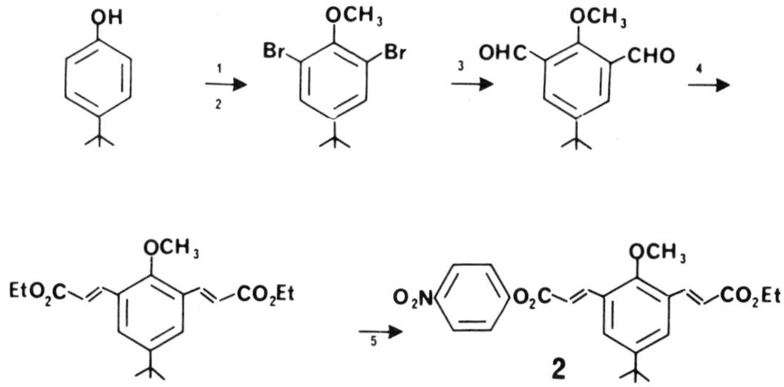

1, Br$_2$ and K$_2$CO$_3$; 2, NaOH and Me$_2$SO$_4$; 3, BuLi, then DMF; 4, Ph$_3$P=CHCO$_2$Et; 5, 1 equiv NaOH, then PCl$_3$, then *p*-nitrophenol in pyridine.

Fig. 10. Synthesis of a substrate with a projection.

group to bind quite so deeply in this case, since the unreactive acrylate arm catches on the rim of the cyclodextrin. This led to an improvement in the rate acceleration for reaction with cyclodextrin, compared with simple hydrolysis, to a value of 4900-fold, three times better than the situation without this projection to adjust the depth of binding.

As we have described above, the adamantane nucleus is too large to penetrate completely into the cyclodextrin cavity, and furthermore its spherical character means that when it binds by occlusion on the cyclodextrin, attached substituents can be moved into good contact with the rim of the cyclodextrin. Two adamantane derivatives were examined, the simple adamantylpropiolic ester **3** and the *t*-butyladamantylpropiolic ester **4** (Fig. 11). The results were complex, and very interesting.

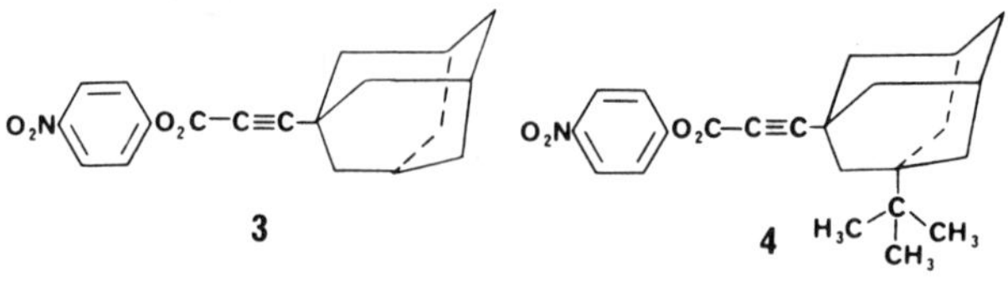

Fig. 11. Two adamantane-based substrates.

The simple adamantylpropiolic ester **3** showed a disappointing 2150-fold acceleration on complexing with β-cyclodextrin, but with the methyl capped cyclodextrin with an intrusive floor the rate acceleration went to 14 000-fold. However, just as striking as this finding of a 7-fold improvement when the methyl capped cyclodextrin was used was an observation that the binding of the substrate decreased by 9-fold. That is, there was simultaneously weaker binding and faster reaction, and by almost the same factor.

The binding of any substrate to cyclodextrin may involve several different geometries in the complex, only one of which is normally suitable for reaction. In this case, a moment's thought makes it clear that the preferred geometry of binding of the substrate should lead to an unreactive complex, in which the side chain is not projected towards the rim but is instead inserted down through the cyclodextrin cavity. The adamantane nucleus can occupy only 65% or so of the cavity, so the propiolic ester chain can contribute to binding if it passes through the rest of the cyclodextrin. However, this carries the ester group out of reach of the secondary hydroxyls, and for that matter the ester cannot even reach the less reactive primary hydroxyl groups of the cyclodextrin in the complex. If this unreactive geometry is the preferred one, but it is in equilibrium with 15% or so of the reactive geometry in which the side chain is not inside the cyclodextrin cavity, then this could explain our results. Putting a floor on the cyclodextrin cavity prevents this unreactive form of complexing, since now the methyl groups of the floor occupy the region through which the side chain would have to pass. Thus the entire complex has the reactive geometry, so the overall reaction rate is 7 times as fast as it was without the floor. At the same time the binding is weaker, because a preferred, if unreactive, mode of complexing has been blocked.

Support for this idea is seen with the related substrate **4** carrying an extra *t*-butyl group. When a *t*-butyladamantane binds to β-cyclodextrin it can completely occupy the cavity, the *t*-butyl group filling the portion not used by the adamantane nucleus. Thus the reactive chain is never buried in an inactive position, and the rate acceleration seen even with simple β-cyclodextrin is 15 000-fold. The alternative geometry in which the chain passes through the cavity and the *t*-butyl group is allowed to project on the secondary side is precluded, according to molecular models, because the *t*-butyl group is so bulky that the adamantane nucleus cannot seat onto the cavity with this alternative geometry. It is interesting that when substrate **4** is examined with the cyclodextrin derivative carrying an intrusive methyl floor, the rate acceleration in the complex decreases slightly, to 9500-fold, but more strikingly the complex now has two cyclodextrins per substrate. There is no longer room in a single cyclodextrin for both the adamantane nucleus and the *t*-butyl group, and models show that one of the modified cyclodextrin units now binds the adamantane segment while a second one can bind to the *t*-butyl group. It is striking that in this extremely bulky system the reaction rate for acylation of a cyclodextrin hydroxyl by the substrate is depressed by only 30%.

In a sense the reason that the tetrahedral intermediates in many of these substrates are bound more poorly than the substrates is that the side chain carrying the reactive carbonyl group comes off at an obtuse angle from the axis of the binding group, and therefore from the axis of the cyclodextrin within the complex. Better rate accelerations were expected if this angle was decreased to perpendicular, and better still if the angle was acute. A perpendicular angle is attainable from a derivative of a metallocene such as ferrocene.

We had found that the ferrocene nucleus binds extremely well to β-cyclodextrin. Molecular models suggest that a suitable side chain projecting from the ferrocene nucleus can react with a hydroxyl group on the secondary side of β-cyclodextrin with only a small upward movement of the nucleus on going to the tetrahedral intermediate. Accordingly, two simple derivatives of the ferrocene system were prepared, the ferrocenylpropiolic ester of *p*-nitrophenol (**5**) and the ferrocenylacrylic ester (**6**) (Fig. 12). Spectacular accelerations were seen in these cases.[10]

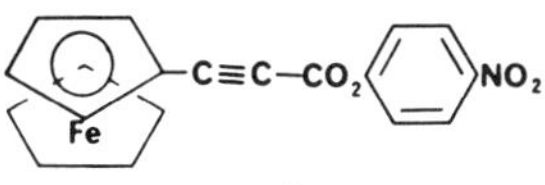

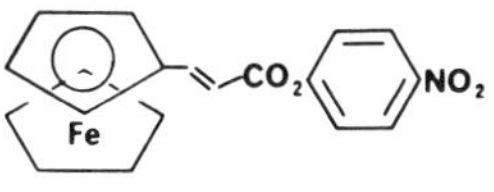

Fig. 12. Two ferrocene-based substrates.

The ferrocenylpropiolic ester **5** showed an acceleration of 140 000-fold for reaction in the cyclodextrin compared with reaction in simple solution at the same pH. In this case, as with all the compounds we have discussed so far except for simple *m-t*-butylphenyl acetate, the reaction is being performed in 60% (v/v) DMSO/water at 30.0° with a buffer which would give a pH of 6.8 in aqueous solution. An even larger acceleration is seen with the ferrocenylacrylic ester (6) which has a 750 000-fold acceleration for reaction with β-cyclodextrin in the complex compared with reaction in the simple DMSO/water solution at the same pH. In both of these cases, molecular models show that the essentially perpendicular geometry of the side chain with respect to the ferrocene nucleus permits formation of the tetrahedral intermediate with only a small loss in depth of binding compared with the geometry of a bound substrate.

Such deductions based on models need experimental confirmation. As one piece of evidence, preliminary work in collaboration with LeNoble on pressure effects on the rate of reaction of 6 with β-cyclodextrin gave values for the volume changes on binding and on proceeding to the transition state. From this work, the transition state is not quite as well bound as the starting material is, so, as the models suggest, the very large acceleration we have observed does not yet correspond to an optimal case. However, the transition state needs to be quite deeply bound in the cavity, so when we use the intrusive floor of the methyl capped β-cyclodextrin with substrate 6 we find an acceleration of only 235 000-fold. The substrate rests on the floor, as is revealed by the fact that binding is more than three times as strong with the floor as it is with simple β-cyclodextrin, but the upward adjustment of the substrate geometry by this intrusive floor is too much, leading to a loss in rate. As suggested by models, the smaller cavity of α-cyclodextrin does not fit substrate 6 well, and the reaction rate with α-cyclodextrin is approximately 2 orders of magnitude less than it is for β-cyclodextrin.

The acceleration seen with substrate 6 is better than the acceleration achieved by the enzyme α-chymotrypsin for the hydrolysis of p-nitrophenyl acetate. The comparison is even more striking when we notice that our reference system is aqueous DMSO, while the enzyme accelerations are normally referred to rates in water at the same pH. Since our solvent mixture can be considered as a mimic of the partially aqueous interior of an enzyme, the full acceleration we achieved including both this solvent change and the acceleration by neighbouring group attack within the complex is 18 000 000-fold. Whether or not one agrees that an extra factor of 25 for the medium effect should be included in the comparison, it is apparent that careful adjustment of geometry has strikingly raised the rate of an intracomplex process with cyclodextrin to the point at which it can be seriously compared with some enzymatic reactions or with some very fast intramolecular processes.

It is surprising that the acrylate ester 6 is a better substrate than the propiolate ester 5, since it looks as if the acrylate ester has more degrees of freedom which must be frozen out in the transition state. Our spectroscopic studies[10] indicated that the strong conjugation between the ferrocene nucleus and the ester group led to highly restricted rotation around the nominal single bonds cf this substrate, so the freedom available to the starting material is much less than appears at first sight. However, we felt that limiting this freedom even further with actual bonds could be helpful to the rate, and this proved to be correct. When a six-membered ring was fused to the system to produce the exocyclic ester 7 (Fig. 13) two rapid rate

Fig. 13. A fused-ring pair of enantiomeric substrates.

processes were observed in the reaction with β-cyclodextrin.[16] The faster of them corresponded to an acceleration, relative to reaction in the aqueous DMSO solvent at the same pH, of 3 200 000-fold; the slower process corresponded to an acceleration of 160 000-fold. We were able to show that these are reactions of the two different enantiomers present in 7, with a 20-fold selection of one enantiomer over the other. Furthermore, since the absolute configuration of 1,2-ferrocenocyclohexen-3-one is known, we were able to show that the more reactive enantiomer in our system was compound 7', not 7". Again this comparison does not yet correct for solvent effects. If we invoke the same 25-fold factor for converting the uncatalysed reaction from aqueous DMSO to water, we have an acceleration for the reactive enantiomer 7' of 80 000 000-fold.

An endocyclic ester incorporating the acrylate system into a six-membered ring was also prepared, but showed smaller accelerations.[16] We were able to ascribe this to a particular steric interaction in the transition state for reaction with cyclodextrin. In subsequent unpublished work[17] we have obtained an even larger acceleration, 5 900 000-fold (150 000 000 including the solvent effect), for the compound corresponding to 7' with a fused cyclopentene rather than cyclohexene ring. Furthermore, the enantiomeric selectivity in this case was a factor of 62, a very striking stereoselectivity. However, even this does not seem to be the optimum substrate, since from models it still has slightly poorer binding of the transition state than of the starting material, while an optimum substrate would reverse that situation. One can expect even higher accelerations with better tailored systems.

5. Cyclodextrins carrying additional reactive groups

Many other laboratories have been concerned with attaching functional groups to cyclodextrin to try to produce better models for the binding and catalytic functions of enzymes.[13] Our studies in this area were initiated about 15 years ago, and addressed several questions. One approach involved attaching known catalytic groups to a cyclodextrin nucleus, including co-enzymes, in order to see the cooperation of a known catalytic process with the binding available in the cyclodextrin cavity. A second study addressed the question of whether the effectiveness of such cooperation was a sensitive function of the position of attachment of a catalytic group to the cyclodextrin nucleus. The third class of studies involved using the cyclodextrins as frameworks on which to mount several functional groups, so as to produce polyfunctional catalysis of a type not easily available with small molecules.

5.1. Metal catalytic groups

Our earliest study involved attaching a metal-binding group to cyclodextrin.[18,19] A pyridinecarboxylic acid group was attached to α-cyclodextrin (Fig. 14) by reaction of the cyclodextrin with the m-nitrophenyl ester of

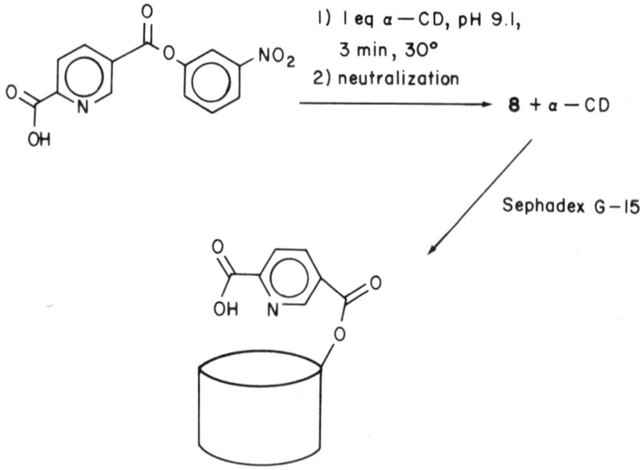

Fig. 14. Preparation of the α-cyclodextrin ester of pyridine-2,5-dicarboxylic acid.

pyridine-2,5-dicarboxylic acid (the ester group was at the 5 position, and acylated the cyclodextrin on its secondary side). Two kinds of catalyses were studied. In one, metals were bound to the pyridinecarboxylate group, and substrates were then hydrolysed by a combination of hydrophobic binding to the cyclodextrin and metal binding and catalysis. As an example, the *p*-nitrophenyl ester of glycine was hydrolysed by the copper complex of the cyclodextrin–ligand system more effectively than if the cyclodextrin was not there. Furthermore, the contribution of the cyclodextrin could be suppressed by adding cyclohexanol, which competitively binds into the cavity. The combination of cyclodextrin binding and metal catalysis was effective with this substrate, but not with another which had the wrong geometry.[19]

In a second more complex example, the pyridinecarboxylate ligand was converted to a nucleophilic catalytic group by reaction with nickel and pyridinecarboxaldoxime. We had earlier shown[20,21] that metal complexes of this oxime are very effective catalysts for attack on substrates which can bind to the metal, but in the present system this requirement for metal binding was replaced by a binding interaction with the cyclodextrin. Thus the effective nickel-oxime reactant was able to attack substrates which could bind into the cavity (Fig. 15), but was not nearly so effective against other

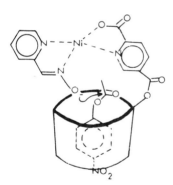

Fig. 15. Attack on bound *p*-nitrophenyl acetate.

substrates which could not. Furthermore, substrate binding could be inhibited by addition of cyclohexanol. This system extended the range of metal catalysed reactions to substrates in which initial metal binding was not strong, but the cooperative effect of the cyclodextrin cavity and the catalytic groups was only modest. Typically, less than a 10-fold acceleration was observed comparing these complex catalysts with simpler ones which did not incorporate the cyclodextrin binding group.

In this work the metal binding group is attached to the secondary side of the cyclodextrin, but we had also prepared a related compound with a metal binding group attached to the cyclodextrin primary side.[22] We found that this was also catalytic toward the same class of substrates. Thus some substrates can orient in either direction in the cyclodextrin cavity, and interact with the groups located on either the primary or the secondary face. A more systematic study of this question led to the same general conclusion.

5.2. Phosphate groups

We have prepared and characterized the pure cycloheptaamylose 2-,3-, and 6-phosphoric acids (Fig. 16).[23] The phosphate group can act as a basic catalyst, while at low pH the corresponding protonated group can act as a general acid. Two processes were examined. Tritium exchange of the methylene hydrogens in p-t-butylphenacyl alcohol shows mild catalysis by β-cyclodextrin itself but more than 100-fold catalysis by the 3-phosphate or

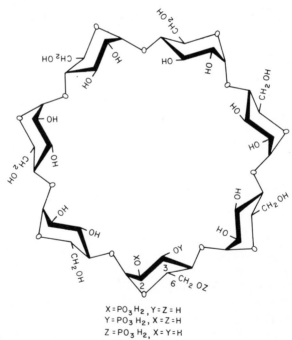

$$X = PO_3H_2, Y = Z = H$$
$$Y = PO_3H_2, X = Z = H$$
$$Z = PO_3H_2, X = Y = H$$

Fig. 16. The three β-cyclodextrin phosphates.

6-phosphate groups, and 80-fold catalysis by the 2-phosphate. This involved binding of the substrate, and was inhibited by cyclohexanol. It was our conclusion that all three positions of placement of basic catalytic groups were useful, and thus that the substrate must bind with almost equal facility so as to expose the methylene group to the catalytic phosphate ions on either face of the cyclodextrin.

One general caution must be noted. As we pointed out earlier, one cannot be sure about the relative proportion of binding of a substrate between productive and non-productive geometries. Thus the finding that a 2-phosphate group is less effective catalytically could mean that it less effectively reaches the active methylene group when the substrate is bound with that methylene group on the secondary side of the cyclodextrin, or it could mean that the 2-phosphate group shifts the equilibrium such that the substrate binds to a lesser extent in the correct geometry and to a greater extent with the methylene group on the opposite side of the cyclodextrin. However, the general conclusion from this, and from the work of Chao,[22] is that the primary face of cyclodextrin is a perfectly suitable location for catalytic groups, and at least in this system no obvious reason is seen for mounting such a group on the secondary side.

By contrast, when we operated at low pH and examined the ability of these cyclodextrin phosphates to acid catalyse the hydrolysis of *p*-nitrophenyl tetrahydropyranyl ether, the phosphate group on C(3) gave a catalyst 5 times as effective as simple cyclodextrin, the phosphate on C(6) was no more effective than cyclodextrin itself, and the phosphate on C(2) was only 1.5 times as effective as was β-cyclodextrin. Thus with this different substrate, and different reaction, there is indeed some preference for mounting the catalytic functional group on the secondary side, and specifically on C(3). However, because the chemistry of secondary side functionalization of cyclodextrins is much more poorly worked out,[24] we have tended to focus on the attachment of functional groups to the primary side. We were greatly encouraged by the fact that in several of our studies primary derivatives of cyclodextrin were just as effective as secondary derivatives in catalysing reactions of bound substrates.

5.3. Pyridoxamine

It is particularly attractive to attach coenzyme groups to cyclodextrins, so as to combine the special catalytic abilities of a coenzyme with the selective binding properties of these cyclodextrins. We described the first example of such a species, in which a pyridoxamine residue is covalently linked to a primary carbon of β-cyclodextrin.[25]

The synthesis involved converting the hydroxymethyl group of pyridoxamine to the bromomethyl substituent. This was then converted to thiopyridoxamine, an analogue carrying a mercaptomethyl group. This thiol was used in a reaction with β-cyclodextrin 6-tosylate to form a thioether link between the cyclodextrin primary side and the pyridoxamine.

Two kinds of special selectivity were examined and found. In one, we tried to pick up the preferential reaction of this pyridoxamine derivative with substrates which could simultaneously react with the pyridoxamine and bind into the cyclodextrin cavity. Molecular models indicated that transamination with ketoacids corresponding to the aromatic amino acids phenylalanine, tyrosine, or tryptophane should be able to utilize this double interaction (Fig. 17). As hoped for from this, a competition between

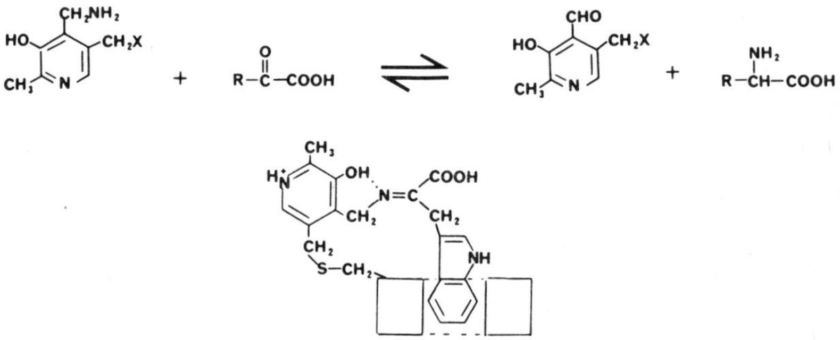

Fig. 17. Transamination by a pyridoxamine-β-cyclodextrin artificial enzyme.

indolepyruvic acid (which forms tryptophane) and pyruvic acid (which forms alanine) showed a 200-fold preference for the indole compound. With simple pyridoxamine, carrying no cyclodextrin binding group, the two ketoacids have essentially the same reactivity. Thus the binding of the indole group into the cyclodextrin ring has created a marked selectivity for this substrate.

Pyridoxamine itself is not chiral, and the amino acids produced by transamination with this molecule in solution are therefore racemic. However, transamination inside the chiral environment of an enzyme leads to highly specific production of optically active amino acids, and we hoped that something similar might be seen in the chiral cavity of β-cyclodextrin. For the synthesis of tryptophane only a modest optical induction was observed, but in the synthesis of phenylalanine from phenylpyruvic acid the L enantiomer was preferentially formed in a 3 to 1 ratio. This optical induction because of the chirality of the cavity can undoubtedly be improved if the environment is preferentially tailored to be very unsymmetrical; work in this area is currently underway in our laboratory. We are also pursuing the attachment of other coenzyme species to cyclodextrin, and have in fact made a molecule with a thiamine residue attached in an appropriate fashion to β-cyclodextrin.[26] One can expect that this molecule will also show special selectivity by simultaneously using the thiamine catalytic group and the cyclodextrin binding ability.

5.4. Mimics for ribonuclease

In spite of its complexity, the coenzyme pyridoxamine is fundamentally a single catalytic entity attached to the cyclodextrin nucleus. Great interest attaches to the possibility that the cyclodextrin framework could be used to mount several different catalytic species, mimicking the polyfunctionality of all real enzymes. As a start in this direction, we decided to attempt to convert cyclodextrin to a mimic of ribonuclease.

The enzyme ribonuclease[27] is one of the smallest and simplest known, and has the attractive feature that its principal catalytic functions are performed by two imidazole rings derived from the amino acids histidine-12 and histidine-119. In addition, an ammonium group of lysine-41 is important in binding the substrate, and probably plays a useful catalytic role. However, it seemed sensible to attempt to mimic the important catalytic function of the two imidazole rings by attaching imidazoles to cyclodextrins. Cramer had made cyclodextrins carrying various imidazole rings, but had reported no interesting catalytic activity in the system.[28] We decided to try to attach two imidazole rings to specific positions on the cyclodextrin and to evaluate it toward phosphate substrates related in character to the natural substrates for ribonuclease.

We had undertaken various studies on the selective functionalization of cyclodextrins, so as to permit the attachment of two groups in particular spots, when a report appeared by Tabushi of the selective bridging of cyclodextrin by a disulphonate system.[29] In the molecule originally reported by Tabushi, diphenylmethane-4,4'-disulphonyl chloride was allowed to react with β-cyclodextrin, producing a bridged derivative. (See Chapter 13, Volume 3). Our subsequent work[30,31] demonstrated that this was actually a mixture, but even so it was useful as a way of attaching imidazole rings. Treatment of the Tabushi compound with an excess of imidazole afforded β-cyclodextrin-bisimidazole (**9**),[31] which we now know was a mixture of positional isomers because of the original mixed bridging by the disulphonate group (Fig. 18).[30] If we label the glucose residues of β-cyclodextrin alphabetically,[30] molecular models indicated that the Tabushi bridge could not possibly reach between adjacent residues, and thus there must be no

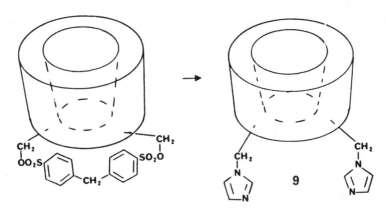

Fig. 18. Synthesis of a cyclodextrin-bisimidazole.

A,B isomer but only an A,C and A,D isomer. NMR studies[30] indicated that indeed this mixture of A,C and A,D isomers was present, as Tabushi confirmed.[32] Our studies used not only the original mixture[31] of positional isomers but also a substance which was principally the A,D isomer.[30] From this work it became clear that both the A,C and A,D bisimidazoles were catalytically active in the process we examined.

As a substrate, we decided to examine a cyclic phosphate. The enzyme ribonuclease cleaves RNA in a two-step process, in which the first step is closure to form a cyclic phosphate. Then in the second step the enzyme catalyses the hydrolysis of the cyclic phosphate, opening the ring. Fundamentally the same mechanism is used in both directions, the water being delivered in the second step by a process which is essentially the reverse of the mechanism by which an alcohol group is lost in the first step. Thus in principle one could examine either or both of these processes in a catalytic system, but we decided to construct a substrate with a cyclic phosphate group, since it is simpler and has a better defined geometry. Furthermore, the enzyme ribonuclease is normally assayed by examining its hydrolysis of appropriate cyclic phosphate compounds.

It would of course be interesting to examine the natural substrates of the enzyme itself, and as will be mentioned later we have actually done so, but we did not expect these to be the optimal systems for our cyclodextrin bisimidazole. The cyclodextrin cavity placement relative to the catalytic groups is not the same as exists within the natural enzyme, so it would be surprising if precisely the same substrate were optimal for the two different systems. Molecular model building made it clear that cyclic phosphates of catechols were the best candidates for the cyclodextrin catalytic system, and specifically the cyclic phosphate derived from 4-*t*-butylcatechol (**10**) seemed to be both available and attractive (Fig. 19).

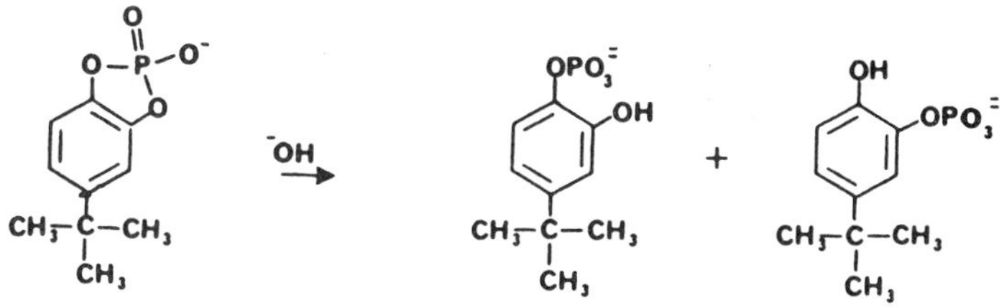

10 **11** **12**

Fig. 19. Random chemical hydrolysis of the substrate.

Hydrolysis of this cyclic phosphate has the other advantage that it can be followed spectroscopically, because of the change in chromophore. Thus we examined this hydrolysis with our cyclodextrin bisimidazole catalysts. Molecular models show that when the *t*-butylphenyl group of compound **10** is hydrophobically bound to the cyclodextrin cavity the phosphate cyclic group is accessible to the imidazole rings of **9**. Specifically, one imidazole ring can act as a general base to deliver water to the phosphorus atom of **10** and the other imidazole ring, protonated on nitrogen, can transfer the proton to oxygen-1 of the cyclic phosphate to assist hydrolysis of the P—O(1) bond. Thus we expected bifunctional catalysis by the cyclodextrin bisimidazole, with a maximum in the plot of rate versus pH because of the requirement that the catalyst must have one neutral and one protonated imidazole group.

As Fig. 20 shows, this prediction was confirmed. In fact, the rate of hydrolysis of substrate **10** by our catalyst **9**, with a large excess of the catalyst at kinetic saturation so we are looking only at the rate within the complex, fits the theoretically calculated curve for bifunctional catalysis over most of the pH region. Only at high pH does the rate exceed that expected for this mechanism, and this is sensible if the acid catalysis by the protonated imidazole group is not absolutely required. The curve in Fig. 20 for catalysis by a simple β-cyclodextrin monoimidazole shows such simple base catalysis, so one would also expect to see it for the bisimidazole catalyst **9**. For the enzyme ribonuclease, a similar curve to that of Fig. 20 is typically seen with its normal substrates, with a rate maximum near pH 7, except that there is a stronger requirement for the acid catalytic group. With our substrate **10**, the leaving group is a phenoxide ion, with a smaller requirement for protonation than would be seen for the alkoxide ions in the normal substrates of the enzyme.

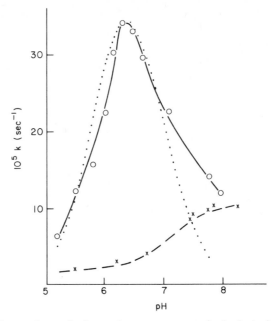

Fig. 20. The observed pseudo-first-order rate constants for hydrolysis of **10** at kinetic saturation with **9** (circles) or with the related 6-β-cyclodextrinylimidazole (crosses) as a function of pH at 25.0° C. The dotted curve is that calculated for bifunctional catalysis by an acidic and a basic group, each with pK_a = 6.3.

Molecular models indicate the situation shown in Fig. 21, in which delivery of the water within the complex can occur in-line with the P—O(1) bond. Unless a five-coordinate intermediate is formed with enough lifetime to undergo pseudorotation (no pseudorotation occurs in the enzymatic process), one would thus expect that this catalyst should hydrolyse the P—O(1) bond, producing as a product **12**, the 2-phosphate of 4-t-butyl-catechol. This is what is observed, with a preference for the formation of this product exceeding 90%. By contrast, ordinary chemical hydrolysis of

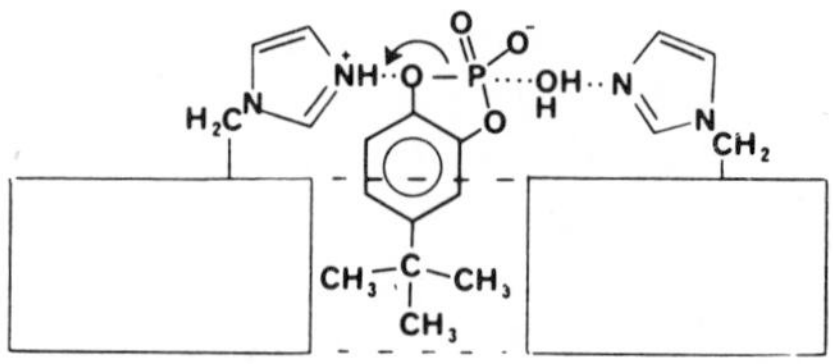

Fig. 21. The specific cleavage of **10** bound in the cavity of **9**.

substrate **10** (Fig. 19) yields a mixture of the 1-phosphate **11** and the 2-phosphate **12**, with a slight preference for the 1-phosphate. Thus not only has catalyst **9** shown bifunctional catalysis in the hydrolysis of a bound substrate, but it also is highly selective in the direction of cleavage. It thus shows many of the properties of the enzyme ribonuclease itself.

As we mentioned earlier, the material first examined as a catalyst was actually a mixture of cyclodextrin bisimidazoles, consisting of both the 6A,6C and 6A,6D isomers. In subsequent work we devised several other bridging groups with different geometric requirements which were able to selectively enrich one of these isomers. On the basis of this work it became clear that both of these isomers were catalytically active, but that the 6A,6C isomer is actually somewhat better. This might at first sight seem strange, since the in-line mechanism for phosphate cleavage seems to require catalytic groups to be located on opposite sides of the cavity from which the substrate projects. However, what is actually required is that the oxygen nucleophile be in-line with the P—O bond being cleaved, but the hydrogen bonds required to catalyse the process will project from the unshared electron pairs of the donor atoms, which are generally not pointed 180° apart. This is illustrated in Fig. 22, showing a top view of the reaction which

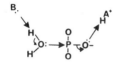

Fig. 22. A top view of the catalysed hydrolysis, showing that linear displacement does not require that the catalyst groups be on opposite sides of the cavity.

rationalizes the finding that somewhat better catalytic activity is found with the 6A,6C isomer of compound **9** in which the catalytic groups are by no means on opposite sides of the protruding substrate molecule.

The other geometric factor in catalyst **9** is the presence of the imidazole catalytic groups close to the cavity, which permits them to deliver water along the line of the P—O(1) bond but not to reach further out and deliver a water in line with the P—O(2) bond. To check these conclusions, and demonstrate that control could be engineered into the molecules, we therefore prepared compound **13** by reaction of 4(5)-mercaptomethylimidazole with the Tabushi β-cyclodextrin bisulphonate (Fig. 23).[30] We also prepared a similar monosubstituted derivative by reacting the mercaptan with β-cyclodextrin 6-tosylate. With both of these catalysts we now found a change in the direction of cleavage, such that with catalyst at 5 m mol dm^{-3} and substrate at 1 mol dm^{-3} the random hydrolysis in solution was completely supressed by binding and only the specific geometry of the catalysed process

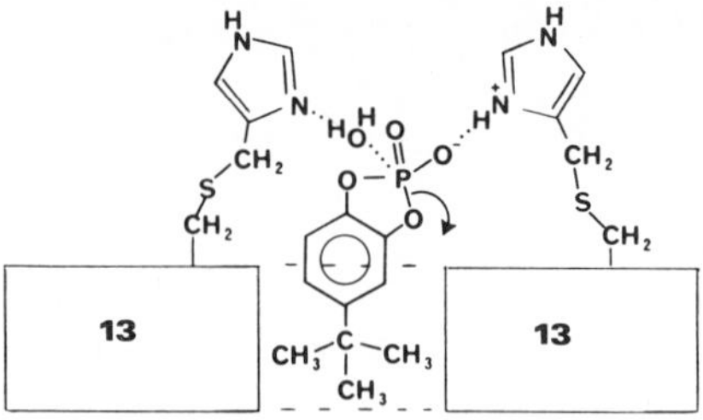

Fig. 23. The specific cleavage of **10** bound in the cavity of catalyst **13**.

was seen. This specific geometry differed from that which was observed previously with catalyst **9**. The product with catalyst **13** or the corresponding monoimidazole derivative also included **11**, the 1-phosphate of 4-*t*-butyl-catechol.

The kinetics of this process also showed a bell-shaped pH rate profile, but with an interesting difference from that seen for catalyst **9** (Fig. 20). With **9** we were able to calculate the pH-rate maximum simply from the titration pK's of the two imidazole rings. This indicated that the pK's of these groups were not strongly perturbed within the complex during the reaction. By contrast, with compound **13** the kinetic curve indicates a reaction pK for the acidic catalytic group almost one unit higher than its titration pK of 6.1. Thus the protonated imidazolium group is stabilized on interaction with the transition state, and probably also with the substrate as is shown in Fig. 23. In this case we therefore invoke an interaction of the imidazolium cation catalytic group with the phosphate anion of the substrate, not with the leaving group oxygen. However, the direction of cleavage of the substrate is dominated by the geometry of delivery of the water molecule by the neutral imidazole catalytic group, and this plays the same role but with a different geometric placement in the two systems.

Other related catalysts have been prepared and examined. For example, β-cyclodextrin has been converted to the heptaimidazole derivative in which all of the imidazole rings are attached to the primary side, C(6).[33] This compound is somewhat more effective as a catalyst than the simple bisimidazole **9**, and shows a rate optimum at a lower pH. In the cyclodextrin heptaimidazole at lower pH, several protonated imidazole rings are present which can imitate the catalytic function not only of a protonated imidazole in the enzyme but also of the lysine cation group. The compound is also a modest catalyst of the hydrolysis of cytidine-2,3-phosphate, a cyclic phosphate substrate frequently used for the assay of the enzyme ribonuclease itself. By contrast, ribonuclease shows no ability to catalyse the hydrolysis of our catechol cyclic phosphate, so the geometric requirements for substrates with the enzyme mimic and the true enzyme are almost mutually exclusive.

Although work is still proceeding on the construction of other cyclodextrin derivatives carrying all three functional groups of the enzyme with well defined placement, a related study should be mentioned. In a sense, the cyclodextrin nucleus in substrate 9 acts both as a binding group for the substrate and as a framework on which to mount the catalytic groups. We have recently explored a simpler situation, in which a framework is used to mount the functional groups so as to incorporate all of the catalytic functionality of the enzyme ribonuclease in more or less the correct geometrical placement, although without a hydrophobic binding cavity as in the molecules we have discussed in this chapter. A catalyst was constructed[34] with a biphenyl framework, based on molecular model building which suggested that it should be able to catalyse the hydrolysis of RNA. In this case we expected polyfunctional catalysis as observed in our models 9 and 13, but with binding achieved only by electrostatic attraction of the catalyst dication for RNA, which is a polyanion. This has indeed proved to be the case, the dication catalysing the hydrolysis of RNA at low catalyst concentrations ($5\ \mathrm{m\ mol\ dm^{-3}}$) over a matter of hours. The appropriate combination of such a structured polyfunctional catalytic system with a binding group, such as can be derived from the cyclodextrins, promises an even more interesting system. In such a combination we can expect better binding, faster rates, and selectivity imposed by the geometric restraints.

Acknowledgment

This work was possible only because of the skillful experimental and intellectual contributions of my co-workers, who are named in the references. Support of this work by the National Institutes of Health is gratefully acknowledged.

References

1. R. Breslow, *J. Am. Chem. Soc.*, 1958, **80**, 3719.
2. (a) R. Breslow, R. Fairweather and J. Keana, *J. Am. Chem. Soc.*, 1967, **89**, 2135, (b) R. Breslow and C. McAllister, *J. Am. Chem. Soc.*, 1971, **93**, 7096, (c) R. Breslow, D. E. McClure, R. S. Brown and J. Eisenach, *J. Am. Chem. Soc.*, 1975, **97**, 194.
3. R. Breslow, *Acc. Chem. Res.*, 1980, **13**, 170.
4. R. Breslow and P. Campbell, *J. Am. Chem. Soc.*, 1969, **91**, 3085.

5. R. Breslow and P. Campbell, *Bioorg. Chem.*, 1971, **1**, 140.
6. R. Breslow, H. Kohn and B. Siegel, *Tetrahedron Lett.*, 1976, 1645.
7. J. Emert and R. Breslow, *J. Am. Chem. Soc.*, 1975, **97**, 670.
8. T. Matsue, M. Fujihara and T. Osa, *Bull. Chem. Soc. Jpn*, 1979, **52**, 3692.
9. R. Breslow and D. Rideout, *J. Am. Chem. Soc.*, 1980, **102**, 7816.
10. R. Breslow, M. F. Czarniecki, J. Emert and H. Hamaguchi, *J. Am. Chem. Soc.*, 1980, **102**, 762.
11. (a) K. Kano, I. Takenoshita and T. Ogawa, *Chem. Lett.*, 1980, 1035, (b) A. Ueno, K. Takahashi and T. Osa, *J. Chem. Soc., Chem. Commun.*, 1980, 921; and earlier references.
12. D. Rideout, unpublished work.
13. M. L. Bender and M. Komiyama, *Cyclodextrin Chemistry*, Springer-Verlag, New York, 1977.
14. R. L. van Etten, J. F. Sebastian, G. A. Clowes and M. L. Bender, *J. Am. Chem. Soc.*, 1967, **89**, 3242, 3253.
15. B. Siegel and R. Breslow, *J. Am. Chem. Soc.*, 1975, **97**, 6869.
16. R. Breslow and G. L. Trainor, *J. Am. Chem. Soc.*, 1981, **103**, 154.
17. G. L. Trainor, unpublished work.
18. R. Breslow and L. E. Overman, *J. Am. Chem. Soc.*, 1970, **92**, 1075.
19. R. Breslow, in *Bioinorganic Chemistry*, Advances in Chemistry Series, American Chemical Society, Washington, 1971.
20. R. Breslow and D. Chipman, *J. Am. Chem. Soc.*, 1965, **87**, 4195.
21. J. Malmin, Ph.D. thesis, Columbia University, 1969.
22. Y. Chao, Ph.D. thesis, Columbia University, 1972.
23. B. Siegel, A. Pinter and R. Breslow, *J. Am. Chem. Soc.*, 1977, **99**, 2309.
24. Although S. Onozuka, M. Kojima, K. Hattori and F. Toda, *Bull. Chem. Soc. Jpn*, 1980, **53**, 3221, report selective tosylation of cyclodextrins on the secondary side, we find that the products are mixtures which include primary tosylates.
25. R. Breslow, M. Hammond and M. Lauer, *J. Am. Chem. Soc.*, 1980, **102**, 421.
26. D. Hilvert, unpublished work.
27. For a review, F. M. Richard and H. W. Wyckoff, in *Enzymes Vol. 4* (ed. P. D. Boyer), Academic Press, New York, London, ch. 24 (1971).
28. F. Cramer and G. Mackensen, *Chem. Ber.*, 1970, **103**, 2138.
29. I. Tabushi, K. Shimokawa, N. Shimizu, H. Shirakata and K. Fujita, *J. Am. Chem. Soc.*, 1976, **98**, 7855.
30. R. Breslow, P. Bovy and C. L. Hersh, *J. Am. Chem. Soc.*, 1980, **102**, 2115.
31. R. Breslow, J. B. Doherty, G. Guillot and C. L. Hersh, *J. Am. Chem. Soc.*, 1978, **100**, 3227.
32. I. Tabushi, Y. Kuroda, K. Yokota and L. C. Yuan, *J. Am. Chem. Soc.*, 1981, **103**, 711.
33. C. L. Hersh, Ph.D. thesis, Columbia University, 1980.
34. R. Corcoran, unpublished work.

15 · ENZYME–SUBSTRATE INTERACTIONS

L. N. JOHNSON
Oxford University, Oxford, UK

1. Introduction

The remarkable specificity of enzymes for their substrates has long been recognised. In 1894 Emil Fischer[1] proposed his famous "lock and key" analogy for the steric relationship between enzyme and substrate. It was envisaged that the sequence and characteristic conformation of the protein resulted in a structure at the active site which allowed a close fit only between the enzyme and its particular substrate. The essence of these simple views has withstood the test of time with slight modification. Nowadays it is recognized that since the substrate changes as it proceeds through various intermediate states on the reaction pathway, the structure of the undistorted enzyme can be complementary to only one of these states. These views have given rise to the concept of a moderately dynamic enzyme structure in which it will be most advantageous for catalysis if the enzyme can adopt a transient rigid structure which is complementary to the structure of the reaction's transition state (e.g. see the discussion by Fersht[2]). Following the determination of enzyme structures by X-ray diffraction methods in the late 1960s, a detailed molecular understanding of the discriminatory interactions

INCLUSION COMPOUNDS III
ISBN 0-12-067103-4

involved in enzyme-substrate complexes became possible. During the period 1959–79, 160 protein structures were solved, with new structures emerging at the rate of 20 per year in later years. Some of these proteins are not enzymes (e.g. haemoglobin, insulin) and for some others detailed analysis is held up because of lack of chemical information or precision in analysis. Nevertheless there is now a wealth of information on enzyme–substrate interactions. It is the purpose of this review to summarize some of the important results in this field and draw together some general conclusions. It will be seen that enzyme–substrate complexes in many cases admirably fit H. M. Powell's[3] early definition of clathrate structures in which different molecules are united through enclosure of one by the other with avoidance of large vacant spaces.

An understanding of enzyme mechanism requires an interdisciplinary study that involves chemical, kinetic and physical methods. X-ray crystallography is the only experimental technique whereby the complete three-dimensional structure of an enzyme can be determined, but the interpretation of these structures and the biological phenomena requires knowledge based on the solution properties of these molecules. In concentrating on the results from X-ray crystallography this review cannot do justice to many of these solution studies.

The examples of enzyme-substrate interactions selected are intended to be illustrative and not comprehensive. Protein–oligosaccharide interactions are discussed because they allow a description of the historical work on lysozyme which clarified many of the ideas on enzyme-substrate interactions. The more recent work on this enzyme has implications for the evolution of active sites of enzymes. Inclusion of the work on phosphorylase, allows a description of the different geometries of the binding sites designed to recognize either $\beta(1-4)$ or $\alpha(1-4)$ linked oligosaccharides. Enzymes which exhibit phosphate recognition sites comprise some 50% of known proteins. From X-ray methods over 30 phosphate binding sites are known. Tabulation and classification of these allows some general observations of the types of interactions expected and their rational. Finally in a book on inclusion compounds, it is appropriate to consider those enzymes that completely—or nearly completely—enclose their substrates. Examples include proteins involved in sugar transport, dehydrogenases, kinases and phosphorylase. They demonstrate the need, in most cases, for the protein to exhibit at least two structures, an "open" structure that allows the substrate to bind and a "closed" structure that encompasses the substrate and allows reaction to proceed.

Of the many enzymes which it is not possible to describe in detail, the serine proteinases form the major omission. These enzymes from both mammalian and bacterial sources contain at their active sites a serine with a special environment which is essential for catalysis. Following the work of Blow *et al.*[4] on bovine pancreatic α-chymotrypsin, the structures of trypsin, elastase, γ-chymotrypsin and a number of inhibited derivatives and the inactive precursors, trypsinogen and chymotrypsinogen, have been determined. In addition two classes of bacterial enzymes have been studied in depth; the first class includes the proteases A and B from *Streptomyces griseus* and α-lytic protease from *Lysobacter enzymogenes* which are homologous in structure and sequence to the mammalian pancreatic enzymes and the second includes subtilisin whose structure is different but which nevertheless contains a similar arrangement of catalytic residues at the active site. These studies have shown how different enzyme specificity can be achieved within the same molecular framework by just a few amino acid substitutions at the binding site; how proposals for enzyme mechanism can be suggested and tested; how with the aid of neutron diffraction the state of ionization of active site residues can be determined;[5] how with the aid of refinement at high resolution the positions and possible role of water molecules can be established;[6] how from comparative studies on the inactive zymogens the concept of activation by a conformational change resulting in a conversion of a flexible to a rigid catalytic binding site has been developed;[7,8] and finally with studies on the naturally occurring protease-inhibitor complexes with the enzyme, the importance of transition state stabilization has been demonstrated.[8] These studies have been well reviewed elsewhere[8–11] and a proper treatment of the results is outside the scope of this review.

The problems posed in enzyme-substrate interactions are not trivial. In the absence of substrate, the protein molecule must adopt a thermodynamically stable structure in which groups on the protein surface interact partly with each other and partly with the surrounding solvent. Nature abhors a vacuum and any large crevasse on the surface must be filled. In order to recognize the substrate there must be sufficient specificity to displace bound water molecules and to discriminate between substrates which may differ by only a few atoms. These specific interactions are weak in comparison to the covalent bond. (Thermodynamic measurements on lysozyme, for example, show that the unitary free energy of binding the substrate is of the order of 39.7 kJ mol^{-1}.[12]) One of the salient features of enzyme molecules is that they are not rigid static molecules (like for example the inert proteins keratin and collagen) but are dynamic with scope for conformational flexibility. These properties together with the observation that enzymes will bind but not catalyze reactions with the wrong substrate, led Koshland[13] to propose an "induced fit" mechanism. Using as an example the enzyme hexokinase, which catalyses the transfer of the γ-phosphate of MgATP (adenosine triphosphate) to the 6-hydroxyl of certain hexoses, Koshland suggested that the reason the enzyme was a good kinase and a poor ATPase was because only the correct sugar substrate could induce the conformational change in the enzyme that is essential for catalysis. Substrate analogues and water molecules are unable to trigger this change.

Most regulatory enzymes of metabolic pathways are oligomeric (composed of two or more subunits) and their reactions are controlled by levels of metabolites which are not related to the substrate. For example it was shown as long ago as 1936 that glycogen phosphorylase b requires AMP (adenosine monophosphate) for activity,[14] although AMP is not a cofactor in the reaction and is related to neither of the substrates, glycogen or glucose-1-phosphate. The two major theories advanced to explain these phenomena[15,16] propose conformational changes both in tertiary and quaternary structures and require the enzyme to exist in at least two distinguishable states. To date haemoglobin is the best understood example of an allosteric protein but structural studies on glycogen phosphorylase, aspartate transcarbamylase and phosphofructokinase are well advanced.

Thus limited flexibility in protein structure is required in order to recognize substrate, to achieve catalysis, to facilitate release of product (often the rate limiting step in an enzyme reaction) and in order to achieve allosteric response. How can such dynamic phenomena be studied by X-ray methods?

2. Scope and limitations of X-ray diffraction methods

2.1. Do crystal lattice forces alter structure?

Protein crystals are usually obtained by bringing a protein solution to supersaturation in the presence of a high concentration of salt (e.g. ammonium sulphate) or by the use of organic solvents.[17] In the early days it was questioned whether these unnatural conditions resulted in a change in the conformation of the protein. Nowadays the question is seldom raised. The generally good correlation of the results with those obtained by other methods in solution together with, in most cases, the superb explanatory powers of the structures, has led to an acceptance of the structure as a meaningful thermodynamic minimum energy conformation of the molecule.

Protein crystals differ from crystals of small molecules in that they contain typically 50% of solvent from which they are crystallized[18] (variation 27%–65%: in certain cases the solvent content may be as great as 75%.[19]) Protein–protein contacts between neighbouring molecules must be sufficient to stabilize the crystal lattice, but these represent only a small proportion of the intra-protein contacts. It has been argued that the close packed arrangement of protein molecules in the crystal is not too different from the close packed arrangement of protein molecules in a cell. (Typical protein concentration in a crystal = 800 mg ml^{-1}; estimate of protein concentration in yeast cytoplasm = 150 mg ml^{-1}.) The large solvent channels in protein crystals enable heavy metals, substrates and other metabolites to be diffused into preformed crystals often in periods of time of the order of 10 min to 6 h depending on the size of the molecule and crystal.[6]

Thermodynamic measurements suggest that most enzyme molecules are only just stable. For example, thermal denaturation studies on lysozyme give a change in Gibbs free energy $\Delta G = -40.5$ kJ mol^{-1} at 25° C on going from the unfolded to the folded state with $\Delta H = -276 \pm 12$ kJ mol^{-1}.[20,21] The large favourable enthalpic term is derived from the many hydrogen bonds and van der Waals interactions that stabilize the folded state. This is almost counter-balanced by the large unfavourable entropic term which reflects the increased order in the system overall and masks the favourable entropic contributions of the hydrophobic bond. The forces that stabilize the crystal lattice are the same order of magnitude as the overall free energy of folding and these do not, in general, appear to affect the conformation.[22,23]

Two examples serve to illustrate:

(a) Comparison has been made of arrangement of atoms at the active sites of two homologous serine proteases, the mammalian enzyme trypsin and the bacterial enzyme from *Streptomyces griseus* (SGPA). Both structures have been refined and they represent two of the most precisely determined structures available. After suitable rotation and translation, the 62 common atom positions at the active centres differed by only 0.39 Å.[24] Sequence analysis reveals that, although there are small regions of high homology between these enzymes, large gaps and insertions are required to align the sequences overall. Nevertheless the remarkable similarity in structure between the mammalian and bacterial enzymes indicates that the structural fold has been more highly conserved than the amino acid sequence and that the arrangement of the crucial atoms at the active site has been maintained without significant alteration at all. Many other examples of conservation in structure of homologous enzymes are also available.[25] Thus it appears that these structural similarities at key positions are not altered by crystallization.

(b) Tuna ferricytochrome c has been refined at a resolution of 1.8 Å with a final crystallographic R factor of 20.8% for the two independent cyto-chromes in the asymmetric unit and 49 water molecules.[26,27] The two molecules of the asymmetric unit are virtually identical in conformation (rms difference in positions of all atoms including the haem = 0.47 Å). The temperature factors are also very similar suggesting that there is little effect of the crystal packing forces on conformation. Tuna ferrocytochrome c crystallizes in a different space group and has been refined at 1.5 Å resolution to a crystallographic R factor of 17.3%. The rms differences in co-ordinates between reduced and oxidized forms is 0.57 Å. In this instance a detailed comparison of the two structures revealed that the reduced form had a slightly less exposed haem and less hydrophilic environment of the haem crevasse, features that would tend to favour the retention of the electron by the Fe^{II} state of the iron. These results both emphasize the precision which can now be obtained in protein structure determination (positional standard deviations in coordinates of less than 0.1 Å) and demonstrate the order of magnitude of conformational changes that may be of significance in biology.

However, there are examples where crystallization and irradiation do change the properties of the molecule. For example, oxy-haemoglobin is converted to aqua-met haemoglobin in the crystal.[28] In chicken triosephosphate isomerase only one of the two sub-units binds substrate and exhibits a conformational change, whereas in the yeast enzyme, for which the lattice contacts are different, both subunits bind substrate and respond with a conformational change.[29] Each case must be examined on its own merits.

2.2. Refinement

Until 1976, almost all protein structures and those of their complexes were the result of a subjective interpretation of electron density maps computed with phases obtained by the heavy atom isomorphous replacement technique. These maps were interpreted with the aid of models carefully machined to represent known stereochemistry. Present analyses[24] suggest that even with an apparently good phase determination (i.e. figure of merit = 0.82 representing a mean error in phase angle of 35°), these phases may differ from those calculated from the refined structure by as much as 57° and some apparently well phased reflections may differ by as much as 180°.

Crystallographic refinement by the conventional least squares method, which is routinely applied to small molecules, is impracticable for all but the smallest proteins (<100 amino acids) largely because of the poor ratio of observed reflections to unknown parameters, the limited resolution, and the computing times required. Initially attempts at improving protein structures employed difference Fourier syntheses, later coupled with least squares refinement. For small proteins such as rubredoxin[30] there was some success at high resolution (~1.2 Å). The introduction by Diamond[31] of real space methods which incorporated idealized models with variable dihedral angles allowed many larger structures to be refined such as in the work of the Munich group on trypsin, the pancreatic trypsin inhibitor and various complexes.[32–34]

However, it was only with the development of restrained parameter least squares refinement that routine and objective assessments of the precision and correctness of protein crystal structures in the resolution range 1.5–2 Å became available. These methods[35,36] and other methods[37,38] utilize the known stereochemistry of the amino acids and their non-bonded contacts to provide additional observations to the measured structure factor amplitudes. By restraining or constraining the structure to conform with known stereochemistry, rapid convergence of the refinement is achieved. In the examples given above, *Streptomyces griseus* protease A,[24] was refined by the method of Konnert and Hendrickson,[36] to yield a crystallographic R value of 0.139 for some 12 662 reflections and some 5912 variable parameters which included positional and individual thermal parameters of the enzyme and some 175 solvent molecules. The final structure differed from ideal bond lengths by an overall root-mean-square deviation of 0.02 Å and the probable mean error in co-ordinates was 0.15 Å. The studies on cytochrome c[26,27] involved the methods of Diamond real space refinement[31] and Jack–Levitt[38] restrained energy and reciprocal space refinement. Some statistics are given above. Both studies exemplify the precision which can now be obtained in protein crystallography, and it is this order of precision which is required to make meaningful interpretations of enzyme substrate interactions.

2.3. Mobility in protein crystals

As a result of the advance in refinement of protein crystal structures, the temperature factors of individual atoms can be assessed. These show that the protein molecules in the crystal lattice possess distinct mobility, which in some cases appears to correlate with biological activity. Because X-ray measurements are averaged over time and crystal space, it is impossible to distinguish, from measurements at a single temperature, between true atomic motion derived from vibrational and rotational modes and static disorder caused by different molecules in the crystal adopting different conformations. Diffraction studies on metmyoglobin[39] at four temperatures between 220 K and 300 K show the structure to be composed of a condensed core around the haem with mean square displacements of the order of 0.04 Å^2 which are temperature sensitive and a semiliquid region towards the outside with mean square displacements of 0.04–0.25 Å^2 which are essentially temperature independent. The movements of the surface residues point to a possible pathway to the haem group.

Preliminary analysis[40] of isotropic temperature factors derived from the refinement of lysozyme at 2 Å resolution reveal an overall mean square displacement amplitude of 0.23 Å^2. If contributions from experimental errors, crystal disorder or imperfections in the refinement are neglected, then the apparent thermal motion is found to be compatible with pairs of molecules that have strongest interactions in the crystal either moving as a rigid body with vibration about a common axis or vibrating in an intramolecular mode. Side chains exposed to the solvent have greater apparent motion than the remainder of the molecule and there is in general greater mobility of the residues that line the active site cleft. The good correlation between these displacements in crystals of human and hen lysozyme,[41] suggests that these intramolecular motions are a property of the lysozyme molecule and that the effects of experimental error and crystal packing are not serious.

In analysis of *Streptomyces griseus* protease A at 1.8 Å resolution, James *et al.*[6] showed that certain regions involved in the substrate binding (residues 167–172, 190–194 and 214–227) have some of the largest mean square amplitudes of vibration in the native enzyme. On formation of a complex with a tetrapeptide substrate, there were small conformational changes (of the order of 0.15 Å) in this region and a concomitant significant decrease in the temperature factors. It has been proposed that these molecular vibrations may provide an important factor in substrate recognition and transition stage stabilization.

While these movements are relatively small, much greater flexibility is observed in some enzymes. In trypsinogen the "activation domain" (residues N-terminus to Gly 19, Gly 142–Pro 152, Gly 184–Gly 193 and Gly 216–Asn 223) is disordered with temperature factors greater than 200 Å^2. On formation of the complex between trypsinogen and the pancreatic trypsin inhibitor (a natural transition state analogue), these residues become ordered.[8] Analysis of the trypsinogen crystals at 173 K and 103 K using synchrotron radiation[42] showed that the overall isotropic temperature factor in methanol water mixture fell from 16.1 Å^2 at room temperature to 11.6 Å^2 at 173 K but there was no further reduction at 103 K. The order of the activation domain of trypsinogen was not increased detectably except in the N-terminal region. These results suggest in this instance that mobility is associated with static disorder.

Thus there is no doubt that in the crystal, protein molecules can exhibit movement which may be considerable in certain local regions of the polypeptide chain. Most substrate binding studies have been carried out by diffusion of substrate analogues into preformed crystals. In certain cases large conformational changes may be observed, such as for example the movement of a loop of chain, in binding substrate to yeast triose phosphate isomerase.[29] However, there is now ample evidence that very large conformational changes may be prevented by the crystal lattice forces, as for example in the cases of hexokinase,[105] oxygenation of haemoglobin[28] and activation of glycogen phosphorylase b by AMP.[99] In these instances a proper appreciation of the gross movements can only be obtained from co-crystallization studies.

2.4. Activity in the crystal

The key question in comparing solution and crystal structures is whether the biological properties of the enzyme are altered on crystallization. Following the pioneer studies of Doscher and Richards,[43] the activity of several enzymes have been assayed in the crystal. From a comparison of many different types of experiments, Rupley[44] concluded that, in general, the equilibrium properties of the enzyme (e.g. the binding of saccharides to lysozyme) and certain kinetic properties (e.g. hydrogen exchange of lysozyme, diffusion of solutes into β-lactoglobulin crystals) were not significantly altered in the crystal. In most cases, however, enzymic activity was significantly diminished by factors varying between 1 and 1000 fold. While even a thousand fold reduction in rate must be set against the overall rate enhancement of the enzymic reaction, which is typically of the order of 10^{10}, an explanation for the significant decreases must be sought.

Firstly, not all enzymes exhibit activity in the crystal (e.g. lysozyme) because neighbouring molecules in the crystal lattice block access to the active site. Secondly, in those enzymes where a conformational change is an obligatory part of the reaction, a reduction in rate may be anticipated if these conformational changes cannot be readily accomplished in the crystal. Thirdly, there is a serious limitation imposed by the rate of diffusion of substrate into, and products out of, the crystal. Quiocho and Richards[45] showed that with carboxypeptidase A_α, crystals of 5 μm or less were required before the specific activity of the enzyme in the crystal became independent of crystal size. Rossi and Bernhard[46,47] studied the deacylation rate of crystallized acylated α-chymotrypsin using a chromophoric substrate. Under conditions where diffusion away of product was not essential for detection of reaction, they showed that deacylation rates were the same in the crystal as in solution.

The problem of crystal reactivity and diffusion limitations has been considered in detail by Makinen and Fink.[48] They provide a simple treatment for crystals approximated as a plane sheet of material which leads to the definition of a limiting crystal thickness λ_c below which kinetic measure-

Table 1. Comparison of reactivities of enzymes and proteins in crystal and solution states[48]

Enzyme or protein	Substrate or ligand	Relative activity solution: crystal[d]	Concentration of enzyme in crystal (mol dm^{-3})	λ_c (μm)	λ Experimental (μm)
α-Chymotrypsin	acetyl-L-tyrosine	1:0.18	0.0311	6.6	crystals of up to 0.2–0.8 mm dimensions used, thickness not stated 15-300
Carboxypeptidase A$_\alpha$	carbobenzoxyglycyl-L-phenylalanine	1:0.33	0.0450	6.0	not stated
Carboxypeptidase A$_\gamma$	carbobenzoxyglycyl-L-phenylalanine	1:0.003	0.0450	6.0	
Elastase	N-benzoyl-L-alanine methylester	1:0.46	0.0294	0.83	1.0
Papain	acetylglycine-ethyl ester	1:1.34	0.0279	8.0	1–6
Phosphorylase	maltoheptaose	1:0.09	0.0068	1.0	2.0[a]
Ribonuclease S	uridine-2',3'-phosphate	active in crystal: ratio not stated	0.0596	0.39	50[a]
Ribonuclease A	cytidine-2',3'-phosphate	1:1;1:10	0.0056	1.3	0.02[b]
Liver alcohol dehydrogenase	NADH	1:0.001[c]	0.0080	0.16	>1
Sperm whale metmyoglobin	azide anion	1:0.048[c]	0.0493	0.90	2.9
Horse methemoglobin	azide anion	1:0.4–0.45[c]	0.0382	0.27	1
Fast reacting deoxyhaemoglobin (horse Hb)	CO	1:1	0.0382	0.11	<1 (but CO already in crystal after flash photolysis)

(a) Average crystal thickness stated by authors.
(b) Estimated crystal thickness.
(c) Ratio of second-order rate constants for binding process.
(d) For references see text and reference 48.

ments of second order rate constants are not affected by rate limiting diffusion processes. Their summary of results for 12 enzymes is reproduced in Table 1. For elastase,[50] papain,[51] ribonuclease A[52] and deoxyhaemoglobin,[53] where the crystal thicknesses are comparable to the critical crystal thickness, reactivities are the same in the crystal and solution. In the case of glycogen phosphorylase b, Kasvinsky and Madsen[49] demonstrated that the K_m values for both substrates, glucose-1-phosphate (37 ± 8 mM) and maltoheptaose (176 ± 20 mM) were the same in the crystal and solution. The 10–100 fold reduction in rate, despite the fact that crystal thickness was only twice the critical thickness, may be partly attributable to the allosteric nature of this enzyme and partly to the fact that the large substrate maltoheptaose (mol. wt. 1152) may not obey the simple diffusion rules in the crystal.

The activity of enzymes in the crystalline state may lead to problems when attempts are made to study enzyme–substrate interactions, especially when one of the substrates is water. Almost all substrate will be converted to product during the time taken to make X-ray measurements (of the order of 2–10 days with conventional X-ray sources). Advances in the use of synchrotron radiation,[54,55] where the high spectral brilliance of the source leads to reduction in exposure times of the order of 10–100 fold, coupled with the use of very low temperatures to slow down the reaction rate (as for example in the studies on elastase[56]), are likely to make possible detailed studies of true enzyme-substrate interaction in the near future. A review of X-ray cryo-enzymology has recently appeared.[57] Most of the results described in this review were obtained with conventional sources and temperatures and utilized substrate analogues, competitive inhibitors or only one pair of a two substrate reaction.

Care may be required in the biological interpretation of the results of such binding studies. For example, the plant protein, concanavalin A, which has been used to study cell surface interactions and cell division, appears to be involved in the binding to receptors containing saccharide on the cell surface. The binding can be inhibited by mono and oligosaccharides related to D-mannose and D-glucose. However, concanavalin A will also bind various molecules that do not contain carbohydrate[58] and this may also be of functional significance. Crystallographic studies on the binding of *o*-iodophenyl-β-D-glucopyranoside failed to detect the sugar binding site but located instead a site that was specific for the hydrophobic aglycone moiety[59,60] at the subunit–subunit interface. Only co-crystallization studies with 2-deoxy-2-iodo-methyl-α-D-mannopyranoside[61] revealed the existence of the sugar binding site situated approximately 13 Å from the Mn^{2+} ion and quite distinct from the site recognized by the iodo-phenyl-glycoside studies.

3. Protein-polysaccharide interactions

3.1. β(1–4) linked polysaccharides: lysozyme

Hen egg white lysozyme was the second protein after myoglobin and the first enzyme whose structure was solved by X-ray diffraction methods.[62] The work has been reviewed many times[12] but is described briefly here because the description of substrate binding deduced from these studies exemplifies many features that are common to all enzyme-substrate interactions. Furthermore recent work has provided information on the water structure,[181] the thermal motion of the atoms[40,41] and evolutionary relationships at the active site.[73–75]

Lysozyme is a relatively small and stable enzyme with a single polypeptide chain of 129 amino acid residues. The first 40 residues from the amino terminus form a compact globular region with a hydrophobic core trapped between two α-helices. Residues 40–85 are hydrophilic in character and placed to form a second domain composed partly of a β-pleated sheet, one surface of which lines the active site cleft. The remainder of the polypeptide chain partially fills the gap between the two domains and then folds around the amino-terminal domain.

Lysozyme catalyses the hydrolysis of β(1–4) linked polymers of *N*-acetyl sugars composed either of *N*-acetylglucosamine or alternate residues of *N*-acetylglucosamine and *N*-acetylmuramic acid. The substrate binding site was established in the crystal with various small oligosaccharides that act as competitive inhibitors of the action. Study of the co-crystallized complex with tri-N-acetylchitotriose at 2 Å resolution[63] revealed that the saccharide binds to the top part of the cleft in the surface of the enzyme. The three sugar subsites are termed A, B and C. The free reducing group points down towards the unoccupied part of the cleft (Fig. 1). The contacts to the enzyme are summarized in Table 2. Site C is the strongest binding site with the NH and carbonyl oxygen atoms of its acetamido side chain making good hydrogen bonds with the CO and NH groups of residues 107 and 59 respectively. The CH₃ group makes van der Waals contact with the indole ring of Trp 108, which shifts slightly away from the sugar in order to optimize the contact. The second sugar B makes many non-polar contacts especially to the indole ring of Trp 62 which moves about 0.75 Å such as to close the cleft. The contacts of the sugar in subsite A are more tenuous; they include some non-polar interactions and a hydrogen bond between the NH of the acetamido group and the side chain of Asp 101. In addition there are hydrogen bonds between the ring oxygen $O(5)$ of one sugar and the $O(3)$ hydroxyl of the adjacent sugar so that the oligosaccharide has a structure

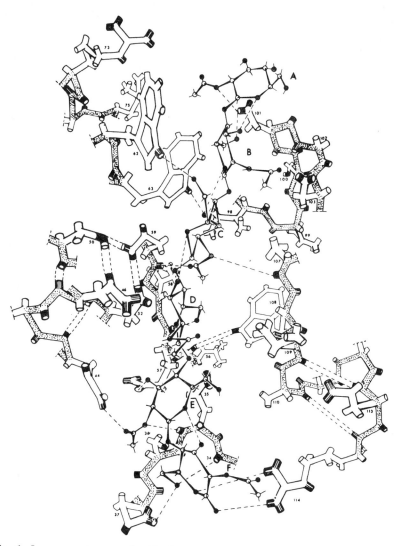

Fig. 1. Lysozyme–hexasaccharide interactions at the active site cleft. The main polypeptide chain is shown speckled and NH and CO are indicated by line and full shading respectively. Sugar residues A, B and C are as observed for the binding of tri-*N*-acetylchitotriose. Residues D, E and F occupy positions inferred from model building. Reproduced with permission from *Proc. R. Soc. London*, 1967, **B167**, 378.[63]

Table 2. Main interatomic contacts between lysozyme and hexasaccharide substrate[12]

Site	Polar contacts	total number of van der Waals <4 Å	ΔG_u (kJ mol^{-1})	ΔH_u (kJ mol^{-1})	ΔS_u (J K^{-1} mol^{-1})	Exposure ratio	Fraction-like contacts
A	NH–Asp101	7	−9.6	−12.1	−9.2	0.6	0.4
B	O(6)–Asp101	11	−11.3	−21.7	−34.3	0.6	0.6
C	O(6)–Trp62	30	−19.2	−25.9	−22.6	0.3	0.5
	O(3)–Trp63						
	NH–CO107						
	CO–NH59						
D	O(6)–CO57	35	+12.1–+25.1	—	—	0.4	0.4
	O(1)–Glu35						
E	O(3)–Gln57	45	−17	—	—	0.3	0.5
	NH–CO35						
F	CO–Asn44	13	−7.1	—	—	0.7	0.5
	O(6)–CO34						
	O(6)–Asn37						
	O(5)–Arg114						
	O(1)–Arg114						

close to that predicted for chitin and the minimum free energy conformation for $\beta(1-4)$ linked glucosyl residues.

The proposed binding of the hexasaccharide substrate was deduced from model building studies and from observations on the specific cleavage patterns of the substrate. The model building revealed the existence of three further sites (D, E and F) in which the sugars in subsites E and F make extensive contacts with the enzyme (Table 2). The sugar in subsite D, while also making extensive contacts, cannot be accommodated in its normal chair conformation because of close contacts between the enzyme and the C(6)–O(6) atoms of the sugar. This overcrowding can be relieved by distorting the sugar to a sofa conformation. Support for this distortion is given by the unfavourable free energy change observed in solution binding studies (Table 2) and experiments summarized in reference 12. Direct observation on binding in site D was provided by studies with a tetrasaccharide lactone complex[64a]—a putative transition state analogue in which the gluconolactone moiety in site D adopts the sofa conformation. Binding in site D has also been observed in a complex of lysozyme with the oligosaccharide NAM–NAG–NAM[64b] (where NAG is *N*-acetyl glucosamine and NAM is *N*-acetyl muramic acid). In this case the NAM residue is less far into the cleft and is undistorted. The complex may resemble the Michaelis complex. In the tetragonal hen egg white lysozyme crystals sites E and F are blocked by neighbouring molecules. However in crystals of turkey lysozyme, the cleft is accessible and direct structural confirmation of the existence of sites E and F has been obtained.[65]

The structural studies on lysozyme led directly to proposals for the catalytic mechanism.[66,12] The essential features involve a glutamic acid (Glu 35) which acts as a general acid to protonate the glycosidic oxygen between the sugars in sites D and E, leading to cleavage of the C(1)–O bond and formation of a carbonium ion at the C(1) position. The carbonium-ion transition state intermediate is both favoured and stabilized by the nearby aspartic acid residue Asp 52 and by the distortion of the sugar in site D to a sofa conformation. Retention of configuration of the product is explained by the steric restrictions at the catalytic site which allowed access of a water molecule only along the line of the leaving group.

Thus the essential features of the lysozyme–substrate interactions involve:

(i) Substrate binding in a cleft in the enzyme surface where two structural domains come together.

(ii) Stabilization of the substrate by weak intermolecular forces (hydrogen bonds, van der Waals interactions and hydrophobic shielding) such that all internal polar groups have their hydrogen bonding capacity satisfied.

(iii) Small conformational changes (e.g. Trp 62) which lead to better contacts between enzyme and substrate.

In addition we note that:

(iv) The strongest binding sites (C and E) are those for which the sugars are most shielded from water. The specificity of lysozyme for N-acetyl sugars is explained by the acetamido binding pocket of site C.

(v) Since the reaction involves hydrolysis, the active site is close to the surface of the enzyme.

(vi) The substrate adopts its preferred conformation except in the region of the bond cleaved.

(vii) The catalytic residues (Glu 35 and Asp 52) have environments which confer on them special features required for catalysis.

Information on the evolutionary origins of lysozyme specificity and catalytic mechanism has come from a comparison of phage T4 lysozyme (T4L) with hen egg white lysozyme (HEWL). Both enzymes catalyse the same reaction but T4 lysozyme is about 250-fold more active than HEWL towards *Escherichia Coli* cell walls, but conversely will not cleave N-acetylglucosamine homopolymers. T4 lysozyme contains 164 amino acids compared with 129 for HEWL and there is no detectable homology between the respective sequences. The determination of the structure of T4L[67] did not at first suggest any structural resemblance with HEWL beyond the fact that both enzymes were bilobal. Subsequently some degree of structural homology was recognized, especially in the region of the active site.[68-70] Recent oligosaccharide binding studies with T4L[71] have showed that N-acetylglucosamine binds preferentially to the equivalent of site C making similar contacts with its acetamido NH and CO atoms and the main chain CO and NH of residues Phe 104 and Leu 32 as in HEWL. Site B is rather more exposed and the saccharide lies against a loop of polypeptide chain at the N-terminal end of an irregular α-helix and there is no interaction complementary to the Trp 62 stacking in lysozyme. Sites D and E were probed by model building studies and like the HEWL, unacceptable bad contacts were found for site D that could be relieved by distortion of the glucopyranose ring to the sofa conformation. In a comparison of the amino acid residues in the region of the glycosidic bond between sites D and E (the bond assumed to be cleaved but not yet confirmed), it was observed that Asp 20 occupied a similar position to that of Asp 52 in HEWL. Support for a vital role of Asp 20 in the T4L mechanism has come from studies with mutants in which conversion of this residue to Glu results in inactivation.[72] The identity of the proton donor is not yet certain. Glu 11 is close to the position occupied by Glu 35 in HEWL but 5 Å away from the bond cleaved. In T4L Glu 11 forms an ion pair with Arg 145, an interaction that might lower its pK and reduce its efficiency as an acid at pH7. The fact that a number of specific interactions between enzyme and substrate, which are known to be important for catalysis, are found to occur in a structurally analogous way for the two enzymes, provides support for the argument that these two lysozymes have arisen by divergent evolution from a common precursor, dispite the lack of homology of the amino acid sequences.[73] The recent DNA sequence determination of the HEWL gene[74] has shown that, in common with many other eukaryotic genes, the HEWL gene is divided into exons which are eventually expressed and the introns whose regions are deleted at the mRNA processing level. The 4 exons of HEWL correspond to residues 18–28 (the signal peptide of prelysozyme and residues 1–28 of mature lysozyme), 28–81/82, 81/82–108 and 108–129. Exon 2 (28–81/82) includes most of the amino acids that bind the substrate and the key catalytic residues. Artymuik et al.[75] have compared the regions of structural homology between T4L and HEWL with this exon pattern and show that the two principal regions of overlap correspond closely to the combination of exons 2 and 3. Exon 1 is absent in T4L and the long C terminal region of T4L has no counterpart in HEWL. These results are consistent with the idea that the region coded by exon 2 could represent a primitive glycosidase on which substrate specificity was conferred by the product of exon 3. The results are also consistent with the view that exonic regions can be rearranged via recombination with introns into new pattenrs in new protein products.[76,77]

3.2. α(1–4) linked oligosaccharides: glycogen phosphorylase

Glycogen phosphorylase exists in two interconvertible states. Phosphorylase b, the form found in resting muscle, is inactive except in the presence of AMP (or analogues of AMP). In response to nervous or hormonal signals, phosphorylase b is converted to phosphorylase a by the enzyme–catalyzed phosphorylation of a single serine residue and is no longer dependent on AMP for activity.[78] X-ray crystallographic studies on phosphorylase b[79] (at 3 Å resolution) were performed on crystals grown in the presence of inosine monophosphate (IMP) a weak activator that produces activity but does not increase the affinity of the enzyme for substrate. X-ray studies of phosphorylase a[80] (at 2.5 Å resolution) were performed on crystals grown in the presence of glucose, an inhibitor. Hence both crystal studies relate, most probably, to the T state of the enzyme (in the nomenclature of Monod et al.[15]).

Glycogen phosphorylase catalyses the sequential phosphorylysis of α(1–4) linked glucosyl sugars releasing α-D-glucose-1-phosphate from the non-reducing end of glycogen. The enzyme digests to within 4 sugars of an α(1–6) branch point and exhibits a much higher affinity for branched substrates, such as glycogen, than linear polymers (see Graves and Wang[78]). Glycogen is present in skeletal muscle as dense spherical granules which are found close to the sarcoplasmic reticulum at the level of the thin filaments. These protein glycogen particles can be isolated and shown to contain significant amounts of enzymes involved in glycogen metabolism.[81] Approximately 17% of phosphorylase in muscle is associated with glycogen particles.[82] Thus there is a glycogen binding site on the surface of phosphorylase distinct from the catalytic site, as first suggested by biochemical experiments[83] and confirmed by X-ray studies.[79,84] A similar site probably exists for those other enzymes of glycogen metabolism associated with the glycogen particle. X-ray studies at 5 Å resolution on porcine pancreatic α-amylase[85] have shown that two binding sites per molecule exist for a modified maltotriose molecule; one site is in a deep cleft and is identified as the active site and the other is on the surface of the molecule, a possible "storage" site—no

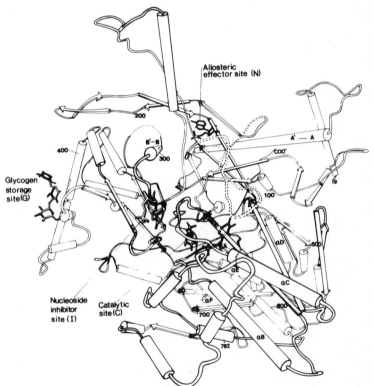

Fig. 2. A schematic drawing of the phosphorylase *b* monomer. α-helices and β-strands are represented by cylinders and arrows respectively. AMP is shown bound to the allosteric effector site (N), and the nucleoside inhibitor site I. G1P and the nearby pyridoxal phosphate are shown at the catalytic site and the major route of access to the catalytic site is shown by the thin line from the label. Maltotriose is shown bound at the glycogen storage site. Reproduced with permission from *J. Mol. Biol.*, 1980, **140**, 565.[91]

second oligosaccharide binding site has been repoted for Taka amylase from *Aspargillus Oryzai*,[86] although in this enzyme there is a covalently linked carbohydrate moiety on the surface.

In the structural studies on both phosphorylase b and phosphorylase a linear oligosaccharides derived from amylose bind preferentially to a site on the surface of the enzyme, termed the glycogen storage site, which has a dissociation constant of approximately 1 mM.[84] The site is some 30 Å from the catalytic site and 39 Å from the allosteric effector site and is not involved in any subunit–subunit contacts (Fig. 2). In contrast to lysozyme which accommodates the ribbon-like $\beta(1-4)$ linked oligosaccharides in a cleft on the surface, the major interaction of the helical $\alpha(1-4)$ oligosaccharide is with a surface α-helix (residues 397–410).

The structural data for the binding of maltoheptaose to glycogen phosphorylase b can best be interpreted in terms of subsites (1–5) and two additional subsites (7–8) (Fig. 3).[87] The oligosaccharide adopts a conformation close to the preferred conformation for $\alpha(1-4)$ linked glucosyl residues in which there is a possible hydrogen bond between the 2-hydroxyl of one sugar and the 3-hydroxyl of an adjacent sugar. (See also the discussion by W. Saenger in Chapter 8, Volume 2). The variety of different helical forms

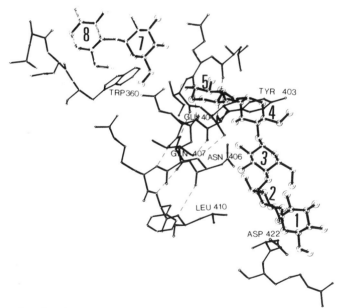

Fig. 3. Phosphorylase b-maltoheptaose interactions at the glycogen storage site. The $\alpha(1-4)$ linked glucosyl residues fill sub-sites 1–5. They adopt a characteristic left-handed α-amylose type helix and interact mostly with amino acid residues 397–412 of the α-helix. Two further sub-sites 7–8 make interactions with this α-helix and also with residues in the vicinity of Trp 360.

of amylose can be described in terms of variations about this preferred structure within the rather broad energy minima recognized from theoretical calculations.[88,89] Binding studies with maltose and maltotriose have shown that sites 2, 3 and 4 are the tightest binding sites. The left-handed amylose-like helix of maltotriose threads by the right handed polypeptide helix making contacts with amino acids residues i, $i+1$, $i+3$, $i+4$, $i+7$ (Tyr 403, Glu 404, Asn 406, Gln 407 and Leu 410). The contacts involve hydrophobic shielding of Tyr 403 and Leu 410 by non-polar parts of the sugar and hydrogen bonds between the hydroxyl groups of the sugars and the polar residues 406 and 407. The hydrogen between the O(2) and O(3) hydroxyls of adjacent sugars are aligned almost parallel with the hydrogen bonds of the α-helix and hence there is scope for a further favourable interaction via anti-parallel dipoles of the oligosaccharide and polypeptide hydrogen bonds. It will be interesting to see if these structural interactions between α-(1–4) linked oligosaccharides and a polypeptide α-helix will be found in other enzymes.

   Pyridoxal phosphate (PLP) has been recognized as an essential co-factor of phosphorylase since 1957. It is linked to the enzyme by a Schiff base with Lys 679. The Schiff base can be reduced with sodium borohydride without loss of activity and hence the role of the co-factor in phosphorylase is quite different from its role in conventional vitamin B_6 dependent enzymes. Accumulated evidence (see Helmreich and Klein[90]) suggests that the 5′-phosphate of PLP is involved in catalysis. The catalytic site has been identified in the structural studies by the presence of PLP and the strong binding site for glucose-1-phosphate (G1P) in phosphorylase b[79,91] and glucose in phosphorylase a.[92,93]

   The site is located at the centre of the molecule where the three domains of this large protein (841 amino acids) come together. The buried nature of the site removed from bulk water is likely to favour phosphorolysis rather than hydrolysis of the glycosidic bond. The sugar moiety of G1P is bound so that the hydrogen bonding capacities of all the peripheral hydroxyls are satisfied. The interactions of the phosphate moiety with the enzyme are weak and involve a rather long contact to Lys 573 and some stabilization by the amino terminus of the α-helix beginning at residue Gly 134 (see Section 4). It is interesting that although G1P binds strongly to crystals of phosphorylase b, it shatters crystals of phosphorylase a and in these the active site has been probed by glucose or 5-thio-glucose-1-phosphate binding studies. Evidently there are structural differences at the two catalytic sites, but no detailed comparisons have been made. The proximity of G1P to the PLP provides direct support for the role of the co-factor in catalysis, although the closest approach of the 5′-phosphate to the glycosidic oxygen is long ($\sim$5–6 Å). Several proposals have been made for the role of the 5′-phosphate groups in catalysis[90,91,94,182–185] based on the X-ray structural results, [31]P

NMR observations and chemical modification studies but further work is required to establish the details of the mechanism. Phosphate binding sites in phosphorylase are discussed in Section 4.

In the crystallographic studies on phosphorylase b and a, binding of the second substrate, glycogen or glycogen analogs, have not been observed at the catalytic site. The reasons for this are not clear. Access to the site is limited. The most obvious route is along a cylindrical channel approximately 15 Å long indicated by the catalytic site arrow in Fig. 2 but other routes may be possible with large conformational changes. Kinetic studies indicate that the apparent affinity of the site for linear oligosaccharides is weak ($K_m \sim 176$ mM). It may well be that before glycogen substrates can bind, large structural changes are required leading first to an open and then to an enclosed conformation as demonstrated in several of the enzymes discussed in Section 5, Enclosed Substrates.

4. Phosphate binding and catalytic sites

Because of the importance of phosphorylated intermediates in many cellular processes, there are a correspondingly large number of enzymes which exhibit affinity for phosphate. The binding sites may be for substrates, (e.g. as in the enzymes of glycolysis or the nucleases), for co-enzymes (e.g. as in NAD (nicotinamide adenine dinucleotide) of the dehydrogenases or pyridoxal phosphate of amino transferases or phosphorolyase) or for phosphorylated control intermediates (e.g. AMP in the regulation of phosphorylase or ADP (adenosine diphosphate) in the regulation of phosphorfructokinase). A summary of the major contacts established by X-ray crystallography for over 30 phosphate binding sites of enzymes is given in Table 3. The list is divided into two parts. Part A gives those sites that are used to locate and orient the substrate, co-enzyme or effector but are not involved in the catalytic reaction. Part B summarizes the known catalytic sites which encompass reactions such as phosphorylation, phosphoryl transfer and hydrolysis of phosphate ester bonds. In some examples, especially those involving ATP, a counter ion is essential, and the positions of the counter ions are also given, where known. In some instances (é.g. glucose-6-phosphate isomerase, aspartatecarbamoyltransferase, p-hydroxybenzoate hydroxylase, gluthathione reductase) the phosphate site geometry is known in considerable detail but lack of sequence data (at the time of writing) prevents a precise interpretation of the intermolecular contacts with the protein. In the case of hexokinase, the sequence is that deduced from the refinement of the structure at high resolution.

Table 3. Phosphate binding and catalytic sites in proteins

Enzyme	Compound bound	Phosphate moiety	Residues at binding sites[c]	Comment	Ref.
A. Binding Sites					
Lactate dehydrogenase (LDH)	NAD (co-enzyme)	pyrophosphate	NH-31, NH-32, Arg 301, Lys 58	Helix dipole	136
Liver alcohol dehydrogenase (LADH)	NADH (co-enzyme)	pyrophosphate	NH-202, NH-203, Arg 47, Arg 369	Helix dipole	141
Glyceraldehyde-3-phosphate dehydrogenase (GAPDH)	NAD (co-enzyme)	pyrophosphate	*B. Stearothermophilus* NH-10, NH-11, Asn 180 Lobster NH-10, NH-11, Ala 180	Helix dipole	144
	Glyceraldehyde-3-phosphate (substrate)	3-phosphate	*B. Stearothermophilus* Thr 179, Asn 181, Arg 231, 02'-nicotinamide ribose Lobster Thr 179, Thr 181, Lys 191, Arg 231, 02'-nicotinamide ribose	Helix dipole Site deduced from tight binding site for sulphate ion	142
Dihydrofolate reductase	NADPH (co-enzyme)	pyrophosphate	*Lactobacillus Casei* amino terminus α-F helix Arg 44, NH-45, Thr 45, NH-100, Thr 126 Avian liver amino terminus α-F and α-C helices NH-56, Thr 118, Thr 146	Helix dipole	95
				Helix dipole Thr 146 probably interacts via a water molecule. Lys 55 (equiv. to Arg 44 (lc) is 4 Å from nicotinamide phosphate	96

Enzyme	Ligand	Phosphate	Interacting residues	Comment	Ref.
Flavodoxin	FMN (co-factor)	2'-phosphate	*Lactobacillus Casei* Arg 43, Thr 63, His 64, NH-64, amino terminus of α-C helix	Phosphate is 5 Å from amino terminus of 2-turn α-C helix	97
			Avian liver Lys 54, Ser 76, Ca^{2+}(?)	Arg 77 (equiv. to His 64(1c) is too distant to participate	98
			Desulfovibrio Vulgaris (01) NH-11, Ser 58 (02) Ser 10, Thr 15 (03) NH-14, Thr 12 *Clostridium MP* (01) NH-8, Ser 54 (02) Ser 7, Thr 12 (03) NH-9, Thr 9, NH-11	Participation of helix dipole from helix 17–27 subsequently recognized by Hol et al.	79, 91
Glycogen phosphorylase	Pyridoxal phosphate (co-enzyme)	5'-phosphate	Phosphorylase b NH-675, NH-676, Lys 567, Lys 573; Phosphorylase a Helix dipole of α-E helix Lys 567, Lys 573(?)	Helix dipole	80, 93
	AMP (allosteric activator)	5'phosphate	Phosphorylase b Arg 309, Lys 247(I); Phosphorylase a Arg 308, Arg 309, Arg 242	Helix dipole	99
	Ser-14-phosphate (covalent activator)	Serylphosphate	Phosphorylase a Arg 69, Arg 43', His 36'		100
				Prime indicates contacts from symmetry related subunit	100, 101
Aspartate amino transferase	Pyridoxal phosphate (co-enzyme)	5'-phosphate	Arg 266, amino terminus of helix (116)	Helix dipole	102

Table 3. Phosphate binding and catalytic sites in proteins—continued

Enzyme	Compound bound	Phosphate moiety	Residues at binding sites[c]	Comment	Ref.
Triose phosphate isomerase	Dihydroxyacetone phosphate (substrate)	phosphate	NH-232, NH-233, Ser 211	Helix-dipole. Identified from sulphate binding site. Movement of loop 169–176 on exchange of substrate with phosphate or substrate	29
Phosphofructokinase (PFK)	Fructose-6-phosphate (substrate)	6-phosphate	His 249, Arg 252, Arg 243', Arg 162'	Prime indicates contacts from symmetry related subunit	103
	Mg–ADP (allosteric activator)	α-phosphate	Mg^{2+} between α and β phosphates Arg 25, Lys 213, Arg 214		
		β-phosphate	Arg 21, Arg 25, Arg 154, NH-59		
Haemoglobin	2,3-diphospho glycerate (allosteric inhibitor)	2-phosphate	α-NH$_3^+$(β_1), His 2 (β_1), His 143(β_2)		104
		3-phosphate	α-NH$_3^+$(β_2), His 2(β_1), His 143(β_1)		
B. Catalytic sites					
Glyceraldehyde-3-phosphate dehydrogenase (GAPDH)	Inorganic phosphate (substrate)		*B. Stearothermophilus* NH-149, NH-150, Ser 148, Thr 208	Helix dipole	144
			Lobster Ser 148, Thr 208	Site identified for weak binding of sulphate ion	142
Glycogen phosphorylase b	Glucose-1-phosphate (substrate)	α-1-phosphate	NH-134, NH-135	Helix dipole	91, 94
	Glucose-1, 2-cyclic phosphate	1,2-cyclic-phosphate	Lys 573, 5'-phosphate of pyridoxal phosphate		94

Enzyme	Substrate	Phosphate	Interactions	Notes	Ref.
Hexokinase (HK)	Co-ATP (substrate)		Co^{21} between β and γ phosphate. Ser 393, Ser 212, main chain NH	Site deduced from model building	105
Phosphoglycerate kinase (PGK)	ATP (substrate)	α-phosphate; β- and γ-phosphate	Horse Lys219 5 Å from amino terminus of α-helix	No other interactions: may need conformational change	106
		α- and β-phosphate; γ phosphate	Yeast Lys 213, Lys 217 NH-371, NH-372 Mn (3 Å)		177
Phosphofructokinase (PFK)	Mg-ATP (substrate)		Mg^{2+} bridge β and γ phosphates; NH-11 and Arg 171 may participate but latter not defined	No definite protein contacts and γ-phosphate not defined	103
Adenylate kinase	Mn-ATP	α-phosphate	Arg 132	No direct binding of ATP or AMP observed. Sites identified from indirect evidence.	107, 108
		β-γ phosphates	Mn^{2+} and β and γ phosphates do not occupy unique positions. γ-phosphate (which is transferred) is close to Asp 93 and His 36	Binding sites lined with number of basic residues which do not appear to be in direct contact with phosphates. Loop 16–22 wraps around phosphates.	
		5'-phosphate			
Phosphoglycerate mutase	3-phosphoglycerate (substrate)	3-phosphate (transferable)	His 8, Thr, Asn, Glu, Ser	His 179 able to participate in reaction. In both substrate and phosphorylated enzyme complexes	109, 110
	phosphorylated enzyme	His-phosphate	His 8, Ser, Glu, main chain NH and CO	conformational changes may increase contacts.	
		non-transferable phosphate site	amino terminus of "loose" α-helix, Arg	Assignments tentative until conformational changes established.	

Table 3. *Phosphate binding and catalytic sites in proteins—continued*

Enzyme	Compound bound	Phosphate moiety	Residues at binding sites[c]	Comment	Ref.
Ribonuclease S	UpcA phosphonate analogue of uridine-5',3', adenylate (substrate analogue)	2'-OH ribose (uridine) 5'-OH ribose (adenine)	His 12 His 119	In binding transition state analogue, Lys 41 moves to within 3 Å of phosphate sites	111, 112, 57
Staphylococcal nuclease	Thymidine-3',5'-bisphosphate-Ca^{2+} (substrate analogue)	5'-phosphate	Arg 35, Arg 87, Glu 43 (via H_2O) Ca^{2+}		113
Ribonuclease T_1 (*Aspergillus oryzae*)	2'-guanylic acid	3'-phosphate 2'-phosphate	Tyr 85 Glu 58 (His 40), His 92, Arg 77		180

[a] ?, residue not firmly located in electron density map or assignment of residue uncertain.

[b] l, long interaction (>3.5 Å).

[c] Where side chain atoms are involved in the contact, the name and number of the residue is given. Where main chain atoms are involved the atoms (e.g. NH or CO) and the residue number are given.

[d] Current recommendations from the nomenclature committee of International Union of Biochemistry suggest differentiating "biphosphate" as in fructose, 1,6-bisphosphate for two phosphate groups, from diphosphate as in adenosine diphosphate for a single group derived from diphosphoric acid. In the table the nomenclature used by the original authors has been retained.

4.1. Binding sites

It is apparent from part A of Table 3 that where the phosphate is involved in binding only, the site has been firmly located by X-ray crystallography. In crystals grown from ammonium sulphate, the site is usually occupied by a sulphate ion (e.g. glyceraldehyde-3-phosphate dehydrogenase, triose phosphate isomerase). Often the tight binding of the anion is achieved by salt linkages to arginine residues (e.g. lactate dehydrogenase, alcohol dehydrogenase, dihydrofolate reductase, phosphofructokinase) or less frequently to a lysine residue (e.g. the 2'-phosphate of the NADPH co-factor of dihydrofolate reductase or the 5'-phosphate of PLP in phosphorylase). The frequent use of arginine at phosphate sites of enzymes has already been commented on.[114,115] The guanidinium group is ideally suited for such interactions by virtue of its planar structure and ability to form multiple hydrogen bonds with the phosphate moiety (Fig. 4). Because of its resonance structure, the guanidinium group is a poor proton donor ($pK_a > 12$). Hence, in general, it would not be expected to function as a general acid catalyst for the hydrolysis of the phosphorylated intermediate, although its role in *Straphylococcal* nuclease (discussed later) may be an exception.

Not all proteins that contain phosphate binding sites exhibit positively charged residues at these sites (e.g. triose phosphate isomerase). For many enzymes the negatively charged phosphate is bound near the amino terminus of an α-helix. The rationale behind this observation has been provided by Hol and his colleagues[116] who drew attention to the fact, well reviewed in the work of Wada,[117] that the alignment of peptide dipoles parallel to the helix axis gives rise to a macrodipole of considerable strength. The individual peptide bond has a dipole moment of about 3.5D (Fig. 5) so that in an α-helix, where some 97% of the peptide dipoles point in the direction of the helix axis, the resulting effect is one in which an isolated half unit of positive charge is located at the N-terminus and a half unit of negative charge separated from it by the full length of the α-helix is located at the C-terminus. The isolation of these charges is unusual because charged groups in protein structures are usually solvated or compensated by a salt bridge. Hol et al.[116] showed that at a distance of 5 Å from the N-terminus of an α-helix the strength of the field increases up to a helix length of about 10 Å (seven residues or two turns of helix) and thereafter further elongation has only a marginal effect. In general the phosphate moieties are found bound at a distance of 3–5 Å from the amino terminus of an α-helix involving an attractive energy contribution of about 0.5 eV or 50 kJ mol^{-1}.

In those proteins that contain several helices running parallel to each other (e.g. the dehydrogenases, triose phosphate isomerase), the combined effect of several helix dipoles may lead to a long range interaction for

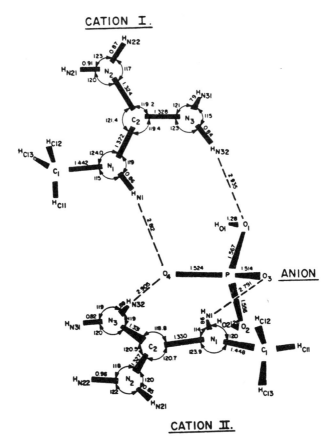

Fig. 4. The crystal structure of a methylguanidinium ion $(CH_3NHC(NH_2)_2)^+$–phosphate (H_2PO_4) complex showing bond lengths, hydrogen bond lengths and bond angles. Reproduced with permission from *J. Am. Chem. Soc.*, 1974, **96**, 4471.[114]

negatively charged substrates. In further instances where there is no apparent anion binding, the helix field may contribute to the catalysis by rendering protein side chains near the N-terminus of an α-helix more nucleophilic and/or by stabilizing transition states along the catalytic pathway. Such mechanisms may operate for papain,[116] subtilisin[116] and triose phosphate isomerase.[29]

Finally, at these tight phosphate binding sites, the phosphate may be held by a network of hydrogen bonds involving serine, threonine and main chain NH atoms. This is the case in flavodoxin where a loop of chain wraps around the phosphate site.

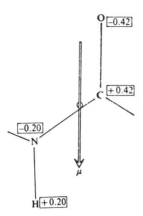

Fig. 5. The dipole moment of the peptide unit. Numbers in boxes give the approximate fractional atomic charge (in units of elementary charge). The dipole moment is 0.72e Å = 3.46D. Reproduced with permission from *Nature*, 1978, **273**, 443.[116]

The interactions of the Ser 14-phosphate in glycogen phosphorylase a deserves comment. Reversible protein phosphorylation is one of the primary regulatory mechanisms in biology.[118,119] This regulation was discovered in the control of glycogen metabolism and it now appears to be ubiquitous throughout metabolism, including neural mechanisms.[119,120] To date phosphorylase is the only phospho-regulatory protein whose structure is known. The single covalent modification catalysed by phosphorylase kinase results in phosphorylation of Ser-14 and the enzyme no longer requires AMP for activity. In the crystal structure of phosphorylase a,[100,101] the N-terminal 16 residues are reasonably well located. The Ser-phosphate is bound at the subunit–subunit interface through two specific ion pairs, one formed within the same monomer (Arg 69) and the other to the symmetry related monomer (Arg 43'). In addition the site is surrounded by a number of positively charged residues from both subunits. In phosphorylase b,[79] the first 20 residues are not defined in the electron density map and are assumed to be mobile. How the conversion of a flexible N-terminal tail to a rigid structure involving ionic links between subunits leads to activation at the catalytic site which is over 30 Å away remains to be established. It may be that the cluster of positive charges constrains the enzyme in the T state (in the nomenclature of Monod, Wyman and Changeux[15]) and that the presence of the seryl phosphate relieves the mutual repulsion of these cationic groups. The first 16 residues (which contain 5 basic and 2 acidic side chains) appear to be essential for control but not for catalysis since in phosphorylase b' (where the first 16 residues have been removed) AMP is still required for activity but there are no heterotropic allosteric effects.[121]

An extremely buried phosphate site has been recognized in another regulatory protein, the bacterial catabolite gene activator protein. This protein binds to specific DNA sequences in the presence of cyclic AMP and regulates transcription of several operons. The regulator, cyclic AMP, is buried within a deep pocket formed by residues from a β-roll and also two long α-helices, one from each subunit of the dimer.[178] The phosphate group of the cyclic AMP interacts with arginine and serine side chains. One of the exciting features of this work is that a comparison of amino acid sequences of the regulatory subunit of the ubiquitous control protein in mammalian systems, cAMP-dependent protein kinase, and the catabolite gene activator protein shows significant homology, especially in the cAMP binding regions.[179] It seems that a common ancestral precursor protein capable of binding cAMP has evolved in both bacteria and mammals. In prokaryotes the cAMP binding domain is attached to a DNA binding domain and regulates gene transcription, while in mammals, the cAMP binding domain is part of a subunit that regulates the activity of a protein kinase, the first enzyme in a cascade system of several metabolic pathways, especially those involved in the metabolism of glycogen.

4.2. Catalytic sites

In contrast (part B Table 3) for those enzymes involved in phosphoryl transfer reactions, the phosphate binding site is weak, as might be anticipated. In almost every case it has been difficult to locate by X-ray crystallographic methods. For example in *Bacillus stearothermophilus* glyceraldehyde-3-phosphate dehydrogenase, the site for the phosphate involved in the phosphorylation reaction has been inferred from the observation that this site is weakly occupied by a sulphate ion in the crystals and is consistent with the position deduced from crystallographic binding studies with the co-factor and part of the substrate. In the lobster enzyme the sites have been identified from comparison of sulphate and citrate soaked crystals in which there are conformational changes of the order of 2 Å and some asymmetry between different subunits. For hexokinase no direct binding of ATP has been observed and its position has been inferred from model building studies. For phosphoglycerate kinase and phosphofructokinase the γ phosphate has been located using the inactive imido analogue of ATP, AMP–PNP. In the kinases (see Section 5.4) it is likely that a conformational change takes place on binding both substrates that results in a shielding of the catalytic site from bulk solvent and may lead to the correct location of amino acid side chains that can both orient the phosphate to be transferred and stabilize the transition state. In the case of the allosteric enzyme glycogen phosphorylase, the crystals of both phosphorylase a and phosphorylase b are grown under conditions in which the enzyme is in the T state with a corresponding low affinity for the substrate, glucose-1-phosphate. Addition of strong activators (e.g. AMP) to phosphorylase b in solution results in a 10 fold reduction of K_m which reflects an increased affinity for the phosphate moiety. It has been suggested that the mode of binding observed in phosphorylase b may not represent the productive enzyme-substrate complex and this is supported by the observation that the potent inhibitor, glucose-1,2-cyclic phosphate has its phosphate site located some 2.4 Å from the phosphate site of glucose-1-phosphate.[94] These observations suggest that in the b form of the enzyme in the absence of strong activators, the phosphate site is not well defined. The conformational changes that occur on activation, are likely to lead to a more precise positioning of the phosphate. In this respect it is interesting to note that a particular arginine, Arg 568, exhibits an increase in its reactivity towards arginine specific reagents on conversion of phosphorylase to its active conformation.[122] In the present structure of phosphorylase, Arg 568 is relatively inaccessible, but a small change in conformation would allow it to contribute to the phosphate binding site and become more accessible.

Ribonuclease provides an example of good correlation between solution and crystallographic studies and illustrates the way in which conformational changes may stabilize the transition state. The enzyme hydrolyses polyribonucleotides in two steps (i) a chain cleavage resulting in a 2':3'-cyclic phosphate and a free 5'-OH on the other side of the bond cleaved and (ii) hydrolysis of the cyclic phosphate to yield the free 3'-phosphate monoester group. The studies with ribonuclease S[111] supported proposals for a mechanism originally made in 1961.[123] The first stage in the reaction is catalyzed by a histidine (His 12) which acts as a base to remove the proton from the 2'-hydroxyl and allows a direct "in line" attack of the 2' alkoxide on the phosphorus atom. Thus the transition state intermediate involves a change in the geometry of the phosphorus from tetrahedral geometry to a penta-coordinated trigonal bipyramidal intermediate. Protonation of the 5'(O)' of the leaving nucleoside by another imidazole in the acid form (His 119) leads to the cyclic intermediate. In the native enzyme His 119 can adopt several different positions. The phosphate binding site must be specific but there must be scope for sufficient flexibility to encompass the change in phosphorus co-ordination as the reaction proceeds through the transition state. Support for the flexible nature of this site has come from recent X-ray studies[57] with ribonuclease A. These have shown that the position of the phosphate in the product 3-CMP (cytosine monophosphate) is about 1 Å different from that in the cyclic intermediate 2′ : 3′-CMP, both binding studies being carried out at −60° C in 70% methanol. On binding the putative transition state analogue uridine–vanadate in which the oxovanadium(IV) ion (VO^{2+}) is thought to form a trigonal bipyramidal structure with the 2′ and 3′ oxygens of the ribose and a water molecule,[124] there is a dramatic shift in the position of the previously mobile side chain of Lys 41 so that its ε-amino group is now within 3 ± 0.2 Å of the vanadate oxygens.[57,112] An independent study using neutron and X-ray scattering has allowed the vanadate interactions to be defined in detail.[110] Chemical fortification of Lys 41 results in inactivation but in the native enzyme this side chain is over 5 Å from the phosphate binding site. Therefore the recent results provide support for the concept that in the native enzyme the phosphate site is relatively weakly defined but that conformational changes may take place which bring additional groups (e.g. Lys 41) into their correct and localized conformation so that they contribute to the stabilization of the transition state. In ribonuclease T_1, the role of Lys 41 could be taken by Arg 77.[180]

Staphylococcal nuclease catalyses the hydrolysis of both DNA and RNA at the 5′-position of the phosphate ester yielding a free 5′-hydroxyl group and a 3′-phosphate monoester. The reaction is dependent upon calcium with a pH optimum of between 8.6 and 10.3. Thymidine-3′ : 5′-bisphosphate (pdTp) is the best known inhibitor of the enzyme ($K_i \approx 10^{-7}$). Since the binding studies in the enzyme crystal were carried out at 1.5 Å resolution on the co-crystallized nuclease-pdTp-Ca^{2+} complex there is good reason to expect that the structure should resemble the actual enzyme–substrate complex. In the complex[113] (Fig. 6) the 5′-phosphate is firmly bound by double hydrogen bonds to two arginine residues and the Ca^{2+}. (A rather similar mode of binding of phosphate involving a single arginine residue (Arg 108) and a Ca^{2+} ion has been proposed for prophospholipase A2[125] although in this enzyme the bond hydrolysed is the ester bond at the 2 position of 1,2 diacyl phosphoglycerides). In Staphylococcal nuclease it has been suggested that the arginine links serve to neutralize the negative charge and to polarize the phosphate group. An "in-line" mechanism has been proposed in which nucleophilic attack at the phosphorus by a water molecule bound between Glu 43 and an oxygen atom of the phosphate is promoted by the interactions of the phosphate with the arginine residues and the Ca^{2+} ion. The structural data for the group involved in the protonation of the 5′-oxygen are indecisive. The guanidinium group of Arg 87 itself is an implausible but not entirely impossible candidate although a considerable environment-induced shift in pK of this residue would be required.

Thus there is growing support for the general hypothesis that phosphate catalytic sites are generally weak and somewhat flexible in the native enzyme structure but that conformational changes can take place on forming the transition state complex that lead to additional stabilization interactions.

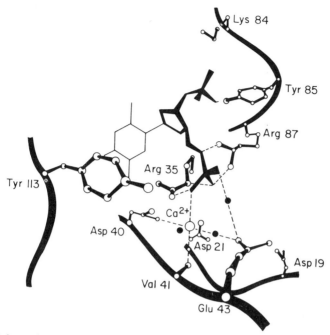

Fig. 6. Schematic view of the binding site in *Straphylococcus* nuclease–pdTp-Ca^{2+} complex. The heavy dots represent water molecules. Reproduced with permission from *Proc. Nat. Acad. Sci. USA*, 1979, **76**, 2551.[113]

In this respect we might expect to find basic residues (such as Lys 41 in ribonuclease, or Arg 568 in phosphorylase) waiting in the wings and ready to move into place to stabilize the transition state. In the co-crystallized complex of nuclease-pdTp-Ca^{2+} we may have an example of this type of stabilization produced by the very specific contacts from two arginine residues to the phosphate oxygens. However enthusiasm for these proposals must be tempered by Cotton *et al.*'s observation[113] that the phosphate in this complex is firmly tetrahedral and not trigonal-bipyramidal and hence resembles more an enzyme-substrate complex rather than an enzyme-transition state analogue.

5. Enclosed substrate binding sites

5.1. Xenon in myoglobin

The xenon binding studies to myoglobin provided the first example of a ligand totally enclosed by the protein.[126] Xenon (54 electrons) is an anaesthetic agent in man. Its mode of action is not understood but, since the electron cloud of xenon is spherically symmetric, narcotic activity cannot depend on any specific structural grouping. Crystallographic studies showed that xenon bound to an interior site in the myoglobin molecule in van der Waals contact with one of the pyrrole rings of the haem group and with the haem linked histidine. In the native structure of myoglobin this cavity is not even large enough to accept a water molecule and there is no open pathway from the solvent to this site. The only means of access therefore must come from slight fluctuations of the protein structure in the crystal which momentarily open up an appropriate channel. It appears that the complex is stabilized by interactions which originate from the polarizability of the xenon atom (namely charge induced–dipole moments, dipole–dipole moments and London forces). Subsequent studies with mercuri-iodide and related compounds that also bind to this site support these ideas.[127] Xenon does not bind to haemoglobin: evidently the interior pocket must be slightly different or fluctuations leading to the route of access blocked.

5.2. Arabinose binding protein from E. Coli

The studies on the arabinose binding protein form the most detailed example to date of an (almost) totally enclosed substrate molecule. The protein is one of a family of Gram-negative bacterial proteins that are involved in high affinity uptake processes in response to osmotic shock. The arabinose binding protein (ABP) mediates the transport of arabinose from the periplasmic space through the cytoplasmic membrane, presumably by interactions with other membrane transport systems. These interactions must involve a discrimination between the sugar-bound ABP complex and the sugar-free ABP which in turn suggests that there should be a structural change between the liganded and unliganded ABP. X-ray crystallographic studies have shown that, when bound, the arabinose molecule is almost inaccessible to solvent. Small angle X-ray scattering experiments indicate a dramatic decrease in radius of gyration of 0.94 Å on binding L-arabinose, consistent with a large conformational change.

The structure of the 306 amino acid ABP has been solved at 2.4 Å resolution.[128] The protein is ellipsoidal (axial ratio 2 : 1) and consists of two globular domains (designated P and Q). Unlike other domain proteins, which show a high degree of neighbourhood correlation, each domain of ABP is composed of two separate polypeptide segments. The two domains are structurally similar (rms difference is $C\alpha$ positions for 92 amino acids = 2.6 Å) and exhibit an α/β topology with a six stranded β-sheet flanked by two α-helices on either side. There is some indication of weak homology in amino acid sequences of the two domains and it appears that the high similarity in conformation may have arisen by gene duplication followed by divergent evolution, a proposal which has also been made for the two structurally equivalent domains in rhodanese.[129]

The sugar binding site is located in the cleft formed by the two domains[130] (Fig. 7). The cleft is 25 Å deep and the opposite walls are separated by as much as 10 Å (α-carbon distances). These walls are lined with predominantly hydrophilic residues (some 14 negatively charged and some 7 positively charged side chains) and the sugar forms a bridge between the walls, evidently stabilizing a closed structure. All the sugar hydroxyls are within hydrogen bonding distance to amino acids (Fig. 7). The structure of

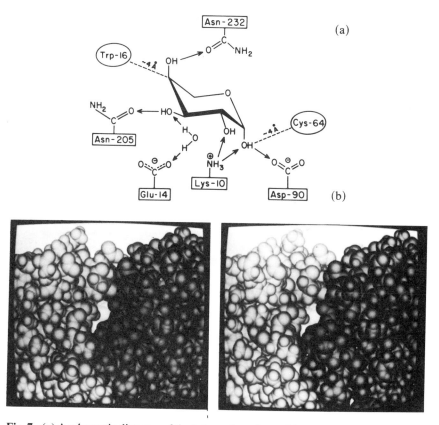

Fig. 7. (a) A schematic diagram of the interaction of L-arabinose to arabinose binding protein. (b) A space filling model of the sugar binding cleft of arabinose binding protein showing two domains (grey and black) and the sugar molecule (white) buried in the cleft. The part of the sugar molecule which can be seen corresponds to the C(2) hydroxyl. Reproduced with permission from *J. Biol. Chem.*, 1982, **256**, 13213.[130]

the sugar itself has weak pseudo-symmetry but although the sugar binds between the two domains that are also related by pseudo-symmetry, the side chains involved in the hydrogen bonds do not include any pairs of structurally equivalent residues. Evidently the recognition site is specific and asymmetric. The results are consistent with the specificity studies that have shown that all four hydroxyls are required for binding. A measure of the inaccessibility of the sugar to solvent is given by the observation that the atom with the greatest exposure is the $C(2)$ hydroxyl which has an accessible surface area[131] of 0.5 Å^2. (A totally exposed glycyl residue in the sequence Ala-Gly-Ala would have an accessible surface area of 82 Å^2.)

The structure of ABP was solved with the sugar already bound and to date there are no crystallographic data on the sugar-free protein. However the buried nature of the sugar site suggests that in the ligand free protein the cleft must be more open to permit access, a feature that could be accomplished by the rotation of one lobe of the bilobate structure relative to the other. Such ideas have received support from low angle X-ray scattering measurements.[132] Similar bilobate structures detected by low resolution studies have been reported for 3 other bacterial transport proteins and it seems therefore that these structures show unifying features. There has been some speculation on the observation that several of the proteins that transverse membranes and are involved in transport have a dimeric structure. It has been proposed that these molecules may be oriented with their axis of symmetry perpendicular to the plane of the membrane so that the subunit–subunit contact can form a channel for translocation.[133] While such ideas may not be generally applicable, the bilobate structure of the arabinose binding protein and similar proteins is consistent to some extent with these views. It might be envisaged that the membrane bound components that recognize the bilobate arabinose binding protein also contain some degree of 2-fold symmetry.

Argos et al.[134] have carried out a comparison of amino acid sequence and secondary structure prediction on the two sugar binding proteins that, in contrast to the arabinose binding protein which is only involved in transport, are involved in chemotaxis. They find that these proteins are likely to be similar in structure to ABP with the exception of a large, well conserved loop, which is close to the sugar binding site and which may indicate a common chemotactic receptor binding site.

5.3. Dehydrogenases

The dehydrogenases for which both crystal structure and chemical sequence are known are listed in Table 4. The common feature of their reactions

Table 4. Dehydrogenases for which both high resolution crystal structures and chemical sequences are known

Form of enzyme		Resolution (Å)	Reference
Lactate dehydrogenase (LDH)			
lactate + $NAD^+ \rightleftarrows$ pyruvate + NADH + H^+			
Apo	–dogfish	2.0	135
Holo	–dogfish		
+NAD	+pyruvate	3.0	
+NAD	+oxalate	3.0	136
+NADH	+oxamate	3.0	
Liver alcohol dehydrogenase (LADH)			
alcohol + $NAD^+ \rightleftarrows$ aldehyde + NADH + H^+			
Apo	–horse	2.4	137
Holo	–horse		
+NADH	+DMSO[a]	2.9	138
+NAD(H)	+trifluoethanol	4.5	139
+H_2NADH	+DACA[b]	2.9	140
Glyceraldehyde-3-phosphate dehydrogenase (GAPDH)			
glyceraldehyde-3-phosphate + NAD^+ + $HPO_4 \rightleftarrows$			
1,3-di-phosphoglycerate + NADH + H^+			
Apo	–lobster	3.0	143
Holo	–lobster	3.0	142
Holo	*B. Stearothermophilus*	2.7	144

[a] Dimethylsulfoxide.
[b] *trans*-4-*N,N*-dimethylamino cinnamaldehyde.

involves the removal of a proton from the hydroxyl group of an alcohol and the transfer of a hydride ion from the C(1) of the alcohol group to the C(4) position of the nicotinamide moiety of NAD^+ (Fig. 8). In glyceraldehyde-3-phosphate dehydrogenase, phosphorylation is coupled to dehydrogenation. These enzymes must therefore recognize and bind NAD in such a way that the nicotinamide moiety is correctly positioned in the active site but must also exhibit some flexibility to allow the change from tetrahedral

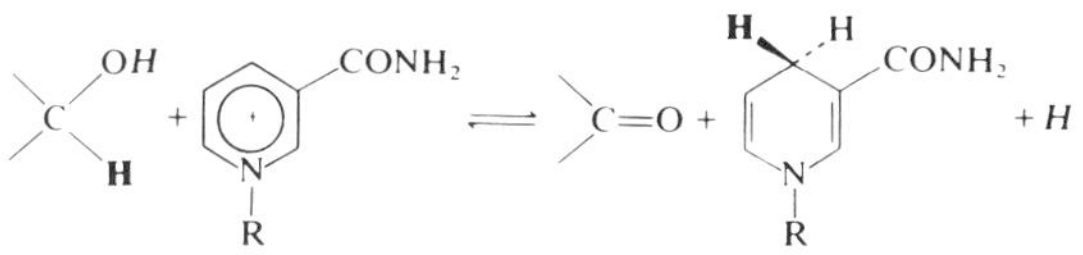

Fig. 8. Generalized scheme of dehydrogenase reaction. Reproduced with permission from A. Fersht, *Enzyme Structure and Mechanism*, W. H. Freeman & Co.[2]

to trigonal geometry at the C(1) atom of the substrate. Further, while substrate binding can best be achieved if the active site is open, hydride transfer will be facilitated by a hydrophobic environment in which water is excluded. Hence it was anticipated from kinetic studies, which indicated an ordered mechanism, and confirmed by structural structures, that a conformational change must accompany substrate binding. This summary of structural comparisons for the dehydrogenases is based on the excellent review by Branden and Eklund.[141]

The structures of the dehydrogenases are composed of two domains; the nucleotide binding domain that binds the co-factor NAD and the catalytic domain that binds substrate. The active site is at the interface of the domains. The catalytic domain structures are based on α/β topology with an anti-parallel β-sheet but exhibit no structural homologies. In contrast the nucleotide binding domains (nbd) show remarkable similarity in structure. The homologies and their evolutionary significance have been extensively compared and discussed.[145,146] The structure, now so familiar because of its ubiquity in other proteins, consists of a 6-stranded parallel β-sheet and four α-helices in which the two units of three strands are related by an approximate 2-fold axis of symmetry. The co-enzyme binds with the adenine part in a non-specific hydrophobic pocket at the carboxy end of the β-sheet between strands βB and βD (Fig. 9). The pyrophosphate moiety binds in a crevice outside the carboxy ends and close to the N-terminal end of helix αB (Table 3) and the nicotinamide is positioned on the other side of the sheet. A similar arrangement for co-enzyme binding is found in the NADP (nicotinamide adenine dinuclotide phosphate) dependent enzyme dihydrofolate reductase[147,148] despite some differences in sheet and topology. A rationalization for the formation of such a binding site in α/β proteins has been provided by Bränden.[148] The right-handedness of the $\beta\alpha\beta$ connection and the reversal of strand order gives rise to a pocket where the loops that connect strands with their respective helices are on opposite sides of the sheet at the carboxy ends of the strands. Differences in orientation, length, and amino acid sequence of these loops can lead to formation of a binding crevice without requiring a change in the basic folding pattern of the domain.

The conformation of NAD when bound is observed to be an extended conformation in which the adenine and nicotinamide rings are about 14 Å apart. This is different from the "closed" structure conformation observed in single crystal studies[149] and detected by NMR experiments in solution (see summary in McDonald et al.[150]). The fact that the structure of NAD is similar in all three dehydrogenases points to the importance of the interactions with the enzyme which evidently stabilize a conformation that is otherwise not favoured. Only in the non-productive complexes of LADH (liver alcohol dehydrogenase) and NAD modified at the nicotinamide is

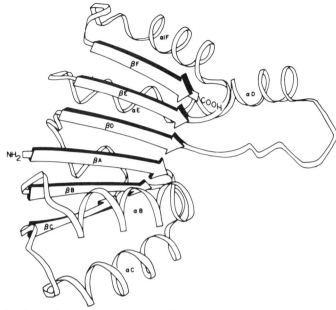

Fig. 9. A schematic drawing of the NAD$^+$ binding region in lactate dehydrogenase. Reproduced with permission from M. G. Rossmann *et al.*[146]

the conformation folded.[153a] The strength of these interactions is also demonstrated by the observations that 8-bromo-adenine derivatives bind with an *anti*-conformation about the *N*-glycosyl bond,[153b,153c] although the preferred conformation observed for 8-bromo-adenosine is *syn*. Although there is no overall homology in sequence between the structurally equivalent residues in the three dehydrogenases, four residues are conserved which, in the LDH (lactate dehydrogenase) numbering system,[140] are Gly 28, Gly 33, Asp 53 and Gly 99: of these a glycine at position 33 is required in order to make a tight bend between βA and αB and the remaining three residues are involved in co-enzyme binding. The 2′ hydroxyl of the adenine ribose hydrogen bonds to Asp 53 at the end of βB and the ribose is so close to the carboxy end of the sheet that the last residue of βA (Gly 28) is invariably a glycine. The O(3′) hydroxyl of nicotinamide ribose forms hydrogen bonds to the main chain carbonyl of residue 98 and contacts to Gly 99 are so close that this residue must be a glycine. A diagrammatic representation of these contacts and those additional ones made for example in LDH[151b] is shown in Fig. 10. The nicotinamide is positioned with one side of the ring facing the substrate bound to the catalytic domain and the other close to a hydrophobic region of the nbd.

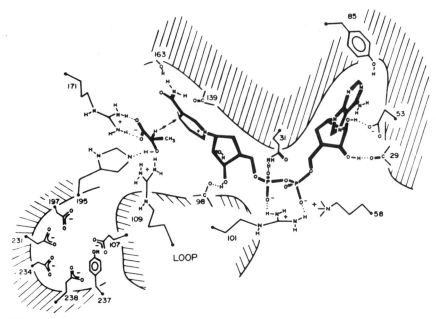

Fig. 10. Diagramatic representation of principle groups in the LDH active centre showing lactate and NAD$^+$ binding. The interaction between residue 31 and the nicotinamide phosphate occurs only in the H$_4$ (heart) isozyme. Reproduced with permission from J. J. Holbrook *et al.*[151a]

In the catalytic mechanism, both LDH and GAPDH (glyceraldehyde 3-phosphate dehydrogenase) utilize a histidine residue (His 195 in LDH) that acts as an acid-base catalyst by forming a hydrogen bond with the carbonyl or alcohol group. In LADH a zinc atom performs a similar role as an electron sink. In all three dehydrogenases conformational changes have been observed between the apo- and the holo-enzymes that are likely to be important in facilitating hydride transfer. The residues involved come from structurally different parts of the molecule. In LDH, the conformational change triggered by co-enzyme binding involves a large movement (up to 10 Å) of a loop (residue 98–114) that includes part of the extended helix α-D and results in a contact between Arg 101 and the pyrophosphate of the co-enzyme. In LADH this loop is not present. Instead co-enzyme binding (especially the nicotinamide moiety) triggers a movement of the whole catalytic domain relative to the co-enzyme binding domain with some concerted changes in each. In GAPDH there is an S shaped loop in the catalytic domain comprising residues 178–201. The arrangement of subunits brings the S-shaped loops into close contact at the centre of the tetramer so that residue Thr 179 makes contact with the adenosine. Low resolution studies on *B. stearothermophilus* GAPDH suggest that when NAD^+ is removed from the enzyme the whole nucleotide binding domain moves away from the active site with concomitant changes in the S loop.[152] However no large changes are seen between apo and holo lobster enzyme[143a] although in this case these may be inhibited by low pH. The "meso"-form of lobster GAPDH (about one NAD molecule per tetramer) is isomorphous with the holo-enzyme.[143b]

The positions of the substrates in lactate dehydrogenase have been deduced from studies on three pseudo-ternary complexes, LDH–NAD–oxalate, LDH–NADH–oxamate and LDH–NAD–pyruvate. In the latter complex there is a covalent link between the C(4) of the nicotinamide and the methyl carbon of pyruvate. All three complexes crystallize isomorphously with the holo-enzyme in which the flexible loop connecting βD and αE has closed the active site. The essential interactions are summarized in Fig. 11. All polar and ionic groups of the lactate are stabilized by correspond-

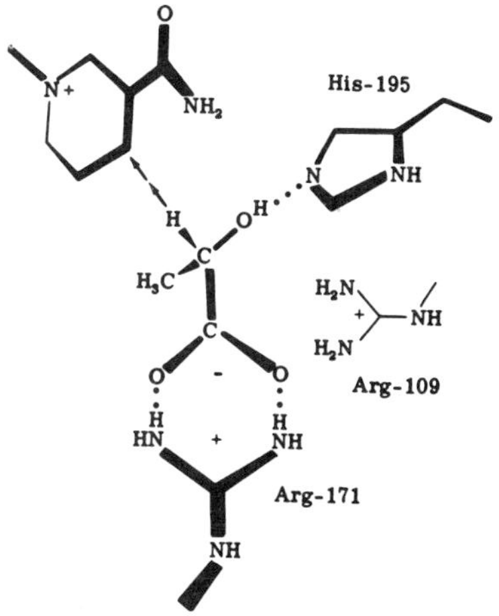

Fig. 11. Diagrammatic representation of the anticipated substrate binding in the active ternary intermediate of LDH. Reproduced with permission from J. J. Holbrook *et al.*[151a]

ing interactions with the protein. The proposed electronic rearrangement, which can now take place at the shielded site, involves abstraction of a hydride ion from the bound lactate by NAD^+. This induces a flow of electrons from the unprotonated histidine, His 195, which serves both to orient the substrate and to act as an acid–base catalyst. The loop of polypeptide chain (98–114), which moves on forming the ternary complex, entraps the co-enzyme and substrate in a hydrophobic environment. This results in a high-energy state for the charged NAD^+ molecule and encourages the flow of electrons which results in the formation of a neutral NADH and pyruvate.[151c]

In alcohol dehydrogenase the crystal structures of several ternary complexes of the enzyme with NAD^+ or NADH and pseudo substrates or inhibitors have been studied in which the substrate or inhibitor is thought to bind in a way which is consistent with a productive binding mode of a substrate. The results are summarized in Fig. 12. The essential Zn^{2+} ion

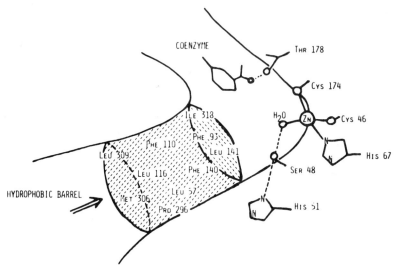

Fig. 12. Schematic representation of substrate binding in liver alcohol dehydrogenase. Reproduced with permission from C.-I. Bränden and M. Eklund, *Dehydrogenases Requiring Nicotinamide Co-Enzyme*, Birkhauser Verlag.[141]

is at the bottom of a hydrophobic pocket formed at the junction of the catalytic and nucleotide binding domains and is bound by Cys 46 and Cys 174 and a nitrogen atom of His 67. The fourth ligand is an ionizable water molecule. On binding substrate this water molecule is displaced and the substrate oxygen becomes liganded to the zinc and also makes a hydrogen bond to Ser 48 which in turn is hydrogen bonded to His 51. The second nitrogen of His 51 hydrogen bonds to the 2′-hydroxyl of the nicotinamide ribose. The hydrogen atom to be transferred points towards the C(4) of the nicotinamide. The rather broad specificity of alcohol dehydrogenase for non-polar primary and secondary alcohols can be explained by the hydrophobic barrel which can accommodate a wide variety of non-polar side chains. The reaction mechanism favoured by the crystallographers and consistent with extensive solution studies involves a co-enzyme induced isomerization on binding NAD^+ that results in a loss of a proton from the zinc bound water molecule via a proton pump mechanism involving Ser 48 and His 51. Thus the proton is transferred from the hydrophobic interior to the surface of the enzyme. In the next step alcohol displaces the zinc bound hydroxyl group and binds directly to zinc as an alcoholate group. A water molecule is formed by the extraction of a proton from the substrate by the zinc bound hydroxyl ion. Interconversion of the ternary complexes by hydride transfer is facilitated by the electrophilic nature of the zinc atom. An alternative mechanism in which the substrate does not displace the zinc bound water molecule but both are bound in a penta-coordinated complex is not supported by the structural studies on the transient intermediate complex of LADH with H_2NADH and an aldehyde substrate.[140] These show that the substrate oxygen is directly liganded to the zinc and the zinc bound water exchanged.

In glyceraldehyde 3-phosphate dehydrogenase, substrate binding sites have been deduced from model building studies on the holo-enzyme (Fig. 13). No direct experimental observations are available. In these studies it is proposed that the 3-phosphate group interacts with the side chains of Arg 231, Thr 179 and the 2'-hydroxyl of the nicotinamide ribose. In the

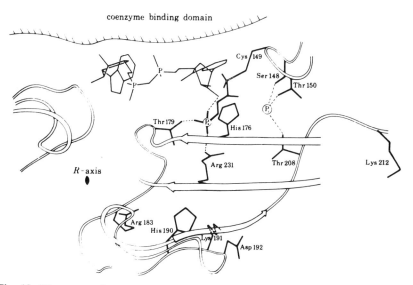

Fig. 13. Diagrammatic representation of the active centre region of the tetrameric molecule of the glyceraldehyde-3-phosphate dehydrogenase from *B. stearother- mophilus* showing the binding of NAD^+ and the disposition of the S-loop of the symmetry related subunit. Reproduced with permission from *Phil. Trans. R. Soc. London*, 1981, **B293**, 105.[152]

crystals obtained from ammonium sulphate this site is occupied by a firmly bound sulphate ion. The C(2) hydroxyl group binds to Ser 148 and the aldehyde is close to the reactive cystine 149 so that in the first step of the reaction a hemithioacetal is formed converting the unreactive carbonyl group into a more reactive alcohol at the C(1) position. The alcohol interacts with His 176 and the substrate is oriented so that the C(1) hydrogen points towards C(4) of the nicotinamide. The dehydrogenation can then occur as described for lactate dehydrogenase. The phosphate for the phosphorylation reaction is bound at a second weaker phosphate binding site (Table 3). In the lobster enzyme the strong substrate and weaker catalytic phosphate binding sites can be spanned by a citrate molecule.[143b] Before phosphorylation of the thio-ester can occur the NADH molecule must be replaced by an NAD$^+$ molecule. The NAD$^+$ possibly serves to polarize the sulphur atom of the hemithioacetal. While such a mechanism is consistent with recent kinetic data, the molecular basis of negative cooperativity (whereby there is tight binding by two equivalents of NAD$^+$ followed by weak binding of two more) remains to be established.[152] Limited molecular asymmetry has been observed in the abortive ternary complex of lobster GAPDH with trifluoroacetone–NAD complex.[152b]

5.4. Kinases

The five kinases, whose structures are known are shown in Table 5. All five structures are based on an α/β secondary structure arrangement. Three of them (hexokinase (HK), phosphoglycerate kinase (PGK) and phosphofructokinase (PFK)) exhibit bilobal structures; pyruvate kinase (PK) has three domains and the smaller adenylate kinase (AK) molecule is partially divided into two parts by a deep cleft. In each structure the substrate binding site is at the region where the domains come together. In order to facilitate the transfer of the phosphoryl group, it is necessary to exclude water from the substrate binding site. For the monomeric kinases, HK, PGK and AK, there is firm evidence either from crystallographic and/or from solution studies that the binding of one or both substrates triggers a conformational change which results in closure of the bilobal structure around the substrate(s) shielding them from water.[157] In the cases of the two oligomeric kinases, PFK and PK there is no direct evidence for a substrate enclosing conformational change. Indeed for PFK it seems that substrates can bind to a relatively open catalytic site that is situated between subunits.

For the adenine binding domains of PGK and, to some extent, AK, there is a topological similarity with the nucleotide binding domain of dehydrogenases which extends to the adenine binding site itself. For hexokinase

Table 5. Kinases whose crystal structures are known

Source/form of enzyme	mol.wt. of subunit	Oligomeric structure	Reaction catalyzed	Resolution (Å)	Reference
1. Hexokinase (HK)					
Yeast (B isozyme)	50 000	Monomer	Glucose + ATP⇌glucose-6-phosphate + ADP	2.1	154
Yeast (A isozyme) + glucose	50 000	Monomer		3.5	155
Yeast (B isozyme)	50 000	Dimer		3.5	156
2. Phosphoglycerate kinase (PGK)					
Equine Muscle	45 000	Monomer	1,3 diphosphoglycerate + $ADP \xrightleftharpoons{Mg^{2+}}$ 3-phosphoglycerate + ATP	2.5	158, 106
Yeast	45 000	Monomer		3.0	159
3. Adenylate kinase (AK)					
Porcine muscle					
A form pH 6.9–8.0	22 000	Monomer	AMP + Mg.ATP⇌Mg.ADP + ADP	3.0	107, 108
B form pH 5.7–5.9				4.7	
4. Phosphofructokinase (PFK)					
B. stearothermophilus	33 900	Tetramer	Fructose-6-phosphate + Mg.ATP⇌ Fructose-1,6-diphosphate +Mg.ADP+H^+	2.4	103, 160
5. Pyruvate kinase (PK)					
Cat muscle	60 000	Tetramer	Phosphoenoylpyruvate + ADP + H^+ $\xrightarrow{Mg^{2+}, K^+}$ Pyruvate + ATP	2.6	161, 162

there appears to be no such topological similarities. While in pyruvate kinase the nucleotide is bound at the carboxy end of a β sheet, as in the dehydrogenases, but it is oriented along the strand direction instead of along the edge of the sheet.

Thus it is more difficult to deduce overall similarities in the kinase structures and mechanisms than in the dehydrogenases. Only in the cases of phosphofructokinase and phosphoglycerate kinase has the transferable γ-phosphate of ATP been located (as discussed in Section 4), but no structures of co-crystallized ternary complexes are available to date. It may be that subsequent studies will reveal unifying principles. A structural and functional similarity has been noted, however, from comparisons of PK with triose phosphate isomerase which is discussed later. As has been pointed out by Lowe,[166] HK, PFK, PK and PGK all catalyse phosporyl transfer with inversion of configuration which is most simply interpreted in terms of an "in line" transfer of the phosphoryl group between substrates in the ternary complex. Hence in the kinases no catalysis takes place unless both substrates are bound. This is in contrast to the mutases where the phosphoryl transfer is to the same substrate and involves first a transfer to the enzyme and then a subsequent rearrangement of substrate (see for example Winn *et al.*[109]).

5.4.1. Hexokinase (HK)

When either sugars or nucleotides are diffused into preformed crystals of the native B isozyme of hexokinase, minor conformational changes occur.[163,164] However much more dramatic differences are observed when the B isozyme structure is compared with the structure of the A isozyme co-crystallized in the presence of glucose.[165a] One lobe of the bilobate molecule, constituting 40% of the structure, rotates by about 12° relative to the other lobe, resulting in movements as much as 8 Å (Fig. 14). These changes are comparable in magnitude to the changes observed in the allosteric transition of haemoglobin. The assumption that these movements are triggered by glucose binding to the enzyme in solution and are not a consequence of the A/B isozyme differences or the crystal lattice forces is supported by low angle X-ray scattering measurements. The radius of gyration of the B isozyme in solution is reduced by 0.95 ± 0.24 Å when glucose is bound and by 1.25 ± 0.28 Å when glucose-6-phosphate is bound.[165b] The former value is consistent with the change computed from the atomic co-ordinates of the two crystal structures.

In the enzyme–glucose complex, the solvent accessible area of glucose is reduced by 94% and only the 6-hydroxyl methyl group is exposed. The cleft has closed so that atoms in the smaller lobe are in contact with those

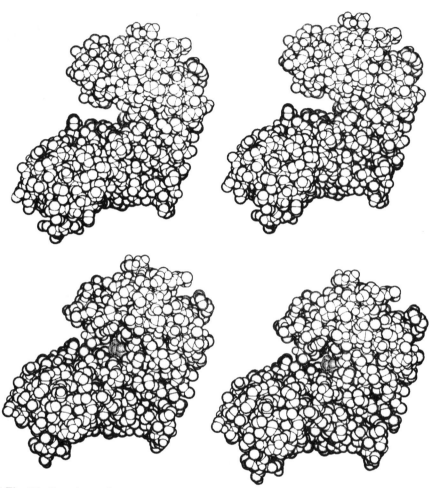

Fig. 14. Drawings of space filling models of yeast hexokinase B in its unliganded form (upper) and of yeast hexokinase A complexed with glucose (lower). The active site cleft that divides the molecule into two lobes is closed by the binding of glucose. Reproduced with permission from *Science*, 1979, **204**, 375.[157]

of the larger lobe and glucose cannot leave without a conformational change. Despite lack of chemical sequence information, considerable progress has been made in identifying residues from the 2.1 Å resolution crystallographic refinement of the native structure.[154] The 1, 3, 4 and 6-hydroxyl groups of the glucose hydrogen bond to polar groups on the enzyme and account for the enzyme's known specificity. The 6-hydroxyl is bound to the carboxylate side chain of Asp 189, a group which may be of importance in catalysis.[163]

Studies on binding of the second substrate ATP are less well advanced because of difficulties associated with crystal cracking and inability to locate the whole molecule. In the current model the γ-phosphate of ATP is 5–6 Å from the 6-hydroxy methyl group of the sugar. Model building in the closed glucose-linked complex, assuming that the ATP binding site remains unaltered shows no significant reduction in this distance. Evidently in order to account for a direct nucleophilic attack of the γ-phosphate on the 6-hydroxyl, further conformational changes must take place on forming the ternary complex.

5.4.2. Phosphoglycerate kinase (PGK)

Horse and yeast PGK have closely homologous structures in which the most striking feature is the division of the molecule into two widely separated domains of equal size, corresponding to the N-terminal and C-terminal portions of the molecule with the final ten residues for the C-terminus packed in the N-terminal domain. The topology of the C-terminal domain is identical to the dinucleotide binding domain of the dehydrogenases and that of the N-terminal domain nearly so.

Substrate binding studies for the horse enzyme were first complicated by the presence of ammonium sulphate in the crystallization medium. Transfer of the crystals to tartrate allowed observations on the binding of MgADP, MgATP and 3-phosphoglycerate.[106] MgADP and MgATP bind to the part of the C-domain that faces the N-domain (Fig. 15). The adenine moiety is located in a narrow hydrophobic slot in a position topologically equivalent to that occupied by the adenine of NAD when bound to lactate dehydrogenase, the ribose involves a contact with a glutamate side chain, while of the three phosphates of ATP only the α phosphate makes a specific contact with the ε-amino group of a lysine. The β and γ phosphates are located about 5 Å from the amino terminus of an α-helix and are presumably stabilized by a helix dipole interaction (Section 4). In ADP the metal is located between the α and β phosphates. Its site in ATP is not clear but appears to be different. Changes which accompany ATP or ADP binding are small and local. In contrast the changes on binding 3-phosphoglycerate are so large that the substrate could not be located with certainty in the difference Fourier synthesis. The most plausible feature was a peak situated close to an arginine-histidine cluster on the N-domain which makes an attractive binding site for the highly charged 3-phosphoglycerate molecule. However this places the substrate 12–15 Å from the γ-phosphate of ATP, a distance rather too large to allow for an in-line attack. It has therefore been proposed that in the ternary structure complex there is a "hinge–bending" motion that brings the two domains together. Support for this

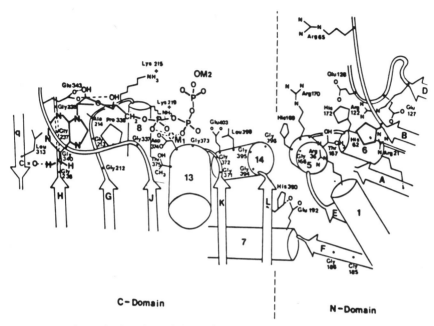

Fig. 15. A schematic drawing of the active site of horse phosphoglycerate kinase with ATP bound. The metal position in the ADP complex is shown by M1. Its position in the ATP complex is not established. Note the cluster of basic residues to the right of the active site. Reproduced with permission from *Phil. Trans. R. Soc. London*, 1981, **B293**, 93.[106]

closing of the molecule has come from low angle X-ray scattering measurements[167] that have shown a decrease in radius of gyration of 1.09 Å on formation of the ternary complex. Such a motion appears structurally feasible.[168]

Details have recently emerged of the substrate binding to yeast phosphoglycerate kinase.[177] The sequence determination has shown that there is a high degree of homology between horse and yeast enzymes (63% of residues identical). In particular there are three regions of a triple glycine sequence which are conserved and which (as described later) may be important in providing regions of inherent flexibility. Crystal soaking experiments with high concentrations of ATP and ADP led to disruption of the crystals, as if the nucleotide phosphates triggered an unfavourable conformational change. Strongest binding was observed with Mn-adenylyl, β,γ, imido diphosphate (Mn-AMP-PNP) at pH 9 (to minimize effects of sulphate ion binding). Almost all of the contacts appear similar to those observed for the horse enzyme (after allowing for a shift of two residues between the

numbering systems of references 177 and 106). The numbers from here on
refer to the yeast enzyme. The γ phosphoryl makes hydrogen bonds with
the main chain nitrogens of residues 371 and probably 372 (a helix dipole
contact). The metal ion appears positioned some 3 Å from the γ phosphate
and 5 Å from the α and β phosphates and also interacts with the ribose
hydroxyl and the carboxylate of Asp 372. Lysines 213 and 217 could also
play a role in stabilizing the phosphates.

The most surprising feature of this work was that the substrate 3-phospho-
glycerate (3-PGA) appears to be strongly bound to the native enzyme,
although no 3-PGA was intentionally added. Its presence is possibly the
result of turnover of metabolites during preparative stages of the enzyme.
The substrate is maximally hydrogen bonded to the enzyme. It is surrounded
by a cluster of 6 basic residues and is in van der Waals contact with the
triple Gly sequence 392–394. One of its carboxyl oxygens is only 4 Å away
and in a suitable position to make an in line attack on the γ phosphoryl
of ATP. Clearly a conformational change has to occur for catalysis and the
results suggest that both substrates ATP and 3-PGA need to be present in
order to release the trigger. Watson and his colleagues note that at the hinge
region there is an ion pair between histidine 388 and glutamic acid 190
(residues 390 and 192 in Fig. 15). They propose that conformational changes,
especially those centred on the triple Gly sequences 369–371 (ATP binding)
and 392–394 (3-PGA binding) could weaken this interaction and thus initiate
domain movement. Clearly it will be most exciting to see the "closed" form
of the enzyme with both substrates bound.

Recent kinetic studies[176] provide evidence for a tight complex between
phosphoglycerate kinase and its substrate 1,3-diphosphoglycerate that is
recognized by glyceraldehyde-3-phosphate dehydrogenase and indicate that
there may be direct transfer of the rather unstable substrate between these
two adjacent molecules in the glycolytic pathway.

5.4.3. Adenylate kinase (AK)
Despite identical space groups and nearly identical cell dimensions, the
crystal structures of the A and B forms of adenylate kinase (Table 5) show
differences. These involve about 15% of the molecule with the largest change
occurring in the loop 16–22. The mid-point in the hysteresis curve in the
$A \rightarrow B$ transition is at pH 6.4, close to the pK of His 36. His 36 is known
from NMR studies to be near to the divalent cation bound to substrate.

AK contains a deep cleft in which the binding sites for AMP and ATP
are located at opposite ends with the phosphate groups lying in the centre
and nearly surrounded by protein. These sites were located by binding
studies in the crystal at 6 Å resolution using a series of analogs and although

the interpretation was not entirely unambiguous the results are consistent with other information. In the A form of the crystals only the ATP site is available. The AMP site is blocked by the loop 16–22. On movement of this loop together with the helix 23–30 in the B form, the AMP specificity binding pocket is opened up. The sequence of the loop 16–22 is Gly. Pro. Gly. Ser. Gly. Lys. The high glycine content allows for flexibility.

Comparison of adenylate kinase with the dehydrogenases[169] show that the ATP binding site of adenylate kinase is topologically similar to the adenine part of the NAD binding site of lactate dehydrogenase. In particular a tyrosine residue (Tyr 95) which interacts with the base is equivalent to a tyrosine (Tyr 85) in lactate dehydrogenase and the mobile loop of adenylate kinase which encircles the phosphate of the AMP binding site is equivalent to a loop in the dehydrogenases which forms contacts with the pyrophosphate of NAD. Further the loop of fifteen residues that moves so dramatically in lactate dehydrogenase between the apo- and the holo-enzyme is equivalent to helix 123–133 which is preceded by a flexible hinge and also moves in adenylate kinase.

The present resolution of the structural studies does not provide sufficient detail for an understanding of mechanism. Further, it is not yet known whether co-crystallization with substrates will produce additional conformational changes beyond those triggered by the pH change. Nevertheless the results at present support the view of an "induced fit" mechanism for adenylate kinase in which the B form represents the open structure of the native enzyme and the A form the closed structure needed to shield the substrates from the water.

5.4.4. Phosphofructokinase (PFK)

In contrast to the other four kinases, phosphofructokinase is an allosteric enzyme whose activity *in vivo* is regulated by the energy requirements of the cell. The bacterial enzyme from *Bacillus stearothermophilus* shows homotropic co-operativity with substrate F6P, allosteric activation by ADP and inhibition by phosphoenolpyruvate. Since the crystals were obtained in the presence of F6P, the form of the enzyme in the crystals is the active form (the R state in the terminology of Monod, Wyman and Changeux[15]), The subunit (Fig. 16) is divided into two domains each of which has a central β-sheet sandwiched between α-helices. Each subunit forms close contacts with only two of the other subunits in the tetramer. The active site lies in an extended cleft between the two domains with the 6-phosphate of F6P making interactions with basic residues from two subunits (Table 3). The F6P molecule is bound by domain 2 and the ATP by domain 1. The Mg^{2+} bridges the α and β phosphates in ADP and moves about 2.8 Å to

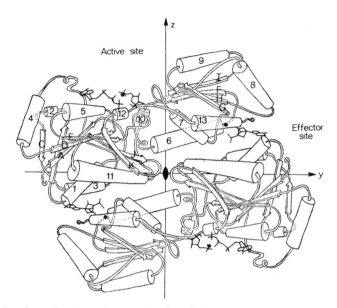

Fig. 16. A schematic view of two subunits of the phosphofructokinase tetramer. α-helices and β-sheet strands are represented by cylinders and arrows respectively. The molecule is divided into two domains. The substrates ATP and F6P are shown in the active site and the activator ADP in the effector site. Reproduced with permission from *Phil. Trans. R. Soc. London*, 1981, **B293**, 53.[103]

bridge the β and γ phosphates in an ATP analogue, but the contacts between the metal and the protein are not clear. The γ phosphate has been established from an AMP.PNP–ADP difference Fourier synthesis at 6 Å resolution. The phosphate lies between ADP and F6P in a position suitable for in line phosphoryl transfer. No large conformational changes are seen in these maps presumably because the structure is already in the R state. The binding of the phosphate of F6P involves two arginines from a neighbouring subunit in addition to a histidine and arginine from its own subunit. Thus any rearrangement of the subunits would alter this site and could explain the co-operativity of substrate binding. In addition the contacts at the activation ADP site also involve contributions from two subunits.

In the present liganded form of the enzyme the inter-subunit contacts around the z-axis contain a number of water molecules, and the catalytic site appears to be relatively open. On the other hand removal of substrate (F6P or P_i) from the crystals causes them to disintegrate, indicative of a conformational change. The structure of the T state has recently been solved independently at 7 Å resolution (P. R. Evans, Private Communication). The change from R to T is broadly described as a rotation of the z-dyad related dimer (Fig. 16) by 8° relative to the other dimer, and a rotation of the small domain relative to the large domain around the long axis of the dimer. Changes in tertiary structure appear to ensure that the subunit contacts are conserved despite the subunit rotations, except that the z-interface is closed up slightly. The beginning of sheet strand I carrying Arg 243 moves into the F6P site and the loop between β strand F and helix 7 (including Arg 171) moves into the F6P site from the other side, thus blocking F6P binding. The molecular details of these interesting changes, which are of the order of 3–4 Å, await higher resolution studies.

5.4.5. *Pyruvate kinase (PK)*

The polypeptide chain of pyruvate kinase is folded into three domains. Domain A (about 220 residues) contains a β-sheet barrel of eight parallel strands linked by α-helices with a topology that is nearly identical to the fold of triose phosphate isomerase. Domain B is inserted between the third strand and third helix of domain A and the course of the polypeptide chain is not well defined although it appears to contain antiparallel sheet and little α-helix. Domain C exhibits similarity to the first part of the nucleotide binding domain of lactate dehydrogenase. In the association of subunits, the inter subunit contacts between the two C domains involve two β sheets that form a single continuous sheet stretching across the subunits in alcohol dehydrogenase.

Substrate binding studies at 6 Å resolution[170] have shown that phosphoenolpyruvate is bound in a cleft between domains A and C. These two domains are connected by only two polypeptide chains which might indicate scope for flexibility. Mn–ATP binds radially perpendicular to the barrel axis so that the end of the density interpreted as the γ-phosphate overlaps the phosphoenolpyruvate site.

Three enzymes have now been observed with a common 8-fold β-barrel-α-helix topology, triose phosphate isomerase,[171], 2-keto-3-deoxy-6-phosphogluconate aldolase,[172] and pyruvate kinase.[161,162] In addition the 5.5 Å resolution structure of glycolate oxidase[19] also shows similarity. These enzymes, despite their different chemical roles in metabolism share a common function: they each activate a C–H bond that is next to a carbonyl group.[173] The first step in the pyruvate kinase reaction involves deprotonation of pyruvate leading to an enoyl-pyruvate intermediate.[174] Deprotonation is decoupled from phosphorylation. In triose phosphate isomerase the deprotonation of glyceraldehyde-3-phosphate leads to the formation of an enediol.[175] Thus the similarity in catalytic functions between triose phosphate isomerase and pyruvate kinase can now be extended to a similarity in structure. At the time of writing the amino acid sequence of pyruvate kinase was not known, but it has recently been determined.[186] The side chain electron densities at the catalytic site suggests that some of these residues might be similar to those of triose phosphate isomerase. Thus these interesting studies have thrown light on the possible evolutionary origins of deprotonation reactions involving enol-intermediates.

6. Conclusions

This limited survey of enzyme–substrate interactions has shown that some general conclusions may be drawn. All the enzyme–substrate complexes are stabilized by weak intermolecular forces involving hydrogen bonds, van der Waals, hydrophobic and charge–charge interactions. These weak forces provide sufficient stabilization to favour the enzyme–substrate complex over the dissociated state. For enzyme–transition state (or transition state analog) complexes, where dissociation constants may be in the range 10^{-9}–10^{-14} M, additional interactions appear to be involved that are related to bond distortion and subsequent electron rearrangement in the substrate and conformational changes in the enzyme.

The survey of phosphate binding sites has shown that, where the sites are involved solely in the binding of substrate, co-factor or effector, well defined charge–charge, helix dipole interactions or a network of hydrogen bonds stabilize the protein–phosphate complex. Where the phosphate site is involved in catalysis, as in phosphoryl transfer or phosphate ester hydrolysis, the interactions appear less specific, as expected, but there is the likelihood of conformational changes that allow additional groups to contribute towards stabilization of the transition state complex.

The active sites of the hydrolytic enzymes (e.g. lysozyme, serine proteinases, ribonucleases) are relatively exposed in a cleft on the surface of the molecule and accessible to water. Small conformational changes, e.g. shifts of <1 Å or a simple rotation of a single residue, take place on binding substrate that lead to closer contracts between the two molecules. For enzymes where it is necessary to remove reactants from bulk water, such as the dehydrogenases, kinases, or phosphorylase, gross conformational changes take place on binding substrate which involve shifts in atoms of the order of 10 Å and relative movements of domains one to another. These changes are so great they cannot be accomplished within the crystal lattice, despite the mobility of protein molecules in the crystal. They result in some instances (such as hexokinase or arabinose binding protein) in the substrate being almost totally enclosed by the enzyme with a reduction in its solvent accessibility of almost 100%. Little information is available on the structural transitions of allosteric proteins, apart from haemoglobin, but the crystal studies on binding of effectors (e.g. H^+ or organic phosphates to haemoglobin; AMP or phosphorylation of Ser-14 in phosphorylase; ADP to phosphofructokinase) show that the effector binding sites are located at subunit–subunit interfaces. Hence relative changes in tertiary structure triggered by effector binding affect quaternary structure (and vice-versa) so that conformational differences between the tertiary and quaternary structures are closely co-ordinated.

Acknowledgements

The author wishes to thank all those who contributed reprints or preprints for the preparation of this review.

References (Literature survey completed April 1982)

1. E. Fischer, *Ber. Dtsch. Chem. Ges.*, 1894, **24**, 2683.
2. A. Fersht, *Enzyme Structure and Mechanism*, W. H. Freeman & Co., Reading, 1977, Ch. 10.
3. H. M. Powell, *J. Chem. Soc.*, 1948, 61.
4. B. W. Matthews, P. D. Sigler, R. Henderson and B. H. Blow, *Nature*, 1967, **214**, 652.
5. A. A. Kossiakoff and S. A. Spencer, *Nature*, 1980, **288**, 414.
6. M. N. G. James, A. R. Sielecki, G. D. Brayer, L. T. J. Delbaere and C.-A. Bauer, *J. Mol. Biol.*, 1980, **144**, 43.
7. R. Huber, *Trends Biochem. Sci.*, 1979, **4**, 271.
8. R. Huber and W. Bode, *Acc. Chem. Res.*, 1978, **11**, 114.
9. J. Kraut, *Ann. Rev. Biochem.*, 1977, **46**, 331.
10. M. N. G. James, *Can. J. Biochem.*, 1980, **58**, 251.
11. G. E. Schulz and R. H. Schirmer, *Principles of Protein Structure*, Springer-Verlag, New York, 1979.
12. T. Imoto, L. N. Johnson, A. C. T. North, D. C. Phillips and J. A. Rupley, in *The Enzymes* (ed. P. D. Boyer), Academic Press, New York, London, 1972, Vol. 7, 3rd ed. p. 665.
13. D. E. Koshland, in *The Enzymes* (ed. P. D. Boyer, H. Lardy and K. Myrbäch), Academic Press, New York, London, 1959, Vol. 1, 2nd ed. pp. 305–346.
14. C. F. Cori and G. T. Cori, *Proc. Soc. Exp. Biol. Med.*, 1936, **34**, 702.
15. J. Monod, J. Wyman and J. P. Changeux, *J. Mol. Biol.*, 1965, **12**, 88.
16. D. E. Koshland and K. E. Neat, *Ann. Rev. Biochem.*, 1967, **37**, 359.
17. T. L. Blundell and L. N. Johnson, *Protein Crystallography*, Academic Press, London, New York, 1976.

18. B. W. Matthews, *J. Mol. Biol.* 1968, **20**, 82.
19. Y. Lindquist and C-I. Bränden, *J. Mol. Biol.*, 1980, **143**, 201.
20. A. J. Sophianopoulos and B. J. Weiss, *Biochemistry*, 1964, **3**, 1920.
21. C. Tanford, *Adv. Prot. Chem.*, 1968, **23**, 121.
22. B. W. Matthews, L. H. Weaver and W. R. Kester, *J. Biol. Chem.*, 1974, **249**, 8030.
23. M. F. Perutz, *J. Mol. Biol.*, 1965, **13**, 646.
24. A. R. Sielecki, W. A. Hendrickson, C. G. Broughton, L. T. J. Delbaere, G. D. Brayer and M. N. G. James, *J. Mol. Biol.*, 1979, **134**, 781.
25. M. G. Rossmann and P. Argos, *Ann. Rev. Biochem.* 1981, **50**, 497.
26. T. Takano and R. E. Dickerson, *J. Mol. Biol.*, 1981, **153**, 95.
27. T. Takano and R. E. Dickerson, *J. Mol. Biol.*, 1981, **153**, 79.
28. M. F. Perutz, *Nature*, 1970, **228**, 726.
29. T. Alber, D. W. Banner, A. C. Bloomer, G. A. Petsko, D. C. Phillips, P. S. Rivers and I. A. Wilson, *Phil. Trans. R. Soc. London*, 1981, **B293**, 159.
30. K. D. Watenpaugh, L. C. Sieker, J. R. Herriott and L. H. Jensen, *Acta Crystallogr.*, 1973, **B29**, 943.
31. R. Diamond, *J. Mol. Biol.*, 1974, **82**, 371.
32. R. Fehlhammer and W. Bode, *J. Mol. Biol.* 1975, **98**, 683.
33. W. Bode and P. Schwager, *J. Mol. Biol.*, 1975, **98**, 693.
34. R. Huber, D. Kukla, W. Bode, P. Schwager, K. Bartels, K. Deisenhofer, and W. Steigemann, *J. Mol. Biol.*, 1979, **89**, 73.
35. J. H. Konnert, *Acta Crystallogr.*, 1976, **A32**, 614.
36. J. H. Konnert and W. A. Hendrickson, *Acta Crystallogr.*, 1980, **A36**, 344.
37. J. H. Sussman, S. R. Holbrook, G. M. Church and S.-H. Kim, *Acta Crystallogr.*, 1977, **A33**, 800.
38. A. Jack and M. Levitt, *Acta Crystallogr.*, 1978, **A34**, 931.
39. H. Frauenfelder, G. A. Petsko and D. Tsernoglou, *Nature*, 1979, **280**, 369.
40. M. J. E. Sternberg, D. E. P. Grace and D. C. Phillips, *J. Mol. Biol.*, 1979, **130**, 231.
41. P. J. Artymiuk, C. C. F. Blake, D. E. P. Grace, S. J. Oatley, D. C. Phillips, and M. J. F. Sternberg, *Nature*, 1979, **280**, 563.
42. J. Walter, W. Steigemann, T. P. Singh, H. D. Bartunik, W. Bode and R. Huber, *Acta Crystallogr.*, 1982, **B38**, 1462.
43. M. S. Doscher and F. M. Richards, *J. Biol. Chem.*, 1963, **238**, 2399.
44. J. A. Rupley in *Biological Macromolecules* (ed. S. N. Timasheff and G. D. Freeman), Vol. 2, Structure and Stability of Biological Macromolecules, Marcel Dekker, New York, 1969, p. 291.
45. F. A. Quiocho and F. M. Richards, *Biochemistry*, 1966, **5**, 4062.
46. G. L. Rossi and S. A. Bernhard, *J. Mol. Biol.*, 1970, **55**, 215.
47. G. L. Rossi and S. A. Bernhard, *J. Mol. Biol.*, 1971, **49**, 85.
48. M. W. Makinen and A. L. Fink, *Ann. Rev. Biophys. Bioeng.*, 1977, **6**, 301.
49. P. J. Kasvinsky and N. B. Madsen, *J. Biol. Chem.*, 1976, **251**, 6852.
50. D. M. Shotton, N. J. White, H. C. Watson, *Cold Spring Harbor Symp. Quant. Biol.*, 19th, **36**, 91.
51. L. A. A. Sluyterman and M. J. M. de Graaf, *Biochim. Biophys. Acta*, 1969, **171**, 272.
52. J. Bella and E. F. Nowoswiat, *Biochim. Biophys. Acta*, 1965, **105**, 325.
53. L. J. Parkhurst and O. M. Gibson, *J. Biol. Chem.*, 1964, **242**, 5762.
54. J. C. Phillips, A. Wlodawer, M. M. Yevitz and K. O. Hodgson, *Proc. Nat. Acad. Sci. USA*, 1976, **73**, 128.

55. H. D. Bartunik, R. Fourme and J. C. Phillips, in *Uses of Synchrotron Radiation in Biology*, (ed. B. B. Stuhrman), Academic Press, London, New York, 1982.

56. T. Alber, G. A. Petsko and D. Tsernoglou, *Nature*, 1976, **263**, 297.

57. A. L. Fink and G. A. Petsko, *Adv. Enzymol.*, 1981, **52**, 177.

58. G. M. Edelman and J. L. Wang, *J. Biol. Chem.*, 1978, **253**, 3016.

59. K. D. Hardman and C. F. Ainsworth, *Biochemistry*, 1973, **12**, 4442.

60. J. W. Becker, G. N. Reeke, J. L. Wang, B. A. Cunningham and G. M. Edelman, *J. Biol. Chem.*, 1975, **250**, 15013.

61. J. W. Becker, G. N. Reeke, B. A. Cunningham and G. M. Edelman, *Nature*, 1976, **259**, 406.

62. C. C. F. Blake, D. F. Koenig, G. A. Mair, A. C. T. North, D. C. Phillips and V. R. Sarma, *Nature*, 1965, **206**, 757.

63. C. C. F. Blake, L. N. Johnson, G. A. Mair, A. C. T. North, D. C. Phillips and V. R. Sarma, *Proc. R. Soc. London*, 1967, **B167**, 378.

64. (*a*) L. O. Ford, L. N. Johnson, P. A. Machin, D. C. Phillips and R. Tjian, *J. Mol. Biol.*, 1974, **88**, 349; (*b*) J. A. Kelly, A. R. Sielecki, B. D. Sykes and M. N. G. James, *Nature*, 1979, **282**, 875.

65. R. Bott and V. R. Sarma, *J. Mol. Biol.*, 1976, **106**, 1037.

66. D. C. Phillips, *Proc. Nat. Acad. Sci. U.S.A.*, 1967, **57**, 484.

67. B. W. Mathews and S. J. Remington, *Proc. Nat. Acad. Sci. USA*, 1974, **71**, 4178.

68. M. G. Rossmann and P. J. Argos, *J. Mol. Biol.*, 1976, **105**, 75.

69. M. Levitt and C. Chothia, *Nature*, 1976, **261**, 552.

70. S. J. Remington and B. W. Matthews, *Proc. Nat. Acad. Sci. USA*, 1978, **75**, 2180.

71. W. F. Anderson, M. G. Grütter, S. J. Remington, L. H. Weaver and B. W. Matthews, *J. Mol. Biol.*, 1981, **147**, 523.

72. Y. Ocada, S. Amagase and A. Tsugita, *J. Mol. Biol.*, 1970, **54**, 219.

73. B. W. Matthews, S. J. Remington, M. G. Grütter, and W. F. Anderson, *J. Mol. Biol.*, 1981, **147**, 545.

74. A. Jung, A. E. Sippel, M. Grez and G. Schütz, *Proc. Nat. Acad. Sci. USA*, 1980, **77**, 5759.

75. P. J. Artymuik, C. C. F. Blake and A. E. Sippel, *Nature*, 1981, **290**, 287.

76. W. Gilbert, *Nature*, 1978, **271**, 501.

77. C. C. F. Blake, *Nature*, 1978, **273**, 267.

78. D. J. Graves and J. H. Wang, in *The Enzymes* (ed. P. D. Boyer), 3rd. edn., Vol. 7. Academic Press, London, New York, 1972, p. 435.

79. I. T. Weber, L. N. Johnson, K. S. Wilson, D. G. R. Yeates, D. L. Wild and J. A. Jenkins, *Nature*, 1978, **274**, 433.

80. S. Sprang and R. J. Fletterick, *J. Mol. Biol.* 1979, **131**, 523.

81. F. Meyer, L. M. G. Heilmeyer, R. M. Haschke and E. M. Fischer, *J. Biol. Chem.*, 1970, **245**, 6642.

82. P. Cohen, *Curr. Top. Cell. Reg.*, 1978, **14**, 117.

83. J. H. Wang, M. L. Shŏnka and D. J. Graves, *Biochemistry*, 1965, **40**, 1340.

84. P. J. Kasvinsky, N. B. Madsen, R. J. Fletterick and J. Sygusch, *J. Biol. Chem.*, 1978, **253**, 1290.

85. F. Payan, R. Haser, M. Pierrot, M. Frey and J. P. Astiev, *Acta Crystallogr.*, 1980, **B36**, 416.

86. Y. Matsura, M. Kusunoki, W. Havada, N. Tanaka, Y. Iga, N. Yasuoka, H. Toda, K. Narita and M. Kakudo, *J. Biochem.* (Tokyo) 1980, **87**, 1555.

87. L. N. Johnson, E. A. Stura, M. S. P. Sansom and Y. S. Babu, *Biochem. Soc. Trans.*, 1983, **11**, 142.

88. G. J. Quigley, A. Sarko and R. H. Marchessault, *J. Am. Chem. Soc.*, 1970, **92**, 5834.
89. D. A. Rees and P. J. C. Smith, *J. Chem. Soc.*, 1975, 836.
90. E. J. M. Helmreich and H. W. Klein, *Angew. Chem. Int. Ed. Eng.*, 1980, **19**, 441.
91. L. N. Johnson, J. A. Jenkins, K. S. Wilson, E. A. Stura and G. Zanotti, *J. Mol. Biol.*, 1980, **140**, 565.
92. J. Sygusch, N. B. Madsen, P. J. Kasvinsky and R. J. Fletterick, *Proc. Nat. Acad. Sci. USA*, 1977, **74**, 4757.
93. R. J. Fletterick and N. B. Madsen, *Ann. Rev. Biochem.*, 1980, **49**, 31.
94. J. A. Jenkins, L. N. Johnson, D. I. Stuart, E. A. Stura, K. S. Wilson and G. Zanotti, *Phil. Trans. R. Soc. London*, 1981, **B293**, 23.
95. D. A. Matthews, R. A. Alden, S. T. Freer, N. Xuong and J. Kraut, *J. Biol. Chem.*, 1979, **254**, 4144.
96. K. W. Volz, D. A. Matthews, R. A. Alden, S. T. Freer, C. Hansch, B. T. Kaufman and J. Kraut, *J. Biol. Chem.*, 1982, **257**, 2528.
97. K. D. Watenpaugh, L. C. Sieker and L. H. Jensen, *Proc. Nat. Acad. Sci. USA*, 1973, **70**, 3853.
98. R. M. Burnett, G. D. Dauling, D. S. Kendall, M. E. LeQuesne, S. G. Mayhew, W. W. Smith and M. L. Ludwig, *J. Biol. Chem.*, 1974, **249**, 4383.
99. L. N. Johnson, E. A. Stura, K. S. Wilson, M. S. P. Sansom and I. T. Weber, *J. Mol. Biol.*, 1979, **134**, 639.
100. R. J. Fletterick, S. Sprang and N. B. Madsen, *Can. J. Biochem.*, 1979, **57**, 789.
101. S. Sprang and R. J. Fletterick, *Biophys. J.*, 1980, **32**, 175.
102. G. C. Ford, G. Eichele and J. N. Jansonius, *Proc. Nat. Acad. Sci. USA*, 1980, **77**, 2559.
103. P. R. Evans, G. W. Farrants and P. J. Hudson, *Phil. Trans. R. Soc. London*, 1981, **B293**, 53.
104. A. Arnone, *Nature*, 1972, **237**, 148.
105. T. A. Steitz, M. Shoham and W. S. Bennett, *Phil. Trans. R. Soc. London*, 1981, **B293**, 43.
106. C. C. F. Blake and D. W. Rice, *Phil. Trans. R. Soc. London*, 1981, **B293**, 93.
107. W. Sachsenheimer and G. E. Schulz, *J. Mol. Biol.*, 1977, **114**, 23.
108. E. F. Pai, W. Sachsenheimer, R. H. Schirmer and G. E. Schulz, *J. Mol. Biol.*, 1977, **114**, 37.
109. S. I. Winn, H. C. Watson, R. N. Harkins and L. A. Fothergill, *Phil. Trans. R. Soc. London*, 1981, **B293**, 121.
110. A. Wlodawer, M. Miller and L. Sjolin, *Proc. Nat. Acad. Sci. USA*, 1983, **80**, 3628.
111. F. M. Richards and H. W. Wyckoff in *The Enzymes* (ed. P. D. Boyer) Academic Press, New York, London, 1971, 647.
112. T. Alber, W. A. Gibert, D. R. Ponzi and G. A. Petsko, *Ciba Symp.*, 1983, **93**, 4.
113. F. A. Cotton, E. E. Hazen and M. J. Legg, *Proc. Nat. Acad. Sci. USA*, 1979, **76**, 2551.
114. F. A. Cotton, V. W. Day, E. E. Hazen, S. Larsen and S. T. K. Wong, *J. Am. Chem. Soc.*, 1974, **96**, 4471.
115. J. F. Riordan, K. D. McElvany and C. L. Borders, *Science*, 1977, **195**, 884.
116. W. G. J. Hol, P. T. van Duijnen and H. J. C. Berendsen, *Nature*, 1978, **273**, 443.
117. A. Wada, *Adv. Biophys.*, 1976, **9**, 1.
118. E. G. Krebs and J. A. Beavo, *Ann. Rev. Biochem.*, 1979, **48**, 923.
119. P. Cohen, *Nature*, 1982, **296**, 613.

120. P. Greengard, *Cyclic Nucleotides, Phosphorylated Proteins and Neuronal Function.* Raven Press, New York, 1978.
121. D. J. Graves, S. A. S. Mann, G. Philip and R. J. Oliveira, *J. Biol. Chem.*, 1968, **243**, 6090.
122. M. Dreyfus, B. Vandenbunder and H. Buc, *Biochemistry*, 1980, **19**, 3643.
123. D. Findlay, D. G. Herries, A. P. Mathias, B. R. Rabin and C. A. Ross, *Nature*, 1961, **190**, 781.
124. R. N. Lindquist, J. L. Lynn and G. E. Lienhard, *J. Am. Chem. Soc.*, 1973, **95**, 8762.
125. J. Drenth, C. M. Enzing, K. H. Kalk and J. C. A. Vessies, *Nature*, 1976, **264**, 373.
126. B. P. Schoenborn, H. C. Watson and J. C. Kendrew, *Nature*, 1965, **207**, 28.
127. R. H. Kretsinger, H. C. Watson and J. C. Kendrew, *J. Mol. Biol.*, 1968, **31**, 305.
128. G. L. Gilliland and F. A. Quiocho, *J. Mol. Biol.*, 1981, **146**, 341.
129. J. H. Ploegman, G. Drent, K. H. Kalk and W. G. J. Hol, *J. Mol. Biol.*, 1978, **123**, 557.
130. M. E. Newcomer, G. L. Gilliland and F. A. Quiocho, *J. Biol. Chem.*, 1982, **256**, 13213.
131. B. Lee and F. M. Richards, *J. Mol. Biol.*, 1971, **55**, 517.
132. M. E. Newcomer, B. A. Lewis and F. A. Quiocho, *J. Biol. Chem.*, 1982, **256**, 13218.
133. M. Klingenberg, *Nature*, 1981, **290**, 449.
134. P. Argos, W. C. Mahoney, M. A. Hermodson, and M. Hanui, *J. Biol. Chem.*, 1981, **256**, 4357.
135. M. J. Adams, G. C. Ford, R. Koekoek, P. J. Lentz, A. McPherson, M. G. Rossmann, I. E. Smiley, R. W. Schevitz and A. J. Wonacott, *Nature*, 1970, **227**, 1098.
136. J. L. White, M. L. Hackert, M. Buehner, M. J. Adams, G. C. Ford, P. J. Lentz, I. E. Smiley, S. J. Steindel and M. G. Rossmann, *J. Mol. Biol.*, 1976, **102**, 759.
137. H. Eklund, B. Nordström, E. Zeppezauer, G. Söderlund, I. Ohlsson, T. Boiwe, B.-O. Sôderberg, O. Tapia, C.-I. Bränden and A. Akeson, *J. Mol. Biol.*, 1976, **102**, 27.
138. H. Eklund and C.-I. Bränden, *J. Biol. Chem.*, 1979, **254**, 3458.
139. B. V. Plapp, H. Eklund and C.-I. Bränden, *J. Mol. Biol.*, 1978, **122**, 23.
140. E. Cedergren-Zeppezauer, J.-P. Samama and H. Eklund, *Biochemistry*, 1982, **21**, 4895.
141. C.-I. Bränden and H. Eklund, in *Dehydrogenases Requiring Nicotinamide Co-Enzyme* (ed. J. Jeffery) Birkhäuser Verlag, Basel, 1980. *Experientia Supplementum* **36**, p. 40.
142. D. Moras, K. W. Olsen, M. N. Sabesan, M. Buehner, G. C. Ford and M. G. Rossmann, *J. Biol. Chem.*, 1975, **250**, 9137.
143. (*a*) M. R. N. Murthy, R. M. Garavito, J. E. Johnson and M. G. Rossmann, *J. Mol. Biol.*, 1980, **138**, 859; (*b*) K. W. Olsen, R. M. Garavito, M. N. Sabesan and M. G. Rossmann, *J. Mol. Biol.*, 1976, **107**, 571.
144. G. Biesecker, J. I. Harris, J.-C. Thierry, J. E. Walker and A. J. Wonacott, *Nature*, 1977, **260**, 328.
145. M. G. Rossmann, D. Moras, and K. W. Olsen, *Nature*, 1974, **250**, 194.
146. M. G. Rossmann, A. Liljas, C.-I. Bränden and L. J. Banaszak, in *The Enzymes*, (ed. P. D. Boyer) 3rd edn. Vol. XI, p. 61 Academic Press, Orlando, New York, London. 1975.

147. D. A. Matthews, R. A. Alden, J. T. Bolin, D. J. Filman, S. T. Freer, R. Hamlin, W. G. J. Hol, R. L. Kislink, E. J. Pastore, L. T. Plante, N. Xuong and J. Kraut, *J. Biol. Chem.*, 1978, **253**, 6946.
148. C.-I. Bränden, *Quart. Rev. Biophys.* 1980, **13**, 317.
149. W. Saenger, B. S. Reddy, K. Mühllegger and G. Weimann, *Nature*, 1977, **267**, 225.
150. G. McDonald, B. Brown, D. Hollis and C. Walter, *Biochemistry*, 1972, **11**, 1920.
151. (a) J. J. Holbrook, A. Liljas, S. J. Steindel and M. G. Rossmann, in The Enzymes (ed. P. D. Boyer) 3rd ed. Vol. XI, p. 191, Academic Press, New York, London, 1975; (b) W. Eventott, M. G. Rossmann, S. S. Taylor, H.-J. Torft, H. Meyer, W. Keil and H. H. Kiltz, *Proc. Nat. Acad. Sci. USA*, 1977, **74**, 2677.; (c) D. M. Park and J. J. Holbrook in *Pyridine Nucleotide-Dependent Dehydrogenases* (ed. H. Sund) W. de Gryter, New York, 1977, p. 485.
152. (a) K. Dalziel, N. V. McFerran and A. J. Wonacott, *Phil. Trans. R. Soc. London*, 1981, **B293**, 105; (b) R. M. Garavito, D. Berger and M. G. Rossmann, *Biochemistry*, 1977, **16**, 4393.
153. (a) J.-P. Samama, A. D. Wrixon and J.-F. Biellmann, *Eur. J. Biochem.*, 1981, **118**, 479; (b) M. A. Abdallah, J. F. Biellmann, B. Nordström and C.-I. Bränden, *Eur. J. Biochem.*, 1975, **50**, 473; (c) K. W. Olsen, R. H. Garavito, M. N. Sabesan and M. G. Rossmann, *J. Mol. Biol.*, 1976, **107**, 577.
154. C. M. Anderson, R. E. Stenkamp and T. A. Steitz, *J. Mol. Biol.* 1978, **123**, 15.
155. W. S. Bennett and T. A. Steitz, *J. Mol. Biol.*, 1980, **140**, 183.
156. T. A. Steitz, W. F. Anderson, R. J. Fletterick and C. M. Anderson, *J. Biol. Chem.*, 1977, **252**, 4494.
157. C. M. Anderson, F. H. Zucker and T. A. Steitz, *Science*, 1979, **204**, 375.
158. R. D. Banks, C. C. F. Blake, P. R. Evans, R. Haser, D. W. Rice, G. W. Hardy, M. Merrett and A. W. Phillips, *Nature*, 1979, **279**, 773.
159. T. N. Bryant, H. C. Watson and P. L. Wendell, *Nature*, 1974, **247**, 14.
160. P. R. Evans and P. J. Hudson, *Nature*, 1979, **279**, 500.
161. M. Levine, H. Muirhead, D. K. Stammers and D. I. Stuart, *Nature*, 1978, **271**, 626.
162. D. I. Stuart, M. Levine, H. Muirhead and D. K. Stammers, *J. Mol. Biol.*, 1979, **134**, 109.
163. C. M. Anderson, R. E. Stenkamp, R. C. McDonald and T. A. Steitz, *J. Mol. Biol.*, 1978, **123**, 207.
164. M. Shoham and T. A. Steitz, *J. Mol. Biol.*, 1980, **140**, 1.
165. (a) W. S. Bennett and T. A. Steitz, *J. Mol. Biol.*, 1980, **140**, 211; (b) R. C. McDonald, T. A. Steitz and D. M. Engelman, *Biochemistry*, 1979, **18**, 338.
166. G. Lowe, P. M. Cullis, R. L. Jarvest, B. V. L. Potter and B. S. Sproat, *Phil. Trans. R. Soc. London*, 1981, **A293**, 75.
167. C. A. Pickover, D. B. McKay, D. M. Engelman and T. A. Steitz, *J. Biol. Chem.*, 1979, **254**, 11323.
168. C. C. F. Blake, F. E. Cohen and D. W. Rice, cited in reference 106.
169. G. E. Schulz and R. H. Schirmer, *Nature*, 1974, **250**, 142.
170. D. K. Stammers and H. Muirhead, *J. Mol. Biol.*, 1975, **95**, 213.
171. D. W. Banner, A. C. Bloomer, G. A. Petsko, D. C. Phillips, C. I. Pogson and I. A. Wilson, *Nature*, 1975, **255**, 609.
172. I. M. Mavridis and A. Tulinsky, *Biochemistry*, 1976, **15**, 4410.
173. I. A. Rose, *Phil. Trans. R. Soc. London*, 1981, **B293**, 131.
174. J. C. Robinson and I. A. Rose, *J. Biol. Chem.*, 1972, **247**, 1096.

175. I. A. Rose, *Brookhaven Symp. Biol.*, 1962, **15**, 293.
176. K. R. Huskins, S. A. Bernhard and F. W. Dahlquist, *Biochemistry*, 1982, **21**, 4180.
177. H. C. Watson, N. P. C. Walker, P. J. Shaw, T. N. Bryant, P. L. Wendell, L. A. Fothergill, R. E. Perkins, S. C. Conroy, M. J. Dobson, M. F. Tuite, A. J. Kingsman and S. M. Kingsman, *EMBO Journal*, 1982, **1**, 1635.
178. D. B. McKay, I. T. Weber and T. A. Steitz, *J. Biol. Chem.*, 1982, **257**, 9518.
179. I. T. Weber, K. Takio, K. Titani and T. A. Steitz, *Proc. Nat. Acad. Sci. USA*, 1982, **79**, 7679.
180. U. Heinemann and W. Saenger, *Nature*, 1982, **299**, 27.
181. C. C. F. Blake, W. C. A. Pulford and P. J. Artymiuk, *J. Mol. Biol.*, 1983, **167**, 693.
182. H. W. Klein, D. Palm and E. J. M. Helmreich, *Biochemistry*, 1982, **21**, 6675.
183. S. G. Withers, N. B. Madsen, B. D. Sykes, M. Takagi, S. Shimomura and T. Fukui, *J. Biol. Chem.*, 1981, **256**, 10759.
184. M. Takagi, T. Fukui and S. Shimomura, *Proc. Nat. Acad. Sci. USA*, 1982, **79**, 3716.
185. S. G. Withers, N. B. Madsen, S. Sprang and R. J. Fletterick, *Biochemistry*, 1982, **21**, 5372.
186. N. Lonberg and W. Gilbert, *Proc. Nat. Acad. Sci. USA*, 1983, **80**, 3661.

16 · CATION TRANSPORT IN LIQUID MEMBRANES MEDIATED BY MACROCYCLIC CROWN ETHER AND CRYPTAND COMPOUNDS

D. W. McBRIDE, Jr., R. M. IZATT,
J. D. LAMB, and J. J. CHRISTENSEN
Brigham Young University, Provo, Utah USA

1. Introduction

The ability to solubilize metal cations in media of low dielectric constant is necessary for many life processes. Indeed, since the beginning of the century a vast amount of effort has been directed towards understanding principles and mechanisms governing permeability, transport, and solubilization of electrolytes and non-electrolytes in both biological and non-biological membranes. During the past decade, we have studied a class of compounds called macrocycles which have the ability to form complexes selectively with metal cations and to solubilize these cations in liquid media of low dielectric constant.

The term "macrocycle" has been applied to a large group of cyclic compounds which can bind metal cations by entrapment within the electron

INCLUSION COMPOUNDS III
ISBN 0-12-067103-4

rich cavity created by inward-oriented donor atoms (i.e. either O, N, S, or some combination of these). Both synthetic and biologically derived macromolecules with widely varying structures have been reported. Among synthetic macrocycles are those of the cyclic polyether ("crown" ether)[1-3] (see Chapter 9, Volume 2) and cryptand[4,5] (see Chapter 10, Volume 2) varieties, which have been used as cation carriers in liquid membranes. Both classes of these synthetic macrocycles are illustrated in Fig. 1.

Owing to their novel reactions with cations, macrocycles have enjoyed much popularity among researchers. This research has led to their use in a steadily increasing number of ways, including the following: ion detection in devices such as cation selective electrodes;[6] phase transfer catalysis;[7,8] selective complexation of cations in studies of reaction mechanism;[8] elucidation of biological transport mechanisms;[9-11] study of solvation and ion pairing effects;[12] solubilization of salts in solvents of low polarity by complexation of the cation and formation of an ion-pair leading to new or improved organic synthesis and increased reaction yields;[7] separation of cations from one another;[13-20] separation of Ca^{2+} isotopes;[21] development of carrier membrane systems with possible application in the removal of heavy metal ions from wastes;[22] and use of drugs to remove cations from, or introduce them into, living systems.[23] In this chapter we shall limit our discussion to their use in liquid membrane transport systems.

Facilitated ion transport through liquid membranes has been investigated by many researchers.[13-20,24-31] These investigations have involved a variety of charged and uncharged mobile carriers. The charged carriers have included both cationic and anionic ion exchange molecules. The cationic exchangers are organophilic molecules containing one or more ionizable groups which, in order to maintain their electroneutrality in the bulk organic phase, may exchange one or more protons for a metal cation.[24] Examples of anion selective ion exchangers are found in the family of quaternary alkylammonium ions.[30] Among the neutral group of carriers are amines, linear polyethers, and macrocyclic crown ethers and cryptands. Thus, macrocycles are not the only mobile carriers able to facilitate ion transport through liquid membranes although they generally demonstrate the highest cation selectivity among possible carriers.

In the following discussion, attention is focused on ion transport mediated by macrocyclic crown ethers and cryptands. A mathematical model is presented for neutral carrier-mediated cation transport in bulk liquid membranes, the experimental procedures followed in these studies are described, and various factors which influence membrane transport and selectivities in single cation systems and in two-cation mixtures are discussed. Then, macrocycle-mediated transport in liquid-surfactant or emulsion type membranes for both unitary cation solutions and binary cation mixtures are

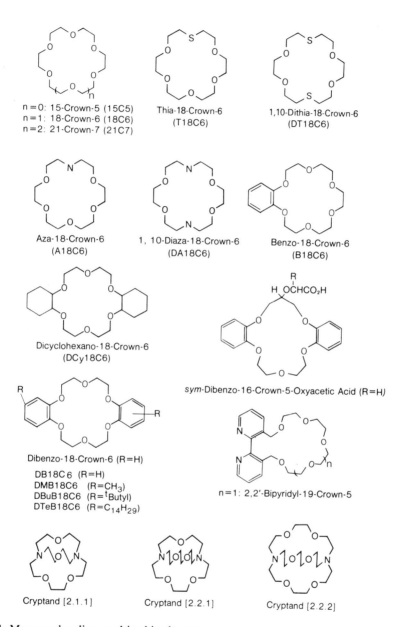

n = 0: 15-Crown-5 (15C5)
n = 1: 18-Crown-6 (18C6)
n = 2: 21-Crown-7 (21C7)

Thia-18-Crown-6
(T18C6)

1,10-Dithia-18-Crown-6
(DT18C6)

Aza-18-Crown-6
(A18C6)

1, 10-Diaza-18-Crown-6
(DA18C6)

Benzo-18-Crown-6
(B18C6)

Dicyclohexano-18-Crown-6
(DCy18C6)

sym-Dibenzo-16-Crown-5-Oxyacetic Acid (R=H)

Dibenzo-18-Crown-6 (R=H)

DB18C6 (R=H)
DMB18C6 (R=CH$_3$)
DBuB18C6 (R=^{t}Butyl)
DTeB18C6 (R=C$_{14}$H$_{29}$)

n = 1: 2,2'-Bipyridyl-19-Crown-5

Cryptand [2.1.1]

Cryptand [2.2.1]

Cryptand [2.2.2]

Fig. 1. Macrocycles discussed in this chapter.

discussed. Finally, mention is made of practical applications of macrocycles in liquid membranes and future prospects for work in this field.

2. Facilitated transport in bulk liquid membranes

A liquid membrane is a liquid or a quasi-liquid phase which separates two other liquid phases in which the membrane is immiscible. A typical membrane consists either of a hydrophobic liquid phase such as chloroform separating two water phases or, less commonly, of an aqueous phase separating two hydrophobic liquid phases. Chemical species may pass from one phase through the membrane to the other phase if they have some solubility in the membrane. This transfer may be accomplished either by simple diffusion or by carrier-facilitated transport wherein species are ushered, often selectively, across the membrane by carrier molecules which reside in the membrane. Net mass transfer of particular species between the separated phases occurs when there is a difference in the chemical potential (or electrochemical potential, if charged species are present) in the two phases. Transfer then occurs down the chemical potential gradient. Possible origins for this gradient include differences in concentration and electrical potential between the two separated phases. Net mass transfer of one species against its chemical potential gradient can also be coupled with the net mass transfer of another species, such as H^+, down its chemical potential gradient.[9,10,20,32,33]

During the past decade many papers on macrocycle-mediated cation transport have appeared. Most of these deal with a limited number of macrocycles such as 15C5,[12] 18C6,[12] DB18C6,[17a,26] crown ether carboxylic acids,[20] calixarenes,[33] and some cryptands[34] interacting with alkali metal cations. There have been few papers reporting macrocycle-mediated cation transport from cation mixtures. We have been engaged in a long-term systematic study of the factors which control or influence cation-macrocycle complex stability[35,36] and macrocycle-mediated cation transport through liquid membranes. In the transport studies, we[27,33,37,38] and others[17a,20,39] have investigated over 50 macrocycles with more than 15 metal and organic cations making nearly 700 combinations. Transport in over 400 combinations of macrocycles and metal cations in two- and three-cation mixtures has been studied also.[13–16,18–20] The results from these studies form a comprehensive set of data available for a detailed study of macrocycle-mediated cation transport. Two membrane systems have been used in these transport studies: 1) bulk liquid membranes which will be discussed in this section and 2) liquid-surfactant or emulsion membranes which will be described in Section

3. In a membrane system with no macrocycle in the organic phase, cation flux from the source phase through the organic phase into the receiving phase is very low. This is a direct consequence of the low solubility of most metal salts in organic solvents of low dielectric constant. However, upon the addition of the appropriate macrocycle to the organic membrane phase the cation flux into the receiving phase may increase several orders of magnitude.

2.1. Theory and model

In going from an aqueous phase to an organic phase, an ion must cross a large electrostatic barrier due to the difference in the bulk phase dielectric constants. The electrostatic barrier can be surmounted only if sufficient energy is available in the system to remove the waters of hydration from the ion. Furthermore, to keep the ion in the organic phase the expended energy to leave the aqueous phase must be regained by suitable electrostatic interactions in the organic phase. The problem is that in the organic phase there is a lack of "suitable electrostatic interactions". To put it another way, the free energy of the ion is larger in the organic phase, due to the lower dielectric constant, than it is in the aqueous phase. This can be seen by examining Equation 1;

$$\Delta G = -\frac{Z^2 e^2}{2r_e}(1 - 1/\varepsilon) \tag{1}$$

where ΔG is the free energy of solvation of the ion, Z is the charge on the ion, e is the electronic charge, ε is the dielectric constant of the solvent, and r_e is the effective ionic radius of the ion. (The value of r_e is the sum of the crystal radius of the ion, r_m, and an empirical constant associated with the solvent). Equation 1 is the Born equation as modified by Latimer, *et al.*[40] Although more elaborate and, from a theoretical standpoint, more sound methods for determining ΔG have been reported,[41,42] Equation 1 is sufficient for the purposes of this chapter. Successful solubilization of ions in the organic phase requires lowering this free energy barrier. This can be accomplished by lowering the charge density of the ion through complexation with a neutral species. For instance, in the discussion which follows, the use of macrocycles to increase markedly the solubility of metal salts in organic solvents is described.

Consider an interfacial region between an aqueous phase and an organic phase. There is not, on an atomic scale, a sharp discontinuity in physical properties in going from one phase to the other in the interfacial region.

Rather, this region is a smooth continuous gradient between the bulk phase properties of both liquids. (The steepness of the gradient, of course, depends on the miscibility of the two phases in each other.) As random thermal motion and mass action due to the large concentration difference move an ion from the bulk aqueous phase into the interfacial region, the ion sees a progressively less polar medium with a progressively lower dielectric constant. The result of such movement is to increase the ion's free energy. Since, from a thermodynamic standpoint, the free energy of the cations must be minimized, most of them will be found in the (predominantly) aqueous portion of the interfacial region thereby maintaining a favourable interaction with water molecules. However, a small number of the cations will be found in the organic portion of the interface. Since these ions are in a low dielectric medium, they have lost much of their screening due to the solvent and can now interact with nearby oppositely charged ions to form ion-pairs, thus mutually lowering their free energies.[12,43-45] Which process will occur, ion pairing or ion solvation in the boundary region between the phases, depends, in part, on the charge densities of the ions involved. High charge density leads to increased hydration and, consequently, lowered organic solubility. On the other hand low charge density results in ion interaction and greater organic solubility.

Macrocycles act to solubilize cations in an organic phase by providing "other suitable electrostatic interactions" for the lowering of the cation's free energy in that phase. These interactions involve the reaction of the cation with the donor atoms which line the cavity of the macrocycle. These cation–cavity electrostatic interactions lower the free energy of the cation in the organic phase due to the decreased (becomes more negative) solvation energy of the complex relative to that of the non-complexed cation. At the same time, the increased radius of the charged complex distributes the cation's charge over a larger volume which increases (becomes less negative) its free energy (see Equation 1). However, such factors as the favourable van der Waals interactions between the hydrophobic exterior surface of the macrocycle and the organic solvent molecules are sufficient to compensate for the increased free energy of the cation complex. The net result is the solubilization of the ion pair.

The preceding discussion is a simplification of the actual mechanisms involved in cation–macrocycle binding. However, it is hoped that it will give some feeling for factors such as macrocycle cavity and cation size diameters, cation charge and type, donor atom type, ligand substituents, and solvent, which may influence the stability of the cation–macrocycle complex.

Before discussing factors which influence macrocycle-mediated cation transport in liquid membranes, a model is presented which has been shown to adequately predict transport behaviour in bulk liquid membranes. A satisfactory model of macrocycle-mediated cation transport in these systems must explain several experimental results. First, if in the source phase the concentration of the counter anion equals the permeant cation concentration, the cation flux is proportional to the square of the cation activity in the source phase at low cation concentrations.[26,46] However, if the concentration of the counter anion is made greater than that of the permeant cation by the addition of a salt containing the same anion but a different and relatively impermeant cation, then the permeant cation flux is proportional to the product of the activities of both the permeant cation and its counter anion.[47] In both of these cases the ion concentration in the receiving phase is assumed negligible. Second, the cation flux is linearly related to the concentration of the ligand in the membrane at low ligand concentrations.[26,46,47] Third, the cation flux is linear in time (again, assuming negligible ion concentration in the receiving phase).[12,26,39,46]* Fourth, the cation flux is dependent on the counter anion present in the source phase.[46-48] Last, the cation flux shows a maximum with respect to log K, the cation-macrocycle interaction, i.e. at both low and high log K values, cation transport is low, while, at intermediate log K values, transport is high.[34,49]

Several carrier mediated membrane transport models have been proposed. Some of these ignore electroneutrality constraints in the membrane.[50-54] While neglecting electroneutrality restraints is reasonable for such systems as biological and artificial lipid bilayer membranes, this is not the case for the bulk liquid membranes discussed here. Other models assume the presence of either charged carriers or ion exchangers in the membrane.[9,10,12,25] Since the macrocycles with which we are concerned are neutral and the membranes used contain no ion exchangers, these models are not applicable. A further limitation of the two types of models mentioned above is that neither one necessitates the concomitant anion transport[55] which is observed experimentally. Other models are used to deduce the electrical properties

---

* Strzelbicki and Bartsch[20] observed non-linear time dependent transport mediated by crown ether carboxylic acids. This time dependence cannot be explained as simply a loss of carrier to the aqueous phase. (The charged crown ether carboxylate ions have considerable solubility in water and some loss for certain ligands is seen.) There are two observations which suggest that the observed non-linear time dependence is not a result of ligand loss from the organic phase. 1) When ligand loss from the membrane to the aqueous phase is minimized by the addition of alkyl chains to the ligand, non-linear time effects are still observed. 2) The nature of the time dependence is such that transport increases with time. Loss of ligand to the aqueous phase would be expected to decrease transport with time. The authors note that this result is at variance with the linear time dependence seen by others.[12,26,39,46] Strzelbicki and Bartsch explain the difference between their results and those of others by stating that equilibrium was not established in their system. However, the fact that aqueous source and receiving phases and the membrane phase are all being stirred at the relatively high rate of 200 rpm and that transport is being measured over a 48 h period (still non-linear after 48 h) indicate that this conclusion may not be valid.

of liquid membrane systems and are not concerned with net mass transfer.[56,57] In the membrane systems discussed in this chapter, an external voltage, used as a driving force is not applied nor is a voltage measured. The principal concern in these systems is net mass transfer and the only "driving force" is a concentration gradient. In 1973, Reusch and Cussler[26] developed a simple and intuitive model and related equation describing cation flux, J_M, though liquid membranes. In this model, the concentration gradient across the membrane was the sole driving force. The transport mechanism proposed by these workers has been tested successfully with respect to cation activity, ligand concentration, and anion type.[48,49] Since their model and subsequent models presented below are steady state models which assume equilibrium with respect both to cation-ligand complexation and to the partitioning of species between phases, the required linear time dependence is also predicted. The flux equation of Reusch and Cussler is given in Equation 2

$$J_M = D_c k_{ip} K L_T M^2 / l \qquad (2)$$

where D_c is the diffusion coefficient for the complex in the membrane, k_{ip} is the ion-pair partition coefficient, K is the equilibrium constant for $(M^{n+}A_n^-)$-macrocycle interaction, L_T is the total ligand concentration, M is the metal ion concentration (activity) in the source phase (the concentration in the receiving phase is considered negligible), and l is the diffusion length. Equation 2 predicts that J_M should vary directly as the equilibrium constant for Reaction 3. In Reaction 3, $(M^{n+}A_n^-)$ represents the cation associated with its anion(s) after partitioning into the membrane from the source phase, L represents the carrier ligand in the membrane, and $(ML^{n+}A_n^-)$ represents

$$(M^{n+}A_n^-) + L = (ML^{n+}A_n^-) \qquad (3)$$

the cation–ligand complex associated with the anion in the membrane.

The model of Reusch and Cussler as described in Equation 2 correctly predicts the exponential rise in J_M in the region of low log K as observed in the experimental points depicted in Figs. 2a and 2b. However, the model fails to predict the observed decrease in cation flux at high log K values. This deficiency can be corrected and the transport of cations from $n : 1$ electrolyte solutions can be evaluated by incorporating additional terms in Equation 2. The rationale behind the development of these equations is now presented.

First, let us consider the model shown in Fig. 3a. Danesi and Chiarizia[58] note that across a planar surface of constant area, liquid–liquid interfacial diffusion-limited mass transfer is characterized by a linear dependence of the transfer on the stirring rate. This is because at low stirring rates the thickness of stagnant interfacial regions is large and is inversely proportional

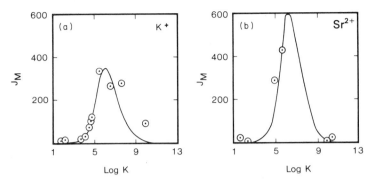

Fig. 2. Plots of log $K(CH_3OH)$ vs. J_M (10^{-8} mol s^{-1} m^{-2}). The solid lines are calculated from Equation 9 using parameters from Lamb *et al.*[49]. Data points are taken from Lamb *et al.*[49]

to the stirring rate. The macrocycle-mediated transport described in this chapter is of the diffusion-limited type.[59] The difference between the concentrations of the cation–ligand–anion complex at the two membrane interfaces is related to J_M by Fick's first law of diffusion:

$$J_M = \frac{D_c}{l_1 + (A'/A'')l_2} \{[MLA]'_{org} - [MLA]''_{org}\}. \tag{4}$$

where A'/A'' is the ratio of the interfacial surface areas for the source and receiving phase interfaces. Expressing the concentration of the complexes in terms of the known quantities: M^{n+}, the bulk aqueous phase cation concentration (assumed to be equivalent, stoichiometrically, to the anion concentration); L_T; k_{ip}; and K results in

$$J_M = \frac{D_c L_T K_{ex}}{l_1 + (A'/A'')l_2} \left(\frac{M'^{n+1}}{1 + K_{ex}M'^{n+1} + K_{ex}^2 M'^{n+1}M''^{n+1}} \right) \tag{5}$$

where

$$K_{ex} = Kk_{ip}n^n \tag{6}$$

and

$$M' = [M^{n+}]'; \; M'' = [M^{n+}]''. \tag{7}$$

In this derivation, it is assumed that $M' \gg M''$. K_{ex} is the extraction coefficient and relates concentration of the cation–ligand–anion complex in the membrane to the aqueous concentration of the cation. While Equation 5 is simple, it is inadequate in describing the data. It does predict a maximum in J_M as a function of K. However, the maximum is much too broad,

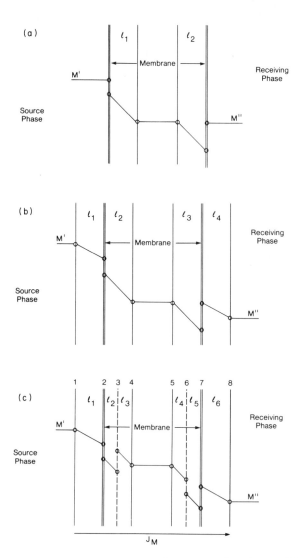

Fig. 3. (a) Transport model in which cation–ligand complexation equilibrium occurs at the interface. No aqueous diffusion layer is present. Membrane boundary layer = l at each interface. (b) Same as (a) except the boundary layers in the aqueous phase, l_1 and l_4, have been added. (c) Transport model in which cation–ligand complexation equilibrium is established at distances l_2 and l_5 from the appropriate interface. Aqueous boundary layers are l_1 and l_6. Membrane diffusion boundary layers are l_3 and l_4.

extending over 3–4 log K units. Furthermore, the predicted flux is much too sensitive to the cation concentration in the receiving phase.

The inclusion of unstirred boundary layers (l values in Fig. 3b) in the aqueous phases results in

$$J_M = \frac{D_c L_T K_{ex}}{l_2 + (A'/A'')l_3} \left(\frac{[M' - (J_M l_1 / D_w)]^{n+1}}{1 + K_{ex}(M' - J_M l_1 / D_w)^{n+1}} \right.$$
$$\left. - \frac{[M'' + (A'/A'')(J_M l_4 / D_w)]^{n+1}}{1 + K_{ex}[M'' + (A'/A'')(J_M l_4 / D_w)]^{n+1}} \right) \quad (8)$$

where D_w is the diffusion coefficient of the cation in water. Equation 8 is no better than Equation 5 in fitting the data and shows some of the same weaknesses as were indicated for Equation 5.

The models represented in Figs. 3a and 3b assume that the relevant reaction equilibria occur at the interface. In the model represented in Fig. 3c, this restriction is lifted. In this model the complexation equilibria are not established until distances l_2 and l_5 are reached away from the membrane interface. This model was proposed by Lamb, *et al.*,[49] but we present here a modified version for $n:1$ electrolytes (the original version was valid only for $n:n$ electrolytes):

$$J_M = \frac{D_c L_T}{l_3 + (A'/A'')l_4} \left(\frac{K_{ex}[M' - (J_M l_1 / D_w)]^{n+1} - K(J_M l_2 / D_I)}{1 + K_{ex}[M' - (J_M l_1 / D_w)]^{n+1} - K(J_M l_2 / D_I)} \right.$$
$$\left. - \frac{K_{ex}[M'' + (A'/A'')(J_M l_6 / D_w)]^{n+1} + K(A'/A'')(J_M l_5 / D_I)}{1 + K_{ex}[M'' + (A'/A'')(J_M l_6 / D_w)]^{n+1} + K(A'/A'')(J_M l_5 / D_I)} \right)$$
$$(9)$$

where the l values are defined in Fig. 3c and D_I is the diffusion coefficient for the ion-pair in the membrane.

Equation 9 is successful in describing cation transport in these liquid membrane systems. This success is seen in Fig. 2 where predicted and experimental points are given for both K^+ and Sr^{2+} with a variety of macrocycles. Note that the sharp maximum peak in J_M as a function of log K is closely reproduced. Similar agreement between the model and experimental data has been found for numerous other M^{n+}-macrocycle systems. The numerical values of the various parameters used to generate Fig. 2 can be found in Reference 49.

One assumption which is used in deriving Equation 9 is that the cation-macrocycle complexation reaction occurs solely in the organic phase. For macrocycles which are poorly soluble in the aqueous phase, this assumption is justified. However, some macrocycles, such as 18C6 and DCy18C6, are much more soluble in the aqueous phase than either the cations or the ion pairs are in the organic phase. In these cases, the assumption that complexation occurs only in the organic phase is most likely incorrect. The fact that log $K(H_2O)$ for metal–ligand complexation is several orders of magnitude lower than the expected log $K(CHCl_3)$ value (see below) can be offset by the combination of a large cation concentration in the aqueous phase and a sufficiently high aqueous solubility of the macrocycle. Under these conditions, the amount of complexed species in the aqueous phase can be appreciable. It would be desirable to evaluate the effect of macrocycle partition coefficients on model predictions. Supporting the possibility that the reaction(s) of interest occur in the aqueous phase is the work of Freiser and his co-workers.[60] These authors, using a high speed stirring and simultaneous sampling apparatus, have presented evidence that, for the solvent extraction from water to chloroform of Cu^{2+} and Ni^{2+} by 2-hydroxy-5-nonylbenzophenone oxime (LIX 65), complexation occurs in the aqueous phase. It would be desirable to incorporate terms for aqueous phase complexation into Equation 9 in order to evaluate these parameters.

Before leaving the subject of modelling transport phenomena, it must be emphasized that the above model assumes that equilibrium exists with respect to the various species distributions across interfaces and with respect to the cation–ligand complexation reaction. In bulk liquid membrane transport, time scales are long, diffusion distances are large, and cation fluxes are small. Therefore, the above assumption is reasonable because there is sufficient time for equilibria to be established across the interface and imaginary planes of the system, and the fluxes are small enough that they do not significantly perturb these equilibria. However, in emulsion or liquid-surfactant membranes (discussed in Section 3) and, to some extent, in solid supported liquid membranes, time scales are short, diffusion distances are small, and, consequently, fluxes are large. For these situations, equilibrium across various boundaries cannot be assumed and one must formulate more complicated kinetic equations. In these cases, the amount of metal transport is expected to be more dependent on kinetic parameters than on equilibrium parameters. The selectivity in these cases can also be affected. For instance, from the data of Liesegang and Eyring[61] the selectivity of monovalent cation complexation with 18C6 in aqueous solution, based on the association rate constant, is $Rb^+ > K^+$, $Cs^+ > Na^+$. This sequence is different than that based on the equilibrium constant for complexation, $K^+ > Rb^+ > Cs^+ > Na^+$.[35] Bartsch and his co-workers[20,62] also note that selectivity sequences based on kinetic parameters may differ from those based on equilibrium parameters. In their studies of transport from cation–mixtures using $CHCl_3$ bulk liquid membranes and crown ether carboxylic acid carriers, they find that, for some carriers, cation selectivity sequences based on transport are different from those based on solvent extraction.

For other carriers, the cation selectivity sequences are the same. Their explanation for the observation that some macrocycles show different cation selectivity sequences depending on whether transport or extraction data are used is that, in the transport system, equilibrium is not established and therefore kinetic parameters dominate, while in the solvent extraction process, equilibrium is attained and therefore equilibrium parameters determine the selectivity sequence. It is apparent that in systems where kinetic parameters control transport, the value of these parameters must be known in order to understand the transport process.

One approach to the study of interfacial phenomena is that employed by Dansesi, *et al.*[28] and by Danesi and Chiriazia[58] in which the interface is immobilized. This technique makes possible the minimization of unstirred aqueous boundary layers since the stirring rate can be increased to a level at which transport may be limited by kinetics rather than by diffusion. These authors have measured the interfacial kinetics of the DB18C6-mediated transfer of K^+-picrate from aqueous solutions into 1,2-dichloroethane.[63] By measuring the dependence of the kinetics on several physical parameters, they were able to postulate a reaction mechanism for the transfer process. In addition, the LIX64N mediated transport of Cu^{2+} in diaphragm-type liquid membranes was explained[28] as being controlled, simultaneously, by diffusion as well as by interfacial chemical reactions. Further use of this technique is expected to yield valuable information concerning the rate determining steps in cation extraction and transport in two-phase systems.

2.2. Bulk liquid membranes

The simplest type of liquid membrane, the "bulk" liquid membrane is illustrated in Fig. 4. This experimental design was used to obtain the bulk membrane results discussed below. Two different sizes of apparatus have been used by us. In the smaller apparatus, a layer of 3 ml of chloroform (the liquid membrane) in a vial (18 mm i.d.) was stirred at 120 rpm using a teflon-coated magnetic stirrer driven by a Hurst synchronous motor. Atop the membrane sat 0.8 ml of an aqueous solution (the source phase) of the cation to be transported and 5.0 ml of distilled deionized water (the receiving phase). The source and receiving phases were separated by a glass tube (8 mm i.d., 10 mm o.d.) (the source phase being on the inside) suspended above the base of the vial but below the surface of the chloroform membrane. The carrier concentration in the membrane was $1.0 \, mmol \, dm^{-3}$, while the cation concentration in the source phase was $1.0 \, mol \, dm^{-3}$ when the salt solubility permitted. Following assembly of the small membrane apparatus, which was closed to prevent evaporation, the membrane was stirred for

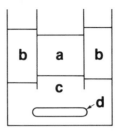

Fig. 4. Diagram of a simple concentric cylinder-type bulk liquid membrane apparatus.[46] (a) Source phase, usually 1 mol dm^{-3} in salt whose flux is to be investigated. (b) Receiving phase, usually distilled water. (c) Membrane phase, usually chloroform, containing macrocycle carrier. (d) Teflon coated magnetic stirrer.

24 h at $25 \pm 1°C$ after which 3 ml of the receiving phase was withdrawn by syringe and analysed for cation either by ion chromatography (Dionex model 10) or by atomic absorption spectrophotometry (Perkin-Elmer model 603), as was appropriate. The larger membrane apparatus differed from the smaller apparatus only in size. When the larger apparatus was used, several samples were withdrawn at regular intervals over the 24 hour period and analysed.[46]

An alternative to the bulk liquid membrane is the diaphram membrane sometimes referred to as a solid supported liquid membrane or an impregnated membrane.[26,47,57,64,65] A typical diaphragm membrane is constructed by soaking a suitable support material such as a filter paper or a porous polymer film (PVC) in the membrane material. The membrane material consists of an organic solvent, e.g. octanol or chloroform, containing the desired cation carrier. The impregnated filter is clamped between two O-ring joints and thus separates two liquid chambers, one containing the source salt solution phase, the other containing the receiving phase. This type of membrane is a thinner, supported version of the bulk liquid membrane described above.

2.3. Factors which influence macrocycle-mediated cation transport

The main components of the membrane transport systems considered here are: the ions to be transported; the mobile carrier; and the solvent used for the membranes. Both the cation and the anion determine whether and to what extent a cation–anion pair (or triplet) will be transported. The cations to be considered include the alkali and alkaline earth cations as well as several monovalent and divalent post-transition metal ions. A wide variety of anions have been studied and their effect on cation flux is discussed in

Section 2.3.2.1. The mobile carriers which have been used include the macrocyclic crown ethers and cryptands listed in Fig. 1. The solvent generally used for the membrane phase is chloroform, although dichloromethane and carbon tetrachloride have also been employed. As noted earlier, solubilizing the desired cation in the organic membrane phase is a necessary but insufficient condition for transport. The factors, related to the three above components of the membrane system, which influence cation transport from single cation solutions in liquid membranes are now presented.

2.3.1. Cation–macrocycle complex stability

Schultz and his co-workers,[66] in a review on work with facilitated transport in impregnated membranes, noted that J_M reached a maximum at intermediate $\log K$ values. Kirch and Lehn[34] first demonstrated that there is an optimum stability constant range for cation-cryptand complexation in which efficient transport occurs in liquid membrane systems. In order to elucidate further the relationship between complex stability and cation flux, Lamb, et al.,[49] measured cation transport using over 90 different cation–ligand combinations. The ligands studied included both crown ethers and cryptands and were chosen so as to give a wide range of $\log K$ values for the M^{n+}-macrocycle complexes. Thus, both cation flux and $\log K$ values were available for the same set of M^{n+}-macrocycle systems. The $\log K$ values, measured in methanol,* ranged from 0.5 to 9.9 for monovalent cations and from 0.5 to 12 for divalent cations. This correlation between $\log K(CH_3OH)$ and J_M showed that there is a limited range or window of $\log K(CH_3OH)$ binding constants which yield significant transport. Intuitively, it can be seen that if the binding is too weak and unfavourable, i.e. small $\log K(CH_3OH)$, insufficient amounts of cations from the source phase will complex in the membrane in order to be transported through it. On the other hand if the binding is too tight and favourable, i.e. large $\log K(CH_3OH)$, then once the complex has been formed the cation will not partition into the receiving phase and there will be a build up of cation–macrocycle complexes in the organic phase at the receiving phase interface. This limited range in $\log K(CH_3OH)$ values has been determined to be 5.5–6.0 for monovalent cations and 6.5–7 for divalent cations.[49]

---

* Since the transport measurements were made using chloroform membranes, the appropriate $\log K$ values should pertain to the complexation reaction in chloroform. Unfortunately, these data are not available nor, due to technical difficulties, are they readily measurable. Based on a linear dependence of $\log K$ on dielectric constant for pure alcohols, Lamb, et al.[49] assumed that $\log K(CH_3OH)$ values for all the complexes would differ by a constant amount from the $\log K(CHCl_3)$ values. Thus, using $\log K(CH_3OH)$ values would be equivalent to a simple shift in the $\log K$ axis and would not influence the shape of the curve. Although this assumption appears to be valid, it has yet to be verified.

Since there is a strong and critical dependence of transport on log $K(CH_3OH)$, we shall discuss factors governing complex stability. Recently, several reviews have appeared concerning the various aspects of cation–macrocycle binding and factors which influence complex stability.[35,36,67–72] The reader interested in a more detailed discussion of cation–macrocycle interactions is invited to consult these reviews. Here, we limit discussion to a few of these factors.

2.3.1.1. Relative cation and ligand cavity sizes. From the inception of crown ether investigation, it was recognized that a correlation often existed between complex stability and the match of metal ion ionic crystal radius and macrocycle cavity radius. Representative cation and macrocyclic cavity radii are given in Table 1. Since cation–macrocycle interaction in the case of neutral macrocycles is assumed to be of the ion–dipole type, it is reasoned that electrostatic bond energies between ligand and cation will be greatest when all donor atoms can fully participate. If the macrocyclic cavity is larger than the metal ion, then the energy needed to desolvate the cation cannot be regained through favourable interaction with donor atoms of the cavity since some of them will be too far away, i.e., the cation will not be favourably coordinated. Furthermore, conformational stress in the macrocycle may preclude a rearrangement of the donor atoms to better coordinate the cation. However, if the macrocycle cavity is sufficiently larger than the cation, then the macrocycle may assume a non-planar conformation

Table 1. Cation,[74] crown ether cavity,[75] and cryptand[5] cavity radii

Cation	Radius (Å)[a]	Crown ether	Radius (Å)	Cryptand	Radius (Å)[b]
Mg^{2+}	0.72	15C5	0.86–0.92[a]	2.1.1.	0.80
Li^+	0.74	18C6	1.34–1.43[a]	2.2.1.	1.10
Ca^{2+}	1.00	21C7	1.7[b]	2.2.2.	1.40
Na^+	1.02	24C8	2.0[b]		
Ag^+	1.15				
Sr^{2+}	1.16				
Pb^{2+}	1.18				
Ba^{2+}	1.36				
K^+	1.38				
Rb^+	1.49				
Tl^+	1.50				
Cs^+	1.70				

[a] From X-ray crystallographic data.
[b] From Corey–Pauling–Koltun models.

which enables it to fit around the cation in order to maximize donor atom–cation interaction. For this reason, it has been suggested[76a] that selectivity due to size compatibility between cation and ligand does not have any meaning for crown ethers larger than 21C7. This is because these ligand cavities are larger than the largest cation studied, Cs^+. Alternatively, large macrocycles may form ring-type coordination with two cations simultaneously.[75] If, on the other hand, the ligand cavity is smaller than the metal ion, then the metal ion can only saddle up to the ligand, not fit inside it. Here, too, the complex stability is diminished because fewer of the cation solvation molecules have been replaced and the cation–ligand bonds are not coincident with the dipole moments of the groups containing the donor atoms. Furthermore, this complex would not be as soluble in low dielectric media as a "good-fit" complex, because of greater cation solvation. However, in cases where the ligand cavity is smaller than the cation, one cation may be sandwiched between two macrocycles forming a 2 : 1, L to M^{n+} complex.

In Fig. 5, the selectivities shown by 15C5, 18C6, and 21C7 for several alkali metal cations are indicated as measured by J_M and log $K(CH_3OH)$. For these systems, the selectivity as measured by J_M is seen to parallel that measured by log $K(CH_3OH)$. Size selectivity for cations as evidenced by log K is even more dramatic for the cryptands as shown in Fig. 6. (A possible reason that cryptands show a stronger cation size dependence than do the cyclic polyethers is that their basket-like cavity is much more rigid and less able to produce the small conformational changes which are necessary to accomodate ions of different sizes.) Also, in Fig. 6, cation transport by these cryptands is illustrated. Unlike the crown ethers, the cryptand transport does not simply parallel the log K selectivity. The explanation for this is as follows. It was mentioned earlier that maximum flux from solutions containing single alkali metal cations occurs in a log $K(CH_3OH)$ range of 5.5–6.0. Below this range, transport increases with log K, while above it transport decreases with log K. For the crown ethers, all log K values for alkali metal cations are below the optimum value, thus transport parallels log K. However, for the cryptands log K values vary from 2–10. Thus, some of the log K values fall in the region where transport increases with log K, while others fall in the region where transport decreases with log K. Hence, for the cryptands, cation flux values from single salt solutions do not simply parallel the log K selectivity.

An interesting result concerning the effect of cavity size and conformation on cation transport is seen in the work of Rebek and his co-workers.[17a,17b] These investigators have synthesized macrocyclic crown ethers of various ring sizes with a 2,2'-bipyridyl function incorporated as a ring member (Fig. 1). These macrocycles offer two cation binding sites: the nitrogen atoms of

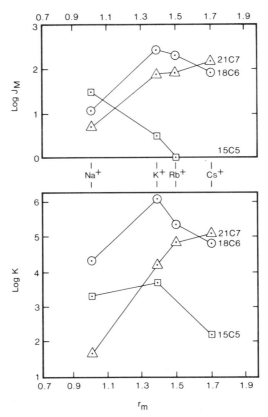

Fig. 5. Cation selectivities of 15C5, 18C6 and 21C7 based on cation flux, J_M, values in a H_2O–$CHCl_3$–H_2O liquid membrane system[27] and on log K (CH_3OH) values at 25° C.[35]

the 2,2'-bipyridyl function and the usual ether–oxygen lined cavity. These two sites are separated and are expected to act independently. Using an NMR technique, these workers[17b] have shown that divalent cations and transition metal complexes preferentially bind to the bipyridyl site and that alkali metal cations preferentially bind to the ether–oxygen lined cavity. Furthermore, they have shown that the occupancy of the bipyridyl site precludes, or greatly diminishes alkali metal cation binding to the oxygen lined cavity. This result was attributed to an allosteric effect of one site on the other. The free crown ether may adopt a variety of conformations in solution, wherein the size of the crown ether cavity and the orientation of the oxygens are related to the dihedral angle defined by the two aromatic rings.[17b] Therefore, when the bipyridyl site is unoccupied, the dihedral angle is not fixed and the cavity is flexible enough to assume a conformation

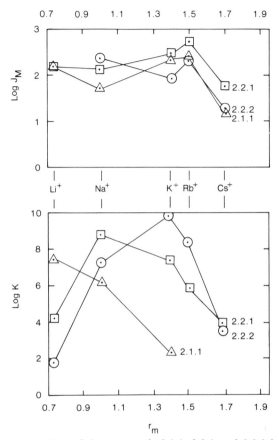

Fig. 6. Cation selectivities of the cryptands 2.1.1, 2.2.1 and 2.2.2 based on cation flux, J_M, values through a H_2O–$CHCl_3$–H_2O liquid membrane system[27] and log K values valid in 95% CH_3OH/5% H_2O (vol/vol) at 25° C.[35]

which favourably coordinates the alkali cation and which maximizes cation–oxygen interaction. However, metal chelation at the bipyridyl site fixes the dihedral angle between the two aromatic rings resulting in decreased macrocycle flexibility, altered macrocycle cavity size, and less than optimal ether oxygen–cation interactions. (The alternative possibility of simple electrostatic respulsion between the two cations was not considered). Thus, the bipyridyl site becomes a modulating site controlling the ability of the macrocycle to bind cations. Consistent with their NMR results, Rebek and Wattley[17a] found that K^+ transport through chloroform membranes mediated by these 2,2′-bipyridyl macrocyclic derivatives is higher when the bipyridyl site is unoccupied than when $W(CO)_4$ is bound to the site.

2.3.1.2. Cation charge. In general, log K for formation of the cation–ligand complex increases with increased cation charge. However, log K is related to the difference in the cation–solvent and cation–ligand interaction energies minus the solvation energy of the cation–ligand complex.[36] Thus, in instances where the cation radius is small, the cation solvation energy is greater than the sum of the other two quantities and complex formation is less favourable. Such is generally the case for Na^+ and Ca^{2+} which are nearly the same size (Table 1) and have relatively high solvation energies. Comparing log K values for various M^{n+}-crown ether complexes,[35] it is seen that complexation is less preferred for the divalent Ca^{2+}, i.e. for 18C6, log $K(CH_3OH)$ values are: Ca^{2+}, 3.86; Na^+, 4.36. This is because, for cations of small radii, the cation solvation term dominates, particularly for divalent cations. On the other hand, consider the larger cations K^+ and Ba^{2+} which also have nearly identical ionic radii (Table 1). For these cations, solvation energies are smaller and do not dominate. This results in an overall stronger complexation for 18C6 with Ba^{2+}, log $K(CH_3OH) = 7.04$ than with K^+, log $K(CH_3OH) = 6.06$. If there is a good fit between ligand cavity size and cation diameter, then the cation–ligand interaction and the cation–ligand complex solvation energies are usually greater than the ion–solvation energy and increasing the charge of the ion results in increased complexation stability. An example of these effects is seen in the preference by the cryptand ligand, 2.2.1, for Ca^{2+} over Na^+.

The effect of cation charge on transport is important in two ways. First, it influences complex stability as discussed above. Second, cation charge affects the ion–pair partition coefficient. This will be discussed in detail later, but suffice it to point out that, given the same aqueous concentrations, the concentration of a monovalent cation in the membrane will be much larger than that of a divalent cation. Thus, from solutions containing a single cation species, the Na^+ flux is greater than the Ca^{2+} flux using the cryptands and, similarly the K^+ flux is greater than the Ba^{2+} flux using 18C6.[27]

2.3.1.3. Cation type. The alkali and alkaline earth cations can be considered to be charged, uniform spheres since loss of their valence electron(s) results in electronic structures with all electron orbitals filled which are identical to those of the noble gas atoms which precede them in the periodic table. On the other hand, post-transition cations with underlying d electron levels completed are polarizable (electron clouds of the ion may be distorted) and usually show specific geometrical coordination requirments. Generally, the polarizability of these non-alkali and non-alkaline earth metal cations makes possible dipole–induced–dipole interactions in the ion–ligand complex which increases complex stability. For example, Tl^+ and Rb^+ are similar in size (Table 1) yet the thermodynamic stability of the Tl^+–18C6 complex, $\log K(H_2O) = 2.27$, is greater than that of the corresponding Rb^+ complex, $\log K(H_2O) = 1.56.$[35] Noble gas and post-transition metal cations can have quite different affinities for macrocycles depending on the donor atom present. For example, Ag^+ exhibits a lower affinity than K^+ for oxygen containing ligands but has a higher affinity for sulphur- and nitrogen-containing ligands.[76a]

2.3.1.4. Donor atom type. Substitution of either sulphur or nitrogen atoms for oxygen donor atoms in macrocyclic ligands affects the stability of the cation–ligand complex as well as the cation flux in liquid membrane systems. These effects are consequences of such factors as differences between the substituted atom and oxygen with respect to their atomic radii and their electronegativities. There are also some specific ion–ligand interactions.

Consider the substitution of a sulphur atom (radius 1.85 Å) for an oxygen atom (radius 1.40 Å).[77a] The larger sulphur atom causes a reduction in the ligand cavity size as well as an increase in the charge separation between the coordinated cation and the donor atom. These effects are expected, generally, to lead to the destabilization of alkali and alkaline earth complexes of these thia–ligands. However, the dipole moment of the thiaether group is expected to be larger than that of the ether group because the dipole moment of diethyl sulphide (1.54 Debye)[77b] is greater than that of diethyl ether (1.15 Debye).[77b] This larger dipole moment should partially offset the destabilization referred to above. The net result of these effects is that sulphur substituted macrocycles generally form less stable cation complexes than do their unsubstituted analogs.[76a] Exceptions are found in the cases of Ag^+ and Hg^{2+}, which show specific affinity for sulphur-containing ligands with resultant increases in complex stability.[76b]

The effect of sulphur substitution on both transport and $\log K$ is seen in Table 2. No transport data have been reported in the case of T18C6. However, $\log K$ values in the case of this macrocycle show that sulphur substitution results in decreased stability of its alkali cation complexes, but increased stability of the corresponding Ag^+ complex. The few $\log K$ data available indicate that these trends continue with DT18C6 where the K^+ complex is less stable than the K^+–T18C6 complex. The $\log K$ value in Table 2 for the Ag^+–DT18C6 complex is valid in H_2O, not CH_3OH. However, the $\log K(H_2O)$ value for the Ag^+–DT18C6 system is nearly as large as the $\log K(CH_3OH)$ value for Ag^+–18C6 interaction. This observation together with the trend in $\log K(CH_3OH)$ for Ag^+–L interaction from 18C6 to T18C6 indicates that the Ag^+–DT18C6 $\log K(CH_3OH)$ value would be much larger than that for Ag^+–T18C6 interaction. It is significant that $\log K(H_2O)$ values in the case of Pb^{2+} change much less in going from 18C6 to DT18C6 than

Table 2. *The effect of sulphur and nitrogen substitutions in 18C6 on J_M and log K*

Cation	18C6 $J_M^{(a)}$	18C6 log $K^{(b)}$	T18C6 log K	DT18C6 $J_M^{(a)}$	DT18C6 log K	A18C6 log K	DA18C6 $J_M^{(a)}$	DA18C6 log K
Na$^+$	26	4.36	2.57$^{(d)}$	4			6	
K$^+$	634	6.06	3.61$^{(d)}$	—$^{(c)}$	1.15$^{(e)}$	3.90$^{(e)}$	9.2	2.04$^{(e)}$
Rb$^+$	492	5.32	3.00$^{(f)}$	—$^{(c)}$			—$^{(c)}$	
Ag$^+$	515	4.58	>5.5^{d}	76	4.34$^{(g)}$	>3.3$^{(e)}$	65	7.8$^{(g)}$
Sr^{2+}	728	>5.5		—$^{(c)}$			290	2.56$^{(g)}$
Ba^{2+}	42 (167)$^{(h)}$	7.04	3.4$^{(d)}$	2 (6)$^{(h)}$			93 (373)$^{(h)}$	2.97$^{(g)}$
Pb^{2+}	310	4.27$^{(g)}$		270	3.13$^{(g)}$		18.3	

(a) J_M values given as 10^{-8} mol s^{-1} m^{-2}; 1 mol dm^{-3} nitrate salts except Ba^{2+} (0.3 mol dm^{-3}).[27]

(b) All log K values are valid in CH$_3$OH unless stated otherwise.[49]

(c) J_M is at the level of the blank, 0.7×10^{-8} mol s^{-1} m^{-2}.

(d) CH$_3$OH.[76a]

(e) CH$_3$OH.[35]

(f) CH$_3$OH.[36]

(g) H$_2$O.[35]

(h) Adjusted for 1 mol dm^{-3}.[27]

do those in the cases of alkali metal ions, alkaline earth metal ions and Ag^+. Cation flux data for these systems are consistent with these log K variations. Transport of the alkali and alkaline earth cations is reduced markedly, that of Ag^+ is less affected and that of Pb^{2+} remains nearly the same. This result is consistent with the preference of alkali and alkaline earth metal cations for "hard" oxygen ligands over "soft" sulphur ligands.[78] The softer cations, Ag^+ and Pb^{2+}, show affinity for sulphur ligands and consequently are effectively transported.

Substitution of nitrogen for oxygen in 18-crown-6 leads to decreased complex stability (Table 2) in the cases of alkali and alkaline earth metal cations for two reasons. First, the van der Waals radius of nitrogen (1.5 Å)[76a] is larger than that of oxygen (1.40 Å) causing increased charge separation between the complexed metal cation and the donor atom. Second, the dipole moment of the carbon–nitrogen–carbon group is smaller than that of the carbon–oxygen–carbon group. This is inferred from the fact that the dipole moment of diethylamine is 0.92 D[77b] and that of diethyl ether is 1.15 D. Both the larger radius of nitrogen and smaller dipole moment of the C–N–C group result in a diminished cation–ligand electrostatic interaction and therefore a weaker complex. The data in Table 2 show that log K for the reaction of K^+ with A18C6 is less than that for the reaction of K^+ with 18C6. Further substitution of nitrogen for oxygen to give DA18C6 results in a still lower log K value in the case of K^+. The log K values in the cases of Sr^{2+} and Ba^{2+} also show sharp decreases in going from 18C6 to DA18C6. The opposite trend in log K is seen in the case of Ag^+. The trends in the J_M values in the series 18C6, DA18C6 are consistent with these log K values. Transport decreases sharply in the cases of K^+ and the other alkali metal cations as log K decreases. In the cases of the alkaline earth cations, transport is appreciable for both macrocycles. For Ag^+, transport decreases because log K in the case of DA18C6 is large enough to prevent appreciable release of Ag^+ into the receiving phase. Thus, DA18C6 seems to effectively discriminate between monovalent alkali metal cations and divalent alkaline earth cations, but shows appreciable Ag^+ transport.

2.3.1.5. *Ligand substituents.* In general, ligand substituents affect cation–ligand complex stability in one or more of the following ways: increase effective radius of the complex, alter electron density in the donor ring, and decrease the macrocycle flexibility. Ligand substituents can also affect the "hydrophobic bulk" of the macrocycle, but this will be dealt with later.

A direct correlation is found between complex solvation energy and complex stability.[36] Since solvation energy is inversely proportional to the radius of the M^{n+}-macrocycle complex, as the macrocycle radius increases the solvation energy becomes less negative, and the stability of the complex decreases. The effect of the macrocycle radius on complex stability can be seen by comparing 18C6 with its bulky analog, DCy18C6 (Table 3). As expected, in every case M^{n+}–DCy18C6 complexes have lower stability constants than do the corresponding M^{n+}–18C6 complexes. Despite the lower log K values, DCy18C6 has consistently higher J_M values than does 18C6. These higher J_M values are most likely due to the lower water solubility of DCy18C6.

Alteration of the electron density of the coordinating atoms in the macrocycle can be accomplished by appropriate substitution of groups on the macrocycle.[69] The effect of ligand substitution can be seen by comparing DB18C6 and DCy18C6 with respect to log K values for cation complexation and J_M values for cation transport. DB18C6, generally, has lower log K values and in all cases lower J_M values than the dicyclohexano derivative (Table 3). It has been suggested[35,69] that the lower log K values result, at least in part, from the electron withdrawing tendency of the aromatic ring. This is further illustrated by the effect of successive substitution of benzo groups on 18C6 (Table 3). The mono-benzo derivative shows a decrease in both stability constant and transport. With the exception of Pb^{2+}, cation transport is diminished. Addition of another benzo group to give DB18C6 results in a further reduction in both complex stability and transport. Furthermore, divalent cation transport by DB18C6 is more affected than is monovalent cation transport.

Except in a few cases where the ligand cavity size and cation diameter exactly match, the cation–ligand complex is greatly stabilized if the

Table 3. *Effect of ligand substituents on cation flux and log K*

	18C6		B18C6		DB18C6		DCy18C6	
Cation	J_M[a]	log K[b]	J_M[a]	log K[b]	J_M[a]	log K[b]	J_M[a]	log K[b]
Na^+	26	4.36	13	4.21[c]	10	4.5	53	4.08
K^+	634	6.06	492	5.29[c]	230	5.0	780	6.01
Rb^+	492	5.32	205	4.48[c]	28	1.08[d]	539	
Cs^+	179	4.79	25		5	3.55	191	4.61
Ag^+	515	4.58	391	4.23[c]	74	1.41[d]	937	
Ca^{2+}	61	3.86	8		[a]		373	
Sr^{2+}	728	>5.5	264	5.12[c]	[a]		1028	3.24[d]
Ba^{2+}	42	7.04	3	5.48[c]	4	4.28	649	3.57[d]
Pb^{2+}	310	6.5[e] 4.27[d]	916	5.49[c]	62	1.89[d]	728	4.95[d]

[a] J_M values given as 10^{-8} mol s^{-1} m^{-2}; 1 mol dm^{-3} nitrate salts except Ba^{2+} (0.30 mol dm^{-3});[27] blank is 0.7×10^{-8} mol s^{-1} m^{-2}.
[b] Log K in methanol from Reference 49, unless otherwise noted. In the case of DCy18C6, value is for the *cis-syn-cis* isomer.[35]
[c] Unpublished results, this laboratory.
[d] H_2O.[35]
[e] 70% MeOH, 30% H_2O,[35]

macrocycle is flexible enough to achieve a conformation which maximizes the coordination of the cation and the electrostatic interactions. Macrocycle derivatives containing benzo-, dibenzo-, and pyridino-groups would be expected to have decreased ring flexibility compared to that shown by the parent macrocycle. In addition, as was mentioned earlier, these groups show electronic induction effects. The relative importance of these effects in determining the observed loss in complex stability has yet to be determined.

2.3.1.6. Solvent. Few transport measurements have been made in solvents other than chloroform. Data for a series of chlorinated methane solvents indicate that cation flux is dependent on membrane solvent (Table 4).[59] An examination of the model represented in Fig. 3c and its associated flux, Equation 9, reveals that there are several parameters whose values depend on solvent and which can affect transport. These are: the thickness of the unstirred boundary layers in the membrane, i.e. the diffusion path length;

Table 4. Effect of solvent on membrane transport

Parameter	CH_2Cl_2	$CHCl_3$	CCl_4
K^+ Transport (DCy18C6)	1270[a]	700[a]	1.6[a]
Ca^{2+} Transport (DCy18C6)	210[a]	130[a]	—[b]
ε, 20° C	9.08	4.81	2.24
$1 - 1/\varepsilon$	0.890	0.792	0.554
Viscosity,[c] 15° C	0.449	0.596	1.038
Molecular weight	84.9	119.4	153.8
Diffusion coefficient[d]	1.48×10^{-5}	1.32×10^{-5}	0.86×10^{-5}
Boundary layer thickness[e] (mm)	0.102	0.101	0.094
Distribution ratio, k_1, for DCy18C6[f]	713	454	108
$k_1/\{1 + x_0(k_1 - 1)\}$[g]	1.997	1.996	1.982
Distribution ratio of K-18C6-picrate[h]	1.13	0.53	<0.001
Relative ion-pair partition coefficient of K^+-NO_3^-, $k_{ip}/k_{ip}(CH_2Cl_2)$[i]	1	2.0×10^{-7}	2.3×10^{-21}

[a] J_M values given as 10^{-8} mol s^{-1} m^{-2}, unpublished results.
[b] J_M at level of blank, 0.7×10^{-8} mol s^{-1} m^{-2}.
[c] Centipoise.
[d] cm^{-2} s^{-1}, calculated using Equation 15 of Reference 49.
[e] Calculated using Equation 16 of Ref. 49.
[f] $[DCy18C6]_{org}/[DCy18C6]_{aq}$, unpublished results except for $CHCl_3$.[77c]
[g] Arbitrary value of x_0 taken to be 0.5.
[h] Organic/aqueous.[79]
[i] Reference.[80]

the diffusion coefficients of all mobile species in these boundary layers; the ion pair partition coefficient; and the log K value for formation of the cation–ligand complex in the organic layer. Additional parameters are the partition coefficient of the macrocycle which is assumed to be infinite in the model and the partition coefficient for the cation–ligand–anion complex. Values for all of these parameters, except log K, for each of the solvents, CH_2Cl_2, $CHCl_3$, and CCl_4, are listed in Table 4. Where experimentally observed values were not available, estimates are given.

Estimates of the boundary layer thickness were made using Equation 16 of Reference 49. These estimates indicate that this layer thickness is essentially equal in the three solvents. On the other hand, diffusion coefficients, estimated using Equation 15 of Ref. 49, showed a decrease upon increasing solvent chlorination. However, this decrease was by a factor of 2, while the decrease in transport is 3 orders of magnitude. Thus, neither boundary layer considerations nor diffusion coefficients is sufficient to explain the solvent effect on transport. Schiffer, et al.,[81] have noted a similar result with respect to membrane selectivity. They point out that differences in diffusion coefficients are unimportant. This is because both the boundary layer thickness and the diffusion coefficient depend on solvent properties such as viscosity, density, and molecular weight. In the more common membrane solvents, the values of these properties and, therefore, boundary layer thickness and diffusion coefficient, are equal to within an order of magnitude.

Unlike the boundary layer thickness and diffusion coefficient, the partitioning of the various species, complexed and uncomplexed, between the organic and aqueous phases as well as log K values for the cation–ligand complexation reactions can be very sensitive to the solvent. At the aqueous source phase-membrane interface, one can consider that a continuous solvent extraction process occurs. From simple equilibrium considerations,* the amount of complexed metal ion in the organic phase is given by Equation 10 where $\bar{L}_T$ is the ligand concentration averaged over both phases, K_{org} is

$$[ML^{n+}A_n^-]_{org} = \frac{\bar{L}_T K_{org} k_{ip} n^n [M^{n+}]_{aq}^{n+1} k_1}{\{1 + x_0(k_1 - 1)\}} \tag{10}$$

---

* Since there is a net cation flux during transport, the system is not at equilibrium, but rather a steady state condition is attained. However, since in bulk liquid membrane transport the cation fluxes are small with respect to the source phase concentration and since we only want to determine the effect of solvent on $[ML^{n+}A_n^-]_{org}$, we assume, for simplicity, that equilibrium exists between the organic and aqueous phases. Even assuming equilibrium, the general expression for $[ML^{n+}A_n^-]_{org}$ can be complicated. Considering equilibrium distributions across a single interface and allowing complexation to occur in both the organic and aqueous phases, Equation 11 may be written

$$[ML^{n+}A_n^-]_{org} = \bar{L}_T \frac{B}{1 + B\{1 + x_0[k_c(n[M^{n+}]_{aq})^n - 1]\}/k_c(n[M^{n+}]_{aq})^n} \tag{11}$$

the complexation constant in the organic phase, k_{ip} is the ion pair partition coefficient, k_l is the ligand partition coefficient and x_0 is the fraction of the total volume occupied by the organic phase. From Equation 10, we see that with the assumption made, i.e., that most of the total ligand is uncomplexed, the amount of metal complexed (and subsequently transported) is related to the product of three terms: K_{org}, k_{ip}, and $k_l/(1 + x_0(k_l - 1))$. A decrease in any one of these three terms would result in a decrease in the cation transport (assuming the log K value falls on the rising portion of the J_M vs. log K plot). Table 4 shows experimental values for k_l in the case of DCy18C6 and calculated values for the term, $k_l/(1 + x_0(k_l - 1))$, using an arbitrary value of 0.5 for x_0. It can be seen that while k_l decreases in the same order as J_M for K^+ and Ca^{2+}, the more important factor, $k_l/(1 + x_0(k_l - 1))$, is essentially constant. In order for k_l to have an appreciable effect on this term, its value must be less than 10. Thus, the solvent effect in the cases of CH_2Cl_2, $CHCl_3$, and CCl_4 cannot be due to differences in k_l. Table 4 also contains relative ion pair partition coefficients for $(K^+-NO_3^-)$ in the three solvents. These k_{ip} values are defined by Equations 14–16 in Section 2.3.2.1 and are calculated using Equations 15 and 16. Values for ΔG_p^C and ΔG_p^A, Equation 16, were calculated using the Abraham–Liszi model for single ion partitioning (see Ref. 80 and the first footnote in Section 2.3.2.2). The solvent dependence of the cation–anion association in the organic phase term in Equation 16, ΔG_p^{CA}, was assumed to depend only on the dielectric constant of the solvent.[45] This dependence was taken to be $\Delta G_p^{CA} = \Delta G_p^{0CA} + 1/\varepsilon$.[45,82] To avoid calculation of the proportionality factor ΔG_p^{0CA}, ratios with respect to k_{ip} (CH_2Cl_2) were taken. Because these ratios are estimations based on theoretical considerations and not on experimental data, they must be regarded as qualitative, rather than quantitative values. The trend in the k_{ip} values does parallel that seen in the J_M values. Thus, the observed effect of solvent on J_M could be related to the effect of solvent on k_{ip}. However, there is a third term to consider, namely, log K_{org}. Unfortunately, data are not available for this parameter in these solvents. However,

---

where

$$B = \frac{K_{org} k_{ip} n^n [M^{n+}]_{aq}^{n+1} k_l}{[1 + x_0(k_l - 1)]},$$

k_c is the partition coefficient for the cation-ligand–anion complex and all other terms have been defined. Equation 11 assumes that a single cation species is present and that $[A^-]_{aq} = n[M^{n+}]_{aq}$. The dependence on K_{aq}, the complexation constant in the aqueous phase, is implicitly included in k_c, as shown in Equation 12.

$$k_c = \frac{[ML^{n+}A_n^-]_{org}}{[ML^{n+}]_{aq}[A^-]_{aq}^n} = \frac{K_{org} k_{ip} k_l}{K_{aq}}. \tag{12}$$

If we assume that most of the total ligand is present in the uncomplexed form, then Equation 11 reduces to Equation 10.

according to Equation 12, the partition coefficient for the ion-pair–ligand complex is related to the product of K_{org}, k_{ip}, and k_l (K_{aq} is not considered since it is independent of the organic solvent). Therefore, $\log K_{org}$ is implicitly included in the measurement of the distribution ratio of the cation-ligand–anion complex. From Table 4, it is apparent that this distribution ratio also parallels, in both direction and magnitude, the trend in transport. (Although this distribution ratio is for the 18C6 complex, we assume that a similar trend and magnitude would exist for the distribution ratio involving the DCy18C6 complex). This still does not give any indication of the trend for the $\log K$ values in the different solvents. More will be said concerning $\log K$ and solvent below, but as a first approximation one would expect $\log K$ to increase with decreasing dielectric constant. This may or may not be the case for the chlorinated methane solvents.* If it is the case, then the $\log K$ values oppose the decrease in transport, and, apparently, the decrease in k_{ip} is sufficient to compensate for the possible increase in $\log K$. The net result would be a decrease in transport in the series CH_2Cl_2, $CHCl_3$, CCl_4.

We have shown that the membrane solvent can greatly affect the ion-pair partition coefficient, and, thereby, transport. However, the solvent can also affect cation selectivity through differences in partition coefficients. In Table 5 are shown selectivity sequences based on the free energy of transfer of an alkali metal cation from water to various solvents. Strictly speaking, the ion partition coefficient for a particular water–solvent system is related to the free energy of transfer from water to the solvent only if the water solubility in that solvent is sufficiently small, i.e., the ions must partition into the solvent unhydrated.[80] The important point is that the solvent can affect the magnitude and selectivity of ion partitioning and judicious selection of the solvent may increase efficiency in cation separation processes.

In the complexation process, macrocycles must compete with solvent molecules for the cations in solution. As a result, the magnitude of K for cation–ligand interaction can be strongly dependent on the solvent. Specifically, it has been observed[35] that greater complex stability is found in solvents of low dielectric constant and cation solvating ability than in those which strongly solvate cations.

It is appealing to attempt to identify a particular property of the solvent which would correlate with measured complex stability constants and even permit one to predict the stability constant of a given M^{n+}–L complex in

---

* In a homologous set of short, linear alcohols, Agostiano, *et al.*[83] found that $\log K$ increases with decreasing dielectric constant (see Fig. 8). The chlorinated methanes used in the study discussed here also form a homologous set of simple solvents. Thus, we assume that in these solvents, $\log K$ also increases with decreasing dielectric constant.

Table 5. *Effect of solvent on cation selectivity sequence*

Solvent	Based on free energy of transfer of the alkali cation from water to the solvent[a]	Based on log K for M^+–DB18C6 interaction in the solvent[b]
Water	—	$K > Rb, Na > Cs^{69}$
		$K > Rb > Cs > Na^{(c)}$
Methanol	$Na > Cs > K > Rb^{84}$	$K > Rb > Na > Cs^{83}$
		$K > Rb > Cs > Na^{(c)}$
(43/57 wt/wt) methanol–water	$Rb > Na > Cs > K^{85}$	$K > Rb > Cs > Na^{(d)}$
Ethanol	$K > Na, Cs > Rb^{85}$	$K > Cs > Na^{83}$
Propanol	$Cs > Na, K > Rb^{84}$	$K > Cs > Na^{83}$
Dimethyl sulphoxide	$Na > Cs > K > Rb^{84}$	$K > Rb > Cs, Na^{69}$
Dimethyl formamide	$Na > Cs > K > Rb^{86}$	$K > Rb > Cs > Na^{69}$
Propylene carbonate	$Cs > Rb > K > Na^{85,86}$	$K > Na > Rb > Cs^{69}$
Acetonitrile	$Cs > Rb > K > Na^{86}$	$Na > K > Rb > Cs^{69}$

[a] Sequence goes from most easily transferred to most difficultly transferred.
[b] Sequence goes from most stable to least stable.
[c] 18C6.[69]
[d] 18C6 in (70/30 wt/wt) methanol–water.[35]

any arbitrary solvent. Unfortunately, as pointed out by Burgess,[86] there is no single solvent property which uniquely determines its ability to solvate ions. To be sure, there are several properties which determine the solvating ability of a solvent. With respect to cation–macrocycle complex stability constants, in some groups of solvents the stability correlates with $1/\varepsilon$, in others it correlates with ε, and in still others the complex stability correlates with the solvent's donor number. (The donor number is related to the ease of donation of the electron pair of the solvent.[86,87])

Frensdorff[88] noted that stability constants for the reaction of metal cations with cyclic polyethers were three to four orders of magnitude higher in methanol than in water. Lehn and Sauvage[71] observed similar behaviour with the cryptands. This effect has been attributed to the difference between the dielectric constants of the two solvents.[35] To further test this hypothesis, stability constants for the interaction of 18C6 with several cations have been measured in a series of methanol–water mixtures.[35,89,90] As a first approximation, one would expect the stability constant to vary with the ion solvation energy as predicted by the simple Born equation.[36] The Born equation (Equation 1) relates solvation energy to the dielectric constant in the following manner:

$$\Delta G \propto 1 - 1/\varepsilon \qquad (13)$$

Thus, if a Born-type relationship is sufficient to explain the variation in stability constant in different water–methanol mixtures, we would expect a linear relationship between $\log K$ and $(1 - 1/\varepsilon)$. Fig. 7 shows that, at least for M^+–18C6 interaction in the water–methanol mixtures, this is the case.

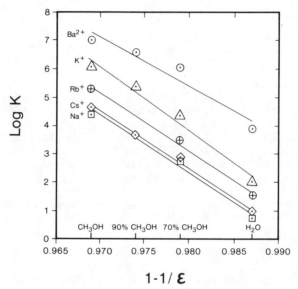

Fig. 7. Plot of $\log K$ vs. $1 - 1/\varepsilon$ for the reaction of 18C6 with several cations in water–methanol mixtures of varying weight fractions.[35,76,89,90]

However, this result may be fortuitous. While the bulk dielectric constant of water–methanol mixtures varies linearly with mole fraction, the ion may experience selective solvation, i.e. in the ion's local environment the solvent mole fraction may be different from that of the bulk phase. Indeed, Abraham and Liszi[80] have shown that when ions partition from water into a solvent in which water is soluble they retain a portion of their hydration sheath and are partially hydrated in the solvent.

Agostiano, et al.[83] measured $\log K$ values for the interaction of DB18C6 with Na^+, K^+, and Cs^+ in a series of pure alcohols. Analysis of their $\log K$ data in these pure solvents along with $\log K$ values in H_2O reveals that the data do not correlate with a simple Born-type solvation model. Rather, an empirical relationship correlating $\log K$ with ε was found as shown in Fig. 8.

To further complicate the solvent situation, Matsuura et al.[91] showed that $\log K$ values for the reaction of DB18C6 with the alkali metal cations in dimethyl sulphoxide (DMSO), dimethylformamide (DMF), and propylene carbonate (PC), do not correlate with dielectric constant, ε. While the

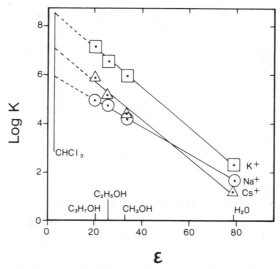

Fig. 8. Plot of ε vs. log K for the reaction of DCy18C6 with K^+, Na^+, and Cs^+ in several pure solvents.[35,83]

dielectric constant decreases in the order PC ($\varepsilon = 65.1$) > DMSO ($\varepsilon = 48.9$) > DMF ($\varepsilon = 36.7$) the log K values increase in the order DMSO < DMF < PC. (If the dielectric constant was the dominating factor in complex stability, one would expect log K values to increase in the same order as the decreasing dielectric constant of the solvents). However, these authors point out that this sequence of increasing log K values correlates with the Gutmann donor numbers (DN) for these solvents which are DMSO (DN = 29.8) > DMF (DN = 26.6) > PC (DN = 15.1). The higher the donor number the better electron donor the solvent is and the better it solvates the cation, making it more difficult for the macrocycle to compete with the solvent for the cation.

Although the Gutmann donor number correlates well with log K values in DMSO, DMF, and PC, it does not correlate with log K values in water, methanol, and ethanol. The Gutmann donor number predicts an increasing log K value in the order of ethanol (DN = 30.0) < methanol (DM = 23.5) < water (DN = 18.0).[86] However, just the opposite occurs, i.e. as mentioned above, log K varies directly with dielectric constant: water ($\varepsilon = 78.5$) < methanol ($\varepsilon = 32.6$) < ethanol ($\varepsilon = 24.3$).

As with ion partition coefficients, cation selectivities based on differences in log K may also vary from one solvent to another. Several examples are noted for DB18C6 in Table 5. In every solvent except acetonitrile, K^+ is preferred. (This is undoubtedly due to K^+ having the cation/ligand cavity ratio nearest to unity.) However, the selectivity order for the other alkali cations does change in several of the solvents.

The above discussion dealt mainly with the effect of the solvent on log K for cation–cyclic polyether complexes. Recently, several studies have been reported on the effect of solvent on log K for formation of cation–cryptand complexes.[92-94] Cox, et al.[92] measured the complex stability for the alkali cations, Ag^+ and Ca^{2+}, with the cryptands [2.1.1], [2.2.1], and [2.2.2] in eight different solvents. It was found that the equilibrium constants for cryptate formation were quite sensitive to solvent variation. For a given cryptate, variations in equilibrium constants of up to 9 orders of magnitude were reported with values in water being consistently lower than those in any of the other solvents. The equilibrium constants in non-aqueous solvents were found, qualitatively, to follow the trends expected from ion–solvent interactions in the solvents. The equilibrium constants are highest in solvents where cation–solvent interactions are relatively weak, e.g., acetonitrile and propylene carbonate. On the other hand, where cation–solvent interactions are large, e.g., dimethylformamide and dimethylsulphoxide, the equilibrium constants are lower. In comparing results in dimethylformamide ($\varepsilon = 36.7$) and N-methylpropionamide ($\varepsilon = 176$), they concluded that the dielectric constant of the solvent does not play a large part in determining cryptate complex stability.

Despite the large differences in equilibrium constants in the various solvents, the selectivity sequence displayed by these cryptands is essentially independent of solvent. This is in contrast to the crown ethers (see Table 5). Cox, et al.[92] suggest that the three dimensional cavity presented by the cryptand results in an effective shielding by the ligand of interactions between the solvent molecules and the cation encapsulated within the cavity. The cavities of small crown ethers differ from cryptand cavities in one important respect. The crown ether contains a planar two-dimensional cavity which allows the entrapped cation to interact with the solvent above and below the plane of the macrocycle. Arnaud-Neu, et al.[94] reached a conclusion similar to that of Cox, et al.[92] concerning the effect of solvent on log K and selectivity for the cryptand complexes of several transition and heavy metal cations.

From the above discussion, it is apparent that the role of the solvent in cation–ligand interactions is not as straightforward as would be hoped. The point is that log K for cation–macrocycle interaction and J_M for cation transport through liquid membranes are both solvent dependent in ways that are not well understood. It would be useful to exploit both solvent and macrocycle cation selectivities to obtain the most efficient separations possible. Furthermore, it would be advantageous to be able to predict, from solvent properties, which solvent would be most effective for a particular separation. At the present time this is not possible. While several instances of success of limited correlations between complex stability and one of the empirical solvent parameters have been noted, an overall correlation has yet to be established. Indeed, as de Jong and Reinhoudt[69] have pointed out, further studies are needed in order to arrive at a model that has any predictive value.

2.3.2. Ion-pair partition coefficients

2.3.2.1. Anion type. In bulk liquid membranes of the type described in this chapter, the diffusion path length of a complex during transport is much larger than the ion's Debye length in the membrane. Therefore, electroneutrality must be maintained in these membranes. In the case of neutral macrocyclic ligands, their cation complexes are positively charged, and in order to maintain the required electroneutrality each complex must be associated with an anion. In the case of either biological cell membranes or artificial lipid bilayer membranes, the restriction of electroneutrality being maintained during ion transport does not apply.[50,95] This is because the membrane thickness (~ 50 Å) is comparable to the Debye length.[96] Thus, in these ultra-thin membranes, cations can be, and indeed are, transported unaccompanied by their co-anions.[11]* Thus, in bulk liquid membranes the anion must accompany the cation–ligand complex and the nature of the anion and of its solvent interactions should be important factors in determining the flux of cations in these membranes.

In order to elucidate the effect of the anion on cation transport, Lamb, et al.,[46,48] performed a series of experiments in which the transport of K^+ by DB18C6 in the presence of various anions was measured. Their results, shown in Fig. 9, show that the flux of K^+ varies over eight orders of magnitude according to the anion present.

The variation of the K^+ flux with anion may be rationalised in terms of the ease with which the anion leaves the water and enters the membrane, i.e., the Gibbs free energy of partitioning between the aqueous and organic phases.[46] If the anion partitions favourably into the organic phase, the

---

* Electroneutrality in the bulk aqueous phases must be maintained, even though the membrane is cation selective. If the aqueous phases are electrically connected via an external electrical path, e.g. a salt bridge or a wire, then charge transfer can occur through this path, thus maintaining bulk electroneutrality. On the other hand, if the two aqueous phases are not electrically connected other than through the membrane itself, i.e. an open circuit condition, then, under appropriate driving conditions, a net cation flux occurs only until a charge separation (charge build up on one side of the membrane) has been established. This charge separation creates an electric field which opposes further net cation flux. The electrical potential thus created is called the Nernst potential.

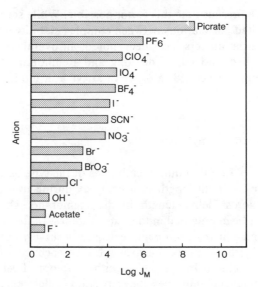

Fig. 9. Log J_M values (arbitrary units) for K^+ in the presence of different anions.[46,48]

concentration of the neutral cation–ligand–anion complex in the membrane will be increased which, in turn, will increase the amount of cation transported through the membrane.

The partition coefficient, k_{ip},

$$k_{ip} = [M^{n+}A_n^-]_{org}/[M^{n+}]_{aq}[A^-]_{aq}^n \tag{14}$$

is related to the Gibbs free energy of partitioning, ΔG_p, between water and membrane by the equation

$$\Delta G_p = -RT \ln k_{ip} \tag{15}$$

The term ΔG_p, for the cation–anion pair, can be divided into its component parts

$$\Delta G_p = \Delta G_p^C + \Delta G_p^A + \Delta G_p^{CA} \tag{16}$$

where the first two terms represent the free energies of partitioning of the cation and anion, respectively, and ΔG_p^{CA} represents the change in free energy of cation–anion interaction in going from water to the membrane phase. If $\Delta G_{g \to M}^A$ represents the free energy of transferring the anion from the gas phase to the membrane phase, and $\Delta G_{g \to w}^A$ that from the gas phase to water, then

$$\Delta G_p^A = \Delta G_{g \to M}^A - \Delta G_{g \to w}^A \tag{17}$$

Substitution of Equation 17 into Equation 16 results in Equation 18,

$$\Delta G_p = \Delta G_p^C + \Delta G_{g\to M}^A - \Delta G_{g\to w}^A + \Delta G^{CA} \tag{18}$$

Of the quantities in Equation 18, Gibbs free energies of hydration, $\Delta G_{g\to w}^A$, are available for most of the anions included in Fig. 9.

Equations 2,* 15, and 18 predict a simple linear correspondence between $\log J_M$ and $\Delta G_{g\to w}^A$ only if it is assumed that all other ΔG terms comprising ΔG_p in Equation 18 either are constant or compensate each other for the anions which are considered. While such an assumption is overly simplistic, Fig. 10 demonstrates that the change in $\log J_M$ with increasing $\Delta G_{g\to w}^A$ is a regular one. This regular trend may arise from the common factors (namely, anion size, charge, and polarisability) which influence the values of all the free-energy terms relating to the anion ($\Delta G_{g\to M}^A$, $\Delta G_{g\to w}^A$, ΔG^{CA}) in Equation 18. Regardless of the reason for this regular behaviour, the data in Fig. 10 show that there is an empirical correlation between cation flux and anion hydration energy.

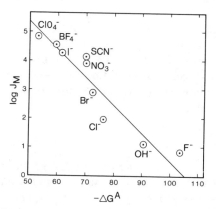

Fig. 10. Plot of $\log J_M$ values (arbitrary units) for K^+ vs. single anion free energy of hydration.[46]

The magnitude of the effect of anion on the facilitated transport of cations across liquid membranes holds significant implications. Interpretation of results of cation-transport experiments where anions move through the membrane requires that the role of the anion(s) be understood. Comparisons of transport carrier effectiveness and system design ought to be made where

---

* Equation 2 has been shown to be inadequate in describing transport behaviour over an extended range of $\log K$ values. However, in this case it may be used with some justification since we are considering only one cation, K^+, whose $\log K$ value falls on the rising portion of the J_M vs. $\log K$ curve in Fig. 2 and we are concerned only with relative effects of the anions.

606   *D. W. McBride, Jr., R. M. Izatt, J. D. Lamb, and J. J. Christensen*

a common anion has been employed. The anion effect may be exploited in that transport of cations across membranes can be turned on or off simply by altering the anion present in the source phase. Also, this anion effect has potential in either separating or detecting anions themselves.

2.3.2.2. Monovalent vs. bivalent cations. As a consequence of the electro-neutrality requirement, the partition coefficient of the anion is important and can play a major role in determining the concentration of the cation–ligand complex in the membrane. Similar effects are noted with cations. These effects are less dramatic than the anion effects, but are consistent with, and can be explained in terms of, the effect of cation charge on the partition coefficient. As was noted earlier, the maximum flux for monovalent cations occurs in the region of 5.5 to 6 $\log K(CH_3OH)$ units while for divalent cations the maximum occurs for $\log K$ values of 6.5 to 7 $\log K(CH_3OH)$ units. This difference in where the maximum flux occurs between monovalent and divalent cations can be rationalized in terms of the effect of the partition coefficient on the concentration of the cation–ligand–anion complex in the membrane.

To compare the relative values of partition coefficients for monovalent and divalent cations, we use Equations 15 and 16. We let $\Delta G_p^{C_1}$ and $\Delta G_p^{C_2}$ be the Gibbs free energy of partitioning for the monovalent and divalent cations, respectively. From a simple Born energy approximation to the difference in electrostatic free energy of the uncomplexed cation between the two phases,* we obtain

$$\Delta G_p^{C_2} = 4\Delta G_p^{C_1} \tag{19}$$

The factor of four comes from squaring the charge on the divalent cation (see Equation 1). Divalent cations have much larger hydration energies than monovalent cations. Consequently, an energy expenditure roughly four times that required for monovalent cations is needed to remove divalent cations from their aqueous environment. (This is only true when cations

---

* Abraham and Liszi[80] have derived more complicated equations than the modified Born equation for ion solvation. Whereas the Born equation assumes the ion to be immediately surrounded by the bulk phase solvent, the Abraham–Liszi model assumes the ion to be surrounded by a number of concentric spherical shells of solvent molecules. The physical properties of each shell differ from those of the bulk phase until the last shell is reached which is the bulk phase. The number of concentric shells between the ion and the bulk phase depends on the solvent, but is generally either one or two. Relevant solvent parameters used in the calculation of ion solvation energy include the bulk phase dielectric constant and the radius of a solvent molecule. In spite of added complexities, the Abraham–Liszi model retains the Z^2 dependence, so Equation 19 is still valid.

are partitioning into organic phases containing neutral extractants such as the crown ethers. If the organic phase contains cation exchangers, then divalent cations extract much better than monovalent cations.)[97] Considering this factor of 4 alone, keeping the other terms in Equation 16 constant, and using Equation 15 lead to

$$k_2 = k_1^4 \tag{20}$$

The terms k_1 and k_2 refer to the partition coefficients for monovalent and divalent cations, respectively. At first glance, Equation 20 might give the impression that $k_2 > k_1$. However, this is not the case for the water–chloroform system. Monovalent cation partition coefficients in this system have been measured to be $\sim 5.0 \times 10^{-6}$ (water, of course, being favoured).[97] Thus, since $k_1 \ll 1$, $k_2 \ll k_1$ in such systems. Furthermore, Equation 20 indicates that whichever phase monovalent cations prefer, divalent cations prefer that phase much more. According to Equation 20 if $k_1 \sim 5.0 \times 10^{-6}$ then $k_2 \sim 6.3 \times 10^{-22}$! This is clearly absurd. If this were the case, divalent cations would never be transported. The point here is that the partition coefficients of divalent cations are smaller than those for monovalent cations.

The double charge of divalent cations affects both their partitioning and the partitioning of anions. If only monovalent anions are present in the aqueous phase, then two anions must partition into the membrane for each divalent cation. This causes a further significant reduction in the ion-pair (or ion-triplet) partitioning into the membrane. Earlier, we discussed the effect of the anion on monovalent cation transport. However, the effect of the anion on divalent cation transport should be even more dramatic. Similar to Equation 19, the effect of the anion in divalent transport should be the square of its effect in monovalent transport.*

We are now in a position to understand the difference in the log K ranges for maximum transport between monovalent and divalent cations. From Equation 4 it is seen that cation transport is related to the concentration of the cation–ligand–anion complex in the organic phase which, in turn, is

---

* Analogous to the situation in which divalent cations are transported with monovalent anions is the situation in which monovalent cations are transported with divalent anions. To satisfy electroneutrality, one anion must accompany two cations. The partition coefficient of the divalent anion is related to that of the monovalent anion according to Equation 20. Hence, we would expect the flux to be much higher for a given cation in the presence of a monovalent anion than for the same cation in the presence of a similar divalent anion. Lamb, *et al.*[46] report that the K^+ flux from $KMnO_4$ or $KClO_4$ solutions is roughly five orders of magnitude greater than the flux from a K_2SO_4 solution. In fact, the flux from a K_2SO_4 solution cannot be distinguished from the flux in the control.

related to the product $Kk_{ip}n^n$.* Therefore, if k_{ip} decreases, as it does for divalent cations, then K for divalent cations must be larger to get the same transport as is found with monovalent cations. (It is assumed that the diffusivities, in the organic phase, are equal for the monovalent and divalent complexes.)

2.3.3. Hydrophobic bulk

In order for facilitated transport of the type described here to be most efficient and to be a candidate for practical use, means must be found to prevent loss of the macrocyclic carrier to the aqueous phases adjacent to the membrane. Substitution of benzo- or cyclohexano-groups into the parent macrocycles contributes somewhat towards accomplishing this end. In addition, we have tested macrocyclic carriers with long aliphatic substituents which ensure extremely low solubility of the carrier in the aqueous phases.[27] The objective was to determine the effect, if any, that the addition of these bulky aliphatic substituents to parent crown ethers would have on cation fluxes and carrier selectivities. In particular, if differences were found, could they be attributed to changes in log K values and/or increased ligand hydrophobicity?

The effect of hydrophobic bulk on cation transport is illustrated in Table 3 for the cases of 18C6 and DCy18C6. Log K values for the interaction of DCy18C6 with cations are less than those for the interaction of 18C6 with the same cations. Thus, DCy18C6 would be expected to have lower transport than 18C6. However, DCy18C6 transports all cations studied better than does 18C6. This unexpected increase in transport for DCy18C6 may be a result of the increased ligand hydrophobicity and subsequent decrease of ligand loss to the aqueous phase. The water solubilities of 18C6, (~ 5 mol dm^{-3}) and DCy18C6 (~ 10 mmol dm^{-3})[23] are consistent with this hypothesis. The log K values for cation interaction with DB18C6 are also less than those with 18C6, except in the case of Na$^+$. However, the increased

---

* This relationship is most easily seen in the model represented in Fig. 3a. This model predicts the following:

$$[ML^{n+}A_n^-]_{org} = L_T \frac{Kk_{ip}n^n[M^{n+}]_{aq}^{n+1}}{1 + Kk_{ip}n^n[M^{n+}]_{aq}^{n+1}} \tag{21}$$

where $[ML^{n+}A_n^-]_{org}$ is the concentration of the cation–ligand–anion complex in the membrane and $[M^{n+}]$ is the cation concentration in the aqueous phase. Other terms have been defined previously. Equation 21 follows directly from Equation 11 if the ligand is limited to the membrane. This is a simple Michaelis–Menten type saturation equation relating $[ML^{n+}A_n^-]_{org}$ to $[M^{n+}]_{aq}$. The appropriate constant relating the two is, in this case, $K_{ex} = Kk_{ip}n^n$. The model presented in Fig. 3a qualitatively predicts transport behaviour and therefore $[ML^{n+}A_n^-]_{org}$. Although a more complicated model is needed to quantitatively predict cation flux, $[ML^{n+}A_n^-]_{org}$ still retains a dependence on K_{ex}, so we use Equation 21 to give a simple, intuitive, and qualitative dependence of $[ML^{n+}A_n^-]_{org}$ on K_{ex}.

hydrophobicity of DB18C6 (aqueous solubility $\sim 0.1\ \text{mmol dm}^{-3}$)[23] apparently is not sufficient to compensate for the decreased log K value since the cation flux is less for DB18C6 than for 18C6.

Aliphatic side chains were added to the benzo-subsituents of DB18C6 (Table 6) to provide increased hydrophobic bulk.[27] In general, cation fluxes are not significantly different either in magnitude or in selectivity for these ligands than for DB18C6, even when a C_{14} chain was added to give DTeB18C6. Thus, selectivity for K^+ among the monovalent cations and for Pb^{2+} among the divalent cations was maintained.

Table 6. Effect of increasing hydrophobic bulk on cation flux[(a)]

Cation	DB18C6	DMB18C6	DBuB18C6	DTeB18C6
Na^+	10	4	7	3
K^+	230	348	220	290
Rb^+	28	30	26	33
Cs^+	5	6	5	7
Ag^+	75	—	92	
Sr^{2+}	—[(b)]	—[(b)]	7	
Ba^{2+}	4	6	7	4
	$(15)^{(c)}$	$(24)^{(c)}$	$(28)^{(c)}$	$(15)^{(c)}$
Pb^{2+}	62	—	80	

[(a)] J_M values given as $10^{-8}\ \text{mol s}^{-1}\ \text{m}^{-2}$; $1\ \text{mol dm}^{-3}$ nitrate salt except Ba^{2+} $(0.30\ \text{mol dm}^{-3})$.
[(b)] Value at blank level, $0.7 \times 10^{-8}\ \text{mol s}^{-1}\ \text{m}^{-2}$.
[(c)] Adjusted to $1\ \text{mol dm}^{-2}$.[27]

Strzelbicki and Bartsch[20] report similar results with crown ether carboxylic acid-mediated alkali cation transport. They added aliphatic side chains of various lengths to *sym*-dibenzo-16-crown-5-oxyacetic acid (Fig. 1). They measured the final ligand concentration in the membrane directly and found that increasing the hydrophobic bulk minimized loss to the aqueous phase. In the case of the unsubstituted ligand, almost half of the ligand originally present in the organic phase was lost. However, with the aliphatic side chains added, no ligand loss was observed. Furthermore, a slight increase in cation flux with increased chain length was observed, although this increase did not correlate quantitatively with the increased ligand concentration in the membrane. These authors also noted that the selectivity sequence was unaltered by increasing ligand hydrophobicity, being $Na^+ > K^+ > Rb^+ > Cs^+ > Li^+$ for all the derivatives. The relative selectivity ratios, on the other hand, did vary, although in a non-systematic manner, with substituent chain length.

There is a point at which increasing the hydrophobic bulk of the macrocycle ceases to change significantly the macrocycle partitioning between the membrane and the aqueous phase. Above this point it is of little use to increase hydrophobic bulk further for the purpose of minimizing loss to the aqueous phase. Thus, since DB18C6 is already poorly soluble in water, little may have been gained in adding aliphatic side chains.

2.4. Cation transport and selectivity from binary cation mixtures

The selectivity of macrocycle-containing membranes may be exploited by selectively transporting cations from cation mixtures. As a first approximation, one would expect the selectivity sequences determined by the magnitude of the flux values in single cation experiments to be reproduced in binary cation mixtures. While this is often true, there are many reversals. (In general, it is not safe to predict behaviour in cation mixtures from that of single cation systems.) Some results for cation transport by DCy18C6 from $Pb^{2+}-M^{n+}$ mixtures are listed in Table 7. It is interesting that selectivity by DCy18C6 for Pb^{2+} over the second cation is high in every case when

Table 7. *Cation fluxes*[(a)] *from two-cation mixtures using bulk liquid membranes containing the listed macrocycles*

Macrocycle	Together	Separate[27]	Log K
DCy18C6			
Pb^{2+}/Li^+	$690/0^{13}$	728/0	$4.95/0.6^{(b)}$
Pb^{2+}/Na^+	$552/2^{13}$	728/53	$4.95/1.21^{(b)}$
Pb^{2+}/K^+	$621/5^{13}$	728/780	$4.95/2.02^{(b)}$
Pb^{2+}/Sr^{2+}	$759/13^{13}$	728/1028	$4.95/3.24^{(b)}$
2.2.1			
Na^+/Li^+	$219/0^{14a}$	310/340	$8.84/4.10^{(c)}$
Na^+/K^+	$182/44^{14a}$	310/664	$8.84/7.45^{(c)}$
Na^+/Rb^+	$230/3^{14a}$	310/1264	$8.84/5.80^{(c)}$
Na^+/Sr^{2+}	$104/1^{14a}$	310/7	$8.84/10.65^{(c)}$
Na^+/Ba^{2+}	$180/3^{14a}$	310/70	$8.84/9.70^{(c)}$
2.2.2			
Na^+/Ca^{2+}	$322/1^{14a}$	552/7	$7.21/7.60^{(c)}$
Na^+/Sr^{2+}	$44/16^{14a}$	552/30	$7.21/11.5^{(c)}$
Na^+/Ba^{2+}	$39/41^{14a}$	552/26	$7.21/12^{(c)}$
Na^+/Pb^{2+}	$5/44^{14a}$	552/44	$7.21/12.4^{(b)}$

(a) J_M values given as 10^{-8} mol s^{-1} m^{-2}; blank value 0.7×10^{-8} mol s^{-1} m^{-2}.
(b) Values in H_2O.[35]
(c) Values in 95% CH_3OH/5% H_2O (vol/vol).[35]

the cations are both present in the source phase, even though the flux values for K^+ and Sr^{2+} are high when these cations are studied alone. Indeed, Sr^{2+} transport is higher than that of Pb^{2+} when the cations are studied alone; but when together in the source phase, Pb^{2+} is selectively transported.[13]

An even more striking reversal of the single cation transport selectivities is seen in Table 7 with Na^+ and the cryptand 2.2.1.[14a] Here both Rb^+ and K^+ have markedly higher transport rates than Na^+ when comparing single cation experiments. However, in mixtures Na^+ is much better transported. When Na^+ is present in binary mixtures with divalent cations, no reversal in selectivity is seen and the selectivity is the same as that predicted from the transport in single cation solutions.

Cation transport from the binary mixtures can be modelled using the scheme represented in Fig. 3c, and allowing two cation species to be present in the aqueous source phase. Each cation species has its own set of parameters. We assume that each species acts independently of the other, i.e. we assume that two "independent" parallel transport processes occur simultaneously, except that the two species are coupled by the constraint that the total amount of ligand is constant:

$$L_T = [L] + [M_1 L^{n_1+} A_{n_1}^-] + [M_2 L^{n_2+} A_{n_2}^-] \tag{22}$$

where M_1 and M_2 represent species 1 and 2, respectively. The flux predicted by the model for a binary cation mixture is given by Equation 23 which is

$$J_{Mi} = B_i \left(\frac{E_i}{1 + K_1 E_1 + K_2 E_2} - \frac{F_i}{1 + K_1 F_1 + K_2 F_2} \right), \quad i = 1 \text{ or } 2 \tag{23}$$

where

$$B_i = \frac{D_{ci} K_i L_T}{l_3 + (A'/A'') l_4},$$

$$E_i = (k_{ip})_i G_i (n_1 G_1 + n_2 G_2)^{ni} - J_{Mi} l_{2i}/D_{Ii},$$

$$G_i = M_i' - J_{Mi} l_1 / D_{wi},$$

$$F_i = (k_{ip})_i H_i (n_1 H_1 + n_2 H_2)^{ni} + (A'/A'') J_{Mi} l_{5i} / D_{Ii}, \text{ and}$$

$$H_i = M_i'' + (A'/A'') J_{Mi} l_6 / D_{wi}$$

a modified version of that found in Christensen, *et al.*[99]

As was mentioned above for transport in single cation experiments, the important parameter is the product $K k_{ip} n^n$, where K is the equilibrium constant, k_{ip} the ion pair partition coefficient and n is the charge on the cation. The concentration of the cation-ligand-anion complex in the membrane is related to this term. Thus, comparison of the separate fluxes of Na^+ and Rb^+ in the presence of cryptand 2.2.1 (Table 7), shows that Rb^+

is favoured. This preference for Rb^+ can be rationalised in terms of the relative log K values[35] for M^+-2.2.1 interaction. The log K value for Rb^+, 5.80, falls within the optimal range for transport, whereas the log K for Na^+, 8.84, is too high for optimal transport. However, when both are present in the source phase, most of the ligand which is complexed is bound to Na^+. Since both Na^+ and Rb^+ are monovalent and since there is a common anion, we can expect the ion–pair partition coefficients to be roughly the same. Thus $(Kk_{ip})_{Na^+}/(Kk_{ip})_{Rb^+} \sim K_{Na^+}/K_{Rb^+} \sim 1000/1$ and the Na^+ complexes outnumber the Rb^+ complexes $1000/1$. So, even though Na^+ is not transported as well as Rb^+ in single cation situations, in mixtures it is the fact that the ligand is complexed predominantly with Na^+ which precludes Rb^+ transport and still allows a respectable Na^+ transport. Thus, since in mixtures involving only monovalent cations the k_{ip} values are approximately equal, the log K value is the deciding factor in determining selectivity. Presumably, the same is true in mixtures of divalent cations. In mixtures involving both monovalent and divalent cations, K, k_{ip}, and n are all important and it is their product which determines selectivity. This fact is illustrated with binary mixtures of Na^+ with either Ca^{2+}, Sr^{2+}, Ba^{2+}, or Pb^{2+} using the cryptand carrier 2.2.2 (Table 7). In all of these cases, the Na^+ flux is larger in the single cation case, but in the mixtures, the pattern is more complex. Na^+ is progressively less preferred in the sequence $Ca^{2+} > Sr^{2+} > Ba^{2+} > Pb^{2+}$. The log K values (in parentheses)[35] of complexation with 2.2.2 increase in the same order, Ca^{2+} (7.60) $< Sr^{2+}$ (11.5) $< Ba^{2+}$ (12) $< Pb^{2+}$ (value valid in H_2O is 12.4, value in CH_3OH is expected to be ~ 3 log K units larger). We assume that the partition coefficients for the divalent cations are approximately equal. Thus, while for Ca^{2+}, $(Kk_{ip})_{Na^+} > (4Kk_{ip})_{Ca^{2+}}$ (resulting in Na^+ being selectively transported), as log K values for the other divalent cations become larger, these cations compete more successfully with Na^+ for the ligand until finally for Pb^{2+} we have $(4Kk_{ip})_{Pb^{2+}} > (Kk_{ip})_{Na^+}$ (resulting in Pb^{2+} being selectively transported, although the total moles transferred has decreased). In the latter case the ligand is predominantly complexed with Pb^{2+} which precludes Na^+ complexation and transport.

3. Facilitated transport in emulsion or liquid-surfactant membranes

Bulk membranes such as those described above are unsuitable for rapidly transporting large quantities of ions because of the low surface area and the large path length between the two aqueous phases. Surface area is greatly increased and path length decreased[100] by use of emulsion type membranes such as those developed by Li and his co-workers.[101] These membrane systems have been used to perform a variety of separations such as the removal of harmful compounds including phenol, citric acid, urea, S^{2-}, NO_2^-, and PO_4^{3-} from waste water[22,102,103] and the rapid and efficient removal of Cu^{2+}, Co^{2+}, and Ni^{2+} from aqueous mixtures.[104] Emulsion membranes have enjoyed increasing popularity in biomedical applications.[102] In this area they have been used for the oxygenation of blood[105] and removal of toxins of uremia from the gastrointestinal tract.[106] Enzymes have also been encapsulated in emulsion membranes and have been shown to retain their catalytic activity.[107–109] In addition, emulsion membranes have promise of being used to encapsulate anti-tumour drugs and to deliver them selectively to a targeted organ or area of the body.[110,111]

Permeation mechanisms involved in emulsion membranes include both simple diffusion and facilitated diffusion. The requirements for simple diffusion are that the permeant species be soluble in the organic phase and be uncharged. Examples of such species are NH_3 and organic molecules such as phenol and acetic acid.[22,103] In facilitated diffusion, a variety of ion exchangers has been used, including both cationic and anionic ion exchangers.[22,32,104,112] Until recently, neutral macrocycles had not been used as permeation mediators in emulsion membranes. This is surprising since macrocycles are used in other related areas and since they exhibit remarkable cation selectivities while ion-exchangers show much less selectivity among cations.[15]

We have incorporated crown ethers into emulsion membrane systems and have used one such system to selectively transport Pb^{2+} from a mixture of that ion with other cations.[16] The emulsion liquid membrane system used by us was patterned after that described by Li and his co-workers[101,103] and is represented in Fig. 11. The emulsion consists of droplets of receiving phase coated by a thin layer of organic membrane. The emulsion is prepared by first vigorously emulsifying (at ~30 000 rpm) the oil phase with the aqueous receiving phase which consists of a 0.1 mol dm^{-3} aqueous solution of a metal pyrophosphate salt. The purpose of the $P_2O_7^{4-}$ is to provide an anion which can form a sufficiently stable complex with Pb^{2+} to strip it from the membrane phase. $Na_4P_2O_7$ was used in the early experiments, but $Li_4P_2O_7$ replaced the Na salt in later experiments because Li^+ is not extracted into the organic membrane. DCy18C6 (mixture of isomers) (Parish Chemical, Orem, Utah) was chosen to be incorporated into the oil phase (0.100 mol dm^{-3}) as the cation carrier because in the bulk $CHCl_3$ liquid membrane systems it demonstrated selectivity for Pb^{2+} among other univalent and bivalent cations.[13,14a] Also, this ligand is one of the few crown ethers soluble in the mixture of S100N oil (Exxon), Indopol L-100 polymer

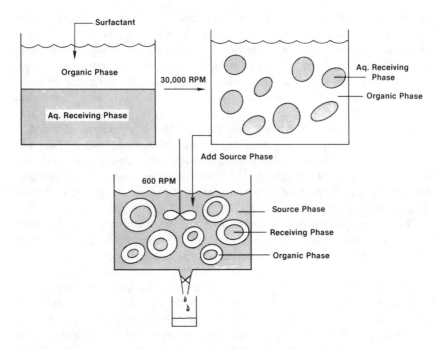

Fig. 11. Diagramatic representation of emulsion membrane system.

(Amoco) and Span 80 surfactant (sorbitan monooleate) (Emulsion Engineering) used as the membrane solvent in our early work. The emulsion, which is stable to appreciable breakage for several hours, is stirred at 600 rpm with the aqueous source phase containing the cation(s) to be transported (nitrate salts). Cation transport from the source phase (outside the membrane spheres) into the receiving phase (inside the membrane spheres) commences immediately and is essentially complete within minutes. The change in cation concentration with time of exposure to the membrane is determined by taking 3-ml samples from the source phase at 0, 3, 6, 10, 15, 20, 25, and 30 min and determining the concentration of each cation by atomic absorption spectrophotometry (Perkin Elmer model 603).

Results are given in Fig. 12 for the transport of Pb^{2+}, both alone and in the presence of Li^+ and K^+, using the S100N/Indopol L-100/Span 80 membrane system.[16] The data in Fig. 12 show that essentially all of the Pb^{2+} is transported within minutes in each case. Little Li^+ is transported, consistent with the lack of affinity of Li^+ for DCy18C6.[35] Transport of K^+ is seen, but at a slower rate than that of Pb^{2+}. These data are consistent with

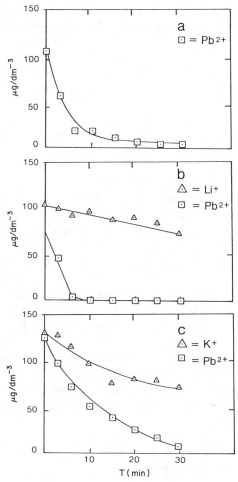

Fig. 12. Cation concentration in aqueous source phase as a function of time for the emulsion membrane system.[16] Each point is the average of points taken from three identical experiments performed simultaneously. The standard deviations were less than 15% in all cases. (a) Pb^{2+} alone, (b) Pb^{2+} and Li^+, both present in the source phase, and (c) Pb^{2+} and K^+, both present in the source phase. Anion $= NO_3^-$. L = DCy18C6.

the relatively higher stability constant for formation of the Pb^{2+} complex.[35] Detailed discussion of these results has been presented.[16]

Experiments were performed which established that the disappearance of Pb^{2+} from the source phase reflected actual transport of Pb^{2+} through the liquid membrane rather than irreversible solvent extraction of cation into the membrane. First, no change in the source Pb^{2+} concentration occurred in the absence of carrier when all other experimental conditions remained unchanged. Second, the crown-containing organic phase was mixed with the aqueous source phase in a solvent extraction-type experiment with no receiving phase present and samples from the source phase were taken at the same time intervals as before. Following an initial small drop in Pb^{2+} concentration due to equilibration with the oil phase, the amount of Pb^{2+} remained constant. Thus, very little solvent extraction of Pb^{2+} by the oil phase occurred. Further evidence for Pb^{2+} transport was obtained by breaking down the emulsion membrane after 30 min and measuring the Pb^{2+} concentration in the receiving phase, which consistently corresponded with the amount of Pb^{2+} which disappeared from the source phase.

Recently, emulsion membranes have been made using toluene mixed with 3% surfactant as the solvent,[15] after the manner of Nakashio and Kondo.[100] Several Pb^{2+}/cation mixtures were used in the source phase in these experiments, with DCy18C6 serving as carrier.[15] In these experiments, 100 ml of 0.020 mol dm^{-3} carrier in toluene were blended at ~30 000 rpm with 100 ml of 0.05 mol dm^{-3} $Li_4P_2O_7$ and 3 ml of Span 80 for 30 s to form the emulsion. 20 ml of the emulsion were placed into each of three stirring vessels and 100 ml of the source phase containing the metal(s) to be tested were added to the emulsion and immediately stirred at 600 rpm. Samples were taken and analysed as described above for the S100N membrane experiments.

Two quantities were determined based on the mass balance for each experiment, the metal enrichment ratio, E, and the separation factor, ϕ (A/B), for metals A and B. These quantities are defined by Equations 24 and 25, respectively,

$$E = (C_f^r / C_i^s) \tag{24}$$

$$\phi(A/B) = (C_f^r / C_f^s)_A (C_f^s / C_f^r)_B \tag{25}$$

where C is the metal concentration in $\mu g\ ml^{-1}$, subscripts i and f refer to the initial and final states, respectively, and superscripts r and s refer to the internal receiving and external source aqueous phases, respectively. Since the ratio of 20 ml of emulsion phase to 100 ml of source phase was used to obtain the results reported below and since the emulsion consisted of equal parts of membrane and aqueous receiving phases, enrichment ratios up to ten are possible with this system.

The stability of the emulsion membrane was determined by making a blank run in which distilled deionized water was used as the source phase. The appearance with time of Li^+ in the source phase was taken to be a measure of membrane breakage since under our experimental conditions

Li^+ is not transported by DCy18C6.[16] At 20 min, the Li^+ concentration in the source phase was 3 $\mu g\ ml^{-1}$. This represents 1% membrane breakage. Even after 180 min the Li^+ concentration in the source phase had increased to only 11 $\mu g\ ml^{-1}$. This corresponds to 4% membrane breakage. Since all runs were completed within 20 min, the amount of membrane breakage during the runs was negligible.

The enrichment ratios for each metal ion studied alone are presented in the second column of Table 8. The enrichment ratios decrease in the order $Tl^+ > Ag^+ > Rb^+ > Cs^+ > K^+ > Na^+$ in the case of the univalent cations and $Pb^{2+} > Sr^{2+} > Ba^{2+} > Ca^{2+}$ in the case of the bivalent cations. A significant change between initial and final concentrations is seen for all metal ions except Na^+, K^+ and Ca^{2+}, while essentially complete extraction is found for Tl^+ and Pb^{2+} within 10 min. The enrichment ratios for Tl^+ and Pb^{2+} are 9.58 and 9.53, respectively. These enrichment ratios are comparable to those obtained for Cu^{2+}, Co^{2+}, and Ni^{2+} by Strzelbicki and Charewicz[104] using emulsion membranes with the carrier LIX 70.

Data for a series of binary systems in which Pb^{2+} was present together with a second cation in the source phase are given in columns 3–5 of Table 8. Comparison of columns 2 and 3 shows that the enrichment ratio for most cations is decreased markedly when Pb^{2+} is present. Indeed, the enrichment ratio for Pb^{2+} in every case is nearly 10, while the enrichment ratio for the other cation is less than 1. This results in separation factors greater than 1000.

In the emulsion membrane system there are several factors which affect cation selectivity and separation factors. Perhaps the most important are the two equilibrium constants: K_{ex}, which includes the ion–pair partition coefficient and log K for the metal–macrocycle reaction; and log K for the M^{n+}–$P_2O_7^{4-}$ reaction. Both of these K values are important, for the former influences metal transport from the source phase across the membrane into the receiving phase, and the second determines the degree to which the metal may be concentrated in the receiving phase. Thus, the preference for Pb^{2+} transport above all the cations tested is consistent with Pb^{2+} having the highest log K values for both the cation–macrocycle and the cation–$P_2O_7^{4-}$ interaction.

The dependence of the cation selectivity and separation factors on the log K value for the "sink reaction" is illustrated by the fact that Ag^+ is preferred over Pb^{2+} when the "sink anion" in the receiving phase is changed from $P_2O_7^{4-}$ to $S_2O_3^{2-}$.[18]

The large membrane surface area and relatively short diffusion path lengths in emulsion membranes result in large cation fluxes. Since the time length (~ 10 min) of the experiment is small, kinetic parameters may be important also in determining selectivity and separation factors. Such factors would include diffusion rates in the membrane and in the receiving phase.

Table 8. Cation enrichment (E) and separation factor (φ) values for Pb^{2+}/M^{n+} transport in emulsion membrane systems from Izatt et al.[15] unless otherwise stated.

Cation	$E_{M^{n+}}$-Single Cation	$E_{M^{n+}}$-Mixture with Pb^{2+}	$E_{Pb^{2+}}$-Mixture with M^{n+}	ϕ $[Pb^{2+}]/[M^{n+}]$	log K DCy18C6	log K $P_2O_7^{4-}$
Na^+	0.40	0.40	9.77	1024	4.08[a] 1.21[b]	2.3
K^+	1.32	0	9.85	—[c]	6.01[a] 2.02[b]	2.3
Rb^+	3.13	0.45	9.93	2856	1.52[b]	
Cs^+	1.53	0	9.92	—[c]	4.61[a]	2.3
Ag^+	4.39	0.38	9.84	1630	2.36[b]	
Tl^+	9.58	0.34	9.84	1739	2.44[a]	1.69
Ca^{2+}	0.73	0.11	9.84	5796		5.6
Sr^{2+}	5.60	0.86	9.92	1321	3.24[b]	5.4
Ba^{2+}	4.00	0.08	9.41	2000	3.57[b]	4.6
Pb^{2+}	9.53	—	—	—	4.95[b]	10.1

[a] Value for cis-syn-cis isomer in CH_3OH.[35]
[b] Value for cis-syn-cis isomer in H_2O.[35]
[c] Undefined.

Since neither of these phases is being stirred directly, stagnant boundary layers could be rate limiting. Currently, mathematical models are being developed to predict macrocycle mediated transport, selectivity, and separation factors together with their time dependence in emulsion membranes.

4. Some practical applications

4.1. Separations

As was discussed earlier, the selective cation complexation exhibited by macrocycles can be exploited to separate selectively a particular cation from a mixture of cations. The choice of macrocycle to be used in a particular separation depends, of course, on the cation to be separated and on other cations in the mixture which might compete with the desired cation for the ligand. For instance, Kimura, *et al.*[113] used DCy18C6 in chloroform to separate, by a liquid–liquid extraction procedure, small amounts of strontium from large amounts of calcium as part of a method for determining radiostrontium in milk ash. In another case, Biehl, *et al.*[16] were able to separate Pb^{2+} from binary mixtures with Li^+ and K^+ using DCy18C6 in emulsion–liquid surfactant membranes. Furthermore, the Pb^{2+} separations were 90% complete in less than 30 min.

Metal isotopes have been separated using crown ethers.[65] The principle involved is that isotopes of a particular metal have very slightly different complexation constants with the same macrocyclic ligand. For example, for the separation of calcium-40 from calcium-44, both isotopes compete for the macrocycle. This reaction can be represented as

$$^{44}Ca^{2+}_{(aq)} + {}^{40}CaL^{2+}_{(org)} \rightleftarrows {}^{40}Ca^{2+}_{(aq)} + {}^{48}CaL^{2+}_{(org)} \qquad (26)$$

where L represents a macrocyclic polyether. This equilibrium is slightly displaced towards the reactants, thereby enriching calcium-40 in the organic phase. Jepson and DeWitt[21] report an equilibrium separation factor of 1.0080 ± 0.0016 for this reaction using DCy18C6. Although this separation factor is near unity, the use of countercurrent extractors to obtain multiple-stage separations can result in significant enrichment and/or separation of the isotopes. In principle, this separation process is applicable to the isotopes of essentially all the metals that form complexes with macrocyclic polyethers. See Chapter 8, Volume 3 for further examples.

Although it has not been done, in principle, any separation listed in Tables 7 and 8 (see Refs. 13, 14, 17–19, and 114 for more complete Tables) could be realized commercially. The large surface areas and small diffusion path lengths in emulsion membrane systems cause these to be the systems of choice where rapid cation separations are desired. Selectivity and separations by emulsion membranes are governed by two factors: the macrocycle and the "anion sink" in the receiving phase. If a suitable anion sink is not available, the permeant species equilibrates among the phases, but enrichment does not occur. However, in these cases a multiple stage separation system could be employed where in successive equilibration steps the source phase is depleted of the permeant species and the receiving phase(s) is enriched in that species. If, on the other hand, an anion sink with the desired complexation selectivity is available, then the above processes may be performed in a single stage. Efficiency in this process requires that the selectivities of the macrocycle and of the anion sink must complement each other. The anion sink is effective only if the cation first permeates the membrane through macrocycle mediated transport. Thus, cation selectivity can be designed into the system by selection of the appropriate macrocycle.

Enantiomeric separations are also possible with certain macrocyclic polyethers. Bradshaw, et al.[115] have shown that chiral recognition exists in the complexation of several chiral alkylammonium cations and the S,S and R,R enantiomers of dimethyldioxopyridino-18-crown-6. Newcomb, et al.[116] also report enantiomer differentiation occurring in the transport of amino ester salts by a chiral host compound. Enantiomeric separations of this type have great biochemical and physiological interest. Pirkle and his co-workers[117] have observed that the increased interest in stereochemistry in many branches of chemistry, biochemistry, and pharmacology has underlined the need for better methods of separating enantiomers and determining their purities. See Chapter 9, Volume 3 for further examples.

Macrocycles may be used in column and thin-layer chromatography and in thin-layer electrophoresis. In column chromatography, monomeric cyclic polyethers can either be adsorbed on silica gel or dissolved in the eluent and used to separate cations and optically active compounds. Applications involving polymeric cyclic polyethers are more widespread and have been reviewed by Blasius and Janzen.[118] Methods have been developed for their use in separating cations, anions, and organic compounds.

4.2. Phase transfer catalysis

Many organic reactions make use of alkali and alkaline earth salts, especially the anionic portion of these salts. Indeed, anions, when unencumbered by strong solvation forces, are potent nucleophiles and bases and provide the basis for development of new and valuable reagents for organic synthesis.[7]

A major problem in the use of such anions is transporting them into the organic phase where the desired reaction with the organic substrate is to take place. A desirable consequence of such transfer is to remove the anion's hydration sheath or, as sometimes stated, to leave the anion "naked" and thereby more accessible for reaction with the substrate. One method which is effective involves the use of a large cation with a lipophilic exterior which, when combined as an ion pair with the anion has high solubility in the organic phase. Macrocycles are effective in forming such cations by reaction with the cations of common salts. (Quaternary ammonium and phosphonium salts may also be used, but will not be discussed here. See Dehmlow[73]). In conjunction with the macrocycle one could use, for instance, the potassium salt of the desired anion. The K^+ would, of course, complex with the macrocycle and to maintain electroneutrality the anion must partition with the resulting cation complex into the organic phase. The K^+-anion pair could be sequestered either from an aqueous solution of the salt (liquid–liquid phase transfer catalysis) or directly from salt crystals in contact with the organic phase (solid–liquid phase transfer catalysis). This technique has proved very useful in allowing reactions to proceed which could not take place because the reactants could not be solubilized in the same phase.[7,8,119] In addition, the procedure avoids unusual and complicated reaction and workup conditions.

4.3. Ion detection and ion selective electrodes

The cation selectivity of macrocycles can also be exploited in the making of ion selective electrodes. These electrodes consist of a reference aqueous phase which is separated from the test solution by an appropriate organic membrane.[98] Two representative examples are shown in Fig. 13. Ion selective liquid membrane microelectrodes are ideally suited for measurement of intracellular ion activities and have been used in biomedical applications.[120,121] As discussed earlier, diaphragm membranes, formed by incorporating macrocycles into solvent impregnated solid supports such as polyvinylchloride, provide systems of high electromotive and mechanical stability with electrode lifetimes longer than a year.[65,122]

Due to the extremely low solubility of ions in the organic membrane, if no carrier is present, the membrane isolates the reference aqueous solution from the test solution. In this case the electrical potential between the two solutions is undefined. However, when a macrocycle is introduced into the membrane a conductive pathway between the two aqueous solutions is established. Thus, a well defined electrical potential exists between the two solutions and, as a first approximation, can be described by an extended

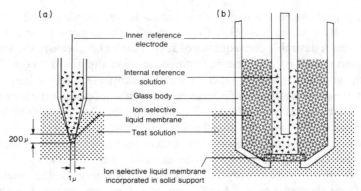

Fig. 13. Diagramatic examples of ion selective electrodes.[65,98] (a) Ion selective microelectrode and (b) ion selective electrode using a solid supported liquid membrane.

Nicolsky equation[57,95,98]

$$E = E_0 + \frac{RT}{F} \ln[a_i + \sum K_{ij}^{Pot}(a_j)^{Z_i/Z_j}] \qquad (27)$$

where E denotes the potential, E_0 is a constant reference potential, a_i is the activity of a primary ion, i, in the test solution, a_j is the activity of an interfering ion, j, in the test solution, K_{ij}^{Pot} is the selectivity factor in the particular membrane–macrocycle system, and RT/F is the usual Nernst factor.

The selectivity factor is a function of u_L, the mobility of the cation–ligand complex for each cation, k_{ip}, the partition coefficient for each ion, and K_L, the cation–ligand complexation constant for each ion and may be expresed[98] as

$$K_{ij}^{Pot} = \frac{u_{jL}(k_{ip})_j K_{jL}}{u_{iL}(k_{ip})_i K_{iL}} \qquad (28)$$

If we assume that for a particular class of cations, the mobility factors and the partition coefficients are approximately equal, then the selectivity factor reduces to the ratio of the complexation constants. Thus, if the complexation constant for a particular ligand–cation pair is much higher than that for other cations for the same ligand, the electrode becomes sensitive to that cation. Mathematically, if $K_{iL} \gg K_{jL}$, then

$$\sum_{j \neq i} K_{ij}^{Pot}(a_j)^{Z_i/Z_j} \sim 0 \qquad (29)$$

and

$$E = E_0 + \frac{RT}{Z_i F} \ln a_i \qquad (30)$$

In principle, ion selective electrodes could be made in this manner for any desired cation by using the macrocycle which binds most strongly to the desired ion. Since bulk transport does not occur in these electrodes, complexation constants too high to permit cation release are not detrimental. On the contrary, larger complexation constants favour selectivity.

5. Future prospects

Much remains to be done to exploit the full potential of the cation selectivity of macrocyclic ligands. Many of the principles described for liquid membranes may be applied to solvent extraction techniques as well. However, the liquid membrane technique has the advantage over solvent extraction that much smaller inventories of complexing agent are needed for a successful process. Hence, more selective, and more expensive complexing agents, such as macrocyclic ligands may be used. However, there is need for the development of cost effective membrane systems into which these selective complexing agents may be incorporated.

The work of Danesi, Horwitz, and their co-workers[28,58] has added much to our understanding of interfacial reaction kinetics and to the determination of rate limiting processes in solvent extraction and liquid membrane systems. The interfacial charge transfer studies of Hofmanová *et al.*[123] are useful in determining where complexation occurs and need to be repeated for many macrocycles. Interfacial surface tensions and the surface activities of macrocycles, both free and complexed, need to be measured. Values for these quantities should provide increased insight concerning the region where the reactions of interest occur. Partition coefficients of macrocycles, inorganic salts, and cation–macrocycle complexes in a variety of H_2O-solvent systems would be useful. The possibility that complexation occurs in the aqueous phase needs to be considered and, possibly, included in mathematical models such as that illustrated in Fig. 3c. The reaction or equilibrium plane distances, l_2 and l_5 (Fig. 3c), need to be determined as well as their dependence on both solvent and ligand.

One of the most interesting questions arising from our work concerns the mechanism of facilitated transport through the emulsion membranes. Carefully controlled studies need to be carried out with these or related model systems in order to elucidate reaction mechanisms. Model systems which might be used include vesicle membranes which can be prepared with well-defined structure, membrane thickness, and size distributions.[124–126] Carrier-mediated transport of Pr^{3+} in vesicle systems has been accomplished by Hunt and his co-workers,[127] Donis, *et al.*,[128] and Grandjean and Laszlo.[129] Several ionophores were used by these workers. Charged carriers such as A-23187, monensin, and lasalocid A were effective in transporting Pr^{3+}. Neutral carriers such as DCy18C6 and DB18C6 were effective only in the presence of an uncoupler such as picrate ion.

There has been great interest in preparing synthetic variations of macrocycles. One promising area of synthesis involves the introduction of acidic groups to the macrocyclic backbone to make possible the formation of neutral cation complexes which may move through membranes without accompanying anions.[20,33] These acid groups are then available to transport protons in the reverse direction from that of the cations. Thus, cations may be caused to move against their concentration gradients by the flow of protons in the opposite direction. Other synthetic areas of current interest include preparation of macrocycles capable of selective transport of organic anions[130] and of combining with more than one cation.[131]

Acknowledgement

The authors acknowledge financial support of this work by the Department of Energy (Contract No. DE-AC02-78ER05016) and helpful discussions with Philip R. Brown and John L. Oscarson.

References

1. C. J. Pedersen, *J. Am. Chem. Soc.*, 1967, **89**, 7017.
2. C. J. Pedersen, in *Synthetic Multidentate Macrocyclic Compounds*, (eds. R. M. Izatt and J. J. Christensen), Academic Press, New York, London, 1978, Ch. 1, p. 1.
3. J. S. Bradshaw, in *Synthetic Multidentate Macrocyclic Compounds*, (eds. R. M. Izatt and J. J. Christensen), Academic Press, New York, London, 1978, Ch. 2, p. 53.
4. J. M. Lehn and J. P. Sauvage, *J. Chem. Soc., Chem. Commun.*, 1971, 440.
5. J. M. Lehn, *Struct. Bonding (Berlin)*, 1973, **16**, 1.
6. W. E. Morf, D. Ammann, R. Bissig, E. Pretsch and W. Simon in *Progress in Macrocyclic Chemistry*, Vol. 1, (eds. R. M. Izatt and J. J. Christensen), Wiley–Interscience, New York, 1979, Ch. 1, p. 1.

7. C. L. Liotta in *Synthetic Multidentate Macrocyclic Compounds*, (eds. R. M. Izatt and J. J. Christensen), Academic Press, New York, London, 1978, Ch. 3. p. 111.
8. W. P. Weber and G. W. Gokel, *Phase Transfer Catalysis in Organic Synthesis*, Springer-Verlag, New York, 1977.
9. E. M. Choy, D. F. Evans and E. L. Cussler, *J. Am. Chem. Soc.*, 1974, **96**, 7085.
10. T. M. Fyles, V. A. Malik-Diemer and D. M. Whitfield, *Can. J. Chem.*, 1981, **59**, 1734.
11. S. G. A. McLaughlin, G. Szabo, S. M. Ciani and G. Eisenman, *J. Membr. Biol.*, 1972, **9**, 3.
12. K. H. Wong, K. Yagi and J. Smid, *J. Membr. Biol.*, 1974, **18**, 379.
13. J. D. Lamb, R. M. Izatt, P. A. Robertson and J. J. Christensen, *J. Am. Chem. Soc.*, 1980, **102**, 2452.
14. (*a*) J. D. Lamb, P. R. Brown, J. J. Christensen, J. S. Bradshaw, D. G. Garrick and R. M. Izatt, *J. Membr. Sci.*, 1983, **13**, 89; (*b*) G. A. Clark, J. D. Lamb, J. J. Christensen, J. S. Bradshaw and R. M. Izatt, *Sep. Sci. Technol.*, in press.
15. R. M. Izatt, M. P. Biehl, J. D. Lamb and J. J. Christensen, *Sep. Sci. Technol.*, 1982, **17**, 1351.
16. M. P. Biehl, R. M. Izatt, J. D. Lamb and J. J. Christensen, *Sep. Sci. Technol.*, 1982, **17**, 289.
17. (*a*) J. Rebek, Jr. and R. V. Wattley, *J. Am. Chem. Soc.*, 1980, **102**, 4853; (*b*) J. Rebek, Jr., J. E. Trend, R. V. Wattley and S. Chakravorti, *J. Am. Chem. Soc.*, 1979, **101**, 4333.
18. J. J. Christensen, S. P. Christensen, M. P. Biehl, S. A. Lowe, J. D. Lamb and R. M. Izatt, *Sep. Sci. Technol.*, 1983, **13**, 353.
19. R. M. Izatt, D. V. Dearden, P. R. Brown, J. S. Bradshaw, J. D. Lamb and J. J. Christensen, *J. Am. Chem. Soc.*, 1983, **105**, 1785.
20. J. Strzelbicki and R. A. Bartsch, *J. Membr. Sci.*, 1982, **10**, 35.
21. B. E. Jepson and R. DeWitt, *J. Inorg. Nucl. Chem.*, 1976, **38**, 1175.
22. T. Kitagawa, Y. Nishikawa, J. W. Frankenfeld and N. N. Li, *Environ. Sci. Technol.*, 1977, **11**, 602.
23. R. M. Izatt, J. D. Lamb, D. J. Eatough, J. J. Christensen and J. H. Rytting in '*Drug Design*' Vol. 8, (ed. E. Ariëns), Academic Press, New York, London, 1979, Ch. 7, p. 355.
24. E. L. Cussler and D. F. Evans, *Sep. Purif. Methods*, 1974, **3**, 399.
25. K. Nomura, A. Matsubara and H. Kimizuka, *Bull. Chem. Soc. Jpn.*, 1981, **54**, 1324.
26. C. F. Reusch and E. L. Cussler, *AIChE. J.*, 1973, **19**, 736.
27. J. D. Lamb, R. M. Izatt, D. G. Garrick, J. S. Bradshaw and J. J. Christensen, *J. Membr. Sci.*, 1981, **9**, 83.
28. P. R. Danesi, E. P. Horwitz, G. F. Vandergrift and R. Chiarizia, *Sep. Sci. Technol.*, 1981, **16**, 201.
29. E. Pefferkorn and R. Varoqui, *J. Colloid Interface Sci.*, 1975, **52**, 89.
30. W. J. Molnar, Jr., M.S. Thesis, Carnegie-Mellon University, 1977.
31. K. Maruyama, H. Tsukube and T. Araki, *J. Am. Chem. Soc.*, 1980, **102**, 3246.
32. E. L. Cussler and D. F. Evans, *J. Membr. Sci.*, 1980, **6**, 113.
33. R. M. Izatt, J. D. Lamb, P. R. Brown, R. T. Hawkins, S. R. Izatt and J. J. Christensen, *J. Am. Chem. Soc.*, 1983, **105**, 1782.
34. M. Kirch and J. M. Lehn, *Angew. Chem. Int.*, *Ed. Engl.*, 1975, **14**, 555.
35. J. D. Lamb, R. M. Izatt, J. J. Christensen and D. J. Eatough, in *Coordination Chemistry of Macrocyclic Compounds*, (ed. G. A. Melson), Plenum Press, New York, 1979, Ch. 3, p. 145.

36. J. D. Lamb, R. M. Izatt and J. J. Christensen in *Progress in Macrocyclic Chemistry*, Vol. 2. (eds. R. M. Izatt and J. J. Christensen), Wiley–Interscience, New York, 1981, Ch. 2, p. 41.
37. R. M. Izatt, B. L. Nielson, J. J. Christensen and J. D. Lamb, *J. Membr. Sci.*, 1981, **9**, 263.
38. P. R. Brown, R. M. Izatt, J. J. Christensen and J. D. Lamb, *J. Membr. Sci.*, 1983, **13**, 85.
39. Y. Kobuke, K. Hanji, K. Horiguchi, M. Asada, Y. Nakayama and J. Furukawa, *J. Am. Chem. Soc.*, 1976, **98**, 7414.
40. W. M. Latimer, K. S. Pitzer and C. M. Slansky, *J. Chem. Phys.*, 1939, **7**, 108.
41. W. Simon, W. E. Morf and P. Ch. Meier, *Struct. Bonding (Berlin)*, 1973, **16**, 113.
42. S. Goldman and R. G. Bates, *J. Am. Chem. Soc.*, 1972, **94**, 1476.
43. M. E. Duffey, D. F. Evans and E. L. Cussler, *J. Membr. Sci.*, 1978, **3**, 1.
44. A. Persoons and M. Van Beylen, *Pure Appl. Chem.*, 1979, **51**, 887.
45. V. van Even and M. C. Haulait-Pirson, *J. Solution Chem.*, 1977, **6**, 757.
46. J. D. Lamb, J. J. Christensen, S. R. Izatt, K. Bedke, M. S. Astin and R. M. Izatt, *J. Am. Chem. Soc.*, 1980, **102**, 3399.
47. F. Caracciolo, E. L. Cussler and D. F. Evans, *AIChE J.*, 1975, **21**, 160.
48. J. J. Christensen, J. D. Lamb, S. R. Izatt, S. E. Starr, G. C. Weed, M. S. Astin, B. D. Stitt and R. M. Izatt, *J. Am. Chem. Soc.*, 1978, **100**, 3219.
49. J. D. Lamb, J. J. Christensen, J. L. Oscarson, B. L. Nielsen, B. W. Asay and R. M. Izatt, *J. Am. Chem. Soc.*, 1980, **102**, 6820.
50. S. M. Ciani, G. Eisenman and G. Szabo, *J. Membr. Biol.*, 1969, **1**, 1.
51. P. Läuger and G. Stark, *Biochim. Biophys. Acta*, 1970, **211**, 458.
52. D. A. Haydon and S. B. Hladky, *Quart. Rev. Biophys.*, 1972, **5**, 187.
53. R. P. Buck, *Anal. Chem.*, 1976, **48**, 23R.
54. J. H. Boles and R. P. Buck, *Anal. Chem.*, 1973, **45**, 2057.
55. W. E. Morf, G. Kahr and W. Simon, *Anal. Lett.*, 1974, **7**, 9.
56. W. E. Morf, P. Wuhrmann and W. Simon, *Anal. Chem.*, 1976, **48**, 1031.
57. A. P. Thoma, A. Viviani-Nauer, S. Arvanitis, W. E. Morf and W. Simon, *Anal. Chem.*, 1977, **49**, 1567.
58. P. R. Danesi and R. Chiarizia, *Crit. Rev. Anal. Chem.*, 1980, **10**, 1.
59. D. W. McBride, Jr., P. R. Brown, R. M. Izatt, J. D. Lamb and J. J. Christensen, unpublished work.
60. H. Watarai, L. Cunningham and H. Freiser, *Anal. Chem.*, 1982, **54**, 2390.
61. G. W. Liesegang and E. M. Eyring in *Synthetic Multidentate Macrocyclic Compounds*, (eds. R. M. Izatt and J. J. Christensen), Academic Press, New York, London, 1978, Ch. 5, p. 245.
62. J. Strzelbicki, G. S. Heo and R. A. Bartsch, *Sep. Sci. Technol.*, 1982, **17**, 635.
63. P. R. Danesi, R. Chiarizia, M. Pizzichini and A. Saltelli, *J. Inorg. Nucl. Chem.*, 1978, **40**, 1119.
64. E. L. Cussler, *AIChE J.*, 1971, **17**, 1300.
65. R. A. Schwind, T. J. Gilligan and E. L. Cussler, in *Synthetic Multidentate Macrocyclic Compounds*, (eds. R. M. Izatt and J. J. Christensen), Academic Press, New York, London, 1978, Ch. 6, p. 289.
66. J. S. Schultz, J. D. Goddard and S. R. Suchdeo, *AIChE. J.*, 1974, **20**, 417.
67. N. S. Poonia in *Progress in Macrocyclic Chemistry*, Vol. 1, (eds. R. M. Izatt and J. J. Christensen), Wiley–Interscience, New York, 1979, Ch. 3, p. 115.
68. E. Weber and F. Vögtle, in *Topics in Current Chemistry–Host Guest Complex Chemistry I*, (ed. F. Vögtle), Springer-Verlag, New York, 1981, Ch. 1, p. 1.
69. F. de Jong and D. N. Reinhoudt, in *Advances in Physical Organic Chemistry*, (eds. V. Gold and D. Bethell), Academic Press, London, New York, 1981. p. 274.

70. R. M. Izatt, J. S. Bradshaw, J. D. Lamb, J. J. Christensen and D. Sen, *Chem. Rev.*, in press.
71. J. M. Lehn and J. P. Sauvage, *J. Am. Chem. Soc.*, 1975, **97**, 6700.
72. E. Kauffmann, J. M. Lehn and J. Sauvage, *Helv. Chim. Acta*, 1976, **59**, 1099.
73. E. V. Dehmlow, *Angew. Chem., Int. Ed. Engl.*, 1974, **13**, 170.
74. R. D. Shannon and C. T. Prewitt, *Acta Crystallogr.*, 1969, **B25**, 925.
75. N. K. Dalley in *Synthetic Multidentate Macrocyclic Compounds*, (eds. R. M. Izatt and J. J. Christensen), Academic Press, New York, London, 1978, Ch. 4, p. 207.
76. (*a*) J. D. Lamb, R. M. Izatt, C. S. Swain and J. J. Christensen, *J. Am. Chem. Soc.*, 1980, **102**, 475; (*b*) R. M. Izatt, R. E. Terry, L. D. Hansen, A. G. Avondet, J. S. Bradshaw, N. K. Dalley, T. E. Jensen, J. J. Christensen and B. L. Haymore, *Inorg. Chim. Acta*, 1978, **30**, 1.
77. (*a*) L. Pauling, *The Nature of the Chemical Bond*, Cornell University Press, Ithaca, New York, 3rd edn., 1960; (*b*) *Handbook of Chemistry and Physics*, CRC Press, Cleveland, Ohio, 1975; (*c*) J. D. Lamb, J. E. King, J. J. Christensen and R. M. Izatt, *Anal. Chem.*, 1981, **53**, 2127.
78. F. Basolo and R. G. Pearson, *Mechanisms of Inorganic Reactions*, 2nd edn., John Wiley and Sons, New York, 1967.
79. T. Iwachido, M. Minami, H. Naito and K. Tôei, *Bull. Chem. Soc. Jpn.*, 1982, **55**, 2378.
80. M. H. Abraham and J. Liszi, *J. Inorg. Nucl. Chem.*, 1981, **43**, 143.
81. D. K. Schiffer, E. M. Choy, D. F. Evans and E. L. Cussler, *AIChE Symposium Series, No. 144*, **70**, 150.
82. E. S. Amis, *Solvent Effects on Reaction Rates and Mechanisms*, Academic Press, New York, London, 1966.
83. A. Agostiano, M. Caselli and M. Della Monica, *J. Electroanal. Chem.*, 1976, **74**, 95.
84. M. H. Abraham and J. Liszi, *J. Chem. Soc., Faraday Trans. 1*, 1978, **74**, 1604.
85. C. M. Criss and M. Salomon, in *Physical Chemistry of Organic Solvent Systems*, (eds. A. K. Covington and T. Dickinson), Plenum Press, New York, 1973, Ch. 2, part 4, p. 253.
86. J. Burgess, *Metal Ions in Solution*, Ellis Horwood Limited, Chichester, England, 1978.
87. V. Gutmann and E. Wychera, *Inorg. Nucl. Chem. Lett.*, 1966, **2**, 257.
88. H. K. Frensdorff, *J. Am. Chem. Soc.*, 1971, **93**, 600.
89. R. M. Izatt, R. E. Terry, D. P. Nelson, Y. Chan, D. J. Eatough, J. S. Bradshaw, L. D. Hansen and J. J. Christensen, *J. Am. Chem. Soc.*, 1976, **98**, 7626.
90. J. D. Lamb, Ph.D. Dissertation, Brigham Young University, Provo, Utah (1978); *Diss. Abs.*, 1979, **39**, 3322B.
91. N. Matsuura, K. Umemoto, Y. Takeda and A. Sasaki, *Bull. Chem. Soc. Jpn.*, 1979, **49**, 1246.
92. B. G. Cox, J. Garcia-Rosas and H. Schneider, *J. Am. Chem. Soc.*, 1981, **103**, 1384.
93. M. K. Chantooni, Jr. and I. M. Kolthoff, *Proc. Natl. Acad. Sci. USA*, 1981, **78**, 7245.
94. F. Arnaud-Neu, B. Spiess and M. J. Schwing-Weill, *J. Am. Chem. Soc.*, 1982, **104**, 5641.
95. W. E. Morf, *The Principles of Ion-Selective Electrodes and of Membrane Transport*, Elsevier, New York, 1981.
96. R. C. Waldbillig and G. Szabo, *Biochim. Biophys. Acta*, 1979, **557**, 295.
97. W. J. McDowell, personal communication.

98. W. Simon, E. Pretsch, D. Ammann, W. E. Morf, M. Güggi, R. Bissig and M. Kessler, *Pure Appl. Chem.*, 1975, **44**, 613.
99. J. J. Christensen, J. D. Lamb, P. R. Brown, J. L. Oscarson and R. M. Izatt, *Sep. Sci. Technol.*, 1981, **16**, 1193.
100. F. Nakashio and K. Kondo, *Sep. Sci. Technol.*, 1980, **15**, 1171.
101. N. N. Li, U.S. Patent 3,410,794 (to Esso Research and Engineering Co.), 1968; *Chem. Abs.*, 1969, **70**, 39550.
102. U. Dayal and B. S. Rawat, *J. Sci. Ind. Res.*, 1978, **37**, 602.
103. R. P. Cahn and N. N. Li, *Sep. Sci.*, 1974, **9**, 505.
104. J. Strzelbicki and W. Charewicz, *Hydromet.*, 1980, **5**, 243.
105. N. N. Li and W. J. Asher, *Chem. Abs.*, 1973, **79**, 108042.
106. W. J. Asher, K. C. Baree, J. W. Frankenfeld, R. W. Hamilton, L. W. Henderson, P. G. Holtzapple and N. N. Li, *Kidney Int.*, 1975, **7**, S-409.
107. R. R. Mohan and N. N. Li, *Biotechnol. Bioeng.*, 1974, **16**, 513.
108. S. W. May and N. N. Li, *Enzyme Eng.*, 1973, 72.
109. G. Gregoriadis, P. D. Leathwood and B. E. Ryman, *FEBS Lett.*, 1971, **14**, 95.
110. G. Gregoriadis, D. E. Neerunjun and R. Hunt, *Life Sci.*, 1977, **21**, 357.
111. J. H. Fendler and A. Romero, *Life Sci.*, 1977, **20**, 1109.
112. A. M. Hochauser and E. L. Cussler, *CEP Symp. Ser.*, 1975, **71**, 136.
113. T. Kimura, K. Iwashima, T. Ishimori and T. Hamada, *Anal. Chem.*, 1979, **51**, 1113.
114. R. M. Izatt, S. R. Izatt, D. W. McBride, Jr., J. S. Bradshaw and J. J. Christensen, *Isr. J. Chem.*, in press.
115. J. S. Bradshaw, B. A. Jones, R. B. Davidson, J. J. Christensen, J. D. Lamb, R. M. Izatt, F. G. Morin and D. M. Grant, *J. Org. Chem.*, 1982, **47**, 3362.
116. M. Newcomb, R. C. Helgeson and D. J. Cram, *J. Am. Chem. Soc.*, 1974, **96**, 7367.
117. W. H. Pirkle, J. M. Finn, J. L. Schreiner and B. C. Hamper, *J. Am. Chem. Soc.*, 1981, **103**, 3964.
118. E. Blasius and K. P. Janzen, in *Topics in Current Chemistry—Host Guest Complex Chemistry I*, (ed. F. Vögtle), Springer-Verlag, New York, 1981, Ch. 4, p. 163.
119. C. M. Starks and C. Liotta, *Phase Transfer Catalysis, Principles and Techniques*, Academic Press, New York, London, 1978.
120. J. L. Walker, Jr., *Anal. Chem.*, 1971, **43**, 89A.
121. G. A. Rechnitz, *Res. Dev.*, 1973, **24**, 18.
122. G. J. Moody and J. D. R. Thomas, *Selective Ion-Sensitive Electrodes*, Morrow, Watford, Herts, UK, 1971.
123. A. Hofmanová, L. Q. Hung and W. Khalil, *J. Electroanal. Chem.*, 1982, **135**, 257.
124. Y. Barenholz, D. Gibbes, B. J. Litman, J. Goll, T. E. Thompson and F. D. Carlson, *Biochemistry*, 1977, **16**, 2806.
125. M. Wong, F. H. Anthony, T. W. Tillack and T. E. Thompson, *Biochemistry*, 1982, **21**, 4126.
126. M. Wong and T. E. Thompson, *Biochemistry*, 1982, **21**, 4133.
127. G. R. A. Hunt, L. R. H. Tipping and M. R. Belmont, *Biophys. Chem.*, 1978, **8**, 341.
128. J. Donis, J. Grandjean, A. Grosjean and P. Laszlo, *Biochem. Biophys. Res. Commun.*, 1981, **102**, 690.
129. J. Grandjean and P. Laszlo, *Biochem. Biophys. Res. Commun.*, 1982, **104**, 1293.
130. I. Tabushi, Y. Kobuke and J. Imuta, *J. Am. Chem. Soc.*, 1981, **103**, 6152.
131. J. Comarmond, P. Plumeré, J. M. Lehn, Y. Agnus, R. Louis, R. Weiss, O. Kahn and I. Morgenstern-Badarau, *J. Am. Chem. Soc.*, 1982, **104**, 6330.

AUTHOR INDEX

The numbers in parentheses are reference numbers. The preceding number is the page number where the reference is first mentioned in the text. The number in bold type is the page number where the full reference can be found.

Gilmore, C. J., 132, **132**, 137, **137**, 139, **139**, 285(48), **294**
Gilson, J.-P., 63(106), **68**
Gleim, W. K. T., 368(181), **387**
Goc, R., 123(119), **128**
Goddard, J. D., 585(66), **626**
Godzik, K., 152(12), **171**
Gokel, G. W., 572(8), **625**
Gold, H. S., 38(115), **68**
Gold, L. W., 90(55), **126**, 91(57), **127**
Gold, V., 586(69), **626**
Goldberg, A., 272(21), **293**
Goldberg, I., 134, **134**, 137, **137**, 130(9), **145**, 288(58), **294**
Goldfarb, T. D., 41(4), **65**
Goldin, V. A., 313(63), **328**
Goldman, S., 575(42), **626**
Goll, J., 624(124), **628**
Golubov, S. I., 152(11), **171**
Gonzalez-Calbet, J., 62(103), **68**
Goodhall, J. I., 59(85), **67**
Gordy, W., 116(104), **128**
Görnerová, T., 245(1), **255**
Gorshkov, V. I., 260(19), **261**
Goto, T., 372(216), **388**
Gough, S. R., 6(9), **34**, 6(12, 15), 7(21, 23), 16(42), **35**, 77(23, 24, 25, 33), 78(28), 85(42), 88(47), 89(48, 49), **126**, 91(58, 59), 108(72), 89(87), **127**, 121(115), **128**
Gourdji, M., 32(74), **36**
Grabowski, Z. R., 194(46, 47), **241**
Grace, D. E. P., 515(40), 516(41), **564**
Gramegna, M. T., 300(11), 301(17), **327**, 309(54), **328**
Grandjean, J., 624(128, 129), **628**
Granicher, H., 91(57), **127**
Grant, D. M., 406(28), **442**, 620(115), **628**
Grassi, M., 307(48), 311(59), **328**
Gravereau, P., 142(19), **146**
Graves, D. J., 525(78), 526(83), **565**, 538(121), **567**
Gray, J. E., 352(110), **385**
Green, B. S., 139, **139**, 140, **140**, 158(23), 169(34), **171**, 272(19), **293**, 278(37), 280(40, 41, 42), **294**
Greengard, P., 537(120), **567**
Grégoire, P., 71(8), **125**, 124(130), **128**
Gregoriadis, G., 613(109, 110), **628**

Grey, N. R., 11(45), **35**, 23(52), **36**
Grez, M., 520(74), **565**
Griffith, O. H., 57(68), 58(70, 71, 72, 73, 74, 75, 77, 80), **67**, 308(49, 50), **328**
Griffiths, D. W., 394(7), **442**, 428(36), **443**
Grishina, L. N., 260(13), **261**
Grosjean, A., 624(128), **628**
Grütter, M. G., 524(71), 520(73), **565**
Guarino, A., 151(6), **170**, 159(24, 25), 160(27, 28), 165(30), 170(36), **171**
Güggi, M., 621(98), **628**
Guibé, L., 31(72), 32(74), **36**
Guillot, G., 459(40), **470**, 501(31), **508**
Günther, I. R., 246(3), **256**
Gustavsen, J. E., 44(11), **65**
Gutmann, V., 599(87), **627**
Gwinn, W. D., 24(54), **36**

Habon, I., 361(138), **386**, 353(161, 173), **387**
Hackert, M. L., 530(136), **567**
Hadni, A., 54(55), **67**
Hafemann, D. R., 6(10, 14), **35**
Halbert, T. R., 60(96), **68**
Halpern, A., 181(15, 16), **241**
Hamada, T., 619(113), **628**
Hamada, Y., 353(159), **387**
Hamaguchi, H., 167(32), **171**, 483(10), **508**
Hamano, M., 345(75), **384**
Hamilton, J. A., 134, **134**, 135, **135**
Hamilton, R. M., 344(36), **383**
Hamilton, R. W., 613(106), **628**
Hamlin, R., 546(147), **568**
Hammond, M., 499(25), **508**
Hamoyose, T., 372(220), **388**
Hamper, B. C., 620(117), **628**
Hanji, K., 574(39), **626**
Hanotier, J., 176(6), **240**, 199(53), **242**, 286(52), **294**
Hanotier-Bridoux, M., 286(52), **294**
Hansch, C., 530(96), **566**
Hansen, L. D., 413(31), **443**, 448(23), **470**, 591(76), 599(89), **627**
Hantos, G., 353(173), **387**
Hanui, M., 544(134), **567**
Harada, A. M., 371(209), **388**

Iwata, S., 372(218), **388**
Izatt, R. M., 572(2, 3, 6, 7), **624,**
572(13, 14, 15, 16, 18, 19, 23, 27),
574(33, 35), **625**, 574(36, 37, 38),
577(46, 48, 49), 579(59), 582(61),
584(65), 580(67), **626**, 586(70, 75),
587(76), 591(77), 599(89), **627,**
611(99), 619(114), 620(115), **628**
Izatt, S. R., 574(33), **625**, 577(46, 48),
626, 619(114), **628**
Izeki, I., 347(88), **385**
Izumi, G., 65(119), **68**

Jack, A., 515(38), **564**
Jacobs, S. M., 54(58), **67**, 105(70), **127**
Jacques, J., 284(46, 47, 53), **294**
Jaffrain, M., 114(95), **128**
Jakus, K. 373(225), **389**
James, M. N. G., 511(6, 10), **563,**
513(24), **564**, 523(64), **565**
James, T. L., 434(43), **443**
Jameson, A. K., 75(17), **126**
Jameson, C. J., 75(17), **126**
Janik, J. A., 114(97), **128**
Janik, J. M., 114(97), **128**
Jansonius, J. N., 531(102), **566**
Janzen, K. P., 620(118), **628**
Járay, M., 335(8), **383**
Jarvest, R. L., 554(166), **568**
Jeffery, J., 530(141), **567**
Jeffrey, G. A., 15(39), **35**, 83(41), **126,**
107(71), **127**, 131, **131**, 134, **134**, 136,
136, 138, **138**, 139, **139**, 140, **140,**
184(21), **241**
Jegoudez, J., 32(76), **36**
Jenkins, J. A., 525(79), **565**, 528(91),
529(94), **566**
Jennings, H. J., 436(45), **443**
Jensen, L. H., 514(30), **564**, 531(97),
566
Jensen, T. E., 591(76), **627**
Jepson, B. E., 260(20), **261**, 572(71),
625
Johari, G. P., 88(45), **126**
Johnson, C. S., 112(85), **127**
Johnson, J. E., 545(143), **567**
Johnson, L. N., 511(12), 512(17), **563,**
520(63), 523(64), 525(79), 527(87),
565, 528(91), 529(94), 517(99), **566**

Johnson, M. R., 135, **135**
Johnson, R. F., 448(25), **470**
Johnston, D. C., 60(96), **68**
Jones, A. L., 175(4), **240**
Jones, B. A., 620(115), **628**
Jones, N. F., 135, **135**
Jones, S. J., 90(55), **126**, 91(57), **127**
Jones, W., 46(117), **68**
Jordan, T. H., 131, **131**, 133, **133**
Jortner, J., 152(12), **171**
Joyner, R. W., 60(89), **67**
Jung, A., 520(74), **565**
Jurczak, J., 239(110), **243**, 374(238),
389

Kadir, Z. A., 57(66), **67**
Kahn, O., 624(131), **628**
Kahn, R., 49(49), **66**
Kahr, G., 577(55), **626**
Kainuma, H., 353(170), **387**, 373(237),
389
Kainuma, K., 334(6), **383**
Kaiser, E. T., 271(17), **293**, 448(19),
469
Kajtár, M., 358(127), **386**
Kakudo, M., 527(86), **565**
Kalk, K. H., 540(125), 543(129), **567**
Kaloustian, J., 7(17, 18, 19), **35**, 77(27),
126
Kamimoto, F., 378(267), **390**
Kampe, W., 395(10), **442**
Kanada, Y., 374(250), **389**
Kanamaru, F., 60(95), **68**
Kaneki, R., 379(273), **390**
Kaneko, Y., 373(227), **389**
Kano, K., 272(69), **295**, 483(11), **508**
Kapon, M., 140, **140**
Kappenstein, C., 246(7), **256**
Kariya, A., 377(261), **389**
Karle, I. L., 123(127), **128**, 133, **133**
Karle, J., 123(127), 124(129), **128**, 133,
133
Kashiwabara, H., 58(78, 79), **67**
Kasvinsky, P. J., 519(49), **564**, 526(84),
565, 528(92), **566**
Katada, M., 60(92), **67**
Katayama, A., 345(55), **384**
Kato, M., 374(245), **389**
Kato, T., 374(244, 245), **389**

SUBJECT INDEX